Research Advancements in Smart Technology, Optimization, and Renewable Energy

Pandian Vasant
University of Technology Petronas, Malaysia

Gerhard Weber
Poznan University of Technology, Poland

Wonsiri Punurai
Mahidol University, Thailand

IGI Global
PUBLISHER of TIMELY KNOWLEDGE

A volume in the Advances in Computer and Electrical Engineering (ACEE) Book Series

Published in the United States of America by
IGI Global
Engineering Science Reference (an imprint of IGI Global)
701 E. Chocolate Avenue
Hershey PA, USA 17033
Tel: 717-533-8845
Fax: 717-533-8661
E-mail: cust@igi-global.com
Web site: http://www.igi-global.com

Library of Congress Cataloging-in-Publication Data

Names: Vasant, Pandian, editor. | Weber, Gerhard, 1960- editor. | Punurai, Wonsiri, 1979- editor.
Title: Research advancements in smart technology, optimization, and renewable energy / Pandian Vasant, Gerhard Weber and Wonsiri Punurai, editors.
Description: Hershey, PA : Engineering Science Reference, an imprint of IGI Global, [2020] | Includes bibliographical references and index. | Summary: "This book explores the recent steps forward for smart applications in sustainability"-- Provided by publisher.
Identifiers: LCCN 2019058334 (print) | LCCN 2019058335 (ebook) | ISBN 9781799839705 (hardcover) | ISBN 9781799850397 (paperback) | ISBN 9781799839712 (ebook)
Subjects: LCSH: Technological innovations. | Smart materials. | Artificial intelligence--Industrial applications. | Wearable technology. | Machine learning. | Renewable energy sources. | Electric power systems.
Classification: LCC T173.8 .R468 2020 (print) | LCC T173.8 (ebook) | DDC 600--dc23
LC record available at https://lccn.loc.gov/2019058334
LC ebook record available at https://lccn.loc.gov/2019058335

This book is published in the IGI Global book series Advances in Computer and Electrical Engineering (ACEE) (ISSN: 2327-039X; eISSN: 2327-0403)

British Cataloguing in Publication Data
A Cataloguing in Publication record for this book is available from the British Library.

All work contributed to this book is new, previously-unpublished material. The views expressed in this book are those of the authors, but not necessarily of the publisher.

For electronic access to this publication, please contact: eresources@igi-global.com.

Advances in Computer and Electrical Engineering (ACEE) Book Series

Srikanta Patnaik
SOA University, India

ISSN:2327-039X
EISSN:2327-0403

Mission

The fields of computer engineering and electrical engineering encompass a broad range of interdisciplinary topics allowing for expansive research developments across multiple fields. Research in these areas continues to develop and become increasingly important as computer and electrical systems have become an integral part of everyday life.

The **Advances in Computer and Electrical Engineering (ACEE) Book Series** aims to publish research on diverse topics pertaining to computer engineering and electrical engineering. **ACEE** encourages scholarly discourse on the latest applications, tools, and methodologies being implemented in the field for the design and development of computer and electrical systems.

Coverage

- Qualitative Methods
- Circuit Analysis
- Applied Electromagnetics
- Programming
- Electrical Power Conversion
- Power Electronics
- Sensor Technologies
- Algorithms
- Optical Electronics
- VLSI Fabrication

IGI Global is currently accepting manuscripts for publication within this series. To submit a proposal for a volume in this series, please contact our Acquisition Editors at Acquisitions@igi-global.com or visit: http://www.igi-global.com/publish/.

The Advances in Computer and Electrical Engineering (ACEE) Book Series (ISSN 2327-039X) is published by IGI Global, 701 E. Chocolate Avenue, Hershey, PA 17033-1240, USA, www.igi-global.com. This series is composed of titles available for purchase individually; each title is edited to be contextually exclusive from any other title within the series. For pricing and ordering information please visit http://www.igi-global.com/book-series/advances-computer-electrical-engineering/73675. Postmaster: Send all address changes to above address.

Titles in this Series

For a list of additional titles in this series, please visit: http://www.igi-global.com/book-series/advances-computer-electrical-engineering/73675

New Methods and Paradigms for Modeling Dynamic Processes Based on Cellular Automata
Stepan Mykolayovych Bilan (State University of Infrastructure and Technology, Ukraine) Mykola Mykolayovych Bilan (Mayakskaya Secondary School, Moldova) and Ruslan Leonidovich Motornyuk (Main Information and Computing Center,Ukraine)
Engineering Science Reference • © 2021 • 200pp • H/C (ISBN: 9781799826491) • US $195.00

Innovations in the Industrial Internet of Things (IIoT) and Smart Factory
Sam Goundar (The University of the South Pacific, Fiji) J. Avanija (Sree Vidyanikethan Engineering College, India) Gurram Sunitha (Sree Vidyanikethan Engineering College, India) K Reddy Madhavi (Sree Vidyanikethan Engineering College, India) and S. Bharath Bhushan (Sree Vidyanikethan Engineering College, India)
Engineering Science Reference • © 2020 • 300pp • H/C (ISBN: 9781799833758) • US $225.00

Advancements in the Design and Implementation of Smart Grid Technology
Ravi Samikannu (Botswana International University of Science and Technology, Botswana) Karthikrajan Senthilnathan (VIT University, India) Balamurugan Shanmugam (Quants IS & CS, India) Iyswarya Annapoorani (VIT University, India) and Bakary Diarra (Institute of Applied Sciences University of Sciences, Techniques and Technologies of Bamako, Mali)
Engineering Science Reference • © 2020 • 300pp • H/C (ISBN: 9781799836575) • US $205.00

Industrial Internet of Things and Cyber-Physical Systems Transforming the Conventional to Digital
Pardeep Kumar (Quaid-e-Awam University of Engineering, Science, and Technology, Pakistan) Vasaki Ponnusamy (Universiti Tunku Abdul Rahman, Malaysia) and Vishal Jain (Bharati Vidyapeeth's Institute of Computer Applications and Management (BVICAM), India)
Engineering Science Reference • © 2020 • 431pp • H/C (ISBN: 9781799828037) • US $225.00

Machine Learning and Deep Learning in Real-Time Applications
Mehul Mahrishi (Swami Keshvanand Institute of Technology, India) Kamal Kant Hiran (Aalborg University, Denmark) Gaurav Meena (Central University of Rajasthan, India) and Paawan Sharma (Pandit Deendayal Petroleum University, India)
Engineering Science Reference • © 2020 • 344pp • H/C (ISBN: 9781799830955) • US $245.00

Advanced Applications of Fractional Differential Operators to Science and Technology
Ahmed Ezzat Matouk (Majmaah University, Saudi Arabia)
Engineering Science Reference • © 2020 • 401pp • H/C (ISBN: 9781799831228) • US $245.00

IGI Global
PUBLISHER of TIMELY KNOWLEDGE

701 East Chocolate Avenue, Hershey, PA 17033, USA
Tel: 717-533-8845 x100 • Fax: 717-533-8661
E-Mail: cust@igi-global.com • www.igi-global.com

Editorial Advisory Board

List of Reviewers

Table of Contents

Detailed Table of Contents

This research investigates user spatial experience transformations that occur in hyperconnected public spaces and transform them to hybrid spaces. Following this target, the authors conduct an experiment in the Municipal Market of Chania, Crete, in which they evaluate user behaviors on a population of 33 participants comparing their spatial experiences before and after the use of ICT. Through qualitative and quantitative methods (the use of the technology Indoor Atlas as well as questionnaires), the authors analyze behavioral change among users with and without access to Crete 3D, an online ICT-based innovative informative platform, aiming to establish a theoretical framework of understanding user interaction with built space. This process enables knowledge transfer in a twofold way: the authors present how to use metrics to evaluate user-building interaction and how users can quickly gain a deep understanding of the building in use.

Chapter 3
Improving the Optimality Verification and the Parallel Processing of the General Knapsack Linear Integer Problem.......... 37
Elias Munapo, North-West University, South Africa

The chapter presents a new approach to improve the verification process of optimality for the general knapsack linear integer problem. The general knapsack linear integer problem is very difficult to solve. A solution for the general knapsack linear integer problem can be accurately estimated, but it can still be very difficult to verify optimality using the brach and bound related methods. In this chapter, a new objective function is generated that is also used as a more binding equality constraint. This generated equality constraint can be shown to significantly reduce the search region for the branch and bound-related algorithms. The verification process for optimality proposed in this chapter is easier than most of the available branch and bound-related approaches. In addition, the proposed approach is massively parallelizable allowing the use of the much needed independent parallel processing.

Chapter 4
New Direction to the Scheduling Problem: A Pre-Processing Integer Formulation Approach.......... 53
Elias Munapo, North-West University, South Africa
Olusegun Sunday Ewemooje, Federal University of Technology, Akure, Nigeria

This chapter presents a new direction to the scheduling problem by exploring the Moore-Hodgson algorithm. This algorithm is used within the context of integer programming to come up with complementarity conditions, more biding constraints, and a strong lower bound for the scheduling problem. With Moore-Hodgson Algorithm, the alternate optimal solutions cannot be easily generated from one optimal solution; however, with integer formulation, this is not a problem. Unfortunately, integer formulations are sometimes very difficult to handle as the number jobs increases. Therefore, the integer formulation presented in this chapter uses infeasibility to verify optimality with branch and bound related algorithms. Thus, the lower bound was obtained using pre-processing and shown to be highly accurate and on its own can be used in those situations where quick scheduling decisions are required.

One of the most interesting and promising areas of development of machine learning is the active involvement of a human in the process of building a model. However, there are problems with the effective integration of humans into a workflow. It is necessary to develop techniques and information technologies that would allow the effective use of human intellectual capabilities, thereby expanding the machine learning tools. This work considers the use of visual analytics with the goal of building a machine learning model by a human and the technique of transferring this model to the machine level. This made it possible to expand the capabilities of machine learning through the active and productive use of human intellectual abilities.

A digital illustration of a novel prevalence of a physical product helps one to gain larger insight into that product's state performance and behavior digital twin, which is an unequivocal advanced copy of an item, method, or control. This living model creates a thread between the physical and digital worlds. A model of a physical object—a 'twin'—enables you to observe its standing, diagnose problems, and take a look at solutions remotely. It's a dynamic virtual illustration of a tool that is unendingly fed with knowledge from embedded sensors and packages. This provides associate degree correct period of time standing of the physical device. Digital twins drive innovation and performance and offer development technicians prognostic analytics that give firms the flexibility to boost client expertise.

Influence maximization in online social networks (OSNs) is the problem of discovering few nodes or users in the social network termed as 'seed nodes', which can help the spread of influence in the network. With the tremendous growth in social networking, the influence exerted by users of a social network on other online users has caught the attention of researchers to develop effective influence maximization algorithms to be applied in the field of business strategies. The main application of influence maximization is promoting the product to a set of users. However, a real challenge in influence maximization algorithms to deal with enormous amount of users or nodes obtainable in any OSN is posed. The authors focused on graph mining of OSNs for generating 'seed sets' using standard influence maximization techniques. Many standard influence maximization models are used for calculation of spread of influence; a novel influence maximization technique, namely the DegGreedy technique, has been illustrated along with experimental results to make a comparative analysis of the existing techniques.

Foreword

The words intelligent and smart have gained great importance in recent times by applying them as a synonym for high technology, that is, technologies incorporated into devices that are able to feel the changes in the environment in which they perform and execute the necessary measures accordingly. to improve its functionality adapting to these changes. With this, minimum response times, low operating costs, high efficiency and in general better performance for its users are achieved. The applications in engineering, physics and medicine are enough, since they incorporate the skills of awareness and reaction integrated into systems with minimal complexity and low cost.

In terms of energy, most countries face strong challenges to meet consumer demand, and the smart grid must provide it with efficiency and quality by regulating the relationship between supply and demand. The use of intelligent technology is indispensable for the generation and distribution of energy throughout the regions that make up the countries in general and particularly those in the process of development using simple but effective systems. For example, Smart Grid technology makes it possible to differentiate the smart network from the traditional network in several ways allowing to address maximum power problems including generation for distribution purposes.

On a domestic level, the concept of the smart home has existed for decades, but smart homes are extremely rare, although digital technology and automated appliances are common things in several regions of the world. Recent technological advances make it possible to efficiently face changes towards a new technological and economically feasible era for the majority of users. In this sense, two aspects associated with the subject are exploited; in the first, the convenience and efficiency of its use should be considered; the second, focuses on a holistic idea about microgeneration and energy distribution, as well as related services that this requires.

This emerging branch of human knowledge integrates several disciplines for its development. From the classic concepts of control theory, through data fusion, smart structures, sensing systems, piezoceramics, sensors and actuators, active control, shape memory alloys, piezoelectric materials, magnetostriction, smart fluid machines, smart biomaterials, natural engineering to modern systems of total control of manufacturing systems (technology 4.0). The field of application of intelligent technology also requires related techniques that support the design and implementation of algorithms for search, optimization and application of practical solutions at low cost and reasonable time giving applied mathematics and computer science new areas of opportunity for deploy their applications and find new lines of research.

Therefore, this book is focused on dealing with the most recent issues in the areas that integrate smart technology, alternative energy sources and the methods and models associated with them through mathematical optimization. Because of its content, it is expected that technologists, mathematicians, physicists and engineers will find in this, a brilliant opportunity to dabble in this fascinating subject through the diverse chapter that offers both as beginners or as experts in the subject.

This material constitutes another effort by the prestigious IGI-Global publishing house and Doctor Pandian Vasant for the dissemination of scientific subjects of original research through the publication of this in a large editorial work. His work helps scientists on providing us with an efficient means of high credibility and reliability to disseminate the results of our research into the most important emerging fields of human knowledge.

Gilberto Pérez Lechuga
Universidad Autónoma del Estado de Hidalgo, Mexico

Preface

As Environmental issues remain at the forefront of Energy research, Renewable Energy is now an all-important field of study. As, furthermore, Smart Technology continues to rise while always becoming refined, its applications broaden and they enhance in their potential impact on leading to paradigm shifts and even revolutionize views and studies about sustainability. This potential can only be fully realized with a thorough understanding of the most recent breakthroughs in the fields of *Renewable Energy and Smart Technology* themselves and in their emerging methodolologies from Mathematics, Statistics, Analytics, Probability Theory and Stochastics, Operational Research and Artificial Intelligence.

Research Advancements in Smart Technology, Optimization, and Renewable Energy is a worldwide collection of innovative research that explores the recent steps forward towards smart applications in sustainability. Featuring the coverage on a wide range of topics including energy assessment, neural fuzzy control, and biogeography, this work is ideally designed for academicians, researchers, and students, ecomomists, managers, advocates, policy-makers, engineers, multiplyers and implementers of solutions, and artists.

The objective of this book project has been to gather along with their considerations, research studies and contributions the global investigators, experts and scholars in the scientific areas of Optimization and Analytics from all over the world, to share their knowledge, experience and newest trends in views and insights on the current research achievements related with *Renewable Energy and Smart Technology*. This book provides to the international research community a "golden opportunity" to familiarize and deepen, to interact and share their novel academic and practical results, findings and most recent discoveries among their friends and colleagues. The book is published by *IGI Global* which gave confidence, encouragement and support at every stage and in every direction of communication, of academic and creative inquiry.

Modern days' challenges of *Renewable Energy and Smart Technology* in natural science and engineering, in economies and societies, in sustainability and social complexity, in environmental, geo and earth sciences, in OR and decision support systems, are becoming more and more recognized and acknowledged in all over the world. Careful multidisciplinary discourse and research is urgently necessitated in order to finding accurate while, at the same time, stable solutions, to make deep insights and impacting contributions which are future-oriented and sustainable. They emanate from real-work motivations and needs, and along excellent, innovative and creative notions, concepts and models they lead towards powerful systems of recommendation and decision help that are prepared to deliver efficient and effective agendas of cultural, social, managerial and politics decisions, locally and globally. This process of analysis, innovation, creativity and construction requires both smartest networks and disciplinary communication dynamics and stimulating interdisciplinary collaborations and networks that are based on curiosity, freedom, mutual respect and responsibility.

At all of these points, modern *Optimization* and *Optimal Control, Data Mining, Machine Learning, AI* and *OR* come into play as *key technologies* of modeling, regularization and careful selection, of pre- and post-processing, of simulation and preparation, of guidance and interest, of continuous respect and concern.

The population of people on earth steadily grows, putting on the agenda numerous and hard questions, so many urgent problems. For instance, there is the need to offer all the necessary food, clothes, housing, services, infrastructure, medical and many further commodities and goods of all kinds. *Optimization* and *Decision-making* has to find, choose and allocate new territories, particularly, in the rural countrysides, and vast amounts of energy, while all of this should be implemented with not more than a moderate complexity by organizations and governments, within an overall atmosphere of care, empathy and freedom. In situations like these, *Smart Technology, Optimization, and Renewable Energy are making a big difference* and emerge as *Key Technologies* of the future, as represented in this book in such impressive ways.

We editors cordially thank all the authors and all the reviewers for their remarkable contributions. Best and high-quality chapters have been encouraged, chosen and reviewed in order to become published in our book *Research Advancements in Smart Technology, Optimization, and Renewable Energy* by IGI Global.

In order to name some of the numerous topics of this book's chapters, we may list the following ones, but there are *many more*:

- *Algorithms,*
- *Biogeography,*
- *Economic Load Dispatch,*
- *Electric Power Sector,*
- *Energy Assessment,*
- *Energy Management Strategies,*
- *Grey Wolf Optimization,*
- *Hybrid Systems,*
- *Neural Fuzzy Control,*
- *System Optimization,*
- *Teaching-Learning-Based Optimization.*

Let us emphasize the high importance of this work for everyone related with its subjects, challenges and promises, interested in them, and for its future readers and appliers. In fact, we trust that in the years to come, *Smart Technology, Optimization, and Renewable Energy* will remain in the core of giving *Decision-Making Help* and *Support* which this book aims at. In this context, we especially name *Stochastic Optimal Control* in the presence of *Impulses* and *Regime Switching* in economic and cultural frames. Here, we are also working on scientifically, towards further involvement of real-life situations and cases, with all their enormous uncertainties and with "human factors", for an *optimal decision-making*.

Based on very careful and rigorous reviewing processes, 17 chapters were accepted for publication and became part of this exclusive collection of chapters – our book *"Research Advancements in Smart Technology, Optimization, and Renewable Energy"*. Short descriptions of these chapters are following subsequently.

In their chapter *"Prediction of Menstrual Cycle Phase by Wearable Heart Rate Sensor"*, *Junichiro Hayano* (Nagoya City University, Graduate School of Medical Sciences, Japan) and *Emi Yuda* (Tohoku University, Graduate School of Engineering, Japan) introduce comprehensive analyses of long-term heart rate data that may be useful for revealing their associations with the menstrual cycle phase. In fact, prediction of menstrual cycle phase and fertile window by easily measurable bio-signals has been a yet unmet need; such technological development will greatly contribute to women's quality of life. Though many studies reported differences in autonomic indices of heart-rate variability (HRV) between follicular and luteal phases, they not yet reached the level that can predict the menstrual cycle phases. Recent development of wearable sensors enabled heart-rate monitoring during daily life; long-term heart rate data obtained herewith include a lot of information, and information that can be extracted by conventional HRV analysis is just a part of it. This chapter ought to mean a significant future advance.

With their contribution *"Enhancing User Experience in Public Spaces by Measuring Passengers' Flow and Perception Through ICT: The Case of the Municipal Market of Chania"*, *Anna Karagianni, Vasiliki Geropanta, Panagiotis Parthenios* (all from Technical University of Crete), *Riccardo Porreca* (UTE University of Quito, Ecuador), *Sofia Mavroudi, Antonios Vogiatzis, Lais-Ioanna Margiori, Christos Mpaknis, Eleutheria Papadosifou* and *Asimina Ioanna Sampani* (all from Technical University of Crete), investigates user-spatial experience transformations in hyper-connected public spaces and transform them to hybrid spaces. They conduct an experiment in the Municipal Market of Chania, Crete, in which they evaluate user behaviors on a population of 33 participants comparing their spatial experiences before and after the use of ICT. Through qualitative and quantitative methods, the authors analyze the behavioral change among users with and without access to Crete 3D, an online ICT-based innovative informative platform, in order to create a theoretical framework on user interaction with built space. This process permits for knowledge transfer in a twofold way: how to use metrics to evaluate user-building interaction and how users can quickly understand the building in use.

Elias Munapo (North West University, South Africa) in his chapter *"Improving the Optimality Verification and the Parallel Processing of the General Knapsack Linear Integer Problem"* presents a new approach to the verification process of optimality for the general knapsack linear integer problem which is very difficult to solve. A solution may be well estimated but it can still be very difficult to verify optimality using branch-and-bound related methods. In his chapter, a new objective function is generated which is also used for a more binding equality constraint that can be shown to significantly reduce the search region for branch-and-bound related algorithms. The verification process for optimality offered is easier than most of the available branch-and-bound related approaches. The proposed approach is widely parallelizable, allowing for the use of the much needed independent parallel processing.

In their chapter *"New Direction to the Scheduling Problem: A Pre-Processing Integer Formulation Approach"*, *Elias Munapo* (North West University, South Africa) (Elias.Munapo@nwu.ac.za) and *Olusegun Sunday Ewemooje* (Federal University of Technology Akure, Nigeria) present a new direction to the scheduling problem by exploring the Moore-Hodgson algorithm used within the context of integer programming to come up with complementarity conditions, extra more biding constraints and a strong lower bound for the scheduling problem. With Moore-Hodgson algorithm, the alternate optimal solutions cannot be easily generated from one optimal solution, but with integer formulation this is not a problem. As integer formulations are sometimes very difficult to handle as the number jobs increases, the authors present the integer formulation using infeasibility to verify optimality with branch-and-bound related algorithms. Hence, the lower bound is obtained using pre-processing and shown to be quite accurate, and it can be employed whenever quick scheduling decisions are required.

In the chapter *"Use of the Neural Network Controller of Sprung Mass to Reduce Vibrations From Road Irregularities"*, *Zakhid Godzhaev, Sergey Senkevich, Viktor Kuzmin* (all from Federal State Budgetary Scientific Institution "Federal Scientific Agroengineering Center VIM", Russia) and *Izzet Melikov* (Dagestan State Agricultural University named after M.M. Dzhambulatov, Russia) propose a new operation regime of the controller based on neuron nets by combining the advantages of the adaptive, radial and basic functions of the neuron net. Its advantages are a learning ability in real time to process indefinite, nonlinear disturbances, and to change the value of the active force in the hydraulic leaf spring by adjusting the weight coefficients of the neuron net or the radial parameters of the basic function. In fact, hydraulic systems which damp active oscillation operate according to a certain non-linear and time-varying algorithm. It is hard to create a controller based on its dynamic model. The author's model is a ¼ hydraulic active sprung mass of a mobile vehicle. The modeling shows that the use of a neuron net controller makes the sprung mass much more efficient.

In his contribution *"The Traveling Salesman Problem, Network Properties, Convex Quadratic Formulation, and Solution"*, *Elias Munapo* (North West University, South Africa) presents a traveling salesman problem, its network properties, convex quadratic formulation and its solution. He shows that adding or subtracting a constant to all arcs with special features in a traveling salesman problem (TSP) network model does not change an optimal solution of the TSP, and that adding or subtracting a constant to all arcs emanating from the same node in a TSP network does not change the TSP's optimal solution. In addition, a minimal spanning tree is employed to detect sub-tours; then sub-tour elimination constraints are generated. From the formulated linear integer model of the TSP network a convex quadratic program is constructed. Finally, interior point algorithms are applied to solve the TSP in polynomial time.

The chapter *"Optimal Sizing of Hybrid Wind and Solar Renewable Energy System: A Case Study of Ethiopia"* by *Diriba Kajela Geleta* (Madda Walabu University, Oromia, Ethiopia) and *Mukhdeep Singh Manshahia* (Punjabi University Patiala, India) employ a nature inspired methodology to optimize hybrids of renewable energy system in the case of Jeldu district of Ethiopia. Indeed, if properly designed and utilized, the Earth has a rich potential of clean energy to satisfy the energy demand of the world. The authors aim to minimize the total cost of the system designed by using appropriate numbers of components based on the pre-designed constraints to satisfy the load demand. MATLAB code was designed and the results were discussed. The authors observed that the proposed approach solved the optimum sizing of the defined problem with high convergence, and energy demand of the village can be optimally satisfied by the use of wind and solar hybrid system. This paper can help countries like Ethiopia to increase access to electricity.

Vladimir Panchenko (Russian University of Transport, Russia) in his chapter *"Prospects for Energy Supply of the Arctic Zone Objects Using Frost-resistant Solar Modules"* aims at the prospect of using frost-resistant solar modules with extended service life of various designs for energy supply of infrastructure facilities of the Arctic zone. He gives the general characteristic of the region under consideration and reflects on energy specifics, directions of energy development based on renewable energy sources. The author proposes frost-resistant planar photovoltaic modules and solar roofing panels with an extended service life for power supply of objects, and frost-resistant planar photovoltaic thermal roofing panels and concentrator solar installation with high-voltage matrix solar modules with a voltage of 1000 V and an electrical efficiency of up to 28% for simultaneous heat and power generation. The addressed solar modules have an extended rated power period due to the technology of sealing solar cells with a two-component polysiloxane compound; they can work effectively at large negative ambient temperatures and large ranges of its fluctuation.

In their contribution *"Development of Self-Organized Group Method of Data Handling (GMDH) Algorithm to Increase Permeate Flux (%) of Helical Shaped Membrane"*, *Anirban Banik, Mrinmoy Majumder, Sushant Kumar Biswal* and *Tarun Kanti Bandyopadhyay* (all from National Institute of Technology Agartala, India) focus on enhancing the permeate flux of helical shaped membrane using Group Method of Data Handling (GMDH) algorithm. Variables such as operating pressure, pore size, and feed velocity were selected as input, and permeate flux as output variable. The uncertainty analysis evaluates the acceptability of the model and it is found that values of Nash-Sutcliffe efficiency (NSE), ratio of the root mean squared error to the standard deviation (RSR), percent bias (PBIAS) are close to the best values, showing the model acceptability. The effect of input parameters on model output is calibrated using sensitivity analysis. It shows that pore size is the most sensitive parameter followed by feed velocity. The optimum values of pore size, operating pressure, and feed velocity are calibrated and found to be 2.21μm, 1.31×10-03KPa, and 0.37m/sec, respectively. The errors in GMDH model are compared with multi-linear regression (MLR) model, showing that GMDH predicts results with minimal error; the predicted variable follows the actual one with good accuracy.

The contribution *"Transmission Risk Optimization in Interconnected Systems: Risk Adjusted Available Transfer Capability"* by *Nimal Madhu M, Jai Govind Singh, Weerakorn Ongsakul* (all from Asian Institute of Technology, Thailand) and *Vivek Mohan* (National Institute of Technology Tiruchirappalli, India) is concerned with Available Transfer Capability as a key indicator of transmission reliability, varying with the variation in power flow pattern through the network. ATC determination addressing uncertainties in renewable generation and demand is of key significance for safe and economic operation of a power system, especially, in a competitive market. A two-stage risk-adjusted stochastic optimal power dispatch is presented minimizing the reduction in ATC due to variation in active power output from renewable energy sources and load. For a combined transmission-distribution system with renewable and conventional energy sources, ATC is estimated combining continuation power flow and power transfer sensitivity factor methods. The joint probability distribution function of ATC is derived. Risk, quantified as the variance of ATC, is minimized using stochastic weight trade-off non-dominated sorting particle swarm optimization, considering various operational objectives of a network operator.

Firuz Ahamed Nahid, Weerakorn Ongsakul, Nimal Madhu M, and *Tanawat Laopaiboon* (all from Asian Institute of Technology, Thailand) in their chapter *"Hybrid Neural Networks for Renewable Energy Forecasting: Solar and Wind Energy Forecasting using LSTM and RNN"* address forecasting of stochastic renewable energy sources as involved in one of the key applications of AI algorithms in power sector. To manage the generation of electricity from solar or wind effectively, accurate forecasting models are imperative. Therefore, a sophisticated hybrid Neural Network formulation is discussed in this chapter. A combination of Long-Short-Term Memory and Recurrent Neural Networks is formulated for very short-term forecasting of wind speed and solar radiation. In intervals of 15-30 minutes, time series forecasts are made that are ahead by multiple steps. Point-wise and probabilistic forecasting approaches are combined for maximum energy harvest. Historic data are collected for solar radiation, wind speed, temperature and relative humidity, and used to train the model. The proposed model is compared with Convolutional and LSTM Neural Networks individually in terms of RMSE, MAPE, MAE and Correlation; it is identified to have better forecasting accuracy.

In their chapter *"Multi-Fuel Power Dispatch Considering Prohibited Operating Zones and Tie-Line Flow Limits Using Ant Lion Optimizer"*, *Ganesan Sivarajan* (Government College of Engineering, Salem, India), *Jayakumar N* (Government Polytechnic College, Uthangarai, India), *Balachandar P* (Government Polytechnic College, Valangaiman, India) and *Subramanian Srikrishna* (Annamalai University,

India) point out that electrical power generation from fossil fuel releases several contaminants into the air, even more if the generating unit is fed by Multiple Fuel Sources (MFS). Stringent environmental regulations have forced the utilities to produce electricity at the cheapest price and the minimum level of pollutant emissions. Restriction in generator operations increases the complexity in plant operations. Cost effective and environmental responsive operations in MFS environment is a multi-objective constrained optimization problem. Ant Lion Optimizer (ALO) has been chosen for solving the MFS dispatch problems. Fuzzy decision-making mechanism is integrated into the search process of ALO to fetch the Best Compromise Solution (BCS). The intended algorithm is implemented on the standard test systems considering the prevailing operational constraints, e.g., valve-point loadings, CO2 emission, prohibited operating zones and tie-line flow limits.

In the chapter *"Oppositional Differential Search Algorithm for the Optimal Tuning of Both Single Input and Dual Input Power System Stabilizer"*, *Sourav Paul* (Dr. B C Roy Engineering College, India) and *Provas Kumar Roy* (Kalyani Government Engineering College, Kalyani, India) are concerned with low frequency oscillation as a major threat in large interconnected power system, as it curtains the power transfer capability of the line. Power System Stabilizer (PSS) helps in diminishing these low frequency oscillations by providing auxiliary control signal to the generator excitation input, thereby restoring stability of the system. The authors incorporate the concept of oppositional based learning (OBL) along with differential search algorithm (DSA) to solve the PSS problem. The proposed technique has been implemented on both single input and dual input PSS, and a comparative study is conducted to show the supremacy of the new techniques. The convergence characteristics authenticate the sovereignty of the considered algorithms as well.

The contribution *"Extreme Value Metaheuristics and Coupled Mapped Lattice Approaches for Gas Turbine-Absorption Chiller Optimization"* by *Timothy Ganesan* (Royal Bank of Canada), *Pandian Vasant* (Universiti Teknologi Petronas, Malaysia), *Igor Litvinchev* (Nuevo Leon State University, Mexico) and *Mohd Shiraz Aris* (TNB Research, Malaysia) focuses on two novel optimization methodologies: extreme value stochastic engines (random number generators) and the coupled map lattice (CML), in the light of the increasing complexity of engineering systems which has spurred the development of highly efficient optimization techniques. This chapter incorporates extreme value distributions into stochastic engines of conventional metaheuristics and the implementation of CMLs to improve the overall optimization. The core idea is to deal with highly complex, large-scale multi-objective (MO) problems. Differential evolution (DE) approach is employed (incorporated with the extreme value stochastic engine) while the CML is used independently (as an analogue to evolutionary algorithms). Then the techniques are applied to optimize a real-world MO Gas Turbine - Absorption Chiller system. The authors carry out comparative analyses among the conventional DE approach (Gauss-DE), extreme value DE strategies and the CML.

Iurii V. Krak (Taras Shevchenko National University of Kyiv Glushkov Cybernetics Institute, Ukraine), *Olexander V. Barmak* and *Eduard A. Manziuk* (both from National University of Khmelnytskyi, Ukraine) in their chapter *"Visual Analytics to Build a Machine Learning Model"* are concerned with the active involvement of a human in the process of building a model, one of the most interesting and promising areas of machine learning. But there are problems with effective integration of humans into a workflow. Therefore, it is necessary to prepare techniques and information technologies that could allow for an effective use of human intellectual capabilities, thereby expanding the machine learning tools. The authors consider visual analytics with the goal of building a machine learning model by a human, and the technique of transferring this model to the machine level. This has made it possible to expand the capabilities of machine learning through an active and productive employment of human intellectual abilities.

In their chapter *"Smart Connected Digital Products and IoT Platform With the Digital Twin"*, *Yuvaraj D* (Cihan University, Duhok, Kurdistan Region, Iraq), *Mohamed Uvaze Ahamed A* (Cihan University, Erbil, Kurdistan Region, Iraq), *Jayanthiladevi A* (Institute of Scientific Research, Bangalore, India), *Balamurugan Easwaran* (University of Africa, Toru-Orua, Nigeria) and *Thamarai Selvi R* (Bishop Heber College, India) address digital illustration with a novel prevalence for a physical product, gain larger insight into that product's state performance and behavior. Digital twin is an unequivocal advanced copy of an item, method or control. This living model creates a thread between the physical and digital worlds. A model of a physical object - a "twin" - enables one to observe its standing, diagnose problems and take a look at solutions remotely. It is a dynamical virtual illustration of a tool which is unendingly fed with knowledge from embedded sensors and packages. This provides to some degree a correct period of time standing of the physical device. Digital twins drive innovation and performance, offering the development of prognostic analytics, and they give firms the flexibility to boost client expertise.

In the chapter *"Taxonomy of Influence Maximization Techniques in Unknown Social Networks: Influence Maximization Techniques"*, *B Bazeer Ahamed* (Al Musanna College of Technology, Oman), and *Sudhakaran Periakaruppan* (TRP Engineering College, India) deal with influence maximization in Online Social Networks (OSNs) which is the problem of discovering few nodes or users in a social network; they are termed as "seed nodes" and can help for spread of influence in a network. With the tremendous growth of social networking the influence exerted by users of a social network on other online users has caught the attention of researchers to develop effective influence maximization algorithms to be applied in the field of business strategies. The main application of influence maximization aims to promote the product to a set of users. However, this poses a real challenge in influence maximization algorithms to deal with a vast amount of users or nodes obtainable in any OSN. The authors focus on graph mining of OSNs for generating "seed sets" by employing standard influence maximization techniques. Many standard influence maximization models are used for calculating spread of influence; a novel influence maximization technique named as DegGreedy technique has been illustrated along with experimental results for comparing with existing techniques.

To all the *authors* of this book, we convey our sincere appreciation and gratitude for having shared their excellence, dedication and idealism with the academic community and, finally, with humankind. Furthermore, we send our gratitude to the publishing house of IGI Global, for making possible and become reality a truly international, excellent book of real-life significance and of potential high impact for the world of tomorrow, for this present generation and for the generations to come.

Now, we would like to wish us all a lot of joy and gain when browsing and reading this interesting work, and we hope that a remarkable benefit is going to be received from it science-wise, personally and societally.

December 2019,

Pandian Vasant
University of Technology Petronas, Malaysia

Valeriy Kharchenko
Federal Scientific Agroengineering Center VIM, Russia

Vladimir Panchenko
Russian University of Transport, Russia

Joshua Thomas
UOW Malaysia KDU, Malaysia

Gerhard-Wilhelm Weber
Poznan University of Technology, Poland & Middle East Technical University, Turkey

Wonsiri Punurai
Mahidol University, Thailand

Acknowledgment

The editors would like to sincerely thank the global research scholars for their wonderful and marvelous help and support in reviewing all the chapters in this book project with full responsibility and dedication. The overall quality, originality, clarity, creativity, superiority and novelty of the chapters have been tremendously improved and enhanced with their incredible contributions.

Chapter 1
Prediction of Menstrual Cycle Phase by Wearable Heart Rate Sensor

Junichiro Hayano
Graduate School of Medical Sciences, Nagoya City University, Japan

Emi Yuda
https://orcid.org/0000-0002-1865-7735
Graduate School of Engineering, Tohoku University, Japan

ABSTRACT

The prediction of the menstrual cycle phase and fertility window by easily measurable bio-signals is an unmet need and such technological development will greatly contribute to women's QoL. Although many studies have reported differences in autonomic indices of heart rate variability (HRV) between follicular and luteal phases, they have not yet reached the level that can predict the menstrual cycle phases. The recent development of wearable sensors-enabled heart rate monitoring during daily life. The long-term heart rate data obtained by them carry plenty of information, and the information that can be extracted by conventional HRV analysis is only a limited part of it. This chapter introduces comprehensive analyses of long-term heart rate data that may be useful for revealing their associations with the menstrual cycle phase.

INTRODUCTION

Although the menstrual cycle is a major biorhythm that governs biological functions of women, it is not easy to know the exact cycle phase like those of circadian and circaseptan (weekly) rhythms (Goodale et al., 2019). The most accurate method to detect the ovulation is transvaginal ultrasound examination (Ecochard, Boehringer, Rabilloud, & Marret, 2001), but it requires clinical visits and considerable cost. Calendar method and basal body temperature have been widely used, but the former is confounded by physiological cycle variation (Fehring, 2005) and the latter is affected by environmental temperature

DOI: 10.4018/978-1-7998-3970-5.ch001

(Shilaih et al., 2018). Cervical mucus monitoring is inconvenient and subjective (Brezina, Haberl, & Wallach, 2011). Although urine-based luteinizing hormone (LH) can be used for detecting LH surge usually followed by ovulation within 2 days (Behre et al., 2000), it already misses the days with the highest conception probability.

A simpler and accurate prediction of the menstrual cycle, especially fertile window and menstrual periods, will further improve the quality of life for women and their partners. The development of useful methods is an important and challenging issue for information technology. This chapter discusses biological signal processing to predict menstrual cycle phases, focusing on heart rate signals that can be obtained from many popular wearable sensors.

HEART RATE VARIABILITY (HRV)

It is well known that the autonomic nervous functions in women are affected by the menstrual cycle. Many studies reported the changes in autonomic indices of heart rate variability (HRV) with menstrual cycle (Bai, Li, Zhou, & Li, 2009; Brar, Singh, & Kumar, 2015; de Zambotti, Nicholas, Colrain, Trinder, & Baker, 2013; Guasti et al., 1999; Leicht, Hirning, & Allen, 2003; Sato & Miyake, 2004; Sato, Miyake, Akatsu, & Kumashiro, 1995; Yildirir, Kabakci, Akgul, Tokgozoglu, & Oto, 2002). These studies commonly suggest that, compared to the follicular phase, the luteal phase is accompanied by relative increase in sympathetic activity to cardiac parasympathetic activity. These studies, however, have not yet reached the level that can predict the phases of the menstrual cycle and they have used short-term HRV obtained under well controlled laboratory conditions.

The analysis of HRV is divided into short-term and long-term HRV (Hayano & Yuda, 2019). Typically, the former uses 5-minute data and the later uses 24-h data, but the difference is not only the length of data. Short-term HRV is analyzed from data obtained in a laboratory under strictly controlled conditions, while long-term HRV uses data recorded by wearable sensors in freely moving subjects. The autonomic nervous system is constantly regulating the body systems in response to various external and internal stimuli. Thus, the autonomic indices of HRV also respond sensitively to measurement conditions. For example, the power of high frequency (HF, 0.15-0.4 Hz) component of HRV that reflects cardiac parasympathetic function increases with supine rest and non-REM sleep (Hayano & Yasuma, 2003) and it decreases with standing (Pomeranz et al., 1985), physical activities(Yamamoto, Hughson, & Peterson, 1991), and food intake (Hayano et al., 1990). The HF power is also affected by respiration independently of autonomic neural activity; it increases with slow and deep breathing (Hayano, Mukai, et al., 1994; Hirsch & Bishop, 1981). Also, low frequency (LF, 0.04-0.15 Hz) component, particularly its ratio to HF power (LF/HF), increased with standing (Hayano et al., 2001).

These indicate that conventional autonomic indices such as HF and LF power and LF/HF observed in long-term HRV do not reflect *autonomic function*, but they reflect *autonomic state* as a result of responses to certain stimuli (Hayano & Yuda, 2019). Even if the power of HF component was measured by long-term HRV analysis, it is impossible to interpret cardiac parasympathetic nerve function without information on the subject's physical and mental states and measurement environment. Even if 24-hour average LF/HF is higher in subject A than subject B, it may not indicate a difference in autonomous function between subjects, but may be the result of subject B lying longer than subject A during recording (Yoshida et al., 2016).

It is therefore reasonable that the most of earlier studies above mentioned (Bai et al., 2009; Brar et al., 2015; de Zambotti et al., 2013; Guasti et al., 1999; Leicht et al., 2003; Sato & Miyake, 2004; Sato et al., 1995; Yildirir et al., 2002) reporting the menstrual cycle of autonomic function used short-term HRV. They report that normalized LF power and LF/HF, measured under controlled conditions, increase in the luteal phase compared to the follicular phase. These research results, however, do not directly apply to long-term HRV acquired by wearable sensors.

LONG-TERM HRV

If the lack of information about subject's activity during recording is the cause of the limitation of autonomic functional assessment by long-term HRV, it may be partially resolved by the inputs of activity data into the analysis. This is a realistic approach because most wearable heart rate sensors have built-in actigraphic sensors (accelerometers).

Figure 1 shows the analysis of long-term HRV together with body movement and postural changes estimated from actigraphic data on day 20 of the menstrual cycle (luteal phase) in a female subject. During nighttime, heart rate appears to decrease and the amplitude of HRV frequency components shows larger peaks. Although they show complicated fluctuations during the day, the fluctuations seem associated partly with body movement and posture.

Figure 1.

Analysis of long-term heart rate variability (HRV) and actigraphic data on day 20 after menstruation in a female subject. HRV indices were analyzed continuously by the method of complex demodulation and averaged over every one minute. HR = heart rate, VLF = amplitude of very-low-frequency component, LF = amplitude of low-frequency component, HF = amplitude of high-frequency component, LF/HF = LF to HF power ratio, PA = physical activity (mG), and Pos = body posture (0-5 = recumbent, 10 = sitting, 30 = standing, and >30 = walking or running). Figure adapted from Figure 1 in reference (Yuda & Hayano, 2018).

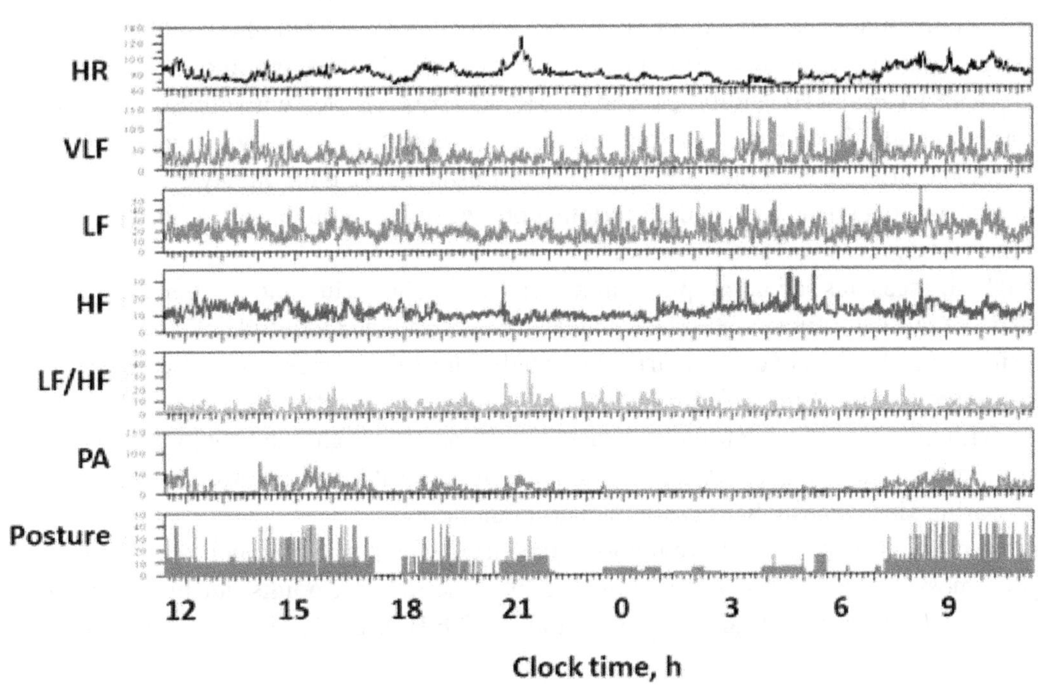

In this analysis, the 24-hour actigraphic data were obtained by built-in triaxial acceleration sensors of a Holter electrocardiographic (ECG) recorder attached on the chest wall. Accelerations in the left-to-right, caudo-cranial, and postero-anterior axes were measured as $x(t)$, $y(t)$, and $z(t)$ values, respectively, and body position was estimated by a model developed by machine-leaning (Yoshida, Yuda, & Hayano, 2019). The time series of $x(t)$, $y(t)$, and $z(t)$ were high-pass filtered and the level of physical activity was estimated as a composite vector value, A(t).

$$\mathrm{A}(t) = \sqrt{x(t)^2 + y(t)^2 + z(t)^2} \tag{1}$$

The changes in HRV indices during 24 hours were analyzed by the method of complex demodulation (CDM) (Hayano, Taylor, et al., 1994; Hayano et al., 1993). CDM is a time-domain method for time series analysis that provides continuous changes in amplitude and frequency of frequency component in the frequency band of interest as the functions of time. For CDM analysis, R-R interval time series were interpolated only using R-R intervals consisting of consecutive sinus beats with a step function and resampled with equidistant time interval at 2 Hz. Then, frequency bands for very low frequency (VLF, 0.0033-0.04 Hz), LF, and HF components were demodulated by CDM. The continuous time series of VLF, LF, and HF amplitude were averaged over every one minute. By the use of CDM, the minute-to-minute changes in HRV components can be matched with physical activity and posture data for the corresponding time.

Figure 2 shows the changes in heart rate and HRV indices averaged over each physical activity category during the menstrual cycle of a healthy female subject in her 30th. She had experienced normal deliveries twice and had a stable menstrual cycle with a period of ~27 days. Using Holter ECG recorder with built-in triaxial acceleration sensors, 24-hour ECGs and actigraphic data were measured before, on the day when a menstruation started, 5, 10, 15, 20, 25 days after the start of the menstruation, and the day when the next menstruation started. According to the level of physical activity, the physical activity state at every one minute was classified into 3 categories; moving, resting (standing or sitting at rest), and sleeping (lying at rest). Minute-by-minute heart rate, standard deviation of normal-to-normal R-R intervals (SDNN), LF, and HF amplitude, and LF/HF data were labeled with physical activity categories for the corresponding time. The data were averaged over physical activity categories for each day of measurement and also for each menstrual phase. For this analysis, data on day 5 and 10 were assumed as follicular phase, data on day 20 and 25 as luteal phase, and data on days 0 at both the beginning and end of menstrual cycle as menstrual period.

Although all measures including physical activity level show fluctuations that are not simple even categorized by physical activity states, heart rate was lower and the amplitude of HRV components were higher during the earlier half of cycle than the latter half. These differences were also detected statistically by evaluating the effects of menstrual phase after adjusting for the effect of time of the day (Figure 3). Compared with the values during follicular phase, heart rate was higher during luteal phase and slightly lower during menstruation period. Also, LF and HF amplitude at rest was lower during luteal phase. HF amplitude during sleep was higher during menstruation period.

Although many earlier studies have reported the changes in autonomic nervous function with menstrual cycle, most of them only compared autonomic function at 2 points during follicular phase and luteal phase by short-term HRV analysis (Bai et al., 2009; Brar et al., 2015; de Zambotti et al., 2013;

Figure 2.
Changes in heart rate and HRV indices categorized by physical activity state during a menstrual cycle in a female subject. SDNN = standard deviation of normal-to-normal R-R intervals. Figure adapted from Figure 2 in reference (Yuda & Hayano, 2018).

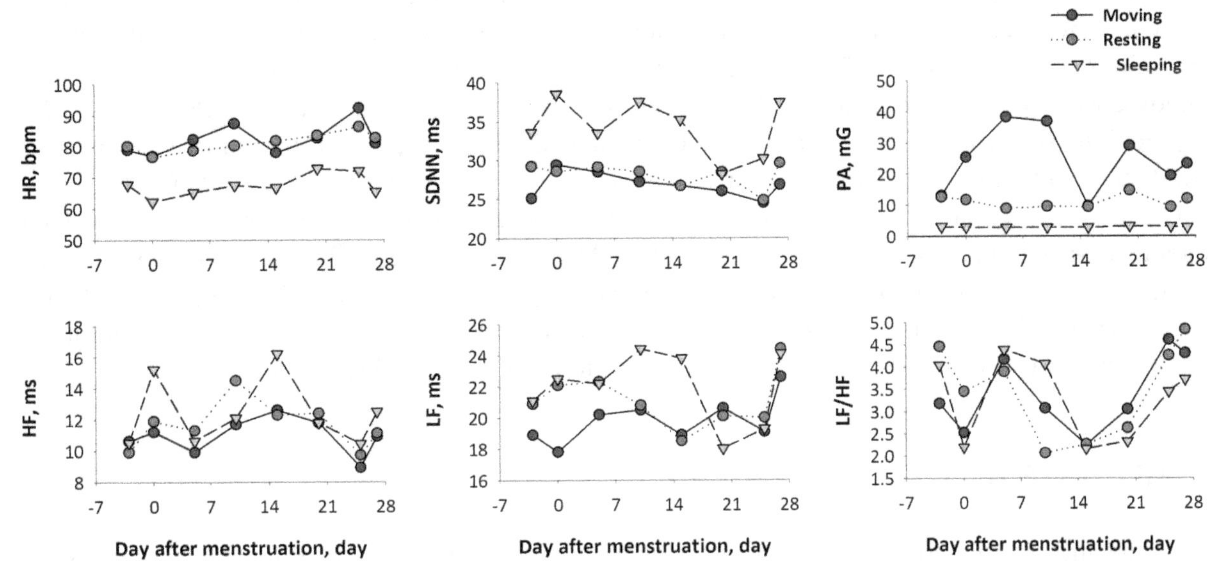

Figure 3.
*Effects of menstrual phase on heart rate and HRV indices categorized by physical activity state. Data are least-square means adjusted for the effects of time of the day. Error bars are standard error. *Values significantly (P <0.05) different from those for follicular phase. Figure adapted from Figure 3 in reference (Yuda & Hayano, 2018).*

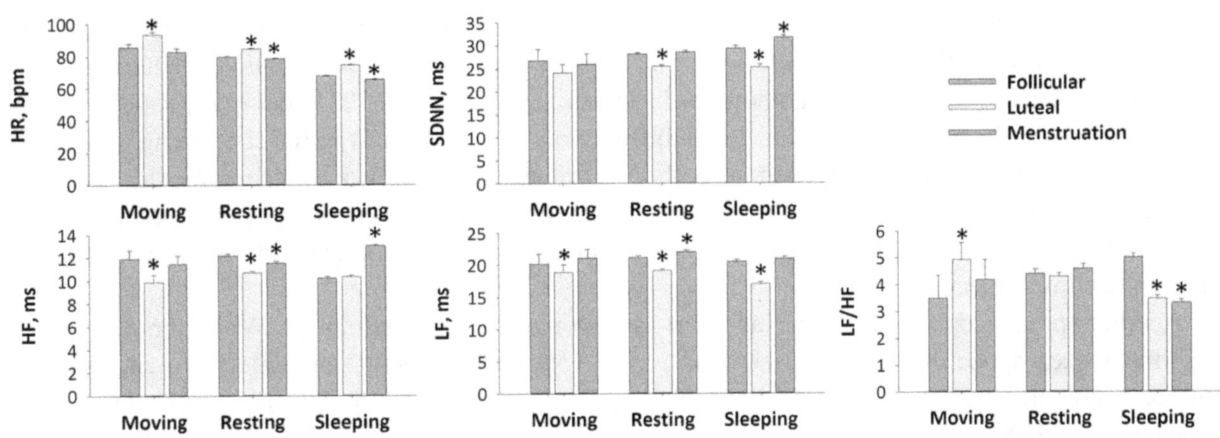

Guasti et al., 1999; Leicht et al., 2003; Sato & Miyake, 2004; Sato et al., 1995; Yildirir et al., 2002). However, the changes in autonomic nervous functions with the menstrual cycle are not binary, but they show complicated fluctuations depending on the kind of autonomic nervous function and the state of physical activity. Additionally, it has been reported that the menstrual cycle affects everyday behaviors such as diet, exercise, and sleep, and that it mediates the apparent relation between menstrual cycle and autonomic nervous function (Tada et al., 2017). These indicate the necessity of simultaneous evaluation of the fluctuation of lifestyle for evaluating the menstrual cycle of autonomic nervous function.

Currently, wearable bio-signal monitors incorporating acceleration and pulse wave sensors are widely used. These techniques are expected to enable longitudinal data collections for weeks or even months under daily life. At present, however, those pulse wave sensors are able to measure pulse wave signal only at rest or during sleep with a signal quality that allows the frequency component analysis (Akiyama et al., 2015; Constant, Laude, Murat, & Elghozi, 1999; Ferrer-Mileo, Guede-Fernandez, Fernandez-Chimeno, Ramos-Castro, & Garcia-Gonzalez, 2015; Gil, Bailon, Vergara, & Laguna, 2010; Gil, Orini, et al., 2010; Hayano, Barros, Kamiya, Ohte, & Yasuma, 2005; Schafer & Vagedes, 2013). Consequently, we presently need to use Holter ECG for continuous data collection including daytime activities, but ECG has limitations due to skin irritation caused by continuous electrodes' attachment and bathing restrictions due to water resistance issue. Thus, 24 hours are the standard length of long-term HRV analysis in most studies so far. Technology is constantly evolving, however. We may be able to evaluate the menstrual cycle from longer-term continuous data and more accurately predict the timing of ovulation and menstruation. The activity-type-specific analysis of HRV during daily life would be useful for them.

HEART RATE RESPONSE TO STIMULI

While the analysis of autonomic nervous function by long-term HRV may be limited by the confusions due to physical and mental activities and the influence of external environment, long-term heart rate monitoring during daily activities provide opportunities to evaluate heart rate responses to spontaneous accidental stimuli that occur during monitoring. One of such stimuli is ventricular ectopic contraction (VPC). VPC usually causes a transient decrease in arterial blood pressure due to insufficient ventricular filling prior to contraction. The decreased arterial pressure stimulates the sino-aortic baroreceptors and causes a suppression of cardiac parasympathetic activity, which results a rapid and transient tachycardia (shortening of several R-R intervals) followed by gradual recovery with an overshoot (lengthening of R-R intervals). This process has been known as heart rate turbulence (HRT) (Bauer et al., 2008; Schmidt et al., 1999). In patients with parasympathetic dysfunction, either a lack of initial shortening of the R-R intervals or a blunted slope is observed, or both. Abnormal HRT reflects baroreceptor reflex dysfunction and is a powerful predictor of increased mortality risk after acute myocardial infarction (Barthel et al., 2003). HRT, however, can be evaluated only when appropriate isolated VPCs exist, which is not often in healthy subjects.

Another spontaneous stimulus during heart rate monitoring is sleep apnea. While the prevalence of clinical sleep apnea that requires treatment is 5 to 15% in the general population (Duran, Esnaola, Rubio, & Iztueta, 2001; Young et al., 1993; Young, Peppard, & Gottlieb, 2002), the episode of sleep apnea itself is observed in >95% of subjects undergoing Holter ECG monitoring (Hayano, Watanabe, et al., 2011). The episodes of sleep apnea are accompanied by characteristic pattern of heart rate response known as cyclic variation of heart rate (CVHR) (Guilleminault, Connolly, Winkle, Melvin, & Tilkian, 1984; Hayano, Watanabe, et al., 2011). CVHR consists of bradycardia during apnea followed by abrupt tachycardia on its cessation. The frequency of CVHR (Fcv) closely correlates with the apnea-hypopnea index that reflects the frequency of sleep apnea episodes (Hayano et al., 2013; Hayano, Watanabe, et al., 2011). CVHR is thought to reflect cardiac autonomic responses to cardio-respiratory perturbation caused by apneic/hypoxic episodes (Zwillich et al., 1982). CVHR is abolished by parasympathetic blocking by atropine (Guilleminault et al., 1984). These indicate that CVHR is a maker of cardiac parasympathetic responsiveness to sleep apnea. Blunted CVHR reflected by reduced amplitude of CVHR (Acv) is observed in high risk patients after myocardial infarction and those with heart failure.

Table 1 shows the comparisons of Fcv and Acv between the follicular and luteal phases in 5 healthy women (age, 37 ± 7 [21-47] y) with a regular menstrual cycle of 28-32 (median, 28) days who underwent 24-hour Holter ECG monitoring in both menstrual phases (5-12 days after the 1st day of menstruation and 5-15 days after the estimated day of ovulation). The time in bed was estimated from actigraphic data and CVHR was detected by an algorithm of auto-correlated wave detection with adaptive threshold (ACAT) (Hayano, Watanabe, et al., 2011). Acv was quantified by signal averaging of R-R interval responses accompanying CHVR (Hayano et al., 2017). Fcv was greater during the follicular phase than luteal phase with a marginal statistical significance (P = 0.05) and the maximum Fcv in 30-minute moving window was significantly greater during follicular phase. These indicate that sleep apnea episodes are more likely to occur during follicular phase than luteal phase. On the other hand, there is no significant difference in Acv with menstrual phase, suggesting that cardiac parasympathetic responsiveness to sleep apnea may not be affected by menstrual phase.

Table 1. Frequency and amplitude of cyclic variation of heart rate (CVHR) in menstrual phases

Variable	Follicular phase	Luteal phase	F	P
Time in bed, min	457 ± 57	404 ± 55	0.59	0.4
Fcv, cycle/h	5.4 ± 1.4	3.4 ± 1.4	4.51	0.05
Max Fcv, cycle/h	29 ± 6.3	21.4 ± 6.2	3.34	0.09
Cycle length of CVHR, s	72.1 ± 2.1	72.4 ± 2	0.01	0.9
Acv, ln(ms)	4.77 ± 0.14	4.74 ± 0.14	0.09	0.7

Acv = amplitude of CVHR, CVHR = cyclic variation of heart rate, Fcv = frequency of CVHR per h. Data adapted from Table 7 in reference (Yuda & Hayano, 2020).

BASAL HEART RATE

Basal heart rate (BHR) that is defined as the minimum heart rate during 24 hours is another index that can be obtained only by long-term heart rate monitoring (Yuda, Yoshida, & Hayano, 2018). To avoid false long R-R intervals (such as due to heart block or R-wave detection failure) recognized as the minimum heart rate, the median R-R intervals in the 3-minute moving windows are calculated first for 24 hours, then the maximum value among the median values during 24 hours was used for calculating BHR. BHR is a robust physiological feature of individual subject (Yuda, Furukawa, Yoshida, & Hayano, 2018). BHR is highest at birth and decreases with age until 20 years, but it shows no significant change with age thereafter. The average time at which the BHR occurs distributes between 02 h and 05 h in both sexes. BHR is thought to reflect the maximum capacity of parasympathetic heart rate control.

Table 2 shows the comparisons of the BHR and the percentile values of heart rate in 24 hours between follicular and luteal phases in 5 healthy women. There is no significant difference in BHR between follicular and luteal phases, indicating the robustness of BHR against menstrual cycle. On the other hand, 50 and 75 percentiles of heart rate are higher during the luteal phase than the follicular phase. These indicate that the increase in heart rate in the luteal phase is not due to a simple rightward shift of the whole heart rate distribution, but to a change in the shape of distribution where the median was shifted to the right.

Table 2. Basal and percentile heart rate in menstrual phases

Variable	Follicular phase	Luteal phase	F	P
100 Percentile (max HR)	126 ± 6	132 ± 6	1.53	0.2
95 Percentile	98 ± 5	108 ± 5	3.31	0.09
75 Percentile	82 ± 2	88.3 ± 2	8.66	0.01
50 Percentile (median HR)	74 ± 2	79.2 ± 2	11.73	0.005
25 Percentile	67 ± 1	69.9 ± 1	2.9	0.1
5 Percentile	58 ± 2	60.4 ± 2	2.43	0.1
0 Percentile (basal HR)	54 ± 2	56.2 ± 2	1.82	0.2

HR = heart rate. Data adapted from Table 6 in reference (Yuda & Hayano, 2020).

NONLINEAR HEART RATE DYNAMICS

Information extracted by the autonomic indices of HRV is only a small part of information carried by the long-term heart rate signal. If it is compared to music, it is just looking at the average volume of each pitch, and from there it is not clear even whether it is music or noise. From there, there is no way to know the song title, emotion, artistry, or performer skills. The attempts to approach them have been made by researches on the nonlinear dynamics of heart rate regulation (Hayano et al., 2018; Huikuri, Perkiomaki, Maestri, & Pinna, 2009; Iyengar, Peng, Morin, Goldberger, & Lipsitz, 1996; Kiyono, Hayano, Watanabe, Struzik, & Yamamoto, 2008; Kiyono et al., 2004; Peng, Havlin, Stanley, & Goldberger, 1995; Pincus, 1991; Yeragani et al., 1998). These studies provided the methods for quantifying the complexity/flexibility of heart rate dynamics, but most of these characteristics have not been studied in relation to the menstrual cycle.

The authors performed four kinds of nonlinear analysis of heart rate dynamics for 24-h data obtained during the follicular and luteal phases. First, approximate entropy (ApEn) was calculated for minute-to-minute heart rate to evaluate unpredictability or irregularity of heart rate (Pincus, 1991; Yeragani et al., 1998). Second, short-term (4-11 beats) and long-term (>11 beats) scaling exponents (α_1 and α_2, respectively) were computed by detrended fluctuation analysis (DFA) (Huikuri et al., 2009; Iyengar et al., 1996; Peng et al., 1995) to evaluate the correlation property of heart rate dynamics. Third, the non-Gaussianity of probability density function of abrupt heart rate changes was evaluated by λ_{25s} (Kiyono et al., 2008; Kiyono et al., 2004). Finally, the ratio between random and regulated components of HRV was estimated from the residual variance of autoregressive (AR) model (ARV) (Hayano, Kiyono, et al., 2011). To evaluate the stationarity of the system to disturbance, the authors also examined if the complex roots of the characteristic equation of the AR model distributed out-side of the unit circle of the complex plane.

Table 3 shows nonlinear indices of 24-hour heart rate dynamics during the follicular and luteal phase in 5 health women. Among nonlinear indices, a reduction in the variance (σ^2) of random component of HRV assessed as AR model residual in the luteal phase was only significant difference between the follicular and luteal phases. In contrast, the analysis of data during sleep period (Table 4) showed significant reductions in ApEn and the σ^2 of random component, and significant increase in the minimum absolute value of characteristic root (RCE) in the luteal phase. These changes indicate that heart rate is more regulated and more stable during the luteal phase than the follicular phase.

Table 3. Nonlinear indices of 24-h heart rate dynamics.

Variable	Follicular phase	Luteal phase	F	P
ApEn of minute HR	1.17 ± 0.05	1.16 ± 0.05	0.28	0.6
DFA α_1	1.28 ± 0.05	1.27 ± 0.05	0.02	0.8
DFA α_2	1.12 ± 0.02	1.16 ± 0.02	3.6	0.08
Non-Gaussianity index λ_{25s}	0.507 ± 0.071	0.53 ± 0.071	0.24	0.6
ARV, %	4.7 ± 1.0	3.7 ± 1.0	4.37	0.06
σ^2 of random component, ms	986 ± 337	837 ± 337	7.96	0.01
σ^2 of regulated component, ms	18044 ± 3252	19763 ± 3196	0.27	0.6
Average of absolute RCE	1.095 ± 0.007	1.105 ± 0.007	1.44	0.2
SD of absolute RCE	0.0133 ± 0.0001	0.0132 ± 0.0001	1.15	0.3
Max absolute RCE	1.106 ± 0.007	1.116 ± 0.007	1.3	0.2
Min absolute RCE	1.048 ± 0.009	1.061 ± 0.009	1.85	0.2

ApEn = approximate entropy, ARV = autoregressive variability, DFA = detrended fluctuation analysis, RCE = root of characteristic equation. Data adapted from Table 4 in reference (Yuda & Hayano, 2020).

Table 4. Nonlinear indices of heart rate dynamics during sleep

Variable	Follicular phase	Luteal phase	F	P
ApEn of minute HR	0.92 ± 0.04	0.84 ± 0.04	5.11	0.04
DFA α_1	1.2 ± 0.07	1.21 ± 0.07	0.05	0.8
DFA α_2	1.01 ± 0.04	1.03 ± 0.04	0.62	0.4
Lambda 25 s	0.513 ± 0.045	0.523 ± 0.045	0.2	0.6
ARV, %	13.2 ± 3.7	10.9 ± 3.7	3.41	0.09
σ^2 of random component, ms	1401 ± 617	1262 ± 616	6.18	0.03
σ^2 of regulated component, ms	9531 ± 2641	9690 ± 2635	0.02	0.8
Average of absolute RCE	10932 ± 2859	10952 ± 2853	0	0.9
SD of absolute RCE	1.100 ± 0.012	1.122 ± 0.0116	4.62	0.05
Max absolute RCE	0.0147 ± 0.0011	0.0160 ± 0.0010	3.91	0.07
Min absolute RCE	1.114 ± 0.013	1.138 ± 0.013	5.07	0.04

Abbreviations are explained in the foot note to Table 3. Data adapted from Table 5 in reference (Yuda & Hayano, 2020).

CONCLUSION

In this chapter, we discussed the possibility to predicting the phase of menstrual cycle from heart rate signals acquired by wearable sensors. The messages of this chapter are summarized as follows:

1. Autonomic function assessed by HRV shows changes in synchrony with the menstrual cycle, but its analysis requires heart rate signals measured under strictly controlled conditions. The HRV autonomic index obtained under daily activities by the wearable sensors provides the state of

autonomic activity, which is the result of responses to various daily activities, from which the autonomic function is unknown.

2. The simultaneous analysis of activity with long-term HRV, however, may be used for detecting the menstrual variations of autonomic function by the separate analyses for each activity state.
3. On the other hand, long-term heart rate signals provide useful measures that are not obtained from short-term heart rate signals; those include Acv that reflects parasympathetic responsiveness to sleep apnea and BHR that reflects the maximum capacity of parasympathetic heart rate control. Although neither BHR nor Acv showed significant menstrual variations, they may be used as the individual personal reference values for heart rate and heart rate responsiveness that are not affected by the menstrual cycle.
4. Finally, long-term heart rate signals provide an analysis of nonlinear heart rate dynamics. In order to measure various characteristics of heart rate dynamics, various nonlinear indices have been introduced, showing a differential association with the follicular and luteal phases.

The long-term heart rate signal obtained by wearable sensors carry various information, which may provide useful features to predict menstrual cycle phase.

REFERENCES

Akiyama, T., Miyazaki, T., Ito, H., Nogami, H., Higurashi, E., Ando, S., & Sawada, R. (2015). Comparable accuracy of micro-electromechanical blood flowmetry-based analysis vs. electrocardiography-based analysis in evaluating heart rate variability. *Circulation Journal*, *79*(4), 794–801. doi:10.1253/circj.CJ-14-1199 PMID:25740056

Bai, X., Li, J., Zhou, L., & Li, X. (2009). Influence of the menstrual cycle on nonlinear properties of heart rate variability in young women. *American Journal of Physiology. Heart and Circulatory Physiology*, *297*(2), H765–H774. doi:10.1152/ajpheart.01283.2008 PMID:19465541

Barthel, P., Schneider, R., Bauer, A., Ulm, K., Schmitt, C., Schomig, A., & Schmidt, G. (2003). Risk stratification after acute myocardial infarction by heart rate turbulence. *Circulation*, *108*(10), 1221–1226. doi:10.1161/01.CIR.0000088783.34082.89 PMID:12939209

Bauer, A., Malik, M., Schmidt, G., Barthel, P., Bonnemeier, H., Cygankiewicz, I., ... Zareba, W. (2008). Heart rate turbulence: standards of measurement, physiological interpretation, and clinical use: International Society for Holter and Noninvasive Electrophysiology Consensus. *Journal of the American College of Cardiology*, *52*(17), 1353–1365. doi:10.1016/j.jacc.2008.07.041 PMID:18940523

Behre, H. M., Kuhlage, J., Gassner, C., Sonntag, B., Schem, C., Schneider, H. P., & Nieschlag, E. (2000). Prediction of ovulation by urinary hormone measurements with the home use ClearPlan Fertility Monitor: Comparison with transvaginal ultrasound scans and serum hormone measurements. *Human Reproduction (Oxford, England)*, *15*(12), 2478–2482. doi:10.1093/humrep/15.12.2478 PMID:11098014

Brar, T. K., Singh, K. D., & Kumar, A. (2015). Effect of Different Phases of Menstrual Cycle on Heart Rate Variability (HRV). *Journal of Clinical and Diagnostic Research: JCDR*, *9*(10), CC01–CC04. doi:10.7860/JCDR/2015/13795.6592 PMID:26557512

Brezina, P. R., Haberl, E., & Wallach, E. (2011). At home testing: Optimizing management for the infertility physician. *Fertility and Sterility*, *95*(6), 1867–1878. doi:10.1016/j.fertnstert.2011.01.001 PMID:21324446

Constant, I., Laude, D., Murat, I., & Elghozi, J. L. (1999). Pulse rate variability is not a surrogate for heart rate variability. *Clinical Science (London, England)*, *97*(4), 391–397. doi:10.1042/cs0970391 PMID:10491338

de Zambotti, M., Nicholas, C. L., Colrain, I. M., Trinder, J. A., & Baker, F. C. (2013). Autonomic regulation across phases of the menstrual cycle and sleep stages in women with premenstrual syndrome and healthy controls. *Psychoneuroendocrinology*, *38*(11), 2618–2627. doi:10.1016/j.psyneuen.2013.06.005 PMID:23850226

Duran, J., Esnaola, S., Rubio, R., & Iztueta, A. (2001). Obstructive sleep apnea-hypopnea and related clinical features in a population-based sample of subjects aged 30 to 70 yr. *American Journal of Respiratory and Critical Care Medicine*, *163*(3 Pt 1), 685–689. doi:10.1164/ajrccm.163.3.2005065 PMID:11254524

Ecochard, R., Boehringer, H., Rabilloud, M., & Marret, H. (2001). Chronological aspects of ultrasonic, hormonal, and other indirect indices of ovulation. *BJOG*, *108*(8), 822–829. doi:10.1111/j.1471-0528.2001.00194.x PMID:11510707

Fehring, R. J. (2005). New low- and high-tech calendar methods of family planning. *Journal of Midwifery & Women's Health*, *50*(1), 31–38. doi:10.1016/j.jmwh.2004.07.001 PMID:15637512

Ferrer-Mileo, V., Guede-Fernandez, F., Fernandez-Chimeno, M., Ramos-Castro, J., & Garcia-Gonzalez, M. A. (2015). Accuracy of heart rate variability estimation by photoplethysmography using an smartphone: Processing optimization and fiducial point selection. *Conference proceedings: ... Annual International Conference of the IEEE Engineering in Medicine and Biology Society. IEEE Engineering in Medicine and Biology Society. Conference, 2015*, 5700-5703.

Gil, E., Bailon, R., Vergara, J. M., & Laguna, P. (2010). PTT variability for discrimination of sleep apnea related decreases in the amplitude fluctuations of PPG signal in children. *IEEE Transactions on Biomedical Engineering*, *57*(5), 1079–1088. doi:10.1109/TBME.2009.2037734 PMID:20142152

Gil, E., Orini, M., Bailon, R., Vergara, J. M., Mainardi, L., & Laguna, P. (2010). Photoplethysmography pulse rate variability as a surrogate measurement of heart rate variability during non-stationary conditions. *Physiological Measurement*, *31*(9), 1271–1290. doi:10.1088/0967-3334/31/9/015 PMID:20702919

Goodale, B. M., Shilaih, M., Falco, L., Dammeier, F., Hamvas, G., & Leeners, B. (2019). Wearable Sensors Reveal Menses-Driven Changes in Physiology and Enable Prediction of the Fertile Window: Observational Study. *Journal of Medical Internet Research*, *21*(4), e13404. doi:10.2196/13404 PMID:30998226

Guasti, L., Grimoldi, P., Mainardi, L. T., Petrozzino, M. R., Piantanida, E., Garganico, D., ... Venco, A. (1999). Autonomic function and baroreflex sensitivity during a normal ovulatory cycle in humans. *Acta Cardiologica*, *54*(4), 209–213. PMID:10511897

Guilleminault, C., Connolly, S., Winkle, R., Melvin, K., & Tilkian, A. (1984). Cyclical variation of the heart rate in sleep apnoea syndrome. Mechanisms, and usefulness of 24 h electrocardiography as a screening technique. *Lancet*, *1*(8369), 126–131. doi:10.1016/S0140-6736(84)90062-X PMID:6140442

Hayano, J., Barros, A. K., Kamiya, A., Ohte, N., & Yasuma, F. (2005). Assessment of pulse rate variability by the method of pulse frequency demodulation. *Biomedical Engineering Online*, *4*(1), 62. doi:10.1186/1475-925X-4-62 PMID:16259639

Hayano, J., Kiyono, K., Struzik, Z. R., Yamamoto, Y., Watanabe, E., Stein, P. K., ... Carney, R. M. (2011). Increased non-gaussianity of heart rate variability predicts cardiac mortality after an acute myocardial infarction. *Frontiers in Physiology*, *2*, 65. doi:10.3389/fphys.2011.00065 PMID:21994495

Hayano, J., Mukai, S., Fukuta, H., Sakata, S., Ohte, N., & Kimura, G. (2001). Postural response of low-frequency component of heart rate variability is an increased risk for mortality in patients with coronary artery disease. *Chest*, *120*(6), 1942–1952. doi:10.1378/chest.120.6.1942 PMID:11742926

Hayano, J., Mukai, S., Sakakibara, M., Okada, A., Takata, K., & Fujinami, T. (1994). Effects of respiratory interval on vagal modulation of heart rate. *The American Journal of Physiology*, *267*(1 Pt 2), H33–H40. PMID:7914066

Hayano, J., Ohashi, K., Yoshida, Y., Yuda, E., Nakamura, T., Kiyono, K., & Yamamoto, Y. (2018). Increase in random component of heart rate variability coinciding with developmental and degenerative stages of life. *Physiological Measurement*, *39*(5), 054004. doi:10.1088/1361-6579/aac007 PMID:29693554

Hayano, J., Sakakibara, Y., Yamada, M., Kamiya, T., Fujinami, T., Yokoyama, K., ... Takata, K. (1990). Diurnal variations in vagal and sympathetic cardiac control. *The American Journal of Physiology*, *258*(3 Pt 2), H642–H646. PMID:2316678

Hayano, J., Taylor, J. A., Mukai, S., Okada, A., Watanabe, Y., Takata, K., & Fujinami, T. (1994). Assessment of frequency shifts in R-R interval variability and respiration with complex demodulation. *Journal of Applied Physiology*, *77*(6), 2879–2888. doi:10.1152/jappl.1994.77.6.2879 PMID:7896636

Hayano, J., Taylor, J. A., Yamada, A., Mukai, S., Hori, R., Asakawa, T., ... Fujinami, T. (1993). Continuous assessment of hemodynamic control by complex demodulation of cardiovascular variability. *The American Journal of Physiology*, *264*, H1229–H1238. PMID:8476100

Hayano, J., Tsukahara, T., Watanabe, E., Sasaki, F., Kawai, K., Sakakibara, H., . . . Fujimoto, K. (2013). Accuracy of ECG-based screening for sleep-disordered breathing: a survey of all male workers in a transport company. *Sleep & breathing = Schlaf & Atmung, 17*(1), 243-251.

Hayano, J., Watanabe, E., Saito, Y., Sasaki, F., Fujimoto, K., Nomiyama, T., ... Sakakibara, H. (2011). Screening for obstructive sleep apnea by cyclic variation of heart rate. *Circulation: Arrhythmia and Electrophysiology*, *4*(1), 64–72. doi:10.1161/CIRCEP.110.958009 PMID:21075771

Hayano, J., & Yasuma, F. (2003). Hypothesis: Respiratory sinus arrhythmia is an intrinsic resting function of cardiopulmonary system. *Cardiovascular Research*, *58*(1), 1–9. doi:10.1016/S0008-6363(02)00851-9 PMID:12667941

Hayano, J., Yasuma, F., Watanabe, E., Carney, R. M., Stein, P. K., Blumenthal, J. A., ... Kodama, I. (2017). Blunted cyclic variation of heart rate predicts mortality risk in post-myocardial infarction, end-stage renal disease, and chronic heart failure patients. *Europace*, *19*(8), 1392–1400. PMID:27789562

Hayano, J., & Yuda, E. (2019). Pitfalls of assessment of autonomic function by heart rate variability. *Journal of Physiological Anthropology*, *38*(1), 3. doi:10.118640101-019-0193-2 PMID:30867063

Hirsch, J. A., & Bishop, B. (1981). Respiratory sinus arrhythmia in humans: How breathing pattern modulates heart rate. *The American Journal of Physiology*, *241*, H620–H629. PMID:7315987

Huikuri, H. V., Perkiomaki, J. S., Maestri, R., & Pinna, G. D. (2009). Clinical impact of evaluation of cardiovascular control by novel methods of heart rate dynamics. *Philosophical Transactions. Series A, Mathematical, Physical, and Engineering Sciences, 367*(1892), 1223-1238.

Iyengar, N., Peng, C. K., Morin, R., Goldberger, A. L., & Lipsitz, L. A. (1996). Age-related alterations in the fractal scaling of cardiac interbeat interval dynamics. *The American Journal of Physiology*, *271*, R1078–R1084. PMID:8898003

Kiyono, K., Hayano, J., Watanabe, E., Struzik, Z. R., & Yamamoto, Y. (2008). Non-Gaussian heart rate as an independent predictor of mortality in patients with chronic heart failure. *Heart Rhythm*, *5*(2), 261–268. doi:10.1016/j.hrthm.2007.10.030 PMID:18242551

Kiyono, K., Struzik, Z. R., Aoyagi, N., Sakata, S., Hayano, J., & Yamamoto, Y. (2004). Critical scale invariance in a healthy human heart rate. *Physical Review Letters*, *93*(17), 178103. doi:10.1103/PhysRevLett.93.178103 PMID:15525130

Leicht, A. S., Hirning, D. A., & Allen, G. D. (2003). Heart rate variability and endogenous sex hormones during the menstrual cycle in young women. *Experimental Physiology*, *88*(3), 441–446. doi:10.1113/eph8802535 PMID:12719769

Peng, C. K., Havlin, S., Stanley, H. E., & Goldberger, A. L. (1995). Quantification of scaling exponents and crossover phenomena in nonstationary heartbeat time series. *Chaos (Woodbury, N.Y.)*, *5*(1), 82–87. doi:10.1063/1.166141 PMID:11538314

Pincus, S. M. (1991). Approximate entropy as a measure of system complexity. *Proceedings of the National Academy of Sciences of the United States of America*, *88*(6), 2297–2301. doi:10.1073/pnas.88.6.2297 PMID:11607165

Pomeranz, B., Macaulay, R. J., Caudill, M. A., Kutz, I., Adam, D., Gordon, D., & … . (1985). Assessment of autonomic function in humans by heart rate spectral analysis. *The American Journal of Physiology*, *248*(1 Pt 2), H151–H153. PMID:3970172

Sato, N., & Miyake, S. (2004). Cardiovascular reactivity to mental stress: Relationship with menstrual cycle and gender. *Journal of Physiological Anthropology and Applied Human Science*, *23*(6), 215–223. doi:10.2114/jpa.23.215 PMID:15599065

Sato, N., Miyake, S., Akatsu, J., & Kumashiro, M. (1995). Power spectral analysis of heart rate variability in healthy young women during the normal menstrual cycle. *Psychosomatic Medicine*, *57*(4), 331–335. doi:10.1097/00006842-199507000-00004 PMID:7480562

Schafer, A., & Vagedes, J. (2013). How accurate is pulse rate variability as an estimate of heart rate variability? A review on studies comparing photoplethysmographic technology with an electrocardiogram. *International Journal of Cardiology*, *166*(1), 15–29. doi:10.1016/j.ijcard.2012.03.119 PMID:22809539

Schmidt, G., Malik, M., Barthel, P., Schneider, R., Ulm, K., Rolnitzky, L., ... Schomig, A. (1999). Heart-rate turbulence after ventricular premature beats as a predictor of mortality after acute myocardial infarction. *Lancet*, *353*(9162), 1390–1396. doi:10.1016/S0140-6736(98)08428-1 PMID:10227219

Shilaih, M., Goodale, B. M., Falco, L., Kubler, F., De Clerck, V., & Leeners, B. (2018). Modern fertility awareness methods: Wrist wearables capture the changes in temperature associated with the menstrual cycle. *Bioscience Reports*, *38*(6), BSR20171279. doi:10.1042/BSR20171279 PMID:29175999

Tada, Y., Yoshizaki, T., Tomata, Y., Yokoyama, Y., Sunami, A., Hida, A., & Kawano, Y. (2017). The Impact of Menstrual Cycle Phases on Cardiac Autonomic Nervous System Activity: An Observational Study Considering Lifestyle (Diet, Physical Activity, and Sleep) among Female College Students. *Journal of Nutritional Science and Vitaminology*, *63*(4), 249–255. doi:10.3177/jnsv.63.249 PMID:28978872

Yamamoto, Y., Hughson, R. L., & Peterson, J. C. (1991). Autonomic control of heart rate during exercise studied by heart rate variability spectral analysis. *Journal of Applied Physiology*, *71*(3), 1136–1142. doi:10.1152/jappl.1991.71.3.1136 PMID:1757310

Yeragani, V. K., Sobolewski, E., Jampala, V. C., Kay, J., Yeragani, S., & Igel, G. (1998). Fractal dimension and approximate entropy of heart period and heart rate: Awake versus sleep differences and methodological issues. *Clinical Science (London, England)*, *95*(3), 295–301. doi:10.1042/cs0950295 PMID:9730848

Yildirir, A., Kabakci, G., Akgul, E., Tokgozoglu, L., & Oto, A. (2002). Effects of menstrual cycle on cardiac autonomic innervation as assessed by heart rate variability. *Annals of Noninvasive Electrocardiology*, *7*(1), 60–63. doi:10.1111/j.1542-474X.2001.tb00140.x PMID:11844293

Yoshida, Y., Furukawa, Y., Ogasawara, H., Yuda, E., Hayano, J., & Group, A. R. (2016). Longer lying position causes lower LF/HF of heart rate variability during ambulatory monitoring. *Proceedings of the 2016 IEEE 5th Global Conference on Consumer Electronics (GCCE).*

Yoshida, Y., Yuda, E., & Hayano, J. (2019). Machine-learning estimation of body posture and physical activity by wearable acceleration and heartbeat sensors. *Signal and Image Processing: an International Journal*, *10*(3), 01–09. doi:10.5121ipij.2019.10301

Young, T., Palta, M., Dempsey, J., Skatrud, J., Weber, S., & Badr, S. (1993). The occurrence of sleep-disordered breathing among middle-aged adults. *The New England Journal of Medicine*, *328*(17), 1230–1235. doi:10.1056/NEJM199304293281704 PMID:8464434

Young, T., Peppard, P. E., & Gottlieb, D. J. (2002). Epidemiology of obstructive sleep apnea: A population health perspective. *American Journal of Respiratory and Critical Care Medicine*, *165*(9), 1217–1239. doi:10.1164/rccm.2109080 PMID:11991871

Yuda, E., Furukawa, Y., Yoshida, Y., & Hayano, J. (2018). Characteristics of basal heart rate during daily life: Influences of age, gender, and seasons. *Artificial Intelligence in Medicine, 18.*

Yuda, E., & Hayano, J. (2018). *Menstrual Cycles of Autonomic Functions and Physical Activities.* Paper presented at the 2018 9th International Conference on Awareness Science and Technology (iCAST), Fukuoka, Japan. 10.1109/ICAwST.2018.8517191

Yuda, E., & Hayano, J. (2020). *Changes in Heart Rate Dynamics with Menstrual Cycles* (Vol. 1072). Cham: Springer. doi:10.1007/978-3-030-33585-4_14

Yuda, E., Yoshida, Y., & Hayano, J. (2018). Impacts of sleeping time during the day on the timing and level of basal heart rate: analysis of ALLSTAR big data. *Wireless Networks,* 1-5.

Zwillich, C., Devlin, T., White, D., Douglas, N., Weil, J., & Martin, R. (1982). Bradycardia during sleep apnea. Characteristics and mechanism. *The Journal of Clinical Investigation, 69*(6), 1286–1292. doi:10.1172/JCI110568 PMID:7085875

KEY TERMS AND DEFINITIONS

Cyclic Variation of Heart Rate (CVHR): The characteristic pattern of heart rate fluctuation accompanying sleep apnea. CVHR appears as repetitive peaks of heart rate occurring at an interval of 25 to 120 seconds, which correspond to the time of cessation of each episode of sleep apnea occurring periodically.

Frequency Components of HRV: Spectral analysis of short-term HRV reveals the presence of high-frequency (HF, 0.15-0.4 Hz) and low-frequency (LF, 0.04-0.15 Hz) components. The HF component reflects respiratory fluctuation of heart rate and is mediated purely by the cardiac parasympathetic nerves, while LF component is mediated by both parasympathetic and sympathetic nerves. Long-term HRV also includes ultra-low frequency (ULF, <0.0033 Hz) and very-low frequency (VLF, 0.0033- 0.04 Hz) components in addition to LF and HF components. ULF and VLF components include heart rate fluctuations caused by many factors including body temperature regulation, renin-angiotensin system, and complex fractal fluctuations caused by brain activities.

Heart Rate Variability (HRV): Physiological fluctuations of inter-beat intervals of the heart. HRV is originated from the brain and mediated through the cardiac autonomic nervous system.

Nonlinear Components of Heart Rate Fluctuation: Mathematically, nonlinear system is defined as that the change in the output is not proportional to the change of the input. In heart rate fluctuations, nonlinear components are observed as those not forming a peak in power spectrum. In healthy peoples, the power of nonlinear components shows $1/f$ distribution, i.e., the power inversely correlated with frequency. This is also known as a characteristic of fractal dynamics, in which values exhibit persistent long-range negative correlations, that is, the larger (smaller) values tend to be followed by smaller (larger) values in all time scales.

Parasympathetic Nervous System: Along with the sympathetic nervous system, one of the two branches of the autonomic nervous system. Although the major role of the cardiac parasympathetic nervous system is to decrease the heart rate, it also has an important role to suppress the occurrence of fatal arrhythmias in the destabilized heart by the sympathetic stimulation.

Chapter 2
Enhancing User Experience in Public Spaces by Measuring Passengers' Flow and Perception Through ICT:
The Case of the Municipal Market of Chania

Anna Karagianni
Technical University of Crete, Greece

Vasiliki Geropanta
https://orcid.org/0000-0003-1232-6174
Technical University of Crete, Greece

Panagiotis Parthenios
Technical University of Crete, Greece

Riccardo Porreca
UTE University of Quito, Ecuador

Sofia Mavroudi
Technical University of Crete, Greece

Antonios Vogiatzis
Technical University of Crete, Greece

Lais-Ioanna Margiori
Technical University of Crete, Greece

Christos Mpaknis
Technical University of Crete, Greece

Eleutheria Papadosifou
Technical University of Crete, Greece

Asimina Ioanna Sampani
Technical University of Crete, Greece

ABSTRACT

This research investigates user spatial experience transformations that occur in hyperconnected public spaces and transform them to hybrid spaces. Following this target, the authors conduct an experiment in the Municipal Market of Chania, Crete, in which they evaluate user behaviors on a population of 33 participants comparing their spatial experiences before and after the use of ICT. Through qualitative and quantitative methods (the use of the technology Indoor Atlas as well as questionnaires), the authors

DOI: 10.4018/978-1-7998-3970-5.ch002

analyze behavioral change among users with and without access to Crete 3D, an online ICT-based innovative informative platform, aiming to establish a theoretical framework of understanding user interaction with built space. This process enables knowledge transfer in a twofold way: the authors present how to use metrics to evaluate user-building interaction and how users can quickly gain a deep understanding of the building in use.

INTRODUCTION

According to Nikos Komninos, intelligent cities are encountered at the intersection of innovation and collaborative digital spaces. "For us, a digital city is a collaborative digital space used to facilitate and augment the activities and functions taking place within the physical space of the city" (Komninos, 2008 p.247). In digital spaces, information acquisition is facilitated by sensors, digital networking and other technologies. But how can we evaluate user experience and user-building interaction in a physical space where digital technologies are not already embedded? The issue of combining wireless ICT with IoT technologies and immersively integrating them into the user environment are the crux of the problem. In this specific case, the objective is to decipher user experience within the public space of the Municipal Market of Chania, by establishing a temporary digital environment within the bustling physical space. Following this objective, the paper has been divided in seven sections. The first chapter analyzes empirically visitor types and their interaction with the building's physical space. The second and third chapter, focus on how ICT and IoT technologies contribute to the decoding of user behavior in this specific case, while the fourth chapter describes the in-situ experiment. Then, qualitative and quantitative data sets extracted by the technology and traditional users prior to their visit are analyzed and compared. The sixth chapter draws upon the entire thesis, tying up the empirical and metric strands, aiming to provide a comprehensive model of user behavior in the framework of new digital technologies. Finally, the conclusion highlights areas for further research.

The methodology of this study is based on empirical analysis as well as qualitative and quantitative data acquisition and comparison. During the first part of the study, the authors visited the building and conducted an empirical documentation of its morphological entities, aiming to extract patterns of flows within the site and highlight the points of maximum interaction between users and built space. As built space, we define: a) the interior space of the building and b) the interstitial spaces that connect the built space of the monument with the urban tissue. During the second part of the study, the authors engaged a group of 33 users, formed by young people that visited the building on a summer day. Two subgroups were formed: the users of the first subgroup entered the building and were asked to circulate freely, while the users of the second subgroup launched the visit after navigating into the Crete3D platform. Crete3D is an online WebGL platform, where users can find replicated morphologies of the main archaeological Cretan monuments presented through a conceptual 3D model. Through the platform, users are able to navigate around the monuments and speculate them at different levels of spatial and contextual detail. The platform allows for switching between seven historical periods and offers a comparative study of their evolution in time. The platform's innovative feature was the potential 'to manage such a large amount of information over the internet, in a transparent, light and simple way for the end user, in addition to offering the ability to compare data over time, during the historical periods.' (Parthenios P. et al, p.1, 2012). Users' location was monitored real-time through the Indooratlas application that was previously installed on every user's smartphone. To facilitate the process, we will refer to the first group of users as 'traditional users' and the second as 'tech users'. After the visit, the team sent a link of the questionnaire in the form of a google survey to both traditional and tech users. The questionnaire consisted of 33 questions, grouped in three sections:

- Level of Knowledge prior to the visit
- Overall Experience Evaluation
- Building Evaluation with emphasis to architectural elements

A total of 33 anonymized questionnaires were completed. The research team analyzed the subjective data, extracted from the questionnaires and juxtaposed them with the objective data, which consist in the real-time location tracking acquired through Indoor Atlas platform. Finally, subjective and objective data was compared among the users that browsed Crete3D application and those that entered without any additional knowledge on the monument.

BACKGROUND

The use of ICT in evaluating user experience is not a new phenomenon (Skarlatidou et al., 2019). It has received considerable attention in the field of Human-Computer Interaction (HCI). Specifically, the areas of product, software and web services rely extensively on user feedback. At the same time, the current advances in UX research have led to an increased interest in evaluating user experience within spatial contexts, both in the light of sustainable design (Jeong Kim, 2013), (Janda, 2009), in Post-Occupancy Building performance (Motsatsi, 2015), and finally in the context of smart buildings design (Jia & Srinivasan, 2015). In that context, researchers have shown an increased interest in developing ICT tools dedicated to user education. These new educational tools have led to the emergence of new learning techniques that transform user into a skilled person, able to be relevant at the fourth industrial revolution (Harrari, 2015).

The juxtaposition of the learning process through ICT within the urban tissue constitute the fundamental pillars of a smart city project (Komninos, 2014) and the identification of a smart / hybrid space (Geropanta V., Cornelio Mari E, 2014). As a matter of fact, from a theoretical point of view, there are three components for a smart city: urban space, ICT and skilled users (Komninos, 2014). These altogether frame the hypothesis that once ICT is applied at urban space, then the predicted flows and behaviors of the traditional setting alter in a more educative and holistic way. This marks the beginning of the passage from the spontaneous way of wandering in the daily social practices to a cognitive layer of information that subtly accompanies the user during the monument visit. This layer of information is continuous and invisible, able to achieve the state of 'calm technology' as defined by Weiser (1995).

THE EXPERIENCE OF THE MUNICIPAL MARKET: A SPATIAL APPROACH

The changing behavioral patterns can show in different ways in the urban space depending on the category of users under study (Karagianni A, et al. 2019 p.606). For instance, it can take the form of an "intense flow in the main corridors of the building with additional stops in the facades of the local products" (Karagianni A, et al. 2019 p.606); it can create larger densities at the entrance, main corridor and exit of the building; bigger densities in the seating areas etc. This presence can also take the form of a 'guided' practice (all cars are parked in a specific place while in the main entrance all social life is gathered) (Karagianni A, et al. 2019 p.607). Elsewhere, we have proposed that the changing behavioral patterns can be understood in relation to a subconscious awareness of the physical and material consistency of the building.

Generally speaking, user experience is regarded as a comprehensive concept describing the subjective experience resulting from the interaction of visitors with the monument as a building. It depends on the different types of users (regular or occasional) and what they expect as well as on the relation between the users and contextual factors such as the physical elements and the architectural and cultural value of the building. As a multifarious concept, is based on distinct human factors and consists of different users' expectations and behavioral reactions. Although they have great research interest as they undertake the notion of user experience from a more human-centric aspect and a very subjective perspective, here the authors observe and classify user experience in more general terms, based on user expectations as well as on physical elements of the building, before the experimentation with ICT. In this specific case, there are two types of users, the regular users or locals (locals who want to shop, locals who wander around the city, locals who socialize,) and the occasional users or visitors (tourists who are unfamiliar with the building). Based on the types of users, their movements and expectations the authors identified the following categories of user experience classifications:

Practical / efficient: The first type of users, locals, expect to perform everyday commercial activities within this building and accomplish practical and quick movement between the two sides of the city while the building is situated just above the new and past city boundaries. This category is characterized by the velocity in their moves, creates either an intense flow in the main corridors of the building or smaller flows with specific stops in certain facades of commercial shops. These users shape a more practical and efficient user experience.

Social: Locals also choose this building as a place to meet other locals, to socially interact and communicate, developing a more social user experience. Apart from the gathering of collectivities, small taverns or coffee shops can serve as spots for privacy, thus large user densities can be observed besides with specific spots of isolation.

Informational: For the second type of users, tourists, a more informative user experience is also shaped. Info points about the history of the building, central tourist kiosk with information about traditional recipes as well as tourist shops with local products provide tourists with all the necessary information and result to a relatively medium flow with large densities around the above points of interests.

At a second level of analysis, user experience is classified based on contextual factors such as the physical elements and the architectural and cultural value of the building.

Emotional: For all user categories, both locals and tourists, the building itself creates a feeling of positive surprise. The dynamic presence of people walking inside the building as well as the architectural elements of the building such as the internal building height, the presence of low intensity light and the limited openings create a unique atmosphere, surpassing tourists' expectations on the one hand reviving memories for the locals on the other. These feelings result in not specific and intense flows, with lots of intermediates stops, and in a more emotional user experience.

Sensory: The specific architectural structure of the monument creates feelings of continuity of space with the surrounding environment and a sense of flowing of activities. Large densities of people, both locals and tourists, are noticed around the 4 wings of the cross where the great openings and the intensity of light give the feeling of an external urban place, connected with the immediate environment.

Motivational: Some spots inside the building serve as info points for the city. The center of the building where the 4 wings intersect has information about the city as well as information about Cretan products while also in another spot old photos of the city and the historical monuments are being exhibited. All these encourage and motivate users to broaden their knowledge about the city and expand their experience by visiting other monuments and relate with the city.

ICT AS A SPACE MAPPING AND KNOWLEDGE GENERATOR TOOL

User experience is presented within all traditional locations in the public space of the municipal market and although ICT is only partially used throughout the experiment, its impact is fundamental since it totally transforms the perception of these places. This catalytic presence of ICT provides a vision of a smart city in concordance with the definition of a smart city theory (Geropanta V., Cornelio Mari E., 2014). Therefore, a deep comprehension of the physical dimension of the ICT here is essential for the desired generation of the enhanced user experience. The authors argue that this analysis will lead them to identify the framework under which experienced is enhanced and becomes the instrument for new education resources.

ICT AS A KNOWLEDGE PLATFORM: CRETE 3D

To evaluate user experience and user interaction with the building, the authors introduced to the group of tech users the online platform Crete3D (www.crete3d.gr), an innovative tool that was designed by architects and archaeologists at the Digital Media Lab, Technical University of Crete. The key idea of Crete3D platform was to provide necessary information of the selected monuments, presented in five levels of detail (Crete, Prefecture, Region, Complex, Monument) and throughout seven different historical periods (Minoan, Classical, Hellenistic, Roman, Byzantine, Venetian, Ottoman, Modern). The platform implemented in WebGL visualizes each cultural monument in five spatial levels of detail representing initially Crete as a whole, then by prefecture, region, complex of monuments and finally focusing on the actual monument (Parthenios P. et al. 2014). Unlike most monument-related visualization methods, the platform is based on the creation of conceptual models instead of providing descriptive details of the monuments. A model is usually constructed in order to discard details from the subject under study and retain only what is essential for some stated purpose (Henderson – Sellers and Gonzalez-Perez, 2010).

The platform depicts 148 monuments in Crete and analyzes fifteen of them in detail. The Municipal Agora of Chania appears in the Region of Chania (Fig. 1), at the third level of detail, under the Prefecture of Chania. The abstract three-dimensional model of the monument appears when visiting the Modern Period at the x-axis and the Region on the y-axis. At the scale of the region, during the Modern Period, users are able to see the building location within the Venetian walls of the city (Fig. 2,3,4). At the same scale, during the Venetian and Ottoman Periods, the Venetian Walls frame the current location of the monument. The user can enter deeper into detail when clicking on the pin of the Monument. At that moment, a more detailed model appears within a part of the Old City of Chania (Fig. 5).

Figure 1. Crete3D Platform Interface: The Region of Chania - Modern Period

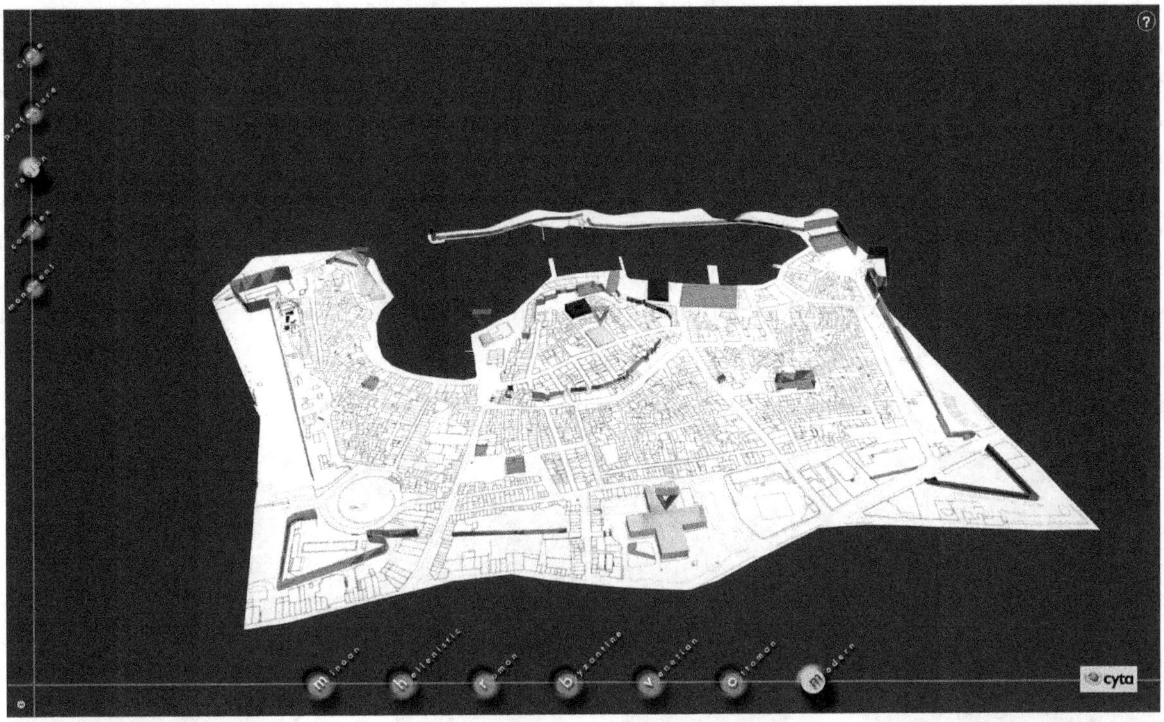

Figure 2. Crete3D Platform Interface: The Region of Chania - Ottoman Period

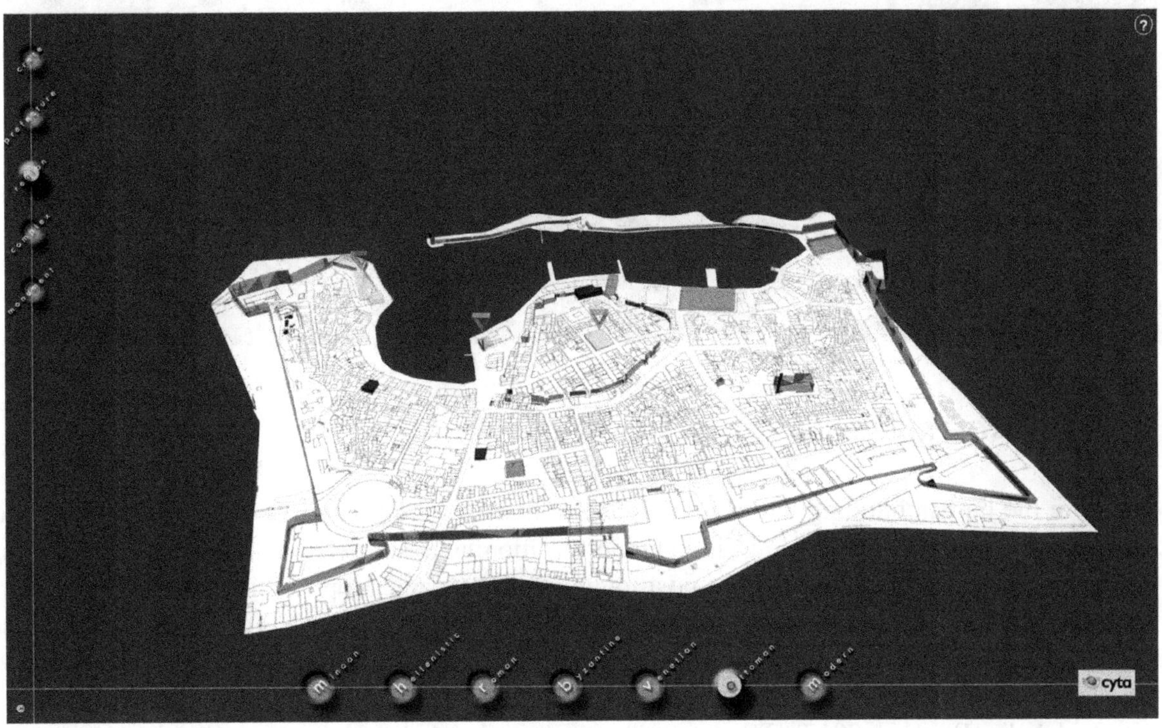

The last level of detail appears at the scale of the monument (Fig. 6). At that point the information is released in a twofold way. On the one hand, the user can gain knowledge regarding the geometry and details of the monument, by navigating into the building or by clicking on the roof and wall (Fig. 7,8), which become animated on mouse-over (Fig. 9). On the other hand, the user can acquire condensed knowledge regarding the monument's history, images, videos, hours of operation, link to google maps and other useful links, by selecting the pop-up menu at the right of the screen (Fig. 10).

The fact that the material described above is available online gave the research team the opportunity to transfer condensed knowledge quickly and through a user-friendly way, a few minutes before the actual visit to the monument. Crete3D does not intend to promote a strictly architecture-base limited image of Crete's past, but a dynamic understanding of its hybrid cultural identity (Parthenios P. et al. 2014).

Figure 3. Crete3D Platform Interface: The Region of Chania - Venetian Period

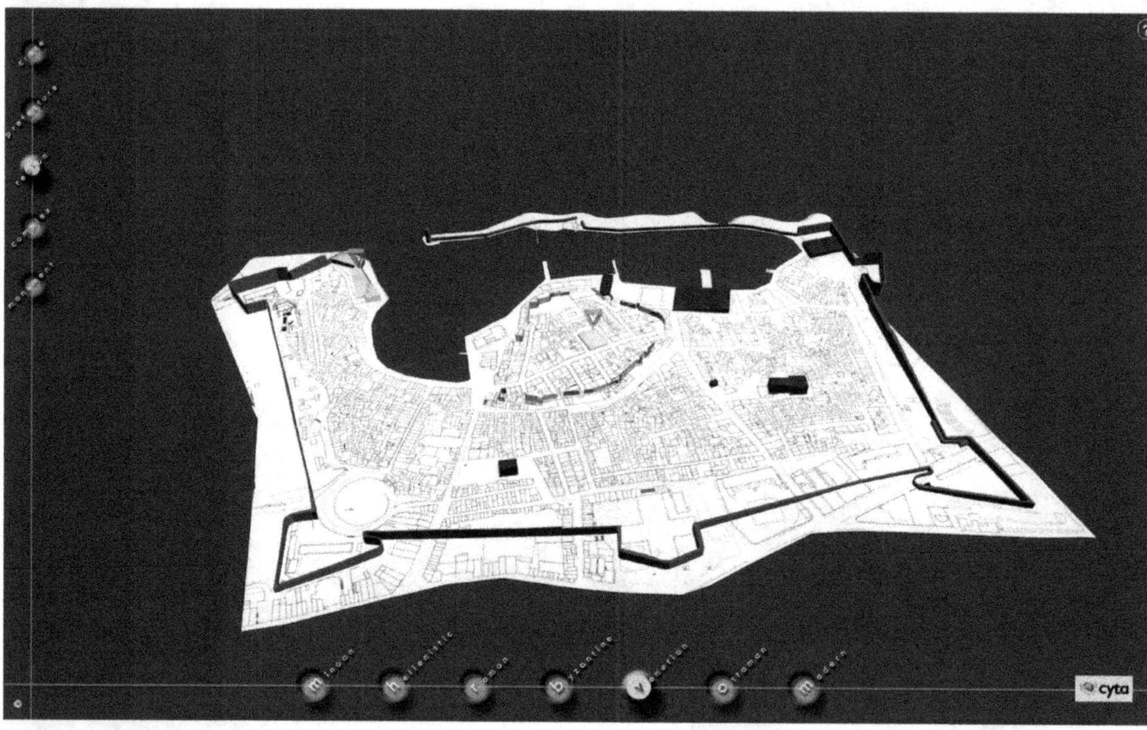

ICT FOR SPACE MAPPING: INDOORATLAS

To measure the human building interaction, the authors used specific ICT and IoT that are based on location tracking data. This technology, is an indoor Positioning System (IPS), a system of network connected devices which is used for wireless positioning of objects and people inside buildings and partly covered areas (Lemmens, 2013). This system, which can be smartphones, tablets or mobile services, uses multiple signals from location devices, as well as sensors etc., calculating accurately the exact location of the device and its specific coordinates (latitude, longitude, and height / number of floors) (Keluža, Marin & Vukelic, Bernard, 2017). These coordinates, co related with the context (Senion,2016) allow to know when the user is located near a certain point, or being able to determine the position of the user regardless of their location in the building. The first refers to a system based on proximity (PS8) and the other is the IPS system (Senion, 2016).

The information obtained by IPS might be essential in the case of a real time analysis where the desired outcome is the identification of user's movement in the building, which zones of the building are more frequently visited, or for more time, and vice versa calculate the quantity of users visiting a particular area at a particular time, determining points of interest and therefore enabling the analysis of user behavior.

In the market there are many IPS products. Some of them namely are the Here Indoor Radio Mapping, the Proximi.io, Anyplace Indoor Service and finally IndoorAtlas MapCreator 2. Briefly, the difference among stands in the way that users depict their geographical location. For instance, in the case of Here Indoor Radio Mapping, users can produce a map by adding buildings and including floor plans and contours to set paths, while in the case of Proximi it is the platform that provides users with accurate location fixes. At Anyplace Indoor Service users can adding floor plans, images, POIs (points of interests), and use the campus mode to signify whether the building is part of a group. Finally, IndoorAtlas MapCreator 2 depends on an app through which the gyroscope and magnetometer on the phone record pathways. The authors chose to work with Indoor Atlas due to (a) prior experience with this app, (b) because it allows for free use plan and an unlimited number of venues with up to 1,000 unique users per account, (c) it requires very easy setup procedure.

Indooratlas depends on a sensor fusion technology that tracks the movement of the user using the built-in inertial sensors in the smartphone. With the integrated magnetic sensor (compass) and other sensor technologies in a smartphone, the software can accurately pinpoint and track the position of a person on the map using the magnetic field inside the building, creating a ''blue dot" on a map. The overall procedure consists of two steps, the setup and the mapping. The setup involves the signing up free at app.indooratlas.com which is the web portal of managing the IndoorAtlas floor plans, maps, API keys and applications. After this step, users upload a floor plan of the venue in the web portal and then make alignment aka Geo-Referenced Floor Plans over the base world map.Mapping (fingerprinting as well) is the method by which indoor positioning can be enabled. It involves collecting signal data from a target location with a mapping software (MapCreator 2) by physically walking through the venue. The research team used a suitable device for mapping (LG nexus 5X and Xiaomi note 5). MapCreator requires hardware magnetometer and gyroscope detectors with access to uncalibrated data. An assessment is made through testing the gyroscope and magnetometer calibrations using MapCreator in order to verify that the sensors function well. The simplest way to test live positioning after a map quality assessment is completed is by the positioning testing feature in MapCreator 2 (Android) or the positioning testing app Indoor Atlas Positioning app in App Store (iOS).

HBI INTERACTION USER EXPERIENCE MEASURING AND ENHANCEMENT

The experiment begins with a visit to the Municipal Agora of Chania. Firstly, the team aligned the IPS coordinates with the GPS coordinates of the building, following the setup – map – positioning process as described in the previous section. The main requirement for the mapping and positioning is strong Wi-Fi or 4G signal. Since weak signal does not allow for accurate data, and inside the agora, the signal was not available at all building surface, the alignment process was repeated several times until being completed.

Figure 4. Crete3D Platform Interface: The Region of Chania – Byzantine Period

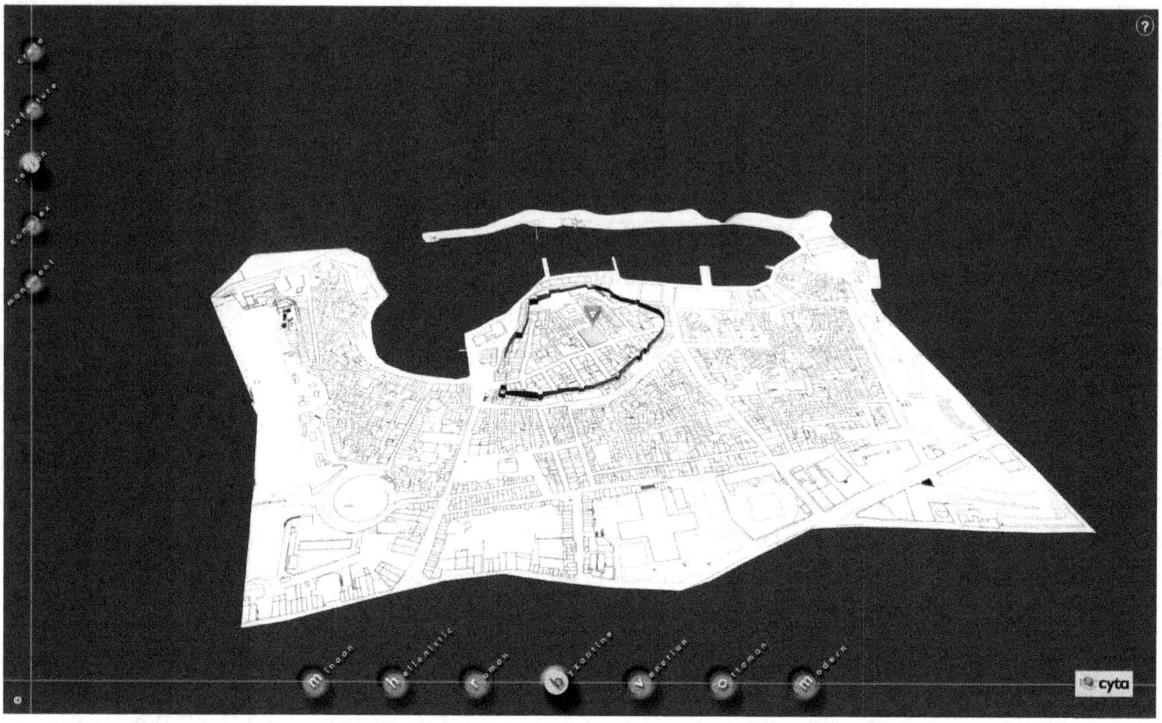

Figure 5. Crete3D Platform Interface: The Complex of Chania – Modern Period

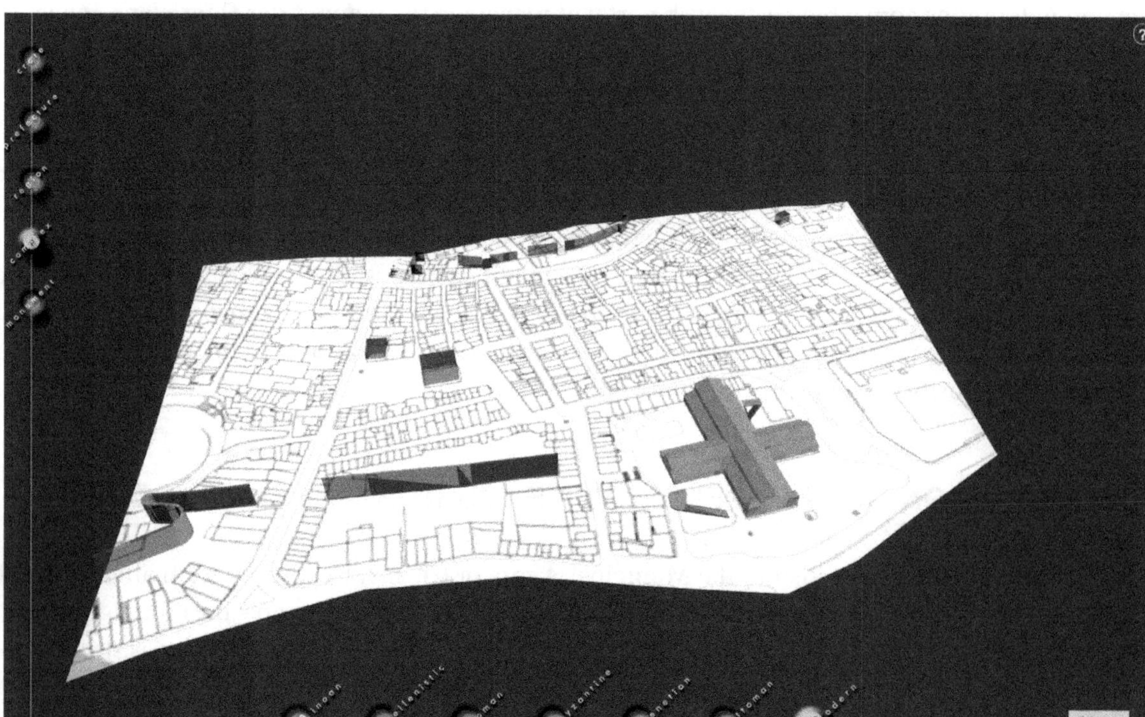

Figure 6. Crete3D Platform Interface: The Monument of Agora – Animated Wall

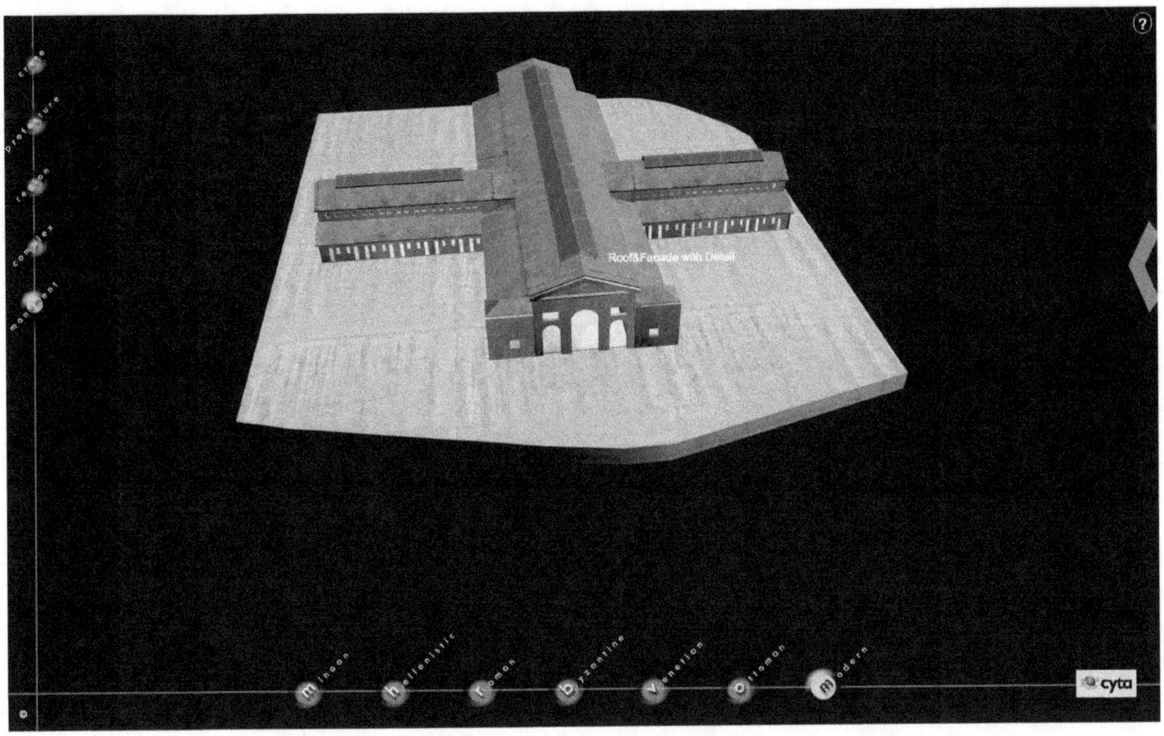

Figure 7. Crete3D Platform Interface: The Monument of Agora – Animated Roof

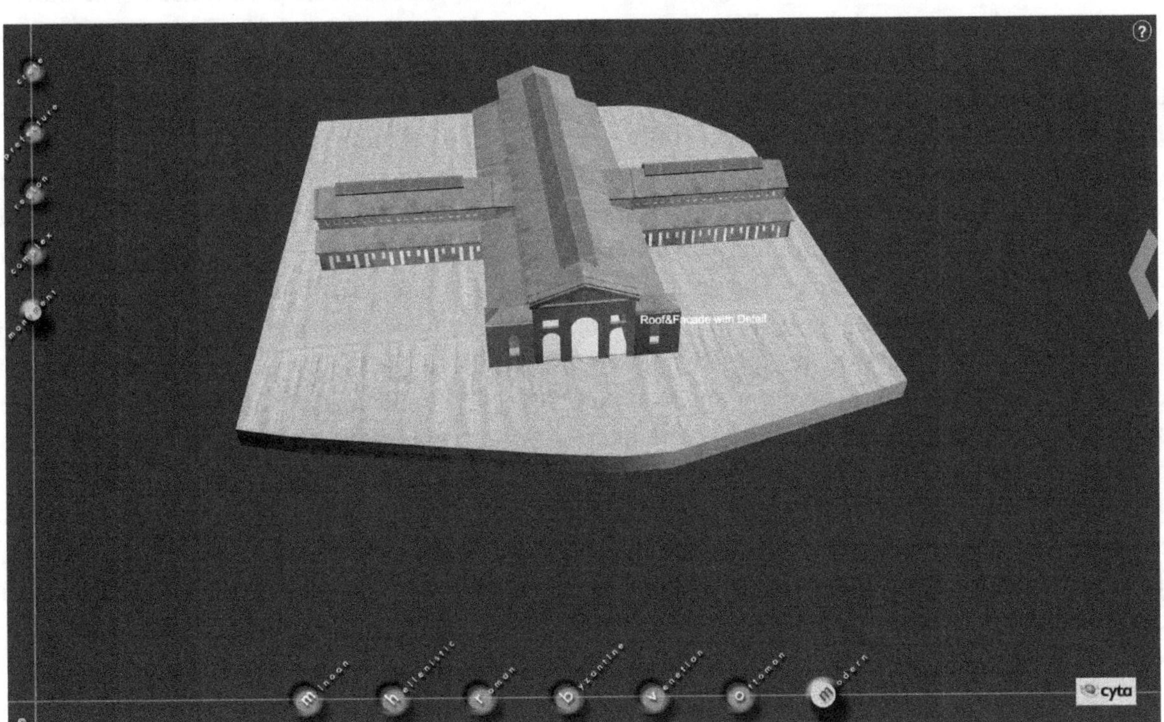

Figure 8. Crete3D Platform Interface: The Monument of Agora – Animated Roof

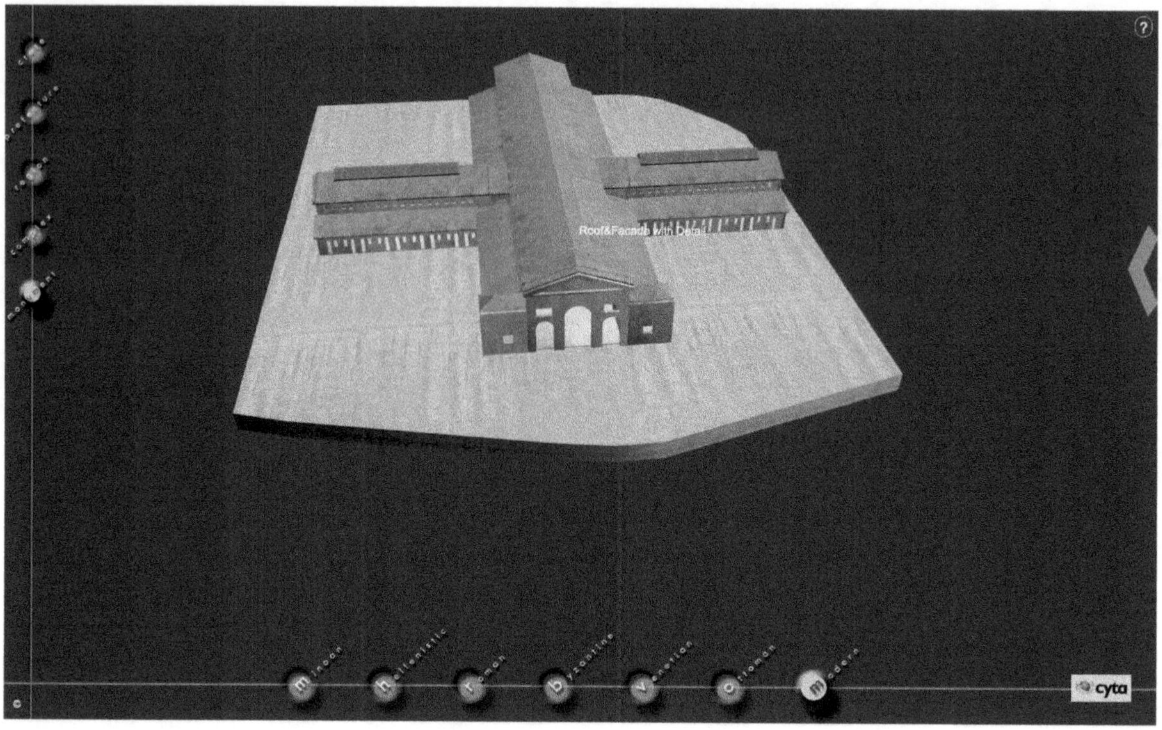

Afterwards, the research team engaged users and led them to the northern entrance of the building where the experiment would start. All participants were aged between 18 and 35. Their age was chosen on the basis of their familiarity with technology, as the experiment required users downloading Indoor-Atlas API or a screen recording application to their smartphones. The time of the visit was deliberately arranged on a summer day that the Market fully functions and hosts tourist groups.

Initially, the entire user population was split into two groups (tech and traditional users). The group of traditional users entered immediately into the building. The only source of information received by these users consisted in bits of info that they heard from tourist leaders as well as an old-fashioned info point in the middle of the building. The group of tech users was given a laptop connected to the internet and were guided to the Crete3D platform. The majority of tech users went through all scales and details of the monument as well as the useful information provided through the pop-up menu, while a few of them clicked on other monuments, regions and complexes. After 15 minutes, the group of tech users entered the monument and made a tour in the interior of the monument. Both groups were advised to return to the same point at the northern entrance when their tour would end.

Despite the fact that Indoor Atlas provides unlimited sessions within the same user account, the research team encountered a sync problem when over ten active users were near the same building point. When this happened, the application was disconnected and the software's accuracy was lost, so the entire process had to be repeated from the beginning. The issue was solved by forming groups of three to five people who were connected to the application with the assistance of the research team.

Figure 9. Crete3D Platform Interface: Pop-up Menu

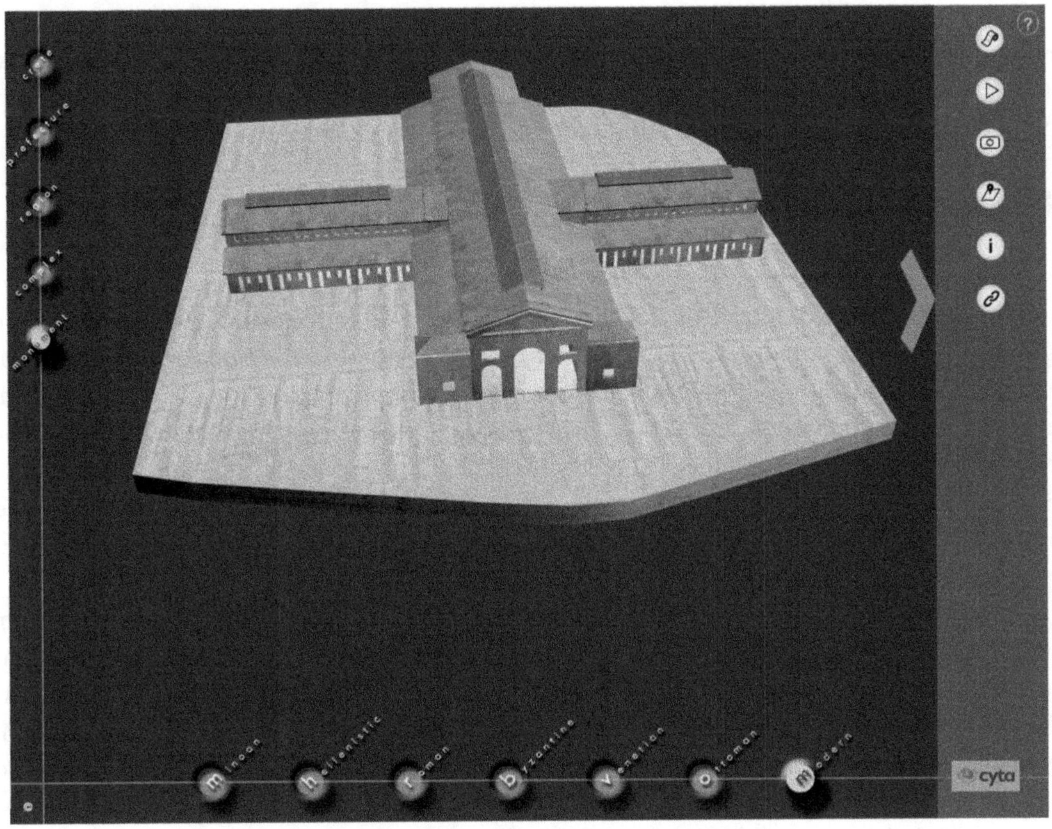

Figure 10. Indoor Atlas Interface: Session ID

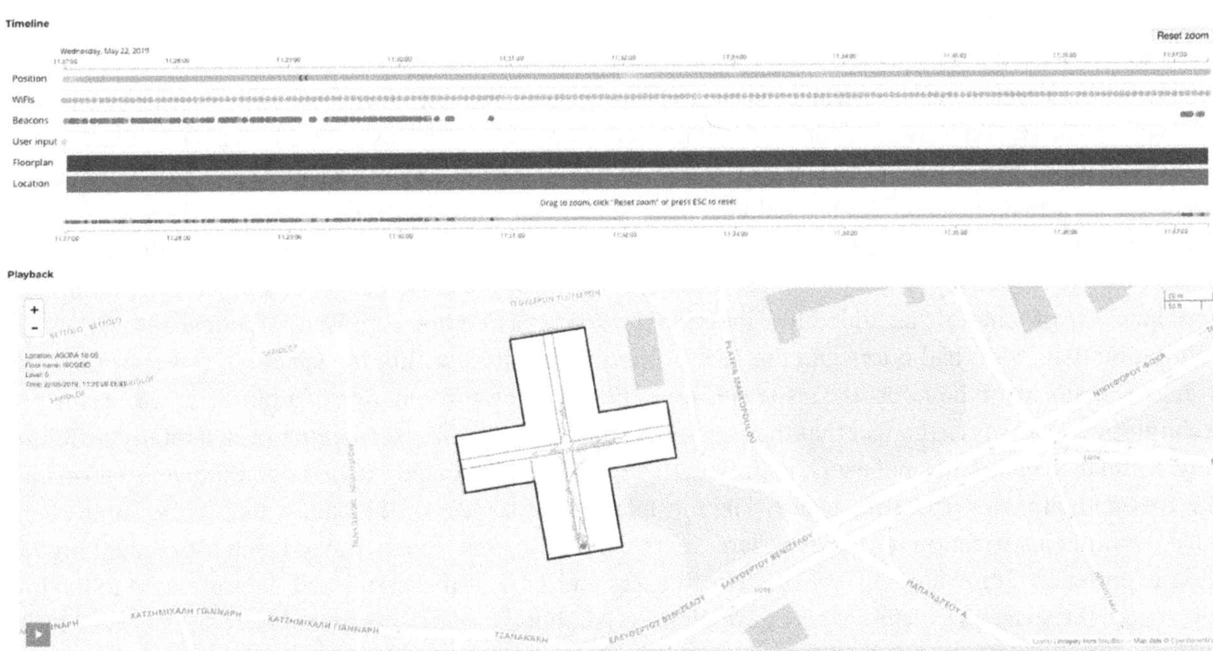

DATA ANALYSIS AND COMPARISON

On completion of the visit to the building, a collection of subjective data through questionnaires was carried out. The research team worked simultaneously on the analysis of the videos of the two-dimensional user flow representation and data acquired through the questionnaires.

Questionnaires / Qualitative Data

The design of the questionnaires was based on a top-down approach of the users' knowledge level. The first group of questions investigated the level of general knowledge regarding the monument and tried to categorize the users in relation to the reason of their prior visit to the building. The second group consisted of evaluating the overall experience of the visit, while the third set of questions focused on the moments of interaction between user and space, aiming to provide a comparative analysis of the building perception between traditional and tech users. By the end of the survey period, data had been collected from 33 individuals, 23 of whom had no previous access to the Crete3D platform and are mentioned here as 'traditional users', while the other 10 had visited the knowledge platform without any time restriction and are called 'tech users'.

Regarding the level of general knowledge, all users had already visited at least once the building, while only one user of the group visited the building for the first time. The reason of their previous building visit was claimed to be: for 60% of tech users and 78,3% of traditional users for tourism while for 30% of tech users and 13% of traditional users for commercial activity. Of the 23 traditional users that responded to the question, only one indicated that his/her prior visit to the building was done for work reasons. Finally, 10% of the work users had used the building as a shortcut to shorten their walking path. In terms of their average previous level of knowledge, it was relatively higher among traditional users (80% of tech users vs. 87% of traditional users rated their level of knowledge between 4-8/10).

The second step consisted in evaluating the overall experience – building, senses, and spatial perception. This step followed the inverted pyramid structure with two set of questions – the first set requiring users to evaluate the visit asking also for their justifications and the second set asking users to give subjective data on the best and worst moments of the interaction with the space. The overall evaluation was very positive. Tech users recorded a significant increase of their evaluation rates compared to the traditional users. Specifically, the evaluation rate of 70% of the tech users was above 4 out of 5, while the correspondent percentage of traditional users was 47.8%. In response to the question: 'Please evaluate the visit (from 1 being Not pleasant to 5 being Very Pleasant)', a range of responses was elicited. Most of the positive responses of traditional users concerned the appreciation of this building as a monument and the flows of tourists that added livelihood to the space, while the negative ones consisted in the bad ventilation that led to bad odors and the lack of thermal comfort within the space. In terms of the tech users, the majority of their positive responses were related to the combination of architecture, tourism and technology as well as to the fact that they had the chance to reinvent the monument in a totally different way. Similar to the traditional users, their negative responses indicated the bad odor and ventilation and the crowded, almost suffocating space. The most surprising aspect of the data is that while some users gave the same justification for the experience evaluation, the users that belonged tech users rated higher their experience. For example, 10% of the tech users and 13% of the traditional ones claimed to having reinvented the space in a different way through this visit, but these tech users evaluated higher their visit.

Within the following set of questions in the second step, users were asked to highlight their best and worst experience during the building visit. Regarding the best experience, there was a significant difference between tech and traditional users. On the contrary, the answers describing the worst experience converge. Specifically, almost two-thirds of the tech users reported that their best moment within the building was when they stopped at the center, where they had the maximum visual contact, as well as the main entrance detail and main axis (north-south). A small minority did not mention any good moment in the building or highlighted the shop experience and communication with people as the best during the visit.

The last question of the second step asked users to choose among multiple choices, up to three spatial qualities that they would like to improve in the building. The question was based on findings from the empirical analysis conducted by the research team. It was considered that this kind of subjective data would usefully supplement and extend the spatial thinking of users. Participants had to choose among the four following features to be improved: ventilation, odors, lighting, materiality. The results were diverse. Tech users voted eight times for ventilation, seven for odors, three for lighting and three for materiality. Traditional users voted nine times for ventilation, eighteen for odors, eleven for lighting and four for materiality.

The third step of the questionnaire required users to give information on the pick of their interaction with the space and on their observations concerning architectural elements of the building. The first question in this set sought to determine the perception that tech users shaped for Crete3D. In response to the question 'What information did you obtain through navigating Crete3D?' Four responses were given. Interestingly enough, the responses to this question were very diverse. Specifically, the majority of users reported that they learnt a lot on this building as a monument and its function. 20% of the users indicated that they gained information on the building's year of construction and current preservation status. Only one individual stated that this platform suggested a new way to explore monuments. On the contrary, 3 users claimed that they did not obtain any new information.

The second, third and fourth questions were dedicated to the identification of the human flows, as perceived by the users. When the participants were asked how many parts of the building they visited, almost all traditional users and 60% of the tech users indicated that they walked to all four parts of the building, while a very small percentage of traditional users and 30% of tech users claimed to have walked to the three parts. Only one tech user reported having walked to two parts. Respondents were asked to indicate how many times they passed through the center. The answers to these questions were similar among tech and traditional users: 60% of tech and 70% of traditional users passed through the center more than 4 times. 30% of tech users and 21% of traditional users passed 3 times, while 10% of tech and 9% of traditional users passes 2 times. Regarding the question 'Which was the main reason of your stops within the building?', there were differences in the ratios of tech and traditional users. 70% of tech users reported that they stopped to observe the architecture of the building, 20% that they stopped to shop or see the shops and 10% to orientate themselves. Interestingly, the ratios of traditional users were evenly distributed: 33% to watch the architecture, 33% to shop or see the shops and 33% to orientate themselves. In that sense, traditional users represent the three categories of the building users described in the first section of this study: the locals, tourists and locals with commercial activity.

The last question of the questionnaire was optional and sought to examine the level of understanding that users had achieved concerning the architectural elements of the building. Interestingly, all tech users answered the question, while 10% of traditional users did not respond. Among tech users, the results were the following:

1) two mentioned the entrance, commenting that 'it gives you the impression that the building is significantly bigger than it actually is',
2) five users highlighted the ceiling, due to its transparency and overall materiality,
3) two users indicated the wall due to the details and overall scale,
4) one user mentioned the overall layout and the scale of the building in relation to the urban tissue.

Regarding the traditional users, the answers were the following:

1) fourteen users mentioned the roof,
2) three users mentioned the central entrance,
3) two users highlighted the cruciform layout and
4) two users indicated the way that the building integrates into the urban tissue.

The most surprising data is that the majority of tech participants mentioned each architectural element 'as described in Crete3D platform' and gave a more detailed description of the element, such as 'the ceiling, due to its transparency and overall materiality', while traditional users responded with one-word answers.

Video Analysis / Quantitative Data

The videos were extracted using two ways: screen recording and the IndoorAtlas location tracking analytics tool. The first way was implemented to participants that used an iPhone, due to technical limitations of the software. Users opened the app interface showing real-time their location into the building and recorded their screen at the same time. The videos produced by the recording were then sent to the research team. The second way consisted in using the IndoorAtlas API, where each user logged in during the building visit and a unique Session ID Number was generated. The Session ID number was also sent to the research team by each participant. In both types of videos, user flow is indicated with a color-coded line, where color is associated with signal accuracy: green for good, yellow for moderate, and red for poor signal. Almost 10% of the overall video duration had poor or no signal and therefore was eliminated from and not considered for data analysis. The only difference between the two extraction techniques was the fact that in screen recording, the periods of time with poor or no signal were manually identified and taken out, while the IndoorAtlas API provided an automated histogram depicting the time in Y axis and the accuracy on the X axis (Fig.11).

The research team defined a series of parameters in order to gain a quantitative understanding of the user flows and stops within the building. The parameters consisted in the direction that the user followed after the first passing through the center, time spent in each of the building's wings, user stops in each wing as well as number of stops and corresponding time spent in the center of the Agora. Other parameters included in the analysis were the inaccurate tracing percentage, the overall user path duration and supplementary comments that varied per case. The results presented here are determined through the quantitative analysis, as well as through the observation of the final video (Fig. 12, 13, 14), where tech user flows and traditional user flows are grouped and then were juxtaposed on a common canvas.

Interestingly, there were enough similarities as well as differences in the paths that that the two groups followed. The first parameter consisted in the direction of users after first passing through the center, In this case, the majority of both tech and traditional users turned left to the west wing of the building when they arrived at the center of the building at the first place. Specifically, contrary to expectations, 50% of tech users and 43,75% of traditional users turned left when first met the central node of the building, despite the fact that at this wing, on the one hand two shops were closed and dark and on the other hand, there were two fisheries and two butcher shops in full function diffusing unpleasant odors. The rest of users moved as follows: 30% of tech users and 25% of traditional users moved straight ahead towards the north and 20% of tech users as well as 31,25 of traditional users turned right. None of the users returned to the north wing after the first passing through the center.

Figure 11 Video Analysis: Tech Users vs. Traditional Users

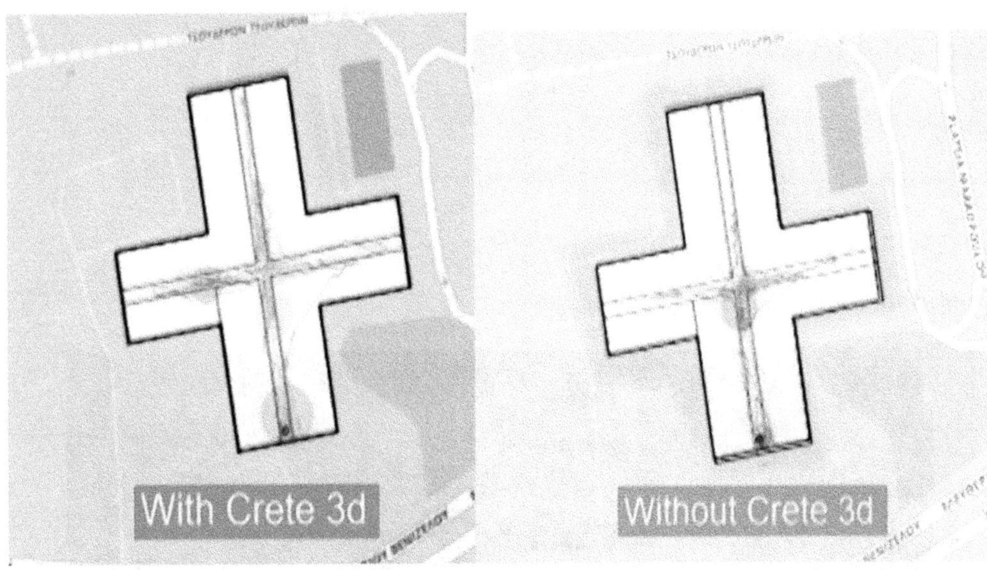

Figure 12. Video Analysis: Tech Users vs. Traditional Users

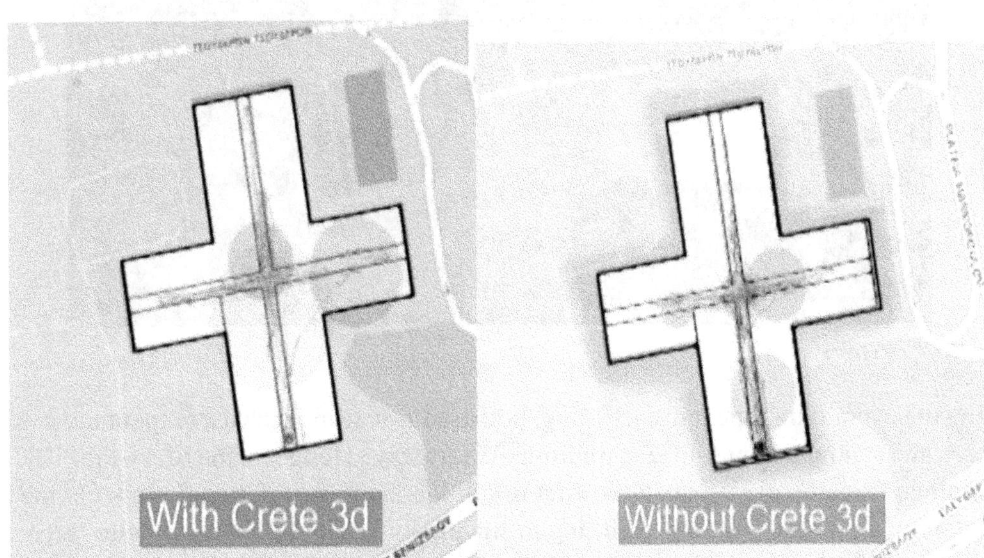

Figure 13. Video Analysis: Tech Users vs. Traditional Users

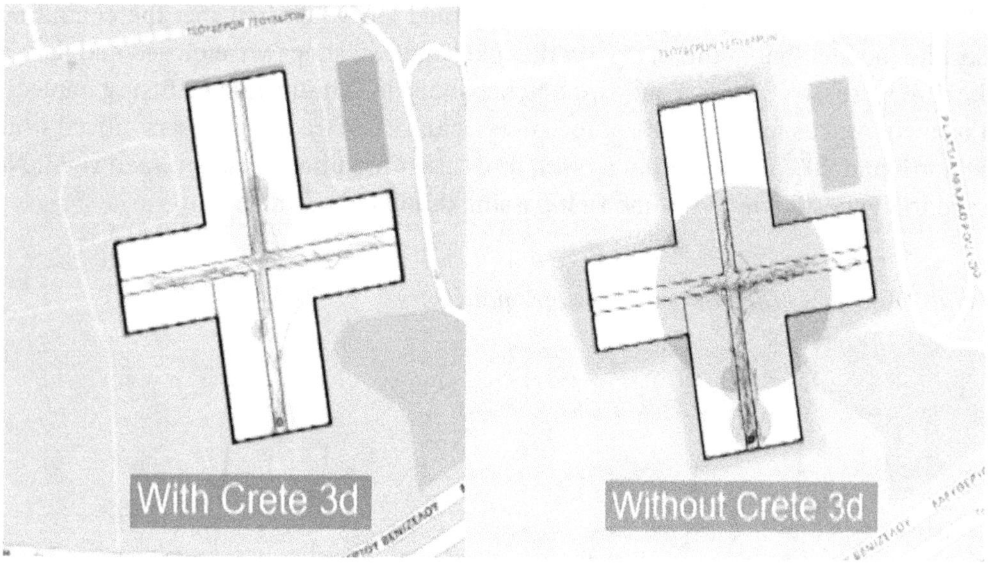

Figure 14. Information Wall in the Agora

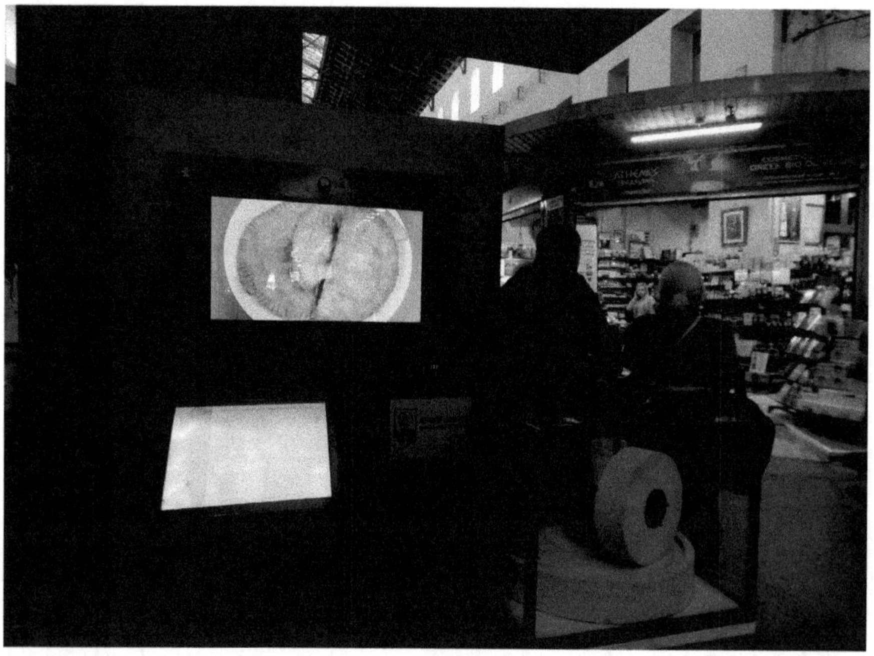

Regarding the overall time spent in each wing, both traditional and tech users spent most of their time in the northern and southern wings. In fact, traditional users stayed longer in the first wing. This correlates with two findings from the overall analysis: first of all, the fact that traditional users claimed that they lacked orientation within the building, resulting in staying longer in the first and more 'secure' wing of the building. Secondly, the fact that Crete3D represents details of the roof on the north-south axis as well as details of the north façade (main entrance), which led the tech group to this specific axis. We argue here that even if users spent more time in the same building section, this happened for different reasons.

Regarding the stops at each part of the building, data was acquired both from the video observation and analysis. Regarding the video observation, it is obvious that tech users traced a longer path and reached all end points of the building. On the contrary, most of the traditional users turned back at the middle of each wing and spent more time in the center. Data analysis shows that tech and traditional users executed the same amount of stops within each wing. What differs is the ratio of the time spent to the number of stops on average of each group. Tech users passed more times through the center, making 5.2 stops on average, while traditional users stayed longer in the center. This might happen due to the fact that traditional users had no knowledge input and tried to gain information through the screen located in the center of the building (Fig.15). In their case, it was the only medium through which they could find history facts and concentrated knowledge regarding the building.

Issues, Controversies, Problems

There are many issues and controversies framing this research. Issues and limitations arise at a macro and microscale. At the macroscale, the existence of a continuous cognitive layer alters the spontaneity of user movement within space. Once informed of being tracked, users no longer feel free to circulate anywhere in the building. The research team observed that most users felt stressed to absorb as much information as possible from the space. They impulsively entered into a state of competing their fellow users. The observed lack of spontaneity was further intensified by the fact that the group of users were young Architecture students, already possessing a level of knowledge on the monument.

At a microscale, the issues emerged were related to the signal inaccuracy, the ICT interface and the sync process. Signal inaccuracy was a constant issue during the experiment and formed a complicating factor at the analysis phase. Moreover, Crete3D is only available on desktop version, rendering the navigation at an open space extremely difficult. Users had to share a laptop prior to the visit, which in many cases interrupted the informational attribute of the visit. Finally, the lack of strong signal and the accumulation of people using the IndoorAtlas API caused sync issues. Specifically, the group of users had to be further distributed into smaller user groups to avoid signal congestion during the experiment.

SOLUTIONS AND RECOMMENDATIONS

To overcome the issues mentioned above, the research team invented solutions at both macro and micro levels, aiming to extract accurate data. At the macro-level, in order for the users to feel that they participated in a collaborative, open process, they were guided by a group of their colleagues. They 'were advised by the research team to feel as flâneurs' (Karagianni et al., p. 605) and get the maximum out of this experience, without trying to memorize any information acquired through the ICT platform.

At the micro-level, the most intriguing part at the analysis phase was the calculation of stops compared to the moments that the signal was lost. This condition was calibrated by an observation: within the Indoor Atlas Session, when user signal was lost, after a couple of seconds it re-appeared at another point of the building, apparently due to the fact that once the end-device connects to the cloud, the system immediately tries to re-connect, depicting user position at the new point. This was crucial for the analysis of user movement and calculation of stops. Finally, regarding the signal congestion issue, the research team firstly reassured the optimal juxtaposition of the IPs to the GPS by conducting the alignment process three times before starting the user visit; secondly, the users formed 5 groups instead of 3 that was initially anticipated, in order to gain control of any signal issues that would arise.

CONCLUSION

In the current study, comparing users with and without ICT-knowledge input during their visit to the Municipal Agora of Chania, showed that ICT triggers an indisputable shift in the user's perception of space. In the empirical analysis, we identified three distinct categories: locals that use the space for commercial reasons, tourists and locals that use it for social reasons. It is interesting to note that when users were asked to point out the main reason that they stopped within the building, traditional users seemed to follow exactly this classification, while tech users followed a distinct path, thus shaping a new typology of people that reinvent the space through technology.

Another important finding was the fact that tech users evaluated higher their visit and identified more architectural and historical elements, although they reported to have lower previous knowledge of the building and its history. However, one unanticipated finding was the fact that traditional users spent a considerable amount of time at the center of the building, where the info point with the screen was located. This further supports the idea that people seek for spatial cognition, not only in terms of understanding the geometry of the space, but also to gain knowledge of the cultural aspects of it. "Building" is a construction of physical elements that creates and protects a space, while each of these two aspects - the physical and the spatial - carry a social value' (Alavi, 2017). It is also known that 'a person's activity space – the set of paths, places, and regions traveled on a regular basis – is an important example of spatial experience that influences people's knowledge of space and place, no matter what their age is' (Montello, 2015). What is crucial to understand is which is the optimal way to gain knowledge in novel spaces of enhanced cultural value.

Together these findings provide important insights into the way that user experience has been transformed by new technologies. This study acts as a knowledge container of spatial analysis, thus creating intermediate-level knowledge, as defined by Kristina Hook and Jonas Lowgreen (2012), knowledge that emerged from interaction design research and lays between general theories and specific instances. It is, in fact, obvious that, through the superimposition of the physical, digital, human and interaction layer, new strong concepts will be identified, articulated and discussed in the field of Human-Centered Design.

FUTURE RESEARCH DIRECTIONS

The work we present here is based on user analysis and research in spatial smart contexts as an ongoing conversation - a process of creating new educational tools and building new spatial knowledge. The relation between user experience, as studied in the field of Human-Computer Interaction is embodied in the Human-Building Interaction approach presented in this study. What seems to be missing, is the articulation of a new language that would integrate the diverse theories and constructs into a new theoretical investigation of User Experience in Smart Environments.

REFERENCES

Alavi, H. S. (2017). The Evolution of Human-Building Interaction: An HCI Perspective. *Interaction Design and Architecture(s) Journal - IxD&A, 32*, 3-6.

Brena, R. F., García-Vázquez, J. P., Galván-Tejada, C. E., Muñoz-Rodriguez, D., Vargas-Rosales, C., & Fangmeyer, J. (2017). Evolution of indoor positioning technologies: A survey. *Journal of Sensors*. doi:10.1155/2017/2630413

Geropanta, V., & Cornelio Mari, E. (2014). Defining smart space: Scales of interaction between architecture and social practices. Journal: Comunicazionepuntodoc, 10, 165-18.

Giovanella, C. (2014) Where's the smartness of learning in smart territories? *Interaction Design and Architecture(s) Journal - IxD&A, 22*, 60-68.

Gonzalez-Perez, C., & Parcero-Oubiña, C. (2011). A conceptual model for cultural definition and motivation. *Proceedings of the 39th Conference on Computer Applications and Quantitative Methods in Archaeology*.

Gozick, B., Subbu, K. P., Dantu, R., & Maeshiro, T. (2011). Magnetic maps for indoor navigation. *IEEE Transactions on Instrumentation and Measurement, 60*(12), 3883–3891. doi:10.1109/TIM.2011.2147690

Henderson-Sellers, B., & Gonzalez-Perez, C. (2010). Granularity in conceptual modelling: application to metamodels. *Proceedings of the 29th international conference on Conceptual modeling*, 219-232. 10.1007/978-3-642-16373-9_16

Janda, K. (2009). Buildings Don't Use Energy: People Do. *PLEA2009 - Proceedings of the 26th Conference on Passive and Low Energy Architecture*.

Jeong Kim, M., & Won Oh, M. (2013). A method for evaluating the performance of green buildings with a focus on user experience. *Energy and Buildings*, (66), 203-210.

Jia, M., & Srinivasa, R. (2015). Occupant Behavior Modeling for Smart Buildings: A Critical Review of Data Acquisition technologies and Modeling Methodologies. *Proceedings of the 2015 Winter Simulation Conference*, 3345-3355. 10.1109/WSC.2015.7408496

Karagianni, A., Geropanta, V., & Parthenios, P. (2019). *Exploring the ICT Potential to Maximize User - Built Space Interaction in Monumental Spaces The case of the municipal agora in Chania, Crete*. https://ecaade.org/conference/downloads/

Keluža, M., & Vukelic, B. (2017). Analysis of an indoor positioning systems. *Zbornik Veleučilišta u Rijeci., 5*(1), 13–32. doi:10.31784/zvr.5.1.2

Kim, Woojin, & Kwang-Ryul. (2017). Indoor positioning system using magnetic field map navigation and an encoder system. *Sensors, 17*(3), 651.

Kim, S.-E. (2012). Indoor positioning system using geomagnetic anomalies for smartphones. In *International conference on indoor positioning and indoor navigation (IPIN)*. IEEE.

Komninos, N. (2008). Intelligent Cities and Globalization of Innovation Networks. *Routledge London, 2008*, 247.

Komninos, N. (2014). *The Age of Intelligent Cities: Smart Environments and Innovation-for-All Strategies*. London: Routledge. doi:10.4324/9781315769349

Lemmens, M. (2013). *Indoor Positioning*. https://www.gim-international.com/content/article/indoor-positioning-2

Li, B. (2012). How feasible is the use of magnetic field alone for indoor positioning? In *International Conference on Indoor Positioning and Indoor Navigation (IPIN)*. IEEE. 10.1109/IPIN.2012.6418880

Li, B., Gallagher, T., Rizos, C., & Dempster, A. G. (2013). Using geomagnetic field for indoor positioning. *Journal of Applied Geodesy*, *7*(4), 299–308. doi:10.1515/jag-2013-0016

Montello, D. (2015). *Spatial Cognition. International Encyclopedia of the Social & Behavioral Sciences*. doi:10.1016/B978-0-08-097086-8.72057-5

Motsatsi, L. (2015). The Development of a Post-Occupancy Evaluation tool for Primary Schools: Learner Comfort Assessment Tool (LCAT). *Proceedings of the Smart And Sustainable Built Environments Conference*.

Parthenios, P., Mania, K., Oikonomou, A., Mallouchou, F., Ragiam, L., Patsavos, N., ... Dimitriou, M. (2014). Using WEBGl to design an interactive 3D Platform for the monuments of Crete. *Proceedings of the 1st CAA GR Conference*.

Piepenrock, C. (2014). Article. *Human Factors*, *56*(5), 942–951. doi:10.1177/0018720813515509 PMID:25141597

Senion. (2016). *Indoor Positioning System*. https://senion.com/indoor-positioning-system/

Vega-Barbas, M., & Pau, I. (2017). Interaction Patterns for Smart Spaces: A Confident Interaction Design Solution for Pervasive Sensitive IoT Services. *IEEE Access: Special Section on Human-Centered Smart Systems and Technologies*.

Weiser, M., & Brown, J. (1995). *Designing Calm Technology*. https://people.csail.mit.edu/rudolph/Teaching/weiser.pdf

KEY TERMS AND DEFINITIONS

Human-Building Interaction: The study and design of interactive opportunities for the occupants to shape the physical, spatial, and social product of their built environments (Alavi et al., 2016).

Smart: The ability to autonomously acquire and apply knowledge.

User: A person that uses something.

Chapter 3
Improving the Optimality Verification and the Parallel Processing of the General Knapsack Linear Integer Problem

Elias Munapo
North-West University, South Africa

ABSTRACT

The chapter presents a new approach to improve the verification process of optimality for the general knapsack linear integer problem. The general knapsack linear integer problem is very difficult to solve. A solution for the general knapsack linear integer problem can be accurately estimated, but it can still be very difficult to verify optimality using the brach and bound related methods. In this chapter, a new objective function is generated that is also used as a more binding equality constraint. This generated equality constraint can be shown to significantly reduce the search region for the branch and bound-related algorithms. The verification process for optimality proposed in this chapter is easier than most of the available branch and bound-related approaches. In addition, the proposed approach is massively parallelizable allowing the use of the much needed independent parallel processing.

INTRODUCTION

The general linear integer programming (LIP) problem has numerous real life applications. Linear integer problems are common as practical problems and include:

(i) traveling salesman,
(ii) assignment and transportation,
(iii) capital budgeting,

DOI: 10.4018/978-1-7998-3970-5.ch003

(iv) facility location,
(v) set packing,
(vi) set covering,
(vii) set partitioning,
(viii) minimal spanning,
(ix) Scheduling and
(x) knapsack.

Knapsack models have many applications in real life and in this paper we select, land conservation, prioritizing conservation projects, optimizing environmental economic policies in a protection zone, maximizing benefits in environmental project selection and plastic bag waste management as examples

A consistent efficient general purpose algorithm for the general LIP is still at large. In other words we are not aware of any efficient consistent general purpose algorithm for this problem as given in Taha(2017). The fact is that this problem is very difficult to solve and heuristics are normally used to approximate solutions in reasonable times (Kumar et. al, 2018). Research on linear integer programming is a very active ongoing process because of the very important applications in real life. The first exact method was a cutting plane approach proposed by Gomory and is given in Gomory (1958), Mitchell (2002) and Salkin (1991). Land and Doig (1960) came up with the branch and bound algorithm. Dakin (1965) improved the branch and bound and this is also explained Bealie (1979). Padberg and Rinald (1991) combined the branch and bound and cuts to come up with a better method which is now called the branch and cut. See Brunetta, Conforti and Rinaldi (1997) and Mitchell (2001) for more on branch and cut algorithms. Salvelsbergh (1997) used pricing within the context of a branch and bound to come up with a good procedure (branch and price) for the linear integer models. See Barnhart et. al (1998) for more on branch and price algorithms. Ladányi, Ralphs and Trotter (2001) combined pricing and cuts and used these in a branch and boud setting. See Barnhart, Hane and Vance (2000) and Fukasawa (2006) for more on branch, cut and price procedure.

THE GENERAL KNAPSACK LINEAR INTEGR PROBLEM

Maximize $Z = c_1x_1 + c_2x_2 + \ldots + c_nx_n$,

Such that $a_1x_1 + a_2x_2 + \ldots + a_nx_n \geq b$, (1)

Where a_j, c_j, and b are integral constants and x_j is an integer $\forall j=1,2,\ldots,n$

The knapsack problem is made up of only one constraint and why this problem NP hard and very difficult to solve? Is it not possible to take advantage of the single constraint to develop an efficient algorithm for this problem?

THE KNAPSACK PROBLEM AND THE KNOWN INTEGER SOLUTION

Suppose the solution of knapsack linear integer problem is known and this solution satisfies (2).

$$c_1x_1 + c_2x_2 + \ldots + c_nx_n = Z_0. \tag{2}$$

Where Z_0 is the optimal value, We can verify the optimality of this solution by solving (3).

Minimize $Z = c_1x_1 + c_2x_2 + \ldots + c_nx_n$,

Such that $a_1x_1 + a_2x_2 + \ldots + a_nx_n \geq b,\ \ c_1x_1 + c_2x_2 + \ldots + c_nx_n = Z_0.$ (3)

It is easier to solve (6) than (3).

CHALLENGE

Unfortunately the optimal solution of a KLIP is not readily available. Infact it is very difficult to solve as explained earlier. This challenge can be easily alleviated by creating a new objective function with a readily available optimal solution which is also optimal to the original prolem.

OBJECTIVE BOUNDING

One thing that appears obvious that must be done to reduce the complexty of the knapsack linear integer problem is to add an objective bound. Suppose the objective bound is given as (4).

$$c_1x_1 + c_2x_2 + \ldots + c_nx_n \leq \ell_o. \tag{4}$$

Where ℓ_o is a constant and is integer. Suppose in in (1), the integer restriction is relaxed and that the optimal solution is given as (5).

$$x_j = x_j^* + f. \tag{5}$$

Where x_j^* is the integral part and f is the fractional part. Then bound ℓ_o is obtained as (6).

$$\ell_o = c_j(x_j^* + 1). \tag{6}$$

Unfortunately an objective bound on its own will not always reduce the the complexity of a knapsack linear integer problem even if it is generated from the optimal solution. We need to do more to significantly reduce the number of branch and bound related iterations.

NUMERICAL ILLUSTRATION

Consider the numerical illustration given in (7).

$$\begin{aligned}\text{Min } Z= \\ &130x_1+140x_2+150x_3+160x_4+180x_5+190x_6+200x_7+210x_8+220x_9+230x_{10} \\ &+240x_{11}+250x_{12}+260x_{13}+270x_{14}+280x_{15}+290x_{16}+300x_{17}+310x_{18}+320x_{19}+330x_{20},\end{aligned} \tag{7}$$

Such that:

$$\begin{aligned}&131x_1+141x_2+151x_3+161x_4+171x_5+181x_6+191x_7+201x_8+221x_9+231x_{10}+241x_{11} \\ &+251x_{12}+261x_{13}+271x_{14}+281x_{15}+291x_{16}+301x_{17}+311x_{18}+321x_{19}+331x_{20}\geq 1217,\end{aligned}$$

Where $x_1, x_2, \ldots, x_{20} \geq 0$ and are integers.

Solving by branch and bound algorithm requires 30771 sub-problems (iterations) to verify one of the the optimal solutions:

$$x_1=7, x_{17}=1, x_2=x_3=\ldots=x_{16}=x_{18=}\ldots=x_{20}=0, Z_0=1210.$$

This is too much in terms of sub-problems given the fact that this solution is found at iteration 599 and we have to go as far as 30771 sub-problems to verify it. Assuming that this optimal solution is known, we add the bound as given in (8).

$$\begin{aligned}&130x_1+140x_2+150x_3+160x_4+180x_5+190x_6+200x_7+210x_8+220x_9+230x_{10}+240x_{11} \\ &+250x_{12}+260x_{13}+270x_{14}+280x_{15}+290x_{16}+300x_{17}+310x_{18}+320x_{19}+330x_{20}\leq 1210.\end{aligned} \tag{8}$$

The new problem becomes as given in (9).

$$\begin{aligned}\text{Min } Z= \\ &130x_1+140x_2+150x_3+160x_4+180x_5+190x_6+200x_7+210x_8+220x_9+230x_{10} \\ &+240x_{11}+250x_{12}+260x_{13}+270x_{14}+280x_{15}+290x_{16}+300x_{17}+310x_{18}+320x_{19}+330x_{20}, \\ \text{Such that:} \\ &131x_1+141x_2+151x_3+161x_4+171x_5+181x_6+191x_7+201x_8+221x_9+231x_{10}+241x_{11} \\ &+251x_{12}+261x_{13}+271x_{14}+281x_{15}+291x_{16}+301x_{17}+311x_{18}+321x_{19}+331x_{20}\geq 1217, \\ &130x_1+140x_2+150x_3+160x_4+180x_5+190x_6+200x_7+210x_8+220x_9+230x_{10}+240x_{11} \\ &+250x_{12}+260x_{13}+270x_{14}+280x_{15}+290x_{16}+300x_{17}+310x_{18}+320x_{19}+330x_{20}\leq 1210.\end{aligned} \tag{9}$$

Where $x_1, x_2, \ldots, x_{20} \geq 0$ and are integers.

We expect problem in (9) to be asier than the original problem that is given in (7). Unfortunately this is not the case. We still need 30229 iterations to verify the optimality of the known solution. This shows a bizzare behavour with the branch and bound algorithm and it also shows that a bound from the objective function on its own is not enough in reducing the complexity of knapsack linear integer problem. There is a need to improve the verification process of the optimal integer solution.

VARIABLE BOUNDING

Variable bounding is very useful in reducing the complexity of a linear integer problem. When searching for feasible integer solutions, the number of sub-problems is minimized if branching starts with the most restricted variable [10]. Let the variable limits $\rho(x_j)$ of a variable x_j be given by (12).

$$\rho(x_j) = \lambda_j \tag{12}$$

Thus the numbers of possible values for the non-basic variables are given in (13).

$$\rho(x_1) = \lambda_1,\ \rho(x_2) = \lambda_2,\ \ldots,\ \rho(x_n) = \lambda_n \tag{13}$$

The variable limits be put in ascending order as shown in (14).

$$\lambda_1 \leq \lambda_2 \leq \ldots \leq \lambda_m \tag{14}$$

So x_1 is the most restricted variable followed by x_2, in that order up to x_n. When constructing the search trees, the following numbers of nodes are visited. The worst case is assumed in the searching process.

STARTING WITH THE MOST RESTRICTED BASIC VARIABLE (x_1)

Stage 1: number of nodes visited = λ_1
Stage 2: number of nodes visited = λ_1,λ_2

...

Stage *n*: number of nodes visited = $\lambda_1,\lambda_2,\ldots,\lambda_n$

Let total number of nodes $T_A = \lambda_1 + \lambda_1\lambda_2 + \ldots + \lambda_1\lambda_2\ldots\lambda_n$.

STARTING WITH THE LEAST RESTRICTED BASIC VARIABLE (x_n)

Stage 1: number of nodes visited = λ_n
Stage 2: number of nodes visited $= \lambda_n\lambda_{n-1}$,

...

Stage *m*: number of nodes visited $= \lambda_n\lambda_{n-1}\ldots\lambda_1$,

Let total number of nodes be $T_B = \lambda_n + \lambda_n\lambda_{n-1} + \ldots + \lambda_n\lambda_{n-1}\ldots\lambda_1$,

Comparing the total number of nodes for we have (15).

$$T_A \leq T_B \tag{15}$$

NUMERICAL ILLUSTRATION-NODES VISITED

Construct a tree of nodes given the possible values as given in (16).

s_1=(0,1,2): $\rho(s_1)$=3,

s_2=(0,1): $\rho(s_2)$=2, (16)

s_3=(0): $\rho(s_3)$=1.

Starting with the most restricted variable we have the search tree given in Figure 1.
Starting with the least restricted variable we have the search tree given in Figure 2.

Figure 1. Search tree when starting with variable s_3

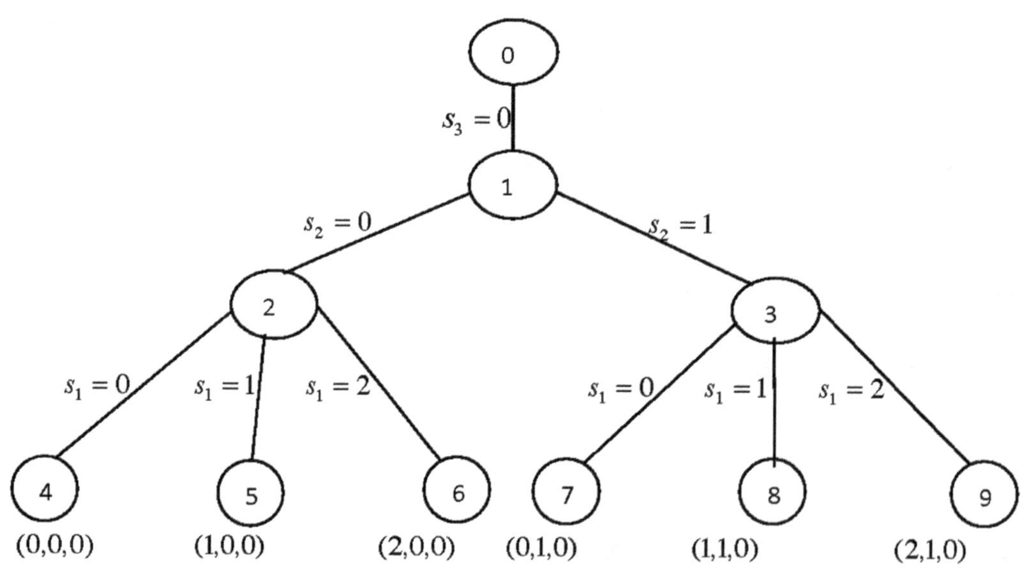

The total number of nodes visited during the searching shown in Figure 1 is 9 (excluding node 0). The total number of nodes visited during the searching presented in Figure 2 is 15 (excluding node 0). Thus it is better to start with the most restricted variable when searching for integer solutions because 15>9. This numerical illustration was taken from Kumar et al.(2007).

Figure 2. Search tree when starting with variable s_1

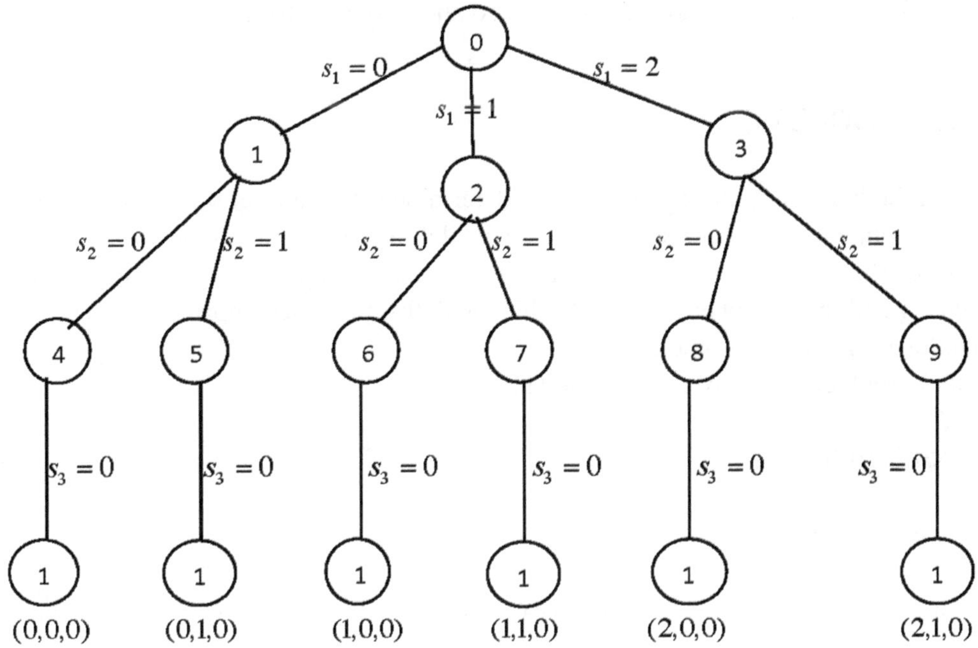

Variable Sum

Since it is is computationally expensive to accurately determine the limits of all the integer variables in a KLIP separately, we rather calculate the limits of the combined variables as given in (10).

$$\ell_1 \leq x_1 + x_2 + \ldots + x_n \leq \ell_k. \tag{10}$$

Where ℓ_1 and ℓ_k are integers. We determine the variable sum limits only when we have a bound ℓ_0. This is done by solving (11) and (12). This is not a challenge as parallel processors can do this at the same time usng the parallel processors.

Maximize $Z=x_1+x_2+\ldots+x_n$,

Such that $a_1x_1 + a_2x_2 + \ldots + a_nx_n \geq b, \ c_1x_1 + c_2x_2 + \ldots + c_nx_n \leq \ell_o$, (11)

Where $x_1,x_2,\ldots,x_n \geq 0$.

Minimize $Z=x_1+x_2+\ldots+x_n$,

Such that $a_1x_1 + a_2x_2 + \ldots + a_nx_n \geq b, \ c_1x_1 + c_2x_2 + \ldots + c_nx_n \leq \ell_o$, (12)

Where $x_1, x_2, \ldots, x_n \geq 0$. The linear problem (LP) in (11) will give us ℓ_1 and the one in (12) will give ℓ_k. Note that variables ina both (11) and (12) are relaxed and easy to solve for.

PARALLEL PROBLEMS

After obtaining the bounds ℓ_1 and ℓ_k we can now determine the k parallel problems that must be soved independently to obtain the optimal solutions. The availability of massively parallel computer processing power makes this procedure very promising. The available versions or variants of the branch and bound related algorithms are not easily parallesible. With this proposed approach each processor can be tasked to handle each parallel problem.

The parallel propblems are given in (13) to (15).

Parallel Problem 1

Minimize $Z=x_1+x_2+\ldots+x_n$,,

$$\text{Such that } a_1x_1 + a_2x_2 + \ldots + a_nx_n \geq b, \;\; c_1x_1 + c_2x_2 + \ldots + c_nx_n \leq \ell_0, \;\; x_1 + x_2 + \ldots + x_n = \ell_1., \tag{13}$$

Where $x_1, x_2, \ldots, x_n \geq 0$ and are integers.

Parallel Problem 2

Minimize $Z=x_1+x_2+\ldots+x_n$,

$$\text{Such that } a_1x_1 + a_2x_2 + \ldots + a_nx_n \geq b, \;\; c_1x_1 + c_2x_2 + \ldots + c_nx_n \leq \ell_0, \;\; x_1 + x_2 + \ldots + x_n = \ell_2. \tag{14}$$

Where $x_1, x_2, \ldots, x_n \geq 0$ and are integers.

Parallel Problem *K*

Minimize $Z=x_1+x_2+\ldots+x_n$,

$$\text{Such that } a_1x_1 + a_2x_2 + \ldots + a_nx_n \geq b, \;\; c_1x_1 + c_2x_2 + \ldots + c_nx_n \leq \ell_0, \;\; x_1 + x_2 + \ldots + x_n = \ell_k. \tag{15}$$

Where $x_1, x_2, \ldots, x_n \geq 0$ and are integers.

WHY VARIABLE SUM OBJECTIVE ROWS ARE SUPERIOR THAN LINEAR FUNCTONS WITH BIGGER COEFFICIENTS?

Note that the objective rows are in variable sum form and objective values in the k parallel problems are readily available. It is easier to verify optimality of a variable sum objective function than any other objective function with coefficients great than one.

Proof

What determines the number of branches when searching for an optimal solution is the number possibilities available. For example if there are k possibles optimal solutions then we have k branches assuming the worst case scenario.

$$Z=x_1+x_2+\ldots+x_n=[0,k] \tag{16}$$

$$Z=c_1x_1+c_2x_2+\ldots+c_nx_n=[0,m] \text{ where } c_1,c_2,\ldots,c_n>1 \tag{17}$$

Figure 3. Branches for [0,k]

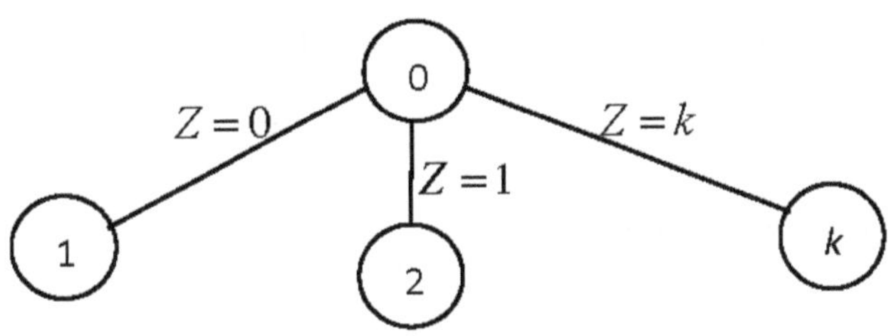

Objective function (17) is the same as:

$$Z=c_1x_1+c_2x_2+\ldots+c_nx_n=x_1+x_2+\ldots+x_n+(c_1-1)x_1+(c_2-1)x_2+\ldots+(c_n-1)x_n$$

Since $c_1,c_2,\ldots,c_n>1$ then $c_1-1\geq 0, c_2-1\geq 0,\ldots,c_n-1\geq 1 \Rightarrow m>k$

Figure 4. Branches for [0,m]

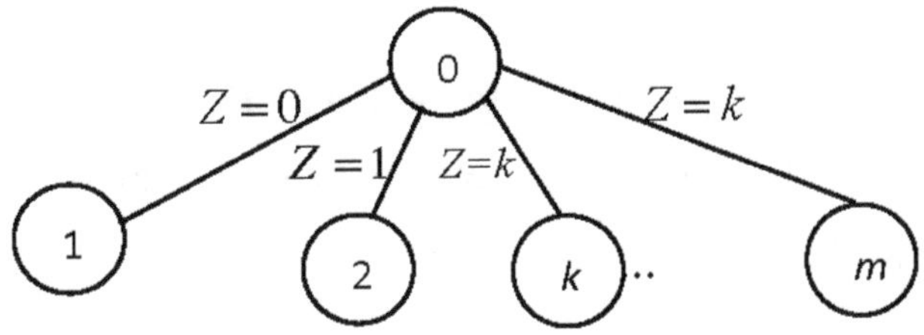

NUMERICAL ILLUSTRATION

Show that the use of variable sum objective function is superior in verifying optimality (18) than the original and given objective function. The optimal solution is known as:

$$x_4 = 1, x_8 = 708, x_1 = x_2 = x_3 = x_5 = x_6 = x_7 = 5678.$$

Minimize $Z = 32x_1 + 12x_2 + 17x_3 + 14x_4 + 43x_5 + 27x_6 + 16x_7 + 8x_8$,

Such that: $43x_1 + 26x_2 + 16x_3 + 27x_4 + 14x_5 + 39x_6 + 32x_7 + 18x_8 \geq 12771,$ (18)

Where $x_1, x_2, \ldots, x_8 \geq 0$ and are integers.

The objective bound is

$$32x_1 + 12x_2 + 17x_3 + 14x_4 + 43x_5 + 27x_6 + 16x_7 + 8x_8 \leq 5678,$$

i.e.

$$32x_1 + 12x_2 + 17x_3 + 14x_4 + 43x_5 + 27x_6 + 16x_7 + 8x_8 = 5678,$$

solving it takes 33 sub-problems to verify optimality.

Using the clique objective row we have

$$x_1 + x_2 + x_3 + x_4 + x_5 + x_6 + x_7 + x_8$$

and that

$$708 \leq x_1 + x_2 + x_3 + x_4 + x_5 + x_6 + x_7 + x_8 \leq 709 \Rightarrow \text{two parallel problems.}$$

The two parallel problems are:

Paralel Problem 1

Maximize $Z = x_1 + x_2 + x_3 + x_4 + x_5 + x_6 + x_7 + x_8$,

$$\text{Such that: } \begin{array}{l} 43x_1 + 26x_2 + 16x_3 + 27x_4 + 14x_5 + 39x_6 + 32x_7 + 18x_8 \geq 12771, \\ 32x_1 + 12x_2 + 17x_3 + 14x_4 + 43x_5 + 27x_6 + 16x_7 + 8x_8 \leq 5678, \\ x_1 + x_2 + x_3 + x_4 + x_5 + x_6 + x_7 + x_8 = 708. \end{array}$$

Paralel Problem 2

Maximize $Z = x_1 + x_2 + x_3 + x_4 + x_5 + x_6 + x_7 + x_8$,

$$\text{Such that: } \begin{aligned} &43x_1 + 26x_2 + 16x_3 + 27x_4 + 14x_5 + 39x_6 + 32x_7 + 18x_8 \geq 12771, \\ &32x_1 + 12x_2 + 17x_3 + 14x_4 + 43x_5 + 27x_6 + 16x_7 + 8x_8 \leq 5678, \\ &x_1 + x_2 + x_3 + x_4 + x_5 + x_6 + x_7 + x_8 = 709. \end{aligned}$$

Solving we note that parallel problem 1 is infeasible and that it takes 17 sub-problems to verify parallel problem 2. Since this is done in parallel it means that the variable sum objective form is superior than the original form of the objective function.

PROPOSED PROCEDURE

Suppose the KLIP is given as:

Maximize $Z = c_1x_1 + c_2x_2 + \ldots + c_nx_n$,

Such that $a_1x_1 + a_2x_2 + \ldots + a_nx_n \geq b$,Where x_j is an integer $\forall_j = 1,2,\ldots,n$

Step 1: Determine the objective bound ℓ_0

$c_1x_1 + c_2x_2 + \ldots + c_nx_n \leq \ell_0$, where ℓ_0 is an integer.

Step 2: Use ℓ_0 to determinine the variable sum bounds ℓ_1 and ℓ_k such that,

$\ell_1 \leq x_1 + x_2 + \ldots + x_n \leq \ell_k$, where ℓ_1 and ℓ_k are integers.

Step 3: Set up the k parallel problems.

Parallel Problem 1

Maximize $Z = x_1 + x_2 + \ldots + x_n$,

Such that $a_1x_1 + a_2x_2 + \ldots + a_nx_n \geq b$, $c_1x_1 + c_2x_2 + \ldots + c_nx_n \leq \ell_0$, $x_1 + x_2 + \ldots + x_n = \ell_1$.,

Where $x_1, x_2, \ldots, x_n \geq 0$ and are integers.

Parallel Problem 2

Maximize $Z=x_1+x_2+\ldots+x_n$,

Such that $a_1x_1+a_2x_2+\ldots+a_nx_n \geq b,\ \ c_1x_1+c_2x_2+\ldots+c_nx_n \leq \ell_0,\ \ x_1+x_2+\ldots+x_n=\ell_2.,$

Where $x_1,x_2,\ldots,x_n \geq 0$ and are integers.

Parallel Problem k

Maximize $Z=x_1+x_2+\ldots+x_n$,

Such that $a_1x_1+a_2x_2+\ldots+a_nx_n \geq b,\ \ c_1x_1+c_2x_2+\ldots+c_nx_n \leq \ell_0,\ \ x_1+x_2+\ldots+x_n=\ell_k.,$

Where $x_1,x_2,\ldots,x_n \geq 0$ and are integers.

Step 4: The optimal solution Z^* is that optimal solution from all the k parallel problems that also minimizes $c_1x_1+c_2x_2+\ldots+c_nx_n$.

FULL EXAMPLE

Use the proposed procedure to solve following KLIP.

Min Z=

$$\begin{aligned}&130x_1+140x_2+150x_3+160x_4+180x_5+190x_6+200x_7+210x_8+220x_9+230x_{10}\\&+240x_{11}+250x_{12}+260x_{13}+270x_{14}+280x_{15}+290x_{16}+300x_{17}+310x_{18}+320x_{19}+330x_{20},\end{aligned} \tag{7}$$

Such that:

$$\begin{aligned}&131x_1+141x_2+151x_3+161x_4+171x_5+181x_6+191x_7+201x_8+221x_9+231x_{10}+241x_{11}\\&+251x_{12}+261x_{13}+271x_{14}+281x_{15}+291x_{16}+301x_{17}+311x_{18}+321x_{19}+331x_{20} \geq 1217,\end{aligned}$$

Where $x_1,x_2,\ldots,x_{20} \geq 0$ and are integers.

Step 1: $\ell_0=1300$
Step 2: Using $\ell_0,\ell_1=6$ and $\ell_k=9$
Step 3: The 4 parallel problems are when,

Parallel problem 1: $\ell_1=6$
Parallelproblem 2: $\ell_2=7$
Parallel problem 3: $\ell_3=8$
Parallel problem 4: $\ell_4=9$

Step 4: Solving with variable sum objective row, objective bound and variable sum equality constraint we have:
Parallel problem 1: Infeasible
Parallel problem 2: x_1=5, x_{10}=x_{20}=1, Z_2=1210
Parallel problem 3: x_1=7, x_{11}=1, Z_3=1210
Parallel problem 4: x_1=7, x_3=2, Z_4=1210

Note that the Z values (Z_2,Z_3 & Z_4) are calculated from the original objective function.

STRENGTHS

(a) Easier to verify optimality.

The procedure is powerful when verifying whether a given bound is optimal or not. A variable sum objective is generated and used to verify optimality.

(b) Minimal risk of combinatorial explosion.

With this procedure we can monitor and manage the size of sub-prpblems. The procedure allows splitting of tasks during the computations.

(c) Allows massively parallel processing

The *k* parallel problems in the procedure can be solved independently allowing the use of massively parallel processing. This gives an edge over most of the available exact approaches of integer prograam-ing such as branch and bound, branch and cut, branch and price and branch cut and price which are are not easily parallelizable. In other words the procedure splits the problem into *k* parts that can be solved independently.

(d) Procedure allows use of other algorithms within its context.

The proposed procedure allows the use of the other approaches in integer programming within the context of itself. Even more efficient and heuristics that are quicker than heuristics can be used to obtain good bounds.

(e) Narrows search region.

With this procedure we search only the most likely points and not the whole feasible space. In other words the complexity of the linear integer problem is significantly reduced.

CONCLUSION

So many improvements have been done on the branch and bound algorithm in terms of addition of cuts to get the branch and cut algorithm, see Brunetta et al. (1997), Mitchell (2001), Padperg and Rinaldi (1991). Pricing was used to improve the the branch and bound so as to come up with the branch and price algorithm, see Barnhart et al. (1998) or Savelsbergh (1997), and also combining these two improved versions to get the branch cut and price hybrid algorithm, see for example Barnhart et al. (2000), Fukusawa et al. (2006) or Ladányi et al. (2001). Objective bounding on its own is not always effective for improving the branch and bound algorithm. Other techiques such as cuts, pricng or or preprocessing as given in Savelsbergh (1994) can significantly reduce the number of sub-problems needed to verify optimality. In this chapter more binding equality constraints are generated and added to the problem. The generated equality constraints can be shown to significantly reduce the search region for the branch and bound related algorithms. The proposed approach is massively parallelizable allowing the use of the much needed independent parallel processing (Akl, 1992 and Sparkias and Gibbon, 1993).

NOTE

TORA Software was used for the numerical illustrions. Please see Taha (2015).

REFERENCES

Akl, S. (1992). *Design and Analysis of Parallel Algorithms*. Prentice Hall Inc.

Barnhart, C., Hane, C. A., & Vance, P. H. (2000). Using branch-and-price-and-cut to solve origin-destination integer multicommodity flow problems. *Operations Research*, *48*(2), 318–326. doi:10.1287/opre.48.2.318.12378

Barnhart, C., Johnson, E. L., Nemhauser, G. L., Savelsbergh, M. W. P., & Vance, P. H. (1998). Branch and price column generation for solving huge integer programs. *Operations Research*, *46*(3), 316–329. doi:10.1287/opre.46.3.316

Bealie, E. M. L. (1979). Branch and Bound Methods for Mathematical Programming systems. *Annals of Discrete Mathematics*, *5*, 201–219. doi:10.1016/S0167-5060(08)70351-0

Brunetta, L., Conforti, M., & Rinaldi, G. (1997). A branch and cut algorithm for the equicut problem. *Mathematical Programming*, *78*(2), 243–263. doi:10.1007/BF02614373

Dakin, R. J. (1965). A tree search algorithm for mixed integer programming problems. *The Computer Journal*, *8*(3), 250–255. doi:10.1093/comjnl/8.3.250

Fukasawa, R., Longo, H., Lysgaard, J., Poggi de Aragao, M., Uchoa, E., & Werneck, R. F. (2006). Robust branch-and-cut-price for the Capacitated vehicle routing problem. *Mathematical Programming Series A*, *106*(3), 491–511. doi:10.100710107-005-0644-x

Gomory, R. E. (1958). Outline of an algorithm for integer solutions to linear programs. *Bulletin of the American Mathematical Society, 64*(5), 275–278. doi:10.1090/S0002-9904-1958-10224-4

Karlof, J. K. (2005). *Integer Programming: Theory and Practice*. CRC Press Inc. doi:10.1201/9781420039597

Kumar, S., Munapo, E., & Jones, B. C. (2007). An Integer Equation Controlled Descending Path to a Protean Pure Integer Program. *Indian Journal of Mathematics, 49*, 211–237.

Kumar, S., Munapo, E., Lesaoana, M., & Nyamugure, P. (2018). A minimum spanning tree based heuristic for the travelling salesman tourm. *OPSEARCH, 55*(1), 150–164. doi:10.100712597-017-0318-5

Ladányi, L., Ralphs, T. K., & Trotter, L. E. Jr. (2001). Branch, cut and Price: Sequential and Parallel. In N. Naddef & M. Jüenger (Eds.), *Computational Combinatorial Optimization* (Vol. 223). Berlin: Springer. doi:10.1007/3-540-45586-8_6

Land, A. H., & Doig, A. G. (1960). An Automatic method for solving discrete programming problems. *Econometrica, 28*(3), 497–520. doi:10.2307/1910129

Mitchell, J. E. (2001). Branch and cut algorithms for integer programming. In C. A. Floudas & P. M. Pardalos (Eds.), *Encyclopedia of Optimization*. Kluwer Academic Publisher. doi:10.1007/0-306-48332-7_215

Mitchell, J. E. (2002). Cutting plane algorithms for combinatorial optimization problems. In Handbook of Applied Optimization. Oxford University Press.

Padberg, M., & Rinaldi, G. (1991). A branch and cut algorithm for the resolution of large-scale symmetric traveling salesman problems. *SIAM Review, 33*(1), 60–100. doi:10.1137/1033004

Salkin, H. M. (1991). A note on Gomory fractional cuts. *Operations Research, 19*(6), 1538–1541. doi:10.1287/opre.19.6.1538

Salvelsbergh, M. W. P. (1997). A branch and price algorithm to solve the generalized assignment problem. *Operations Research, 45*, 381–841.

Savelsbergh, M. W. P. (1994). Preprocessing and Probing Techniques for Mixed Integer Programming Problems. *ORSA Journal on Computing, 6*(4), 445–454. doi:10.1287/ijoc.6.4.445

Sparkias, H., & Gibbon, A. (1993). *Lecture Notes on Parallel Computations*. Cambridge University Press.

Taha, H.A. (2015). *TORA Software*. Windows Version 4.

Taha, H. A. (2017). *Operations Research: An Introduction* (10th ed.). Pearson Educators.

KEY TERMS AND DEFINITIONS

Algorithm: A procedure or set of rules to be followed in problem-solving operations, specifically by a computer.

Branch and Bound Method: This is a solution method which partitions the feasible solution space into smaller subsets of solutions.

Combinatoral Explosion: Is the rapid growth of the complexity of a problem is affected by the number variables, constraints, and bounds of the problem.

Knapsack Problem: Is a problem in combinatorial optimization with an objective function and only one constraint.

Objective Function: Is a mathematical expression that minimizes or maximizes a linear or nonlinear problem.

Optimal Solution: A feasible solution at the point where the objective function reaches its maximum/ minimum value.

Parallel Processing: Is a method of simultaneously breaking up and running program tasks on multiple microprocessors, thereby reducing processing time.

Variable: Is a quantity that needs to be determined in order to solve the linear or nonlinear problem.

Chapter 4
New Direction to the Scheduling Problem:
A Pre-Processing Integer Formulation Approach

Elias Munapo
North-West University, South Africa

Olusegun Sunday Ewemooje
https://orcid.org/0000-0003-3236-6018
Federal University of Technology, Akure, Nigeria

ABSTRACT

This chapter presents a new direction to the scheduling problem by exploring the Moore-Hodgson algorithm. This algorithm is used within the context of integer programming to come up with complementarity conditions, more biding constraints, and a strong lower bound for the scheduling problem. With Moore-Hodgson Algorithm, the alternate optimal solutions cannot be easily generated from one optimal solution; however, with integer formulation, this is not a problem. Unfortunately, integer formulations are sometimes very difficult to handle as the number jobs increases. Therefore, the integer formulation presented in this chapter uses infeasibility to verify optimality with branch and bound related algorithms. Thus, the lower bound was obtained using pre-processing and shown to be highly accurate and on its own can be used in those situations where quick scheduling decisions are required.

INTRODUCTION

The Moore-Hodgson algorithm was proposed by Moore (1968) and the algorithm was attributed to Thom Hodgson. It is for this reason that in literature this algorithm is often referred as the "Moore-Hodgson Algorithm". The scheduling problem assumes that all jobs are processed on a single machine, the processing times of the machine are known with certainty and that the release time for each job is

DOI: 10.4018/978-1-7998-3970-5.ch004

zero. The Moore-Hodgson was developed to improve the earliest due date (EDD) sequence by minimizing tardiness. Here, this algorithm is use within the context of integer programming to come up with complementarity conditions, extra more biding constraints and a strong lower bound for the scheduling problem. The complementarity conditions, extra more binding constraints together with the lower bound can significantly reduce the complexity of the scheduling problem. Alternate optimal solution are readily available with integer formulation while with the Moore-Algorithm algorithm the problem size does not increase at every iteration. The lower bound proposed in this approach is highly accurate and can be used in those circumstances where very quick scheduling decisions are required. Even though heuristics such as Genetic Local Search (Kim et al., 2018) can quickly find a near optimal solution for the scheduling, but the difference between a near optimal solution and an exact solution may be in millions of dollars as exact solutions are necessary for the scheduling problem. TORA software (Taha, 2011a; 2011b) is used in determining the number of branch and bound sub-problems (Land and Doig, 1960) in this paper.

STATEMENT OF THE PROBLEM

Problem: Using EDD, how can the jobs 1,2,3,…,*n* be sequenced so that tardiness is minimized? Suppose the scheduling problem is given as Table 1.

Table 1. Scheduling problem based on n jobs

Job (J_i)	1	2	3	…	N
Processing time (P_i)	p_1	p_2	p_3	…	p_n
Due Date (D_i)	d_1	d_2	d_3	…	d_n

Assumptions: All jobs are processed on a single machine, processing times are known with certainty and that the release time of each job is 0.

Suppose;

(i) (i) EDD is used i.e.

$$d_1 \le d_2 \le d_3 \le \ldots \le d_n. \quad (1)$$

(ii) L_i is lateness, C_i is the cumulative processing time and T_i is tardiness.

Then the complete table is obtained as shown on Table 2.

We do not have tardiness or lateness in the jth job if

$$p_1 + p_2 + p_3 + \ldots + p_j - d_j \le 0. \quad (2)$$

Table 2. Completed table using Earliest Due Date (EDD)

J_i	P_i	D_i	C_i	L_i	T_i
1	p_1	d_1	$c_1=p_1$	$-l_1=c_1-d_1$	0
2	p_2	d_2	$c_2=p_1+p_2$	$-l_2=c_2-d_2$	0
3	p_3	d_3	$c_3=p_1+p_2+p_3$	$-l_3=c_3-d_3$	0
...	...	...	...	...	...
J	p_j	d_j	$c_j=p_1+p_2+p_3+\ldots+p_j$	$l_j=c_j-d_j$	l_j
...	...	...	...	...	...
N	p_n	d_n	$c_n=p_1+p_2+p_3+\ldots+p_j+\ldots+p_n$	$l_n=c_n-d_n$	l_n

MATERIALS AND METHODS

Moore-Hodgson Algorithm

This algorithm yields a schedule with a minimum number of late jobs. The algorithm was proposed by Moore (1968).

Algorithm

(i) Sort jobs in order of increasing due date: $d_j \uparrow$;
(ii) Start with scheduled job set $J_0 = \varnothing$, load $\lambda = 0$;
(iii) For $j = 1, \ldots, n$, if $\lambda + p_j \leq d_j$, then $J_j = J_{j-1} \cup \{j\}$; $\lambda = \lambda + p_j$; otherwise, let $j_{\max} \in J_j-_1 \cup \{j\}$ have largest processing time; set $J_j = J_j-_1 \cup \{j\} \setminus \{j_{\max}\}$; $\lambda = \lambda + p_j - p_{j\max}$.
(iv) Schedule jobs in J_n in order of due date; discard jobs not in J_n or schedule them in any order after the jobs in J_n.

Proof that Moore-Hodgson Algorithm Yields a Schedule with a Minimum Number of Late Jobs

This is a proof by contradiction and is given in Moore (1968), Sidney (1973) and Pinedo (2002). Assume the jobs are already in due date order. Then the claim is equivalent to: (*) for each $k = 1,\ldots,n$ and J_k is a maximum cardinality feasible subset of $S(k) := \{1,\ldots,k\}$. We prove a slightly stronger statement: (**) for each $k = 1,\ldots,n$ and J_k is a maximum cardinality feasible subset of $S(k) := \{1,\ldots,k\}$, and among all maximum cardinality feasible sets, J_k has smallest total length.

To prove (**), let $N(k)$ denote the true maximum cardinality of a feasible subset of $S(k)$, and let F_k denote such a maximum cardinality set of minimum total length. Here *feasible* means that the subset can be scheduled in time, which can be tested by checking the EDD schedule for F_k. Assuming that (**) is not true, consider the smallest counter-example. Evidently, $n > 1$, since for a single job, $J_1 = \varnothing$ if and only if $N(1) = 0$ if and only if $p_1 > d_1$.

By minimality we have that $|J_{n-1}| = N(n-1)$ and $p(J_{n-1}) = p(F_{n-1})$. Note that $|J_{n-1}| \leq |J_n| \leq |J_{n-1}| + 1$, and similarly, $N(n-1) \leq N(n) \leq N(n-1) + 1$, and furthermore $|J_n| \leq N(n)$. As (**) is not true we must have

(a) $|J_{n-1}| = |J_n|$ and $N(n) = N(n-1) + 1$, or

(b) $|J_n| = N(n)$ but J_n is not of minimum total length.

If we are in case (a), then F_n is of size $N(n)$ and contains job n. But then $F_n \setminus \{n\}$ has size $N(n-1)$ and has total length at least that of J_{n-1}. F_n is feasible, hence its EDD schedule is feasible. It ends with job n, which means that $J_{n-1} \cup \{n\}$ is also feasible. Hence $J_n = J_{n-1} \cup \{n\}$ contradicting (a).

If we are in case (b), $N(n) = N(n-1) + 1$, then $n \in J_n$, and $n \in F_n$ with $|F_n| = |J_n|$, and $p(F_n) < p(J_n)$. But then $p(F_n \setminus \{n\}) < p(J_{n-1})$, contradicting the minimum length of J_n-_1. Since, we are in case (b) and $N(n) = N(n-1)$, then $|J_n| \leq |F(n)| = |J_{n-1}|$. So insertion of n was followed by deletion of some job k, so that $p(J_n) \leq p(J_{n-1}) = p(F_{n-1})$. Now it follows from $p(F_n) < p(J_n)$, that $n \in F_n$. Now, let $j_{\max} = \text{argmax}\{p(j)|j \in J_{n-1} \cup \{n\}\}$, and let $j_1 = \max\{j|j \in J_{n-1} \setminus F_n\}$. Then $p(F_n \cup \{j_1\} \setminus \{n\}) = p(F_n) + p(j_1) - p(n) < p(J_n) + p(j_1) - p(n) \leq p(J_n) + p(j_{\max}) - p(n) = p(J_n-_1) = p(F_n-_1)$.

Note that by definition of j_1, the set J^0 of all jobs in J_{n-1} higher than j_1 belongs to F_n as well. From the schedule for F_n remove job n and jobs J^0, process remaining jobs as early as possible, next process job j_1 and then jobs J^0 in EDD order. Then the latter jobs complete earlier than they do in the schedule for J_{n-1}, as $p(F_n) + p(j_1) - p(n) < p(J_{n-1})$. So the set $F_n \cup \{j_1\} \setminus \{n\}$ is feasible, contradicting the minimality of $p(F_{n-1})$. See Sidney (1973) and Pinedo (2002) for more on this algorithm.

Numerical Illustration: Moore Hodgson Algorithm

Use More-Hodgson algorithm to sequence the following 8 jobs given in Table 3.

Using EDD and calculating tardiness we have Table 4,

Removing Job 7 and putting it at the bottom we have Table 5.

Removing Job 1 and putting it at the bottom we have Table 6.

Table 3. Scheduling problem based on eight (8) jobs

Job (J_i)	1	2	3	4	5	6	7	8
Processing time (P_i)	20	6	8	16	20	12	14	6
Due Date (D_i)	30	12	18	46	40	60	10	70

Table 4. Moore-Hodgson iteration 1

Job i	P_i	D_i	C_i	L_i	T_i
7	14	10	14	4	4
2	6	12	20	8	8
3	8	18	28	10	10
1	20	30	48	18	18
5	20	40	68	28	28
4	16	46	84	38	38
6	12	60	96	36	36
8	6	70	102	32	32

Where C_i is the cumulative processing time, L_i=C_i-D_i and T_i is the tardiness.

Table 5. Moore-Hodgson iteration 2

Job i	P_i	D_i	C_i	L_i	T_i
2	6	12	6	-6	0
3	8	18	14	-4	0
1	20	30	34	4	4
5	20	40	54	14	14
4	20	46	70	24	24
6	16	60	82	22	22
8	6	70	88	18	18
7	14	10	102	92	92

Max{6,8,20}=20

Table 6. Moore-Hodgson iteration 3

Job i	P_i	D_i	C_i	L_i	T_i
2	6	12	6	-6	0
3	8	18	14	-4	0
5	20	40	34	-6	0
4	16	46	50	4	4
6	12	60	62	2	2
8	6	70	68	-2	0
7	14	10	82	72	72
1	20	30	102	72	72

Max{6,8,20,16,6}=20

Removing Job 5 and putting it at the bottom we have Table 7.
Job 7 was previously moved from the top. Optimal solution is now available.
The sequence (2,3,4,6,8,7,1,5) will minimize tardiness.

Table 7. Moore-Hodgson iteration 4

Job i	P_i	D_i	C_i	L_i	T_i
2	6	12	6	-6	0
3	8	18	14	-4	0
4	16	46	30	-16	0
6	12	60	42	-18	0
8	6	70	48	-22	0
7	14	10	62	52	52
1	20	30	82	52	52
5	20	40	102	62	62

Linear Integer Formulation

Minimizing Total Tardiness

The formulation is based on the fact that if $p_1 + p_2 + p_3 + ... + p_j - d_j \leq 0$, then there is no tardiness. In other words, we need to,

Maximize $x_1 + x_2 + x_3 + ... + x_n$;

Such that: $p_1x_1 - d_1x_1 \leq 0 + My_1$;

$$\begin{aligned} &p_1x_1 + p_2x_2 - d_2x_2 \leq 0 + My_2; \\ &... \\ &p_1x_1 + p_2x_2 + p_3x_3 + ... + p_nx_n - d_nx_n \leq 0 + My_n; \\ &(x_1 + x_2 + x_3 + ... + x_n) + (y_1 + y_2 + y_3 + ... + y_n) = n. \end{aligned} \tag{3}$$

Where $x_i, y_j = \{0,1\}$ $\forall i \& j$ and M is a very big constant relative to all other constants.

This is a linear integer problem and it can be simplified by;

Maximizing $x_1 + x_2 + x_3 + ... + x_n$;

Such that: $p_1x_1 - d_1x_1 \leq My_1$;

$$\begin{aligned} &p_1x_1 + p_2x_2 - d_2x_2 \leq My_2; \\ &... \\ &p_1x_1 + p_2x_2 + p_3x_3 + ... + p_nx_n - d_nx_n \leq My_n; \\ &x_1 + x_2 + x_3 + ... + x_n + y_1 + y_2 + y_3 + ... + y_n = n. \end{aligned} \tag{4}$$

This can be solved by the available methods for integer programming but there is need for pre-processing to significantly reduce the complexity of the problem. Let

$$p_1x_1 - d_1x_1 = \rho_1x_1;\ p_2x_2 - d_2x_2 = \rho_2x_2;\ ... p_nx_n - d_nx_n = \rho_nx_n. \tag{5}$$

This is further simplified as:

$$\rho_1x_1 \leq My_1;\ p_1x_1 + \rho_2x_2 \leq My_2;\ ... p_1x_1 + p_2x_2 + p_3x_3 + ... + \rho_nx_n \leq My_n. \tag{6}$$

This forms a triangle array of numbers with properties that can be used in solving the problem.

Complementarity

Let the complementarity condition for the scheduling problem be given as (7).

$$x_j+y_j=1 \tag{7}$$

With the complementarity condition the constraint, $x_1+x_2+x_3+...+x_n+y_1+y_2+y_3+...+y_n=n$, becomes redundant. The integer formulation with complementarity conditions become:

Maximize $x_1+x_2+x_3+...+x_n$;

Such that:

$$\rho_1x_1 \le My_1;\ \ p_1x_1+\rho_2x_2 \le My_2;\ ...\ p_1x_1+p_2x_2+p_3x_3+...+\rho_nx_n \le My_n; \tag{8}$$

$$x_1+y_1=1;\ \ x_2+y_2=1;\ ...\ x_n+y_n=1.$$

The formulation in equation (8) has a smaller solution space than equation (4). In other words, complementarity conditions reduce the complexity of the problem.

Pre-Processing

Complementarity reduces the complexity of the problem but there is need for further complexity reduction. Integer models are highly unpredictable when solving by branch and bound related algorithms. Another way of reducing the complexity of an integer model is by pre-processing. In this paper pre-processing is used to determine a strong lower bound for the scheduling problem. More n pre-processing can be obtained from Savelsbergh (1994).

Pre-Processing Procedure

Suppose the inequalities in the rectangular array are numbered as given in equation (9).
$\rho_1x_1 \le My_1;\ \ p_1x_1+\rho_2x_2 \le My_2;\ ...\ p_1x_1+p_2x_2+p_3x_3+...+\rho_nx_n \le My_n;$ (9)

Step 1: Solve $\rho_1x_1 \le My_1$ to get $x_1=x_1^*$ & $y_1=y_1^*$.
Step 2: Use the solution in the previous step to determine the unknown variable in the next inequality.
Step 3: Repeat step 2 until all variables are determined.

The lower bound is given as:

$$x_1+x_2+...+x_n \ge x_1^*+x_2^*+...+x_n^*. \tag{10}$$

Extra More Binding Constraints

The extra more binding constraints can be generated from the constraints on the triangle array of numbers. Suppose the constraint is,

$$p_1x_1 + p_2x_2 + p_3x_3 + \ldots + \rho_j x_j \le My_j.$$

By rearranging the coefficients $p_1, p_2, \ldots, p_3, \ldots, \rho_r$, in ascending order and using the fact that

$$p_1x_1 + p_2x_2 + p_3x_3 + \ldots + \rho_j x_j \le 0,$$

the extra more binding constraint can be generated as:

$$x_1 + x_2 + x_3 + \ldots + x_j \le \ell. \tag{11}$$

Where $\ell \le j$, is the largest number of variables that satisfy

$$p_1x_1 + p_2x_2 + p_3x_3 + \ldots + \rho_j x_j \le 0.$$

It may not be possible or necessary to generate the extra more binding constraints from all the rows. In this paper we illustrate by using only 25% of the rows.

Pre-Processed Scheduling Problem

The pre-processed scheduling problem becomes:

Maximize $x_1 + x_2 + x_3 + \ldots + x_n$;

Such that:

$$\rho_1 x_1 \le My_1;\; p_1x_1 + \rho_2 x_2 \le My_2;\; \ldots\, p_1x_1 + p_2x_2 + p_3x_3 + \ldots + \rho_n x_n \le My_n; \tag{12}$$

$$x_1 + y_1 = 1;\; x_2 + y_2 = 1;\; \ldots\; x_n + y_n = 1;$$

$$x_1 + x_2 + \ldots + x_n \ge (x_1^* + x_2^* + \ldots + x_n^*) + 1;\; x_1 + x_2 + x_3 + \ldots + x_j \le \ell.$$

If equation (12) is infeasible implies that $(x_1^*, x_2^*, \ldots, x_n^*)$ is optimal. This integer formulation is better than the one given in equation (8) in terms complexity.

Numerical Illustration: Integer Formulation

Using (i) plain integer formulation,
(ii) integer formulation with complementarity conditions and
(iii) integer formulation with pre-processing

to sequence the 8 jobs in Table 3, we have:

Plain Integer Formulation

Let M=10000 The plain integer formulation is given as:

Maximize $Z_o = x_1 + x_2 + x_3 + x_4 + x_5 + x_6 + x_7 + x_8.$

Such that

$$4x_7 \leq 10000y_7;$$

$$-6x_2 + 14x_7 \leq 10000y_2;$$

$$-10x_3 + 6x_2 + 14x_7 \leq 10000y_3;$$

$$-10x_1 + 8x_3 + 6x_2 + 14x_7 \leq 10000y_1;$$

$$-20x_5 + 20x_1 + 8x_3 + 6x_2 + 14x_7 \leq 10000y_5;$$

$$-30x_4 + 20x_5 + 20x_1 + 8x_3 + 6x_2 + 14x_7 \leq 10000y_4;$$

$$-48x_6 + 16x_4 + 20x_5 + 20x_1 + 8x_3 + 6x_2 + 14x_7 \leq 10000y_6;$$

$$-64x_8 + 12x_6 + 16x_4 + 20x_5 + 20x_1 + 8x_3 + 6x_2 + 14x_7 \leq 10000y_8;$$

$$x_1 + x_2 + x_3 + x_4 + x_5 + x_6 + x_7 + x_8 + y_1 + y_2 + y_3 + y_4 + y_5 + y_6 + y_7 + y_8 = 8. \quad (13)$$

Solving using the automated branch and bound algorithm (Savelsbergh,1994; Sidney, 1973), we have:

$$Z_o = 5;$$
$$x_2 = x_3 = x_4 = x_6 = x_8 = 1; \qquad (14)$$
$$x_1 = x_5 = x_7 = 0.$$

Note that there are alternate optimal solutions and using the automated branch and bound algorithm we find the solution at sub-problem 11 and it requires 425 sub-problems to verify this optimal solution.

The optimal sequence is: (2,3,4,6,8,7,1,5), i.e. using EDD to sequence the basic variables first and then the non-basic variables.

Integer Formulation With Complementarity Conditions

The integer formulation with complementarity conditions is given as:

Maximize $Z_o = x_1 + x_2 + x_3 + x_4 + x_5 + x_6 + x_7 + x_8;$

Such that:

$$4x_7 \le 10000 y_7;$$

$$-6x_2 + 14x_7 \le 10000 y_2;$$

$$-10x_3 + 6x_2 + 14x_7 \le 10000 y_3;$$

$$-10x_1 + 8x_3 + 6x_2 + 14x_7 \le 10000 y_1;$$

$$-20x_5 + 20x_1 + 8x_3 + 6x_2 + 14x_7 \le 10000 y_5;$$

$$-30x_4 + 20x_5 + 20x_1 + 8x_3 + 6x_2 + 14x_7 \le 10000 y_4;$$

$$-48x_6 + 16x_4 + 20x_5 + 20x_1 + 8x_3 + 6x_2 + 14x_7 \le 10000 y_6;$$

$$-64x_8 + 12x_6 + 16x_4 + 20x_5 + 20x_1 + 8x_3 + 6x_2 + 14x_7 \le 10000 y_8;$$

$$x_1 + y_1 = 1, x_2 + y_2 = 1, x_3 + y_3, = 1x_4 + y_4 = 1, x_5 + y_5 = 1, x_6 + y_6 = 1, x_7 + y_7 = 1, x_8 + y_8 = 1. \qquad (15)$$

Also, there are alternate solutions and solving using the branch and bound algorithm, we have:

$$Z_o = 5;$$
$$x_3 = x_4 = x_5 = x_6 = x_8 = 1; \quad (16)$$
$$x_1 = x_2 = x_7 = 0.$$

Using the automated branch and bound algorithm, we find the solution at sub-problem 24 and it requires 63 sub-problems to verify this optimal solution.

Integer Formulation With Pre-Processing

In pre-processing, the inequalities are numbered as:

$$4x_7 \leq 0 + 10000y_7; \quad (17.1)$$

$$-6x_2 + 14x_7 \leq 0 + 10000y_2; \quad (17.2)$$

$$-10x_3 + 6x_2 + 14x_7 \leq 0 + 10000y_3; \quad (17.3)$$

$$-10x_1 + 8x_3 + 6x_2 + 14x_7 \leq 0 + 10000y_1; \quad (17.4)$$

$$-20x_5 + 20x_1 + 8x_3 + 6x_2 + 14x_7 \leq 0 + 10000y_5; \quad (17.5)$$

$$-30x_4 + 20x_5 + 20x_1 + 8x_3 + 6x_2 + 14x_7 \leq 0 + 10000y_4; \quad (17.6)$$

$$-48x_6 + 16x_4 + 20x_5 + 20x_1 + 8x_3 + 6x_2 + 14x_7 \leq 0 + 10000y_6; \quad (17.7)$$

$$-64x_8 + 12x_6 + 16x_4 + 20x_5 + 20x_1 + 8x_3 + 6x_2 + 14x_7 \leq 0 + 10000y_8. \quad (17.8)$$

Pre-processing using complementarity, we have:

From (17.1) we have; x_7=0 & y_7=1 (18)

From (17.2) and (18) we have; x_2=1 & y_2=0 (19)

From (17.3), (18) & (19) we have; x_3=1 & y_3=0 (20)

From (17.4), (18), (19) & (20) we have; x_1=0 & y_1=1 (21)

From (17.5), (18), (19), (20) & (21) we have; x_5=1 & y_5=0 (22)

From (17.6), (18), (19), (20), (21) & (22) we have; x_4=0 & y_4=1 (23)

From (17.7), (18), (19), (20), (21), (22) & (23) we have; x_6=1 & y_6=0 (24)

From (17.8), (18), (19), (20), (21), (22), (23) & (24) we have; x_8=1 & y_8=0 (25)

From the pre-processing;

(i) Current solution is given as:

$$Z_{LB} = 5;$$
$$x_2 = x_3 = x_5 = x_6 = x_8 = 1; \qquad (26)$$
$$x_1 = x_4 = x_7 = 0.$$

(ii) The lower bound is obtained as:

$$\ell b = x_2 + x_3 + x_5 + x_6 + x_8 = 1+1+1+1+1 = 5. \qquad (27)$$

Pre-Processing Inequality

The pre-processing inequality is given as:

$$x_1 + x_2 + x_3 + x_4 + x_5 + x_6 + x_7 + x_8 \geq 5+1 = 6. \qquad (28)$$

Extra More Binding Constraints

The extra more binding constraints are:

From 17.1, we have: $4x_7 \leq 0 \Rightarrow x_7 \leq 0 \Rightarrow x_7 = 0.$ (29)

From 17.4, we have: $x_1 + x_2 + x_3 \leq 2.$ (30)

(iii) The integer formulation with pre-processing and complementarity conditions is given as:

Maximize $Z_o = x_1 + x_2 + x_3 + x_4 + x_5 + x_6 + x_7 + x_8;$

Such that:

$4x_7 \leq 10000y_7;$

$-6x_2 + 14x_7 \leq 10000y_2;$

$-10x_3 + 6x_2 + 14x_7 \leq 10000y_3;$

$-10x_1 + 8x_3 + 6x_2 + 14x_7 \leq 10000y_1;$

$-20x_5 + 20x_1 + 8x_3 + 6x_2 + 14x_7 \leq 10000y_5;$

$-30x_4 + 20x_5 + 20x_1 + 8x_3 + 6x_2 + 14x_7 \leq 10000y_4;$

$-48x_6 + 16x_4 + 20x_5 + 20x_1 + 8x_3 + 6x_2 + 14x_7 \leq 10000y_6;$

$-64x_8 + 12x_6 + 16x_4 + 20x_5 + 20x_1 + 8x_3 + 6x_2 + 14x_7 \leq 10000y_8;$

$x_1 + y_1 = 1, x_2 + y_2 = 1, x_3 + y_3, = 1x_4 + y_4 = 1, x_5 + y_5 = 1, x_6 + y_6 = 1, x_7 + y_7 = 1, x_8 + y_8 = 1;$

$$x_1 + x_2 + x_3 + x_4 + x_5 + x_6 + x_7 + x_8 \geq 6;\ x_7 \leq 0;\ x_1 + x_2 + x_3 \leq 2. \quad (31)$$

The solution is infeasible, implying that the current solution given in equation (32) is one of the optimal solution.

$$Z_o = 5; \\ x_2 = x_3 = x_5 = x_6 = x_8 = 1; \\ x_1 = x_4 = x_7 = 0. \quad (32)$$

Hence, it requires only 1 sub-problem of the branch and bound algorithm to verify infeasibility.

Some Practical Applications of Scheduling

(i) Scheduling is very important in real life. In big hospitals where there are very large numbers of workers and patients, there is need for proper planning. Since workers cannot work continuously without resting, then the workers must be scheduled so that they have enough time to rest and at the same time patients get the assistance they need throughout their stay at the hospital. Assignment models which are special binary integer models are used to handle these scheduling problems.

(ii) At very busy airports such as the one in London, Dubai, Tokyo, US and Chinese towns, requires very efficient algorithms to schedule the flights. Delays in scheduling the flights is costly to big airlines operating in these big cities.
(iii) Big universities which enrol large numbers of students such as those in China and India require proper planning when scheduling examinations and venues. At these universities there is limited number of resources such as writing venues, laboratories, invigilators etc. The Scheduling problem is formulated as an integer model, solved quickly using heuristics to get a near optimal solution if time is not available for the examination organizers. If the examinations are months away, an exact optimal solution can be obtained by branch and cut related approaches for integer models.
(iv) Production processes require scheduling so as to maximize use of company resources, minimize costs and maximise profits. Production managers require efficient scheduling techniques so as to meet deadline targets, use all the ordered raw materials before expiry dates and also to produce high quality products.

Limitation

Formulating the scheduling problem has its own weaknesses. Complexity of integer models increase with an increase in the size of the problem. It is usually difficult to solve practical problems as integer models in reasonable times.

CONCLUSION

The scheduling problem has real life applications and deserves further research. The rate of research on the scheduling problem show that it is an ongoing viable research area (Chen, 2007; Tönissen et al., 2017; van den Akker et al., 2018). The Moore-Hodgson Algorithm is shown to be efficient and with this algorithm the problem does not change in size at every iteration and there are no chances of any combinatorial explosion. The identified problem with Moore-Hodgson Algorithm is that alternate optimal solutions cannot be easily generated from one optimal solution while with integer formulation, this is not a problem. Unfortunately, integer formulations are sometimes very difficult to handle as the number jobs increases. However, the integer formulation presented in this paper uses infeasibility to verify optimality. Infeasibility is easy to verify with branch and bound related algorithms. Hence, the lower bound was obtained using pre-processing and shown to be highly accurate, and on its own can be used in those situations where quick scheduling decisions are required.

STRENGTHS

With the proposed approach alternate optimal solutions can be explored using sensitivity analysis. Some optimal solutions obtained using Moore-Hodgson may not necessarily be feasible in real life. After solving a problem, one of the machines may be found to be faulty. In the case of Moore-Hodgson, the problem has to be resolved which consumes time.

AREA FOR FURTHER RESEARCH

In future there is need to compare the proposed approach with known algorithms on benchmark models. There is need for comparison in the form of computational experiments and computational times.

REFERENCES

Chen, W.-J. (2007). An efficient algorithm for scheduling jobs on a machine with periodic maintenance. *International Journal of Advanced Manufacturing Technology*, *34*(11-12), 1173–1182. doi:10.100700170-006-0689-x

Kim, J., Jeon, W., Ko, Y.-W., Uhmn, S., & Kim, D.-H. (2018). Genetic Local Search for Nurse Scheduling Problem. *Advanced Science Letters*, *24*(1), 608–612. doi:10.1166/asl.2018.11770

Land, A. H., & Doig, A. G. (1960). An Automatic method for solving discrete programming problems. *Econometrica*, *28*(3), 497–520. doi:10.2307/1910129

Moore, J. M. (1968). An n job, one machine sequencing algorithm for minimizing the number of late jobs. *Management Science*, *15*(1), 102–109. doi:10.1287/mnsc.15.1.102

Pinedo, M. (2002). *Scheduling: Theory, Algorithms, and Systems*. Prentice Hall.

Savelsbergh, M. W. P. (1994). Pre-processing and Probing Techniques for Mixed Integer Programming Problems. *ORSA Journal on Computing*, *6*(4), 445–454. doi:10.1287/ijoc.6.4.445

Sidney, J. B. (1973) An Extension of Moore's Due Date Algorithm. *Lecture Notes in Economics and Mathematical Systems, 86*, 393–398.

Taha, H. A. (2011). *Operations Research: An Introduction* (9th ed.). Pearson Educators.

Taha, H.A. (2011). *TORA Software*. Windows Version 1.00.

Tönissen, D. D., Van den Akker, J. M., & Hoogeveen, J. A. (2017). Column generation strategies and decomposition approaches for the two-stage stochastic multiple knapsack problem. *Computers & Operations Research*, *83*, 125–139. doi:10.1016/j.cor.2017.02.009

van den Akker, J. M., Hoogeveen, H., & Stoef, J. J. (2018). Combining two-stage stochastic programming and recoverable robustness to minimize the number of late jobs in the case of uncertain processing times. *Journal of Scheduling*, *21*(6), 607–617. doi:10.100710951-018-0559-z

KEY TERMS AND DEFINITIONS

Algorithm: A procedure or set of rules to be followed in problem-solving operations, specifically by a computer.

Branch and Bound Method: This is a solution method which partitions the feasible solution space into smaller subsets of solutions.

Complementarity Conditions: A way to model a constraint that has a combinatorial nature e.g. it implies that either A or B must be 0 or both may be 0 as well.

Earliest Due Date: A rule where jobs are scheduled according to the earliest due date given and to minimize the total tardiness of the whole jobs.

Iteration: A repetition of a process in order to generate a sequence of outcomes as a means of obtaining successively closer approximations to the solution of a problem.

Optimal Solution: A feasible solution at the point where the objective function reaches its maximum/minimum value.

Tardiness: A quality or fact of being late in completing a job or project.

Chapter 5
Use of the Neural Network Controller of Sprung Mass to Reduce Vibrations From Road Irregularities

Zakhid Godzhaev
Federal State Budgetary Scientific Institution "Federal Scientific Agroengineering Center VIM", Russia

Sergey Senkevich
https://orcid.org/0000-0001-6354-7220
Federal State Budgetary Scientific Institution "Federal Scientific Agroengineering Center VIM", Russia

Viktor Kuzmin
Federal State Budgetary Scientific Institution "Federal Scientific Agroengineering Center VIM", Russia

Izzet Melikov
Dagestan State Agricultural University Named After M.M. Dzhambulatov, Russia

ABSTRACT

Hydraulic systems that damp active oscillation operate according to a certain non-linear and time-varying algorithm. It is difficult to create a controller based on its dynamic model. This chapter proposes a new operation regime of the controller based on neuron nets by combining the advantages of the adaptive, radial, and basic functions of the neuron net. Its undoubted advantages are a learning (tilting) ability in real time to process indefinite, nonlinear disturbances, and to change the value of the active force in the hydraulic leaf spring by adjusting the weight coefficients of the neuron net and/or the radial parameters of the basic function. The model is a ¼ hydraulic active sprung mass of a mobile vehicle. The modeling shows that the use of a neuron net controller makes the sprung mass much more efficient.

DOI: 10.4018/978-1-7998-3970-5.ch005

INTRODUCTION

Most of the operating time, wheeled agricultural tractors have to work in fields with different soil backgrounds, along dirt or rough roads or even off roads (Senkevich S. et al, 2019; Senkevich S.E. et al, 2019). These conditions require sprung mass which allows supporting a regulated smooth ride and an effective protection of the operator from vibrations. The sprung mass must also keep the tires in contact with the road surface in order to support good stability and controllability of the vehicle (TC). This is especially important when a vehicle turns, brakes or accelerates (Frolov, 1995). The parts of sprung mass are usually located in the corners of the vehicle and connect the transmission and the frame with the wheels. Typically, the sprung mass includes elements of such two main types as elastic elements and dampers installed in parallel. Elastic elements smooth dynamic loads caused by moving along rough road. When the sprung mass is transmitted a dynamic disturbance from the irregularities of the roads, the elastic elements are deformed and accumulate potential energy, which is then released. When the vehicle moves forward and back, the suspension dampers dissipate the oscillation energy. Similarly, elastic elements and dampers keep controlled movement of the vehicle frame during the turns (Derbaremdiker,1985). Most of the vehicles use steel elastic elements in the sprung mass, which are divided into 3 types: coil, torsion, leaf springs. Some vehicles use pneumatic and hydro-pneumatic elastic elements. As dampers in the vehicle sprung mass hydraulic shock absorbers are basically used (Derbaremdiker, 1985; Senkevich, S.E. et al, 2020; Senkevich S. et al, 2020a; Senkevich S. et al, 2020b).

BACKGROUND

Smooth Ride

The primary task of the vehicle sprung mass is to ensure a regulated smooth ride. This indicator is also very important for agricultural tractors, in which the value of vibration accelerations under the operator's seat is higher than that of other vehicles during operation (Deboli, R., Calvo, A. & Preti, C., 2017). In addition, agricultural tractor operators spend a lot of time in the field, when dynamic effect from the soil surface is significantly larger than when driving along the road (Gurhan & Cay, 2008). The resulting noise and vibration affect the operator's work efficiency and his health (Marjoram, et al, 2008; Duke, 2007). Traditionally, the sources of oscillations that affect the noise and vibration load on the operator's workplace are divided into two classes: inboard and road. The inboard sources are all rotating components of the tractor including an engine, wheels, transmission, etc. The fluctuations caused by the sources range from 25 to 20000 Hz, which are perceived by a human ear as noise. The second category of sources is a rough road. The oscillation frequency usually ranges from 0 to 25 Hz. This range includes frequencies that are uncomfortable for a human body. For this reason, vibrations from road bump are the most negative in terms of smooth vehicle ride and effect on a vehicle operator (Silayev, 1972). The upper limit of the frequency range, within which the sprung mass operates, usually does not exceed 25 Hz. The smooth vehicle ride largely depends on the dynamic behavior of the vehicle body (i.e. the suspension linkage), which is exposed to a combination of vertical, longitudinal-angle and cross-angular vibrations during operation (Hansson, 2002). The vibration effect on a human body, especially on passengers of the vehicle, was considered in many studies (Marjoram, et al, 2008; Pobedin, et al, 2016; Sukhorukov, 2003). At the same time, a human body reacts differently to different vibrations. It depends on the direction and frequency of vibrations. According to studies (Sukhorukov, 2003), the human body is more sensitive to horizontal vibrations than to vertical ones. In the vertical direction, the 4-8 Hz range which corresponds to the resonant frequencies of the abdominal organs is the most sensitive. Sensitivity in horizontal directions is the highest in the 1-2 Hz range. The 0.5-0.75 Hz range causes 'seasickness'. Thus, the most 'inconvenient' frequencies for a person, and those that should be weakened by the sprung mass, are of 0.5-10 Hz range. Based on this, in this paper the tractor sprung mass was evaluated in this frequency range (Sukhorukov, 2003).

The vibrations of the frame should be measured vertically and horizontally. The most commonly used measuring method is the measurement of rms acceleration (RMS). It is calculated as follows:

$$RMS = \sqrt{\frac{1}{T}\int_{0}^{T} a^2(\mathrm{t})\,\mathrm{dt}} \tag{1}$$

where $a(t)$ is an immediate acceleration, m/s^2; and T is a period of vibrations, s.

To simplify the experiment, only the magnitude of the vertical vibration acceleration is often measured. In this study, the measurement was carried out only on the vertical component. Summing up, the formula (1) is used to determine the magnitude of the rms acceleration and to evaluate the sprung mass in terms of smooth ride.

There are certain standards for measuring the magnitude of smooth vehicle ride and the maximum allowable values of vibration accelerations perceived by a vehicle operator. Among them, there are usually two ISO standards:

- ISO 2631-74 regulates the main issues related to the measurement of the vibration level and the vibration effect on human health. The standard uses the generalized vibration load coefficient, which depends on the magnitude of the vertical and horizontal vibration accelerations. On the basis of this coefficient, the standard establishes time limits for work.
- ISO 5008-74. This standard describes a technique for measuring and analyzing the vibration load of a vehicle operator. This standard regulates the speed of the tests (5 and 12 km/h), as well as, based on the ISO 2631-74 standard, the information on the maximum permissible vibration frequency is given. On the basis of all these data, the value of rms acceleration is formed, which is used in further work.

Vehicle Handleability

Handleability is a characteristic of the vehicle, which shows the amount of stability and traffic safety that can be created due to the steady contact between the tires and the road. The indicator of handleability is especially critical during vehicle maneuvers, such as turning, braking or speeding. In these extreme situations, poor wheel contact with the road can greatly reduce handling, which in turn can affect the safety of passengers. In this regard, the most important purpose of vehicle sprung mass is to ensure a good and constant contact of the propulsion unit with the road. Controllability is directly related to the amount of wheel contact with the road, which, in turn, depends on two factors: vibration of the wheel and vehicle. The vertical movement of the wheel mainly depends on the road roughness, and the vehicle body moves mainly during the change in the direction of the vehicle movement (turns, speeding, braking).

It is considered that body movements, as a result of vehicle turning speeding or braking are the most problematic. For example, when turning, the centrifugal force pushes the vehicle. It is opposed by the resistance force resulting from the contact of the tire with the road. At the same time, centrifugal force causes load shift on the tires from one side to the other, which results in a significant reduction of the tire lifetime. Similarly, when the vehicle brakes or rapidly accelerates, the load on the different axes of the vehicle becomes different.

The carrying capacity of the vehicle is influenced by its various characteristics. The main one is the characteristic of the vehicle sprung mass. A good sprung mass provides necessary reactions and prevents excessive weight transfer between the sides and vehicle axles. In addition, an effective sprung mass can control the amount of vertical vibration of the wheel, which depends on the degree of road roughness and the speed of the vehicle.

As mentioned above, controllability and smooth ride are two main functions of the vehicle sprung mass. To evaluate the sprung mass, these two characteristics must be measured and analyzed. Unlike smoothness, as mentioned above, there is no standard for quantitative estimation of vehicle controllability. This makes the processing of experimental data more complicated, but, nevertheless, there are some techniques for indirect measuring the amount of vehicle controllability. Good handleability can be ensured by stable contact loads on the wheel. The result of a change in contact (dynamic) force between the tires and the road can be used for quantitative estimation of the controllability. The smaller the dynamic loading of the wheel, the higher the vehicle controllability, and vice versa. In field trials, it is quite difficult to directly measure the magnitude of the dynamic load, therefore, the tire strain indicator is used as an alternative. Since the tires are elastic, the amount of strain is proportional to the vertical loading of the wheel.

The controllability indicator is especially important for agricultural tractors. If we compare agricultural tractors with a car, it is easy to see that tractors have a much higher center of gravity, a much higher weight, and there is no sprung mass for at least one axle. As a result, the tractor has poor resistance to tilting. In extreme situations, poor controllability can lead to inefficient work of the steering and braking systems of the vehicle. On the other hand, from year to year the working speed of the tractor increases, and since a high speed multiplies the efficiency of the tractor operation, the requirements to it increase as well. Therefore, the improvement of the sprung mass is one of the ways to improve the vehicle controllability. The mobile vehicle sprung mass (MV) should effectively dampen vibrations from rough roads, namely, reduce the vertical oscillations of the leaf springs of the MV, as well as ensure proper contact of the tire with the road. Hydraulic and pneumatic components are widely used in semi-active or active MV sprung mass. These systems operate nonlinearly and dynamically vague (Feuerhuber, Lindert, & Schlacher, 2013). In order to control dynamic systems under uncertainty, there were proposed the traditional control systems for adaptive mode of the sprung mass. However, it is necessary to obtain additional information about the operation of the system to create a control algorithm. As a result, the theme of developing a model with an adaptive control structure is of great importance today. To create controllers of the sprung mass with an active control, there were mainly used complex dynamic systems with unclear logic (Cherry & Jones, 1995) and Neural Network control (Kuzmin, Fedotkin & Kryuchkov, 2017), as well as their combinations. For the combined approaches a complex tilting (learning) mechanism, or a specific base of effective solutions appeared by hit-and-miss method are required (Kuzmin, Fedotkin & Kryuchkov, 2017). In the present study, a model of a new sliding mode controller was developed on the basis of radial basis functions in the Neural Network (RBFNN). The controller is applied for the ¼ hydraulic active sprung mass of a mobile vehicle. The control algorithm is based on radial basis functions and combines the advantage of an adaptive control system and sliding control. The adaptation rule is used to regulate the main functions based on information about a given sliding surface in real time.

Since such an approach is able to learn, its implementation can be started without any initial values of the RBFNN.

Active Scheme and Model of The Sprung Mass

In the design of passive sprung mass, as mentioned earlier, there should be found a compromise between smooth ride and controllability. The priority of a particular criterion depends on the operating conditions.

For example, if the vehicle moves mainly on a rough road, smooth ride should be considered as a priority. If the vehicle moves on a smooth winding road, priority is given to controllability. Therefore, the sprung mass, which can change its elastic-damping characteristics (EDC) depending on traffic conditions, is of great advantage.

Such a system is called adaptive. For each road profile, it works as passive with the selected EDC for the given profile, when the profile changes, the system analyzes this change and accordingly changes the EDC (Yu F., Crolla D., 1998).

In the simplest types of adaptive systems, the change in the road profile is recognized directly by the operator of the vehicle, and the change in the suspension EDC is done manually. The system allows the operator to choose one of three modes: 'soft', 'medium' and 'hard'. These options are actually three options for the damping level of the sprung mass. For example, the operator may select the soft suspension mode when the vehicle runs along a rough road. This provides better smooth ride. While driving on a highway at high speed, a hard type of suspension can be selected for better stability and control.

The advantage of this system is the ability to set suspension stiffness according to the subjective perceptions of the operator, but with the help of such a manual system it is impossible to provide quick control and correct handling for short periods of time. Therefore, in advanced adaptive sprung mass, automotive computerized control systems are used instead of manual control.

In vehicle adaptive control systems, the operating condition of the leaf spring is recognized by the sensor information, after which the controller makes the necessary changes to the sprung mass parameters. There are various ways to determine the operating condition of the sprung mass. In some systems, these are taken into account the operator's effects on the gas pedal, brakes and a steering rack. The received signals from the sensors are processed and after that there are made necessary changes to the parameters of the sprung mass.

For example, intense and frequent commands to change the position of the steering rack can be recognized as a 'winding road' where a more rigid regime of the sprung mass operation is required to provide better controllability (Venhovens, 1994).

In other types of adaptive systems, the working condition is determined by analyzing the vehicle vibration accelerations, which are measured by its own accelerometers, then the AFC analysis is performed by the controller and he makes a decision to change the EDC. The correct changes in the damping value should be the output parameters of the adaptive sprung mass. The adaptive sprung mass requires a narrow bandpass of the controller, since the adaptive system responds only to a stable form of changes in the operating condition of the suspension, for example, when the vehicle leaves the hard road surface on the side road.

In new adaptive systems, the controller's bandpass increases in order to give the leaf spring ability quickly respond to its input parameters, such as an impact or sudden braking. However, the capacity of these systems is still lower than the natural vibration frequency of the vehicle body, and their performance is limited.

Wider bandpasses are used in other types of sprung mass, for example, in semi-active or fully active systems (Chalasani, 1986).

A conventional passive sprung mass works by preserving the vibration energy with a spring and dissipating it with a damper, while a fully active sprung mass can add an active effect to the system itself. Such an operation is performed using a power drive, which is placed directly between the sprung and unsprung mass of the vehicle.

Contrary to the passive systems, in the active systems a control effect does not depend on the amplitude or speed of the elements of the sprung mass; it is set by a controller operating according to a special algorithm (Williams, 1994).

An active sprung mass is designed to be controlled by the spring in the entire range of the mass. In particular, this means that this spring has the best response in the range of the natural frequencies of the vehicle and wheels. The natural frequencies of the vehicle usually range in 1-3 Hz, and the natural frequencies of the wheels are in 5-15 Hz range.

These systems completely control the vibration process. They provide good indicators of smooth ride and controllability of the vehicle. However, their big drawback is significant energy consumption, up to 10 kW of additional load for passenger cars. High power consumption increases fuel consumption by 10–15% (Williams, 1994).

Another disadvantage of the system is the need for fast and accurate system components. This results in a significant complication and rise in price of the sprung mass as a whole. Moreover, high-speed actuators of such systems generate unwanted noise, which generally affects vehicle smooth ride. Since, in the active sprung mass (SM), the change in the EDC is made by an operating actuator, power of its driving gear plays a significant role in the characteristics of the SM. The actuator's failure can seriously affect the SM operation as a whole.

For example, a controller's error to determine the required control action can result in a serious failure when changing the damping or stiffness of the system. This can significantly worsen ride quality. On the other hand, insufficient driving gear power may result in loss of wheel contact with the road surface, which will critically affect the stability and controllability of the vehicle. Such a situation may lead to the dangerous regimes for vehicles and passengers.

Taking into account these drawbacks, fully active SM, despite their high efficiency, are usually not used in conventional vehicles, but are only used in models of the premium segment or special vehicles.

At the same time, there is a more practical option of a fully active SM, it's SM with a narrow latitude bandpass. In this system, the power drive unit is successively connected with the passive elastic element and, in some cases, successively with the passive damper (Figure 1).

Unlike a fully active SM, this system does not work in the entire range of vibration frequencies. On the contrary, its purpose is to improve the performance of the SM only near the natural frequencies of a human body (1-3 Hz), since this range is decisive in smooth ride. At higher frequencies, the power drive works inefficiently, and the vibrations are limited mainly to passive elements of SM.

In practice, despite the lower throughput capacity, the working characteristics of an active SM with low transmissive capacity are relatively close to a fully active SM with good quality of vehicle ride and handling.

Figure 1. The scheme of the completely active sprung mass compared with the passive sprung mass

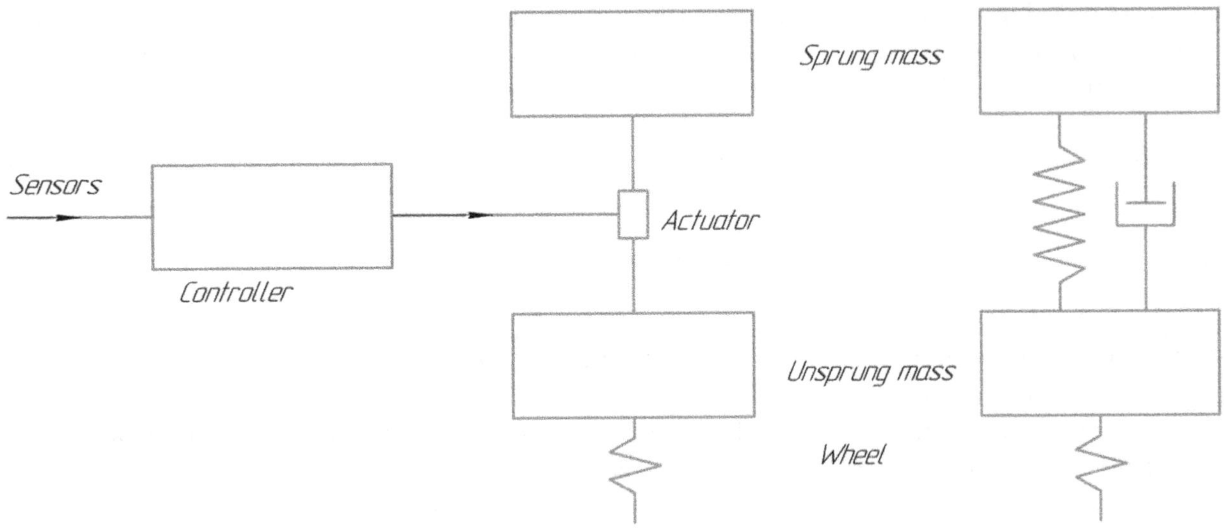

At the same time, the efficiency of these systems is achieved with a significantly lower energy consumption than the systems with high throughput capacity. In addition, since in these systems the passive elements of SM are used along with the active elements, failure cases do not become critical in the operation of SM. Due to their advantages, such systems are being actively implemented by such huge vehicle manufacturers as Jaguar, BMW, Mercedes Benz, Toyota, etc.

There have been conducted many researches to evaluate the efficiency of fully active SM. One of these researches is the fundamental paper (Chalasani, 1986), which studied an active SM with a low throughput capacity on a vehicle model of ¼ scale. The paper shows that such SM reduces the root-mean-square acceleration of the sprung mass on 20%, compared to the passive SM.

The power consumption of fully active SM with low throughput capacity is much lower than in the systems with high capacity, however this consumption is still significant, especially on heavy vehicles.

In a research paper (Deakin, 1997), there was studied power consumption of a military heavy vehicle equipped with an active low transmissive capacity SM. The result of this study showed that up to 25% of engine power was spent on the SM work. However, on asphalt roads, this index is only 3%.

In the work of Gohrle there was studied an active SM with a low transmissive capacity of a military heavy utility vehicle (Gohrle, 2014). The tests have shown that the active SM consumed up to 30 hp of a 150-horsepower engine. This resulted in significant fuel consumption.

Fully active SMs provide high sprung mass quality, but they are not used on agricultural tractors because too much energy is used to ensure their operation. However, for secondary sprung mass (for example, suspension of the tractor cab or operator's seat), such systems are successfully used.

We have prepared a mathematical model of an active vehicle sprung mass with two degrees of freedom. The model is designed to study the dynamic characteristics of the spring. The system includes a control unit, a hydraulic actuator, a module for simulating road bumps, a sprung unit, a data input/output interface unit, and a control unit.

There have been accepted further assumptions: the tire is always in contact with the road surface, the wheel damping is not taken into account (the tire function is modeled by the spring with constant stiffness k_t and the unsprung mass m_u). A diagram of the two active sprung mass is shown in Figure 2.

There are obtained dynamic equations of this sprung mass:

$$m_s\ddot{z}_s = -k_s(z_s - z_u) - b_s(\dot{z}_s - \dot{z}_u) + F_a - F_f \tag{2}$$

$$m_u\ddot{z}_u = k_s(z_s - z_u) + b_s(\dot{z}_s - \dot{z}_u) + k_t(z_r - z_u) - F_a + F_f \tag{3}$$

where z_r is vibrations of road surface; F_a and F_f are the resistance force of the active hydraulic system and the hydraulic friction force, respectively.

The correlation between the servo valve shift *x(t)*, the hydraulic flow rate $Q_L(t)$ and the fullness of the chamber hydrocylinder is presented below (Sukhorukov et al., 2003):

$$Q_L(t) = k_g(t)x_v(t) - k_c P_L(t) = A_p(\dot{z}_x(t) - \dot{z}_u(t)) + C_{lt}P_L(t) + (V_t / 4\beta)\dot{P}_L(t) \tag{4}$$

where k_c is a coefficient of servo valve pressure ratio;

$P_L(t)$ is cylinder pressure difference;
k_g is a coefficient of servo valve flow increase, changing in time;
A_p is cross-sectional area of the cylinder;
C_{lt} is total leakage rate from the hydrocylinder;
V_t is total amount of compression;
β is a module of system volume expansion.

The correlation between the servo valve coil shift and the control voltage is described as $x_v(t)=k_v u(t)$, where k_v is the servo actuator ratio. There has been obtained a time derivative of the effective force of the hydraulic sprung mass:

$$\dot{F}_a(t) = \dot{P}_L(t)A_p = A_p(4\beta / V_t)[k_g(t)k_v u(t) - C_T P_L(t) - A_p(\dot{z}_s(t) - \dot{z}_u(t))] \tag{5}$$

The combined dynamic equation of the sprung mass can be written as:

$$\begin{aligned} m_s\dddot{z}_s = {} & -(4\beta C_T k_s / V_t)z_s - [k_s + 4\beta(C_T b_s + A_p^2)]\dot{z}_s - (b_s + 4\beta C_T m_s / V_t)\ddot{z}_s + (4\beta A_p k_g(t)k_v / V_t)u(t) \\ & -(4\beta C_T F_f / V_t + \dot{F}_f) + [b_s\ddot{z}_u + (k_s + 4\beta A_p^2 / V_t + 4\beta C_T b_s / V_t)\dot{z}_u + 4\beta C_T k_s z_u / V_t \end{aligned} \tag{6}$$

The dynamic equation of this hydraulic servo system has a non-linear characteristic and a multi-output connection that changes over time (Kuzmin, Fedotkin & Kryuchkov, 2017).

Figure 2. A scheme of the active vehicle sprung mass: where m_s, m_u are sprung and unsprung mass respectively; k_s, k_t are stiffness of the element of sprung and unsprung mass, b_s is sprung mass damping, z_s, z_u, z_r are variables representing the displacement of the wheel mass and the road respectively (Pobedin, A.V., Dolotov, A.A. & Shekhovtsov, V. V., 2016).

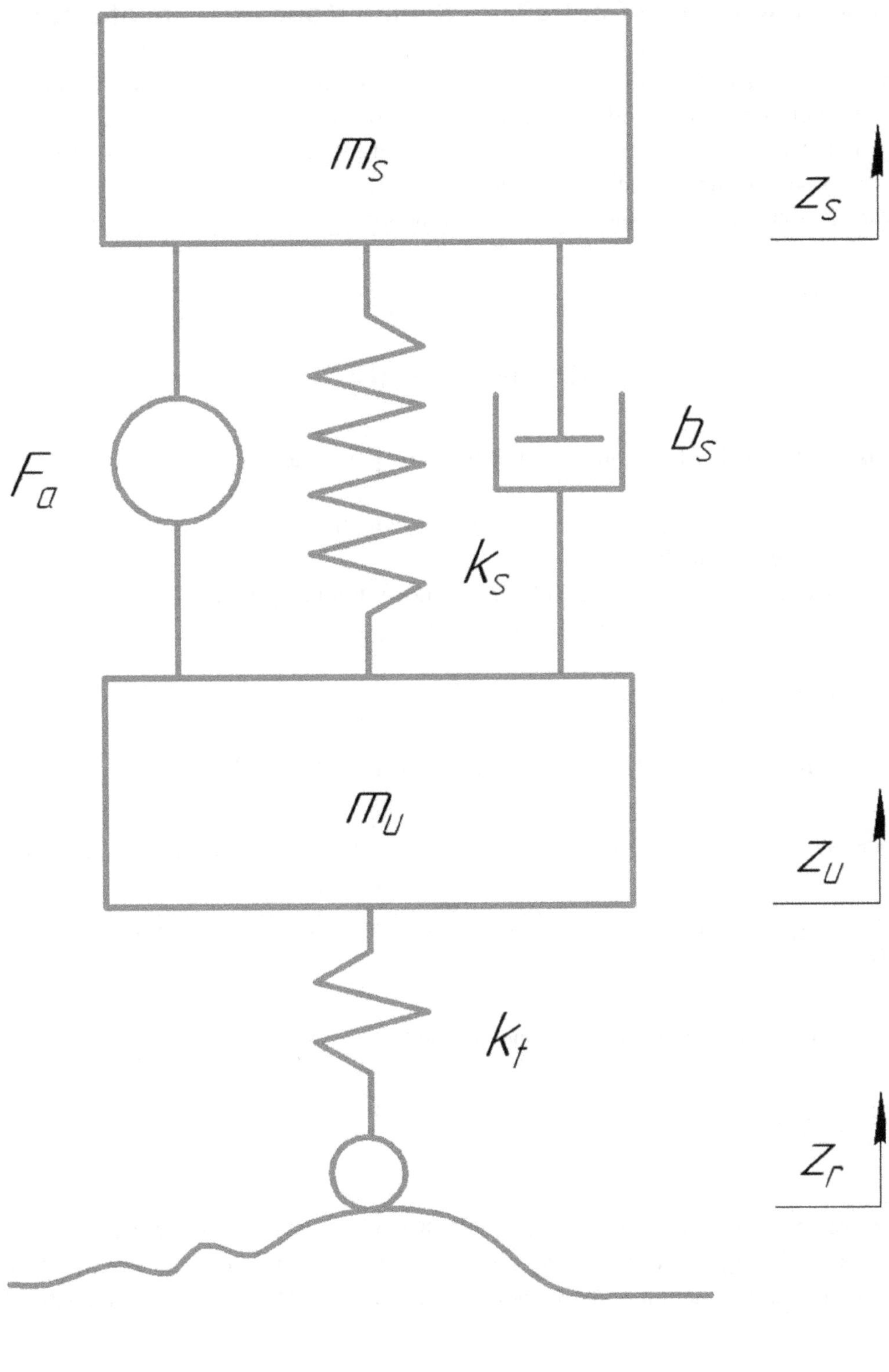

It is quite difficult to estimate these system parameters and use this dynamic equation to design a controller. Consequently, it is necessary to develop a modeless intelligent control scheme used for developing a sprung mass controller.

Radial Base Function of a Controller's Sliding Mode (RBFSM of a controller)

The dynamic equations (4.6) of the hydraulic sprung mass have a multivariable dynamic relationship with a random vibration on the road surface. It has been proposed to use the neural network to control the sliding mode parameters. To develop a nonlinear controller, two variables z_s, z_u are required based on a nonlinear time-varying dynamic 3rd order equation (6).

Other variables are considered as time-varying functions.

$$\begin{aligned}
&\dot{x}(t) = \dot{z}_s(t) = x_2(t)\\
&\dot{x}_2(t) = x_3(t)\\
&\dot{x}_3(t) = -a_1(t)x_1(t) - a_2(t)x_2(t) - a_3(t)x_3(t) + f_u(t) + b(t)u(t) - u_d(t)
\end{aligned} \tag{7}$$

where $a_i(t)$ is a time-varying function, which depends on parameters of sprung mass;

$f_u(t)$ is a function of some non-measurable variables z_u, which are limited by sprung mass;
$b(t)$ is a positive time-varying function;
$u_d(t)$ is a disturbance due to friction force change in the hydraulic system;
$e_1=x_{1d}-x_1$, $e_2=x_{2d}-x_2$ и $e_3=x_{3d}-x_3$ are state variable errors.

Then the equation (4) can be rewritten as:

$$\ddot{e}_3(t) = \dddot{x}_{3d}(t) + a_1(t)x_1(t) + a_2(t)x_2(t) + a_3(t)x_3(t) - b(t)u(t) - f_u(t) + u_d(t) \tag{8}$$

If all the above time-varying functions are known, then the equation of sprung mass control will be:

$$u_{eq}(t) = \frac{1}{b(t)}[\dddot{x}_{3u}(t) + a_1(t)x_1(t) + a_2(t)x_2(t) + a_3(t)x_3(t) - f_u(t) + u_d(t) - \ddot{e}_3(t) + \dot{s}(t) + \lambda s(t)] \tag{9}$$

where $s(t)$ is a sliding surface on a phase plane, defined as

$$s(t) = \left(\frac{d}{dt} + \lambda\right)e_1(t) = e_2(t) + \lambda e_1(t) \tag{10}$$

Based on previous studies (Cherry & Jones, 1995; Kuzmin, Fedotkin & Kryuchkov, 2017), the neural network is a powerful algorithm for developing a non-linear dynamic model. The RBFSM is used to simulate the nonlinear interaction of the variable sliding surface $s(t)$ and the system control law $u(t)$.

The gaussian functions are used as functions to activate each neuron in the hidden layer of this controller. The magnitude of the disturbance of the gaussian functions is a distance between the input value of the sliding variable *s(t)* and the average position of the gaussian function.

$$\theta_j = (s - c_j)^2 \tag{11}$$

where c_j is an average position of the neuron *j*.

The weight coefficients w_j between the neurons of the input layer and the neurons of the hidden layer are indicated as a 1.0constant. The weight coefficients w_k between the neurons of the hidden layer and the neurons of the output layer are corrected in accordance with the adaptation rule. The output of the RBFSM is:

$$g(s) = \sum_{j=1}^{n} w_j \varphi_j (s - c_j) \tag{12}$$

where $\varphi_j(s) = \exp(-\frac{(s - c_j)^2}{\sigma_j^2})$ is a gaussian function and neuron *j* of the output layer;

σ_j and c_j are root-mean-square difference and maths expectation of the gaussian function, respectively; *n* is a number of neurons which is an input value of RBFSM .

In order to take advantage of the sliding mode and adaptive control schemes in the RBFNS, the variable of the sliding surface is set as an input value of the RBFSM, the adaptation rule is introduced to control the weights between the hidden and output neural layers.

For the case of one input and one output, the control input of the neural network controller is

$$u = \sum_{j=1}^{n} w_j \exp\left(-\frac{(s - c_j)^2}{\sigma_j^2} \right) \tag{13}$$

The approximated law of RBFSM control *u* may differ from the control law u_{eq}, from equation (8).

$$\dot{s}(t) = \dot{e}_2 + \lambda \dot{e}_1 = -\lambda s(t) + b(t)[u_{eq}(t) - u(t)] \tag{14}$$

$$s(t)\dot{s}(t) = s(t)(-\lambda s(t) + b(t)[u_{eq}(t) - u(t)]) \tag{15}$$

Based on the Lyapunov's theorem, $s \cdot \dot{s} < 0$ is the condition to reach a sliding surface (Kuzmin, Fedotkin & Kryuchkov, 2017). If you choose the control input signal *u* to fulfill this condition, the system will converge to the beginning of the phase plane.

The weight coefficients of the RBFSM are regulated if $s \cdot \dot{s} < 0$ is achieved. The adaptation rule is used to adjust the weights and search for their optimal values, as well as to achieve stable convergence.

The adaptation rule is derived from the short-term descent rule to minimize the $s \cdot \dot{s}$ value with respect to w_j. Then the equation of the weight coefficients is:

$$\dot{w}_j = -G\frac{ds(t)\dot{s}(t)}{du(t)}\frac{du(t)}{dw_j(t)} = Gb(t)s(t)\frac{du(t)}{dw_j(t)} = \gamma s(t)\exp\left(-\frac{(s-c_j)^2}{\sigma_j^2}\right) = \gamma s(t)\varphi_j(s) \quad (16)$$

The adaptation speed parameter G and the system input parameter $b(t)$ are combined as a tilting (learning) speed parameter γ. The weight coefficients between the hidden and output layers of the neural network can be adjusted in real time.

From equation (6) it can be concluded that the variable $b(t)$ is always a positive value for this sprung mass. From equation (14) it can be seen that $\dot{s}$ is increased with decreasing u and vice versa. If s> 0, an increase in u due to an increase in w_j will lead to $s \cdot \dot{s}$ decrease. When the condition s <0, $s \cdot \dot{s}$ will decrease with decreasing u through decreasing w_j. Therefore, rule (16) is achievable.

The radial basic function (RBF) (Rolland L.Hardy, 1971) and the adaptive structure of the sliding controller (radial-basis function of the sliding mode) are shown in Figures 3 and 4, respectively.

Figure 3. Gaussian function of RBF

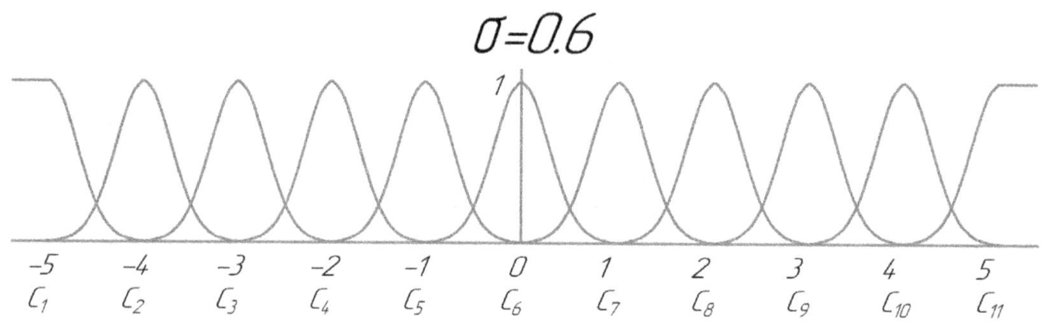

Figure 4. Structure of the adaptive neural network controller

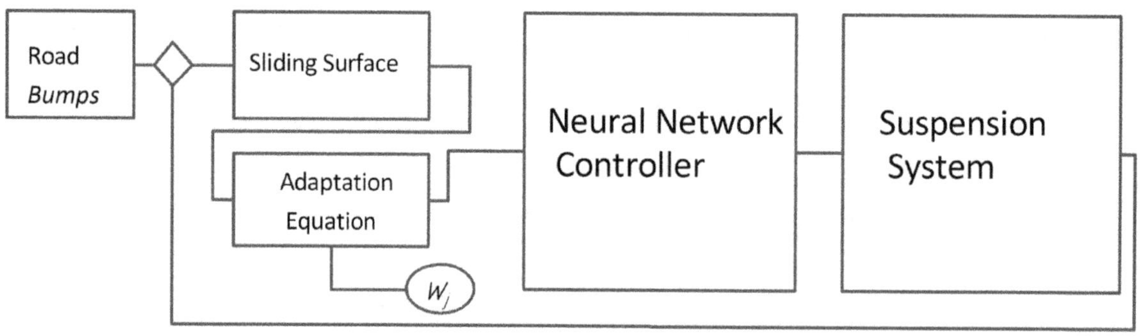

The parameters of the Gaussian functions σ_j and c_j in some cases can be given as constants. The distribution coefficient is $u_j = 0.6$, and the mean value of the functions c_j is given as integral constants from -5 to +5 for simplicity. The scale coefficient g_s is selected from the range to display the sliding surface variable s.

Theoretically, RBFSM can be used to simulate and approximate any nonlinear controller. The coefficient $\tilde{W} = \bar{W} - \widehat{W}$ shows the difference between the optimal value of the weight coefficient and its current value. $\widehat{W}$ is a current value of the neural network weight coefficient. Let's choose the Lyapunov function as:

$$V = \frac{1}{2}s^2 + \frac{b}{2\gamma}\tilde{W}^T\tilde{W} \quad (17)$$

It could be said that the Lyapunov function gradually decreases, and the control system is stable. The sliding surface variable $s(t)$ converges along the boundary layer $s = 0$ in the phase plane, but $\tilde{W}$ remains limited.

Results of Modelling

To control the active vehicle sprung mass, there is proposed an adaptive controller with radial base function of its sliding mode. The adjustable parameter for RBFSM of a controller is selected as $g_s = 8.5$ to cover the range of gaussian functions.

The parameter λ in the equation of sliding mode (7) is 0.5, this value determines the slope of the sliding surface in the phase plane. The value of the tilting (learning) speed parameter γ was chosen to be 0.6. The coefficient of the equation of sliding mode control based on the neural network g_u is equal to 0.8.

The distribution coefficient of the gaussian functions is given as a constant of 0.6. To study the parameters of the control signal based on the spring dynamics, there were conducted the following experiments. The sampling rate in the experiments is 150 Hz.

The dynamic change in the spring position is shown in Figure 5, when the vehicle moves along a sinusoidal road surface with a vibration amplitude of 40 mm, using the RBFNS control. The full line graph shows the road profile, the dash-and-dot line shows the passive movement of the sprung mass, and the thin line indicates the change in the spring mass position with the active sprung mass.

The maximum shift of the sprung mass is 7.5 mm for the initial learning period. After that, the vibration of the vehicle sprung mass is less than 1.5 mm. This slight vibration is caused by a dynamic change in tire deformation. Figure 6 shows the mass acceleration of a spring with a different number of RBFs, in order to estimate the effect of the number of nodes of this neural network.

The root-mean-square values of the vibration acceleration amplitudes are 16.61 and 10.31 mm/c^2 for 5 and 11 nodes of RBF, respectively. The position and the dynamic characteristics of the of the sprung mass acceleration while the vehicle is moving along a rough road with a random amplitude are shown in Figure 7 and 8, respectively.

You can see that the amplitude of the acceleration vibrations is within 18 mm/c^2, excluding the initial learning period. The root-mean-square values of movement and acceleration are 1.04 mm and 8.52 mm/c^2, respectively.

Figure 5. Dynamic movement of the sprung mass in time

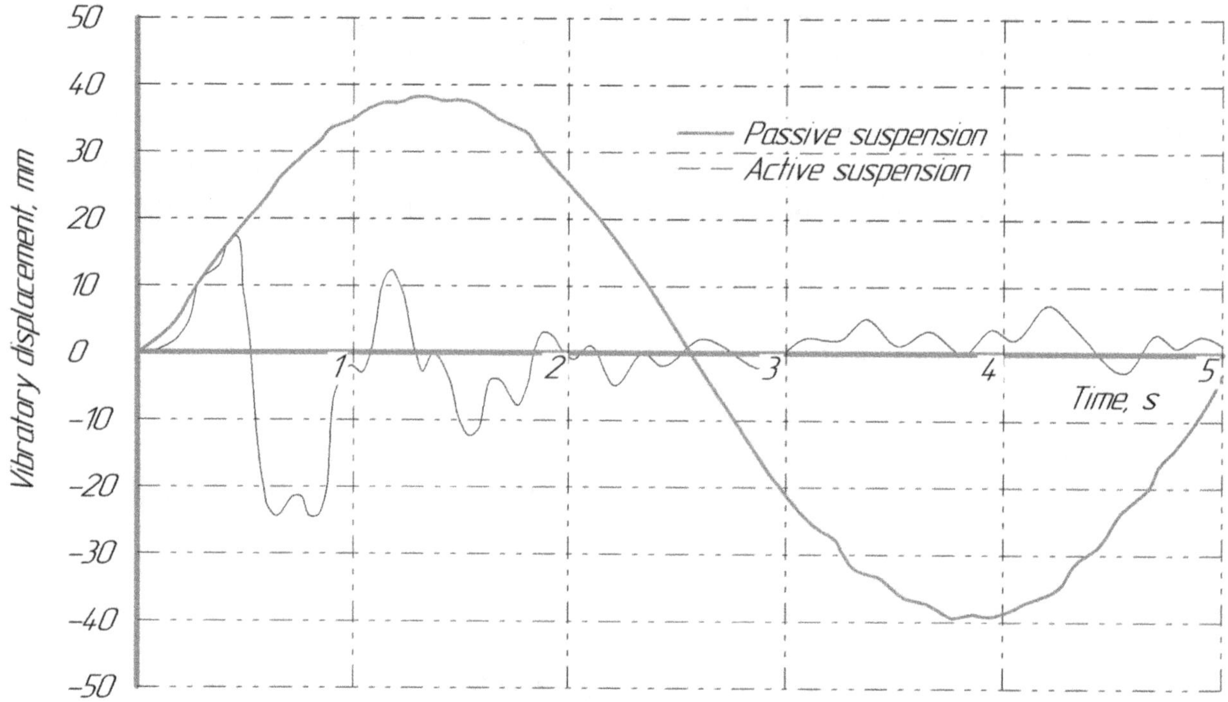

Figure 6. Change of the sprung mass acceleration with 5 and 11 nodes of RBF

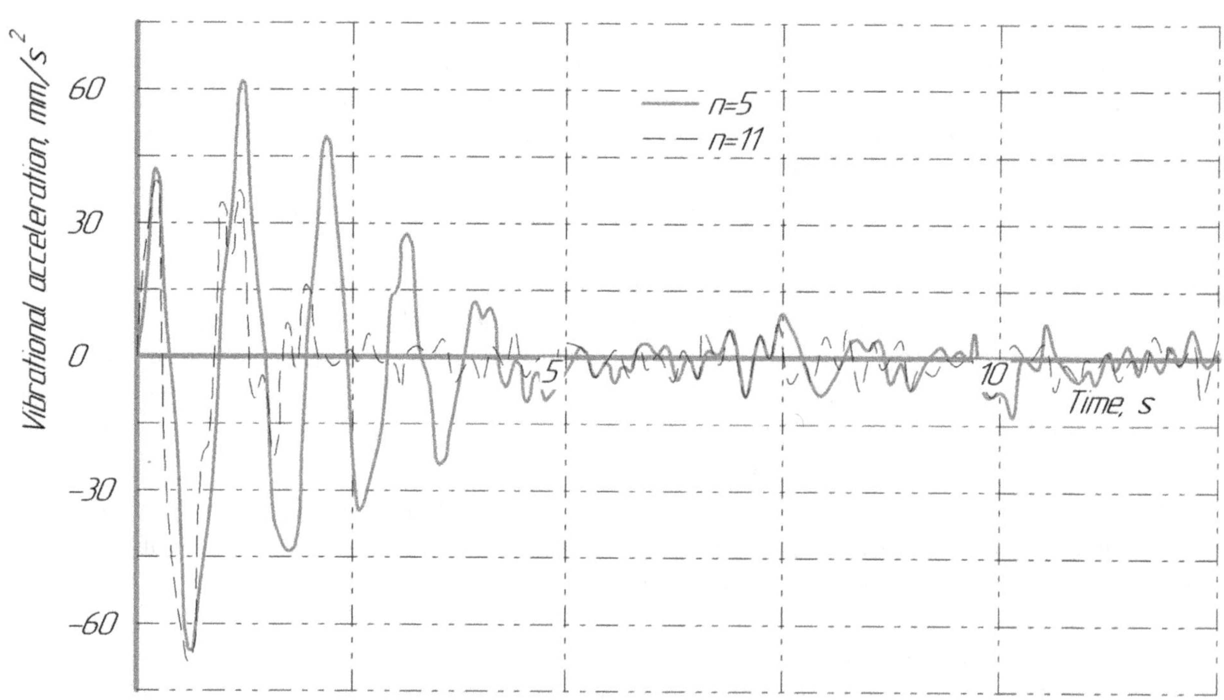

Figure 7. Change of the sprung mass position in time

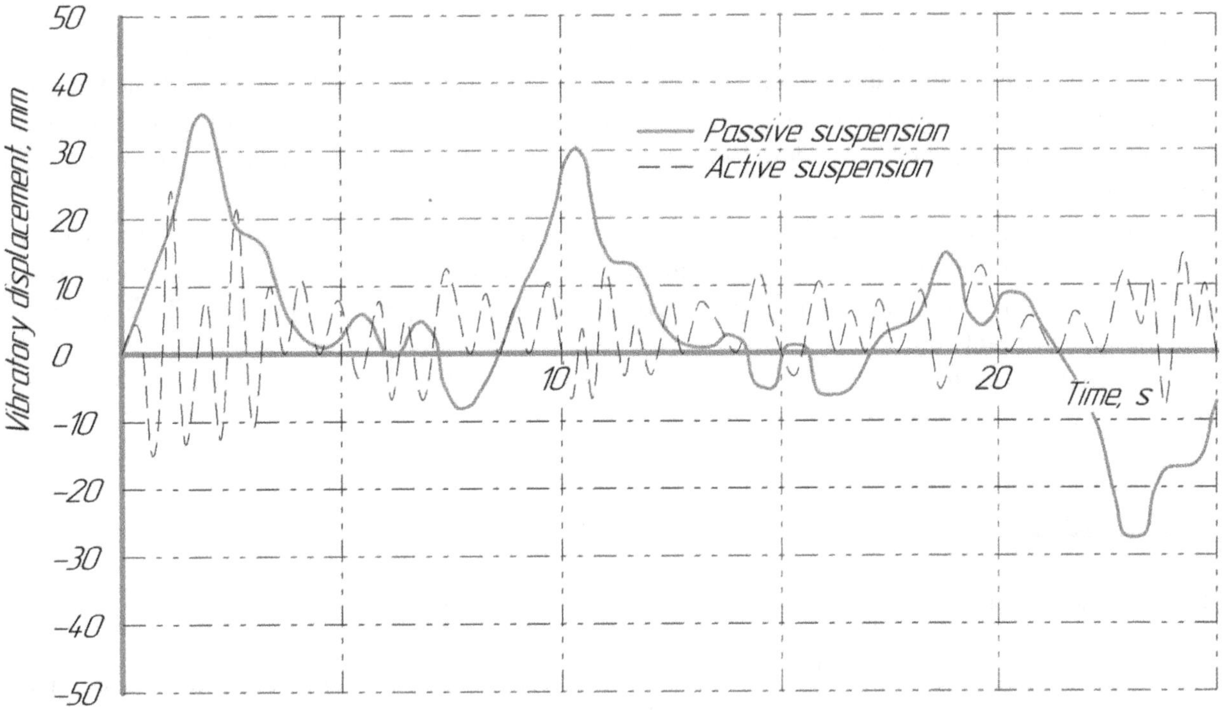

Figure 8. Sprung mass acceleration in time

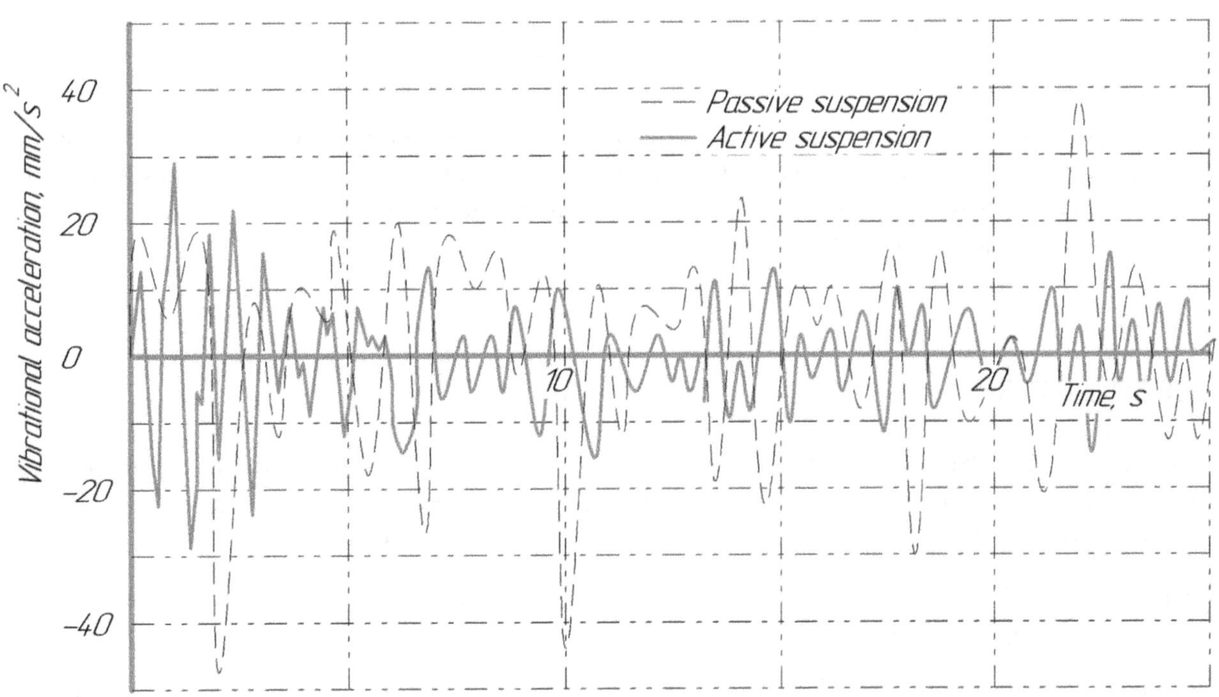

FUTURE RESEARCH DIRECTIONS

Further development of this work will be aimed at solving a number of tasks that are related to the search for new constructive and technical solutions to increase the efficiency of the system and reduce its cost.

First of all, a software control algorithm should be developed for all elements of the system, which depends on the input controller parameters for analysis of the impact of the suspension system on the vehicle handling.

It is necessary to develop all versions of the work, up to emergency and extreme modes. This is necessary to create a ready-made software product suitable for monitoring and automatic control of the system.

The next task is to conduct a comparative assessment of the suspension system on the example of several tractor variants.

Great attention should be paid to the topic of a detailed study of the operation of each system element in real operating conditions.

CONCLUSION

Based on the conducted study, the following conclusions can be drawn:

1. There has been proposed a new controller of sliding mode for a hydraulically active sprung mass based on a neural network. This control algorithm combines the methods of adaptation and the method of sliding mode control, sets the corresponding weights of the RBF by learning the algorithm in real time.
2. The simulation results show that due to the use of this neural network controller, the sprung mass effectively reduces the vibrational activity of the vibration system of the studied vehicle caused by road irregularities.
3. The developed algorithm and the control method used with a neural network controller can be successfully applied in the development and upgrade of the sprung mass of the vehicles and energy resources. This will improve the accuracy of the calculated indicators of the mass and reduce the time for the estimation of these indicators.

REFERENCES

Chalasani, R. M. (1986). Ride performance potential of active sprung weigts - part II: Comprehensive analysis based on a full-car model. American Society of Mechanical Engineers. *Applied Mechanics Division, AMD, 80*, 205–234.

Cherry, A. S., & Jones, R. P. (1995). Fuzzy logic control of an automotive sprung weights. *IEE Proc. Control Theory Applications, 142*(2), 149–160.

Deakin, A., Crolla, D., Shovlin, A., & Charlton, N. (1997). *Power Consumption in Ride of a Combat Support Vehicle Slow-Active Suspension* (SAE Technical Paper No. 973205). Warrendale, PA: SAE International. doi:10.4271/973205

Deboli, R., Calvo, A., & Preti, C. (2017). Whole-body vibration: Measurement of horizontal and vertical transmissibility of an agricultural tractor seat. *International Journal of Industrial Ergonomics*, *58*, 69–78. doi:10.1016/j.ergon.2017.02.002

Derbaremdiker, A. D. (1985). *Amortizatory transportnykh mashin* [Shock absorbers for vehicles]. Moscow, Russia: Mashinostroyeniye.

Duke, M., & Goss, G. (2007). Investigation of tractor driver seat performance with non-linear stiffness and on-off damper. *Biosystems Engineering, 94*(4), 477 – 486. WOS:000245890600005

Feuerhuber, K., Lindert, S. O., & Schlacher, K. (2013). *Vibration attenuation by semi-active dampers.* Paper presented at 14th International Conference on Computer Aided Systems Theory (EUROCAST), Las Palmas, Spain. 10.1007/978-3-642-53862-9_35

Frolov, K. V. (1995). *Vibratsii v tekhnike: Spravochnik* [Vibrations in the technique: Reference Mechanical Engineering]. Vol. 6). Moscow, Russia: Mashinostroyeniye.

Gohrle, C., Schindler, A., Wagner, A., & Sawodny, O. (2014). Design and Vehicle Implementation of Preview Active Suspension Controllers. *IEEE Transactions on Control Systems Technology*, *22*(3), 1135–1142. doi:10.1109/TCST.2013.2272342

Gurhan, R., &Cay, CI. (2008). Performances of tractor's driver seats with passive damping system under simulated test conditions. *TarimBilimleriDergisi-Journal of Agricultural Sciences, 14*(4), 401 – 408. WOS:000207576500013

Hansson, P. A. (2002). Working space requirement for an agricultural tractor axle suspension. *Biosystems Engineering*, *81*(1), 57–71. doi:10.1006/bioe.2001.0007

Hardy, R. L. (1971). Multiquadric Equations of Topography and Other Irregular Surface. *Journal of Geophysical Research*, *76*(8), 1905–1915. doi:10.1029/JB076i008p01905

Kuzmin, V. A., Fedotkin, R. S., & Kryuchkov, V. A. (2017). Iskusstvennaya neyronnaya set' dlya obosnovaniya parametrov khodovykh sistem traktorov [Artificial neural network to justify the parameters of the tractor's propulsion systems]. *Sel'skokhozyaystvennyye Mashiny i Tekhnologii*, (4), 24–30.

Marjoram, R., St-Clair, K., McMahon, B., Goelz, A., & Achen, A. (2008). Integrated MR cabin suspension for reduction of human exposure to whole-body vibration in heavy-duty vehicles. Bremen, Germany: Messebremen-hvghanseatischeveranstaltungs-gmbh. WOS:000270938400110

Pobedin, A. V., Dolotov, A. A., & Shekhovtsov, V. V. (2016). *Decrease of the vibration load level on the tractor operator working place by means of using of vibrations dynamic dampers in the cabin suspension.* Paper presented at 2nd International Conference on Industrial Engineering (ICIE), Chelyabinsk, Russia. 10.1016/j.proeng.2016.07.136

Senkevich, S., Kravchenko, V., Duriagina, V., Senkevich, A., & Vasilev, E. (2019). Optimization of the Parameters of the Elastic Damping Mechanism in Class 1,4 Tractor Transmission for Work in the Main Agricultural Operations. In P. Vasant, I. Zelinka, & G. W. Weber (Eds.), *Intelligent Computing & Optimization. ICO 2018. Advances in Intelligent Systems and Computing* (Vol. 866). Cham: Springer; doi:10.1007/978-3-030-00979-3_17

Senkevich, S. E., Sergeev, N. V., Vasilev, E. K., Godzhaev, Z. A., & Babayev, V. (2019). Use of an Elastic-Damping Mechanism in the Tractor Transmission of a Small Class of Traction (14 kN): Theoretical and Experimental Substantiation (Chapter 6). Handbook of Advanced Agro-Engineering Technologies for Rural Business Development. – Hershey, Pennsylvania (USA): IGI Global – P. 149-179. Doi:10.4018/978-1-5225-7573-3.ch006

Senkevich, S. E., Lavrukhin, P. V., Senkevich, A. A., Ivanov, P. A., & Sergeev, N. V. (2020). *Improvement of Traction and Coupling Properties of the Small Class Tractor for Grain Crop Sowing by Means of the Hydropneumatic Damping Device. Handbook of Research on Energy-Saving Technologies for Environmentally-Friendly Agricultural Development* (pp. 1–27). Hershey, Pennsylvania, USA: IGI Global; doi:10.4018/978-1-5225-9420-8.ch001

Senkevich, S., Kravchenko, V., Lavrukhin, P., Ivanov, P., & Senkevich, A. (2020a). *Theoretical Study of the Effect of an Elastic-Damping Mechanism in the Tractor Transmission on a Machine-Tractor Unit Performance While Sowing. Handbook of Research on Smart Computing for Renewable Energy and Agro-Engineering* (pp. 423–463). Hershey, Pennsylvania, USA: IGI Global; doi:10.4018/978-1-7998-1216-6.ch017

Senkevich, S., Duriagina, V., Kravchenko, V., Gamolina, I., & Pavkin, D. (2020b). Improvement of the Numerical Simulation of the Machine-Tractor Unit Functioning with an Elastic-Damping Mechanism in the Tractor Transmission of a Small Class of Traction (14 kN). In P. Vasant, I. Zelinka, & G. W. Weber (Eds.), *Intelligent Computing and Optimization. ICO 2019. Advances in Intelligent Systems and Computing* (Vol. 1072, pp. 204–213). Cham: Springer; doi:10.1007/978-3-030-33585-4_20

Silayev, A. A. (1972). *Spektral'naya teoriya podressorivaniya transportnykh mashin* [Spectral theory of the suspension of transport vehicles]. Moscow, Russia: Mashgiz.

Sukhorukov, A. V. (2003). Upravleniye dempfiruyushchimi elementami v sisteme podressorivaniya bystrokhodnoy gusenichnoy mashiny (Diss. ... kand. tekhn. nauk. Spetsial'nost' 05.05.03 - Avtomobili i traktory). [*Control of damping elements in the sprung weigt of a high-speed tracked vehicle* (Diss. ... candidate of technical sciences. Specialty 05.05.03 - Automobiles and tractors)].

Venhovens, P. J. T. (1994). The Development and Implementation of Adaptive Semi-Active Suspension Control*. *Vehicle System Dynamics, 23*(1), 211–235. doi:10.1080/00423119408969057

Williams, R. A. (1994). Electronically controlled automotive suspensions. *Computing & Control Engineering Journal, 5*(3), 143–148. doi:10.1049/cce:19940310

Yu, F., & Crolla, D. A. (1998). State Observer Design for an Adaptive Vehicle Suspension. *Vehicle System Dynamics, 30*(6), 457–471. doi:10.1080/00423119808969461

Zheng, E. L., Fan, Y. D., Zhu, R., Zhu, Y., & Xian, J. Y. (2016). Prediction of the vibration characteristics for wheeled tractor with suspended driver seat including air spring and MR damper. *Journal of Mechanical Science and Technology, 30*(9), 4143–4156. doi:10.100712206-016-0826-x

KEY TERMS AND DEFINITIONS

Active Suspension: Is an active vehicle suspension system that controls the vertical movement of the wheels relative to the body.

Adaptive Suspension: (Another name is semi-active suspension) Is a type of active suspension in which the degree of damping of shock absorbers varies depending on the condition of the road surface, traffic parameters and driver's requests.

Hydroelastic Suspension: Is a type of suspension that provides control of the level of the body relative to the road through the use of hydraulic elastic elements.

Neural Network: Is a mathematical model, as well as its software or hardware embodiment, built on the principle of the organization and functioning of biological neural networks.

Nonlinear Characteristics: (Nonlinear system) Is a dynamic system in which processes are described by nonlinear differential equations. The properties and characteristics of nonlinear systems depend on their state.

Sprung Mass: Is called the mass of the elements that are held by the suspension of the tractor (or car) over the road.

Unsprung Mass: Is a concept applicable to ground vehicles with a suspension, which means the mass, including the mass of the wheels and other parts of the structure, separated by the suspension from the frame or body.

Chapter 6
The Traveling Salesman Problem:
Network Properties, Convex Quadratic Formulation, and Solution

Elias Munapo
North-West University, South Africa

ABSTRACT

The chapter presents a traveling salesman problem, its network properties, convex quadratic formulation, and the solution. In this chapter, it is shown that adding or subtracting a constant to all arcs with special features in a traveling salesman problem (TSP) network model does not change an optimal solution of the TSP. It is also shown that adding or subtracting a constant to all arcs emanating from the same node in a TSP network does not change the TSP optimal solution. In addition, a minimal spanning tree is used to detect sub-tours, and then sub-tour elimination constraints are generated. A convex quadratic program is constructed from the formulated linear integer model of the TSP network. Interior point algorithms are then applied to solve the TSP in polynomial time.

INTRODUCTION

The traveling salesman problem (TSP) was believed to be difficulty until now (Munapo, 2020) or (Papadimitriou, 1977) or (Berman & Karpinski, 2006). The TSP has so many applications which include scheduling, sequencing, vehicle routing, engineering, electronics and genetics. In this paper it is shown that adding or subtracting a constant to all arcs with special features in a traveling salesman problem (TSP) does not change an optimal solution of the (TSP). It is also shown that adding or subtracting a constant to all arcs emanating from the same node in a TSP network does not change the TSP optimal solution. A minimal spanning tree (Wolsey, 1989) is used to detect sub-tours and then sub-tour elimination constraint generated (Bektas & Gouveira, 2006). A convex quadratic program (Jensen & Bard, 2003) is constructed from the formulated linear TSP integer model. Interior point algorithms (Gondzio, 2012) are then applied to solve the TSP in polynomial time.

DOI: 10.4018/978-1-7998-3970-5.ch006

The TSP Model

The objective is to start from an origin node and return to it, traversing the nodes in such a way that every node is visited once and that the total distance travelled is minimized (Applegate et al. 2006) or (Gutin & Punnen, 2006). It is also assumed that all nodes have at least two arcs emanating from them.

Figure 1. TSP model

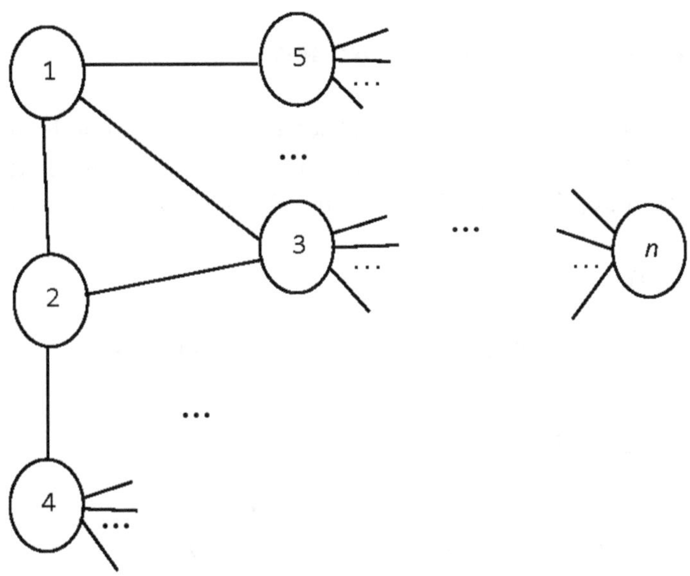

AVAILABLE METHODS

Exact Approaches

There are several exact approaches that are currently available for the TSP. These include:

a) Brute force. In this approach we try all possible routes and see which one is the cheapest and the worst case complexity is *O*(n!) and is not practical for large number of nodes.
b) Linear programming based branch, cut and price methods. See Padberg & Rianld (1991), Mitchell (2001), or Nadef (2002) for more on branch and cut methods.
c) Branch and bound method in which assignment sub-problems are used (Winston, 2004).
d) Dynamic programming techniques (Winston, 2004).

A lot of effort has been injected into these exact approaches and a consistent efficient method is now available.

Heuristics

These are methods that quickly give solutions that are close to optimal solution (Winston, 2004). The problem with these heuristics is that the difference between the exact solution and the near optimal solution is sometimes very big. In terms of money it means millions of dollars for the large real life problems.

DIRECT LINEAR PROGRAMMING FORMULATION AND ITS CHALLENGES

Direct Linear Programming Based Approach

It is easy to directly formulate the TSP as an LP and solve using the available interior point algorithms to obtain an optimal solution in polynomial time. Unfortunately direct LP formulation has a serious challenge. The challenge is that the optimal solution contains sub-tours as illustrated in the following example.

Numerical Illustration

Solve the following TSP in network form by formulating it as an LP and then solve.

Figure 2. TSP numerical illustration 1

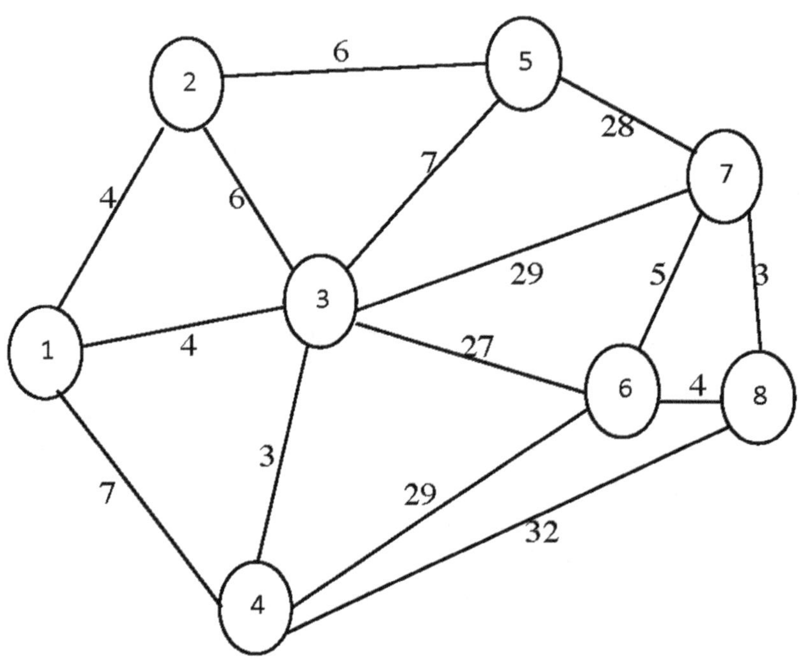

Formulation as A Linear Programming Problem

The problem can be formulated as a linear program as given in (1).

Min

$$4x_{12} + 4_{13} + 7x_{14} + 6x_{23} + 6x_{25} + 3x_{34} + 7x_{35} + 27x_{36} + 29x_{37} + 29x_{46} + 32x_{48} + 28x_{57} + 5x_{67} + 4x_{68} + 3x_{78}.$$

Such that:

Node 1: $x_{12} + x_{13} + x_{14} = 2.$

Node 2: $x_{12} + x_{23} + x_{25} = 2.$

Node 3: $x_{13} + x_{23} + x_{34} + x_{35} + x_{36} + x_{37} = 2.$

Node 4: $x_{14} + x_{34} + x_{46} + x_{48} = 2.$

Node 5: $x_{23} + x_{35} + x_{57} = 2.$

Node 6: $x_{36} + x_{46} + x_{67} + x_{68} = 2.$

Node 7: $x_{37} + x_{57} + x_{67} + x_{78} = 2.$

Node 8: $x_{48} + x_{68} + x_{78} = 2.$ (1)

Where $x_{12}, x_{13}, \ldots, x_{78} \geq 0$ and binary.

Solving as a linear programming problem we have the optimal solution given in (2).

$$Z_0 = 39, x_{12} = x_{14} = x_{25} = x_{34} = x_{35} = x_{67} = x_{68} = x_{78} = 1,$$
$$x_{13} = x_{23} = x_{36} = x_{37} = x_{46} = x_{57} = 0. \quad (2)$$

In the form of a TSP network diagram the optimal solution is given in Figure 3.

The solution given in Figure 3 is not optimal as there is a sub-tour. This is the weakness of direct LP formulation and this challenge can be alleviated. In this chapter we call the constraints that are formed by direct LP formulation *standard constraints*. This chapter presents a way of detecting the sub-tours before LP formulation and a technique to eliminate them. Before we present this sub-tour elimination technique we present some useful properties of a TSP network.

Figure 3. Solution to numerical illustration 1- existence of sub-tours

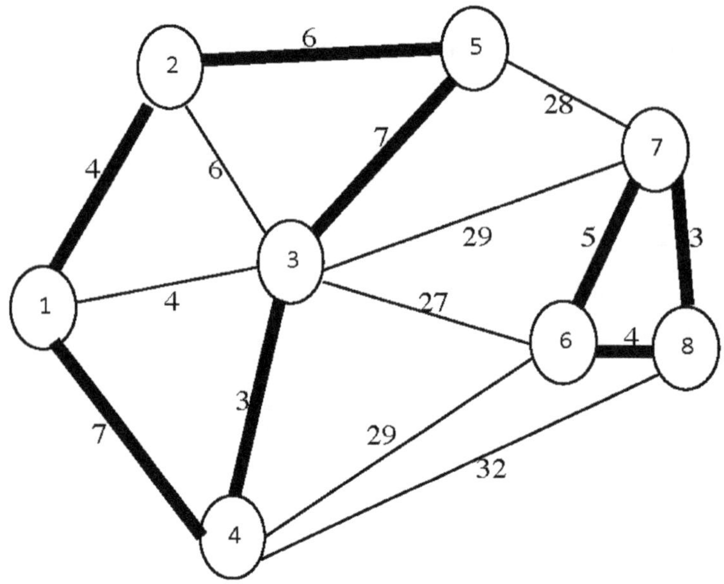

PROPERTIES OF TSP NETWORK DIAGRAM

TSP Network Property 1

Subtracting or adding a constant to all arcs emanating from a given node in a TSP network does not change the TSP optimal solution.

Proof

Suppose there are j arcs emanating from node m as given in Figure 4.

Figure 4. Arcs emanating from the same node m.

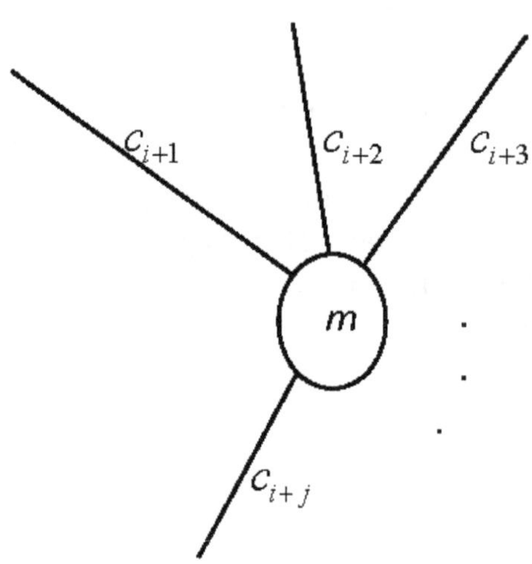

In a TSP network model every node is visited once which implies that we need two arcs to visit this node. The possible lengths of the two arcs are:

$$c_{i+1}+c_{i+2}\quad c_{i+1}+c_{i+3}\ \dots\ c_{i+1}+c_{i+j}\quad c_{i+2}+c_{i+3}\ \dots\ c_{i+3}+c_{i+j} \tag{3}$$

Subtracting π from all arcs emanating from node *m* we have Figure 5.

Figure 5. Subtracting π from all arcs emanating from the same node m.

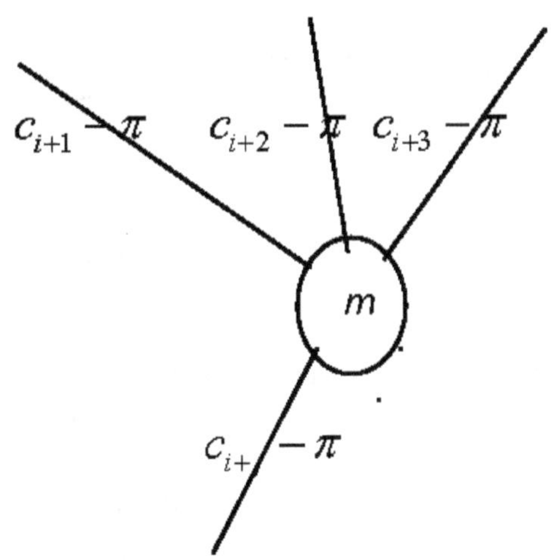

In other words the possible sums of the two arcs are given in (19).

$$c_{i+1}+c_{i+2}-2\pi\quad c_{i+1}+c_{i+3}-2\pi\ \dots\ c_{i+1}+c_{i+j}-2\pi\quad c_{i+2}+c_{i+3}-2\pi\ \dots\ c_{i+3}+c_{i+j}-2\pi \tag{4}$$

The objective function changes by a constant implying there is no change in the optimal solution.

TSP Network Property 2

Adding or subtracting a constant to all arcs with special features does not change an optimal solution of a TSP. Arcs with special features are given in Figure 6. In other words these are arcs in a TSP network that are positioned in such a way that a line can be drawn in from outside the TSP and crosses all these arcs and then leaves the TSP network without touching a single node. The arcs with special features and the line γ_1 that comes from outside TSP network are shown in Figure 6.

Definitions

a) Let Set A be made up of all the nodes on the l.h.s. of the line γ_1 while Set B is that set of all nodes on the r.h.s.

b) Let the arc joining a node in Set A to a node in Set B be represented by x_{i+j} where $j=1,2,\dots\ell 1$.

Figure 6. The arcs with special features

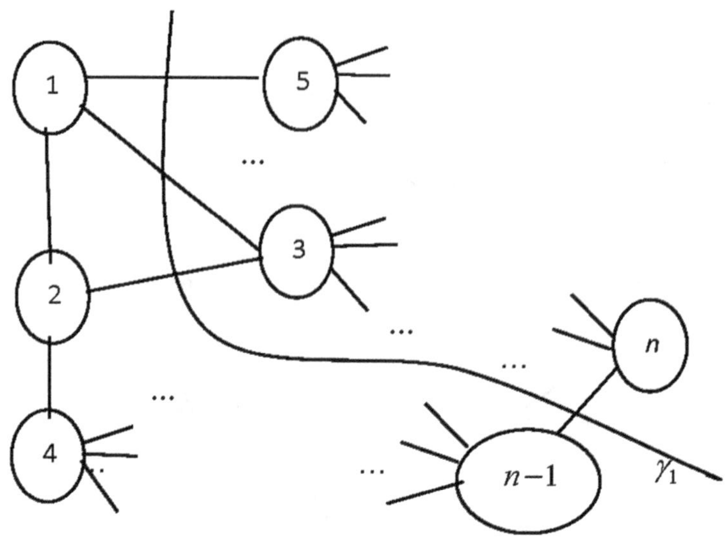

c) Come back principle. Suppose we start from an arc in set A, it means we can go to set B visit some nodes and come back. If we fail to cover all nodes in Set B then we can return there $\ell 1$ times until all nodes are visited. The come back sense comes from the fact that for us to have a tour we must move from A to Set B and come back to Set A and it does not matter the number of time.

d) Adding a constant (π_1) to all arcs with special features does not affect the nature of part of the optimal solution in Set A, neither will it affect what is in Set B. Suppose at optimality the arcs with special features are as given in (5).

$$Arc(x_{i+1}),\ Arc(x_{i+2}),\ \ldots\ Arc(x_{\ell 1}). \tag{5}$$

Since at optimality the values are known then we have (6).

$$x_{i+1} + x_{i+2} + \ldots + x_{\ell 1} = \tau_1. \tag{6}$$

Where τ_1 is an integral constant.

Proof of Property 2

Linear programming can be used and in linear programming form the formulated problem is given in (7).

$$\text{Min}\, c_1x_1 + c_2x_2 + \ldots + \{(c_{i+1} + \pi_1)x_{i+1} + (c_{i+2} + \pi_1)x_{i+2} + \ldots + (c_{\ell 1} + \pi_1)x_{\ell 1}\} + \ldots + c_nx_n;$$

Such that:

Node i:

$$x_{i+1}+x_{i+2}+\ldots+x_{\ell 1}=\overline{\tau}_1 \;\ldots\; x_1+x_2+\ldots+x_r=2. \quad (7)$$

Where $\overline{\tau}_1$ is an integral constant. The objective function simplifies to (8).

$$\text{Min}\, c_1x_1+c_2x_2+\ldots+c_{i+1}x_{i+1}+c_{i+2}x_{i+2}+\ldots+c_{\ell 1}x_{\ell 1}+\ldots+c_nx_n+(\pi_1+\pi_1+\ldots+\pi_1).$$

i.e.

$$\text{Min}\, c_1x_1+c_2x_2+\ldots+c_{i+1}x_{i+1}+c_{i+2}x_{i+2}+\ldots+c_{\ell 1}x_{\ell 1}+\ldots+c_nx_n+K. \quad (8)$$

Where $K=\pi_1+\pi_1+\ldots+\pi_1$, is an integral constant. The objective function changes by a constant implying there is no change in the optimal solution.

TSP Network Property 3

A minimal spanning tree (MST) algorithm will always give a minimum spanning tree.

Proof

Let S be the minimum spanning tree, C_k be the nodes connected after iteration k of MST has been completed, $\overline{C}_k$ be the nodes not connected after iteration k of MST has been completed, A_k be the set of arcs in minimum spanning tree after k iterations of MST algorithm have been completed. Suppose that the MST algorithm does not yield a minimum spanning tree. Then the arc chosen at iteration k (a_k) is not in S, i.e., $a_k \notin S$. All arcs in A_{k-1} are in S, i.e., $A_{k-1} \in S$ This implies $\overline{a}_k \in S$ and $\overline{a}_k$ leads from node in C_{k-1} to a node $\overline{C}_{k-1}$. Replacing $\overline{a}_k$ by a_k we obtain a shorter spanning tree than S. This contradicts the fact that S is the minimum spanning tree. Therefore the MST algorithm yields a minimum spanning tree.

NUMERICAL ILLUSTRATION

Arcs with special features are shown in Figure 7. Subtracting a constant 2 from all arcs emanating from Node 3 is shown in Figure 8. The MST algorithm yields a minimum spanning tree in all cases is illustrated in Figure 9.

Note that the line γ_2 enters network diagram from outside, crosses all arcs without touching a single node and then exits.

MST = 4+6+28+3+4+29+3+4 = 81

Figure 7. Arcs with special features

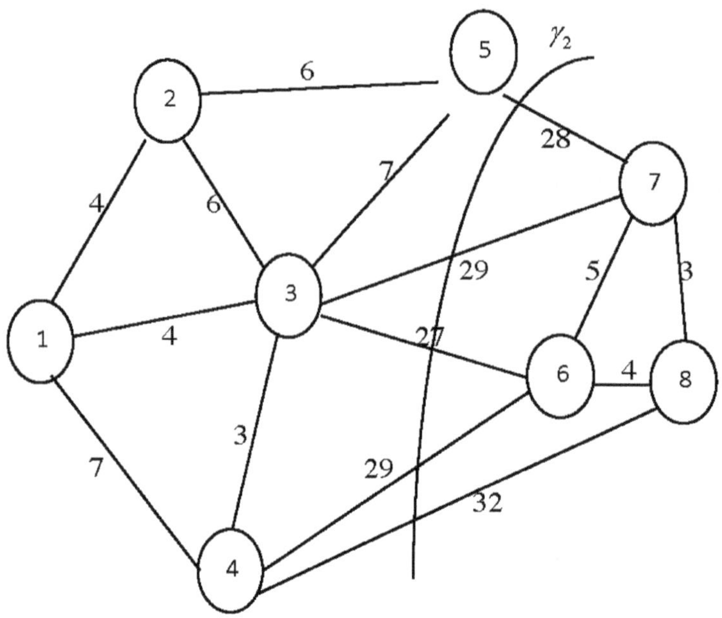

Figure 8. Subtracting 2 from all arcs emanating from Node 3 does not change the optimal solution

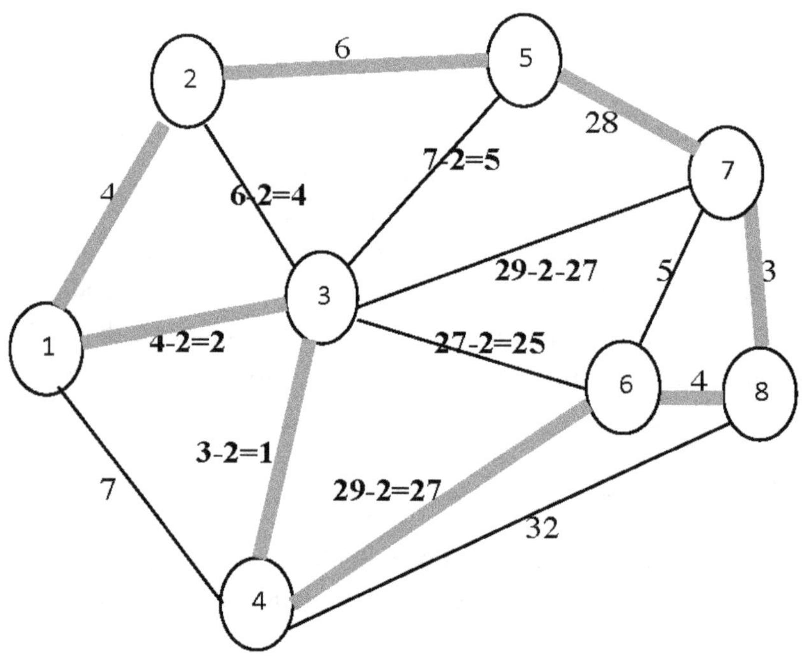

Figure 9. Minimal spanning tree

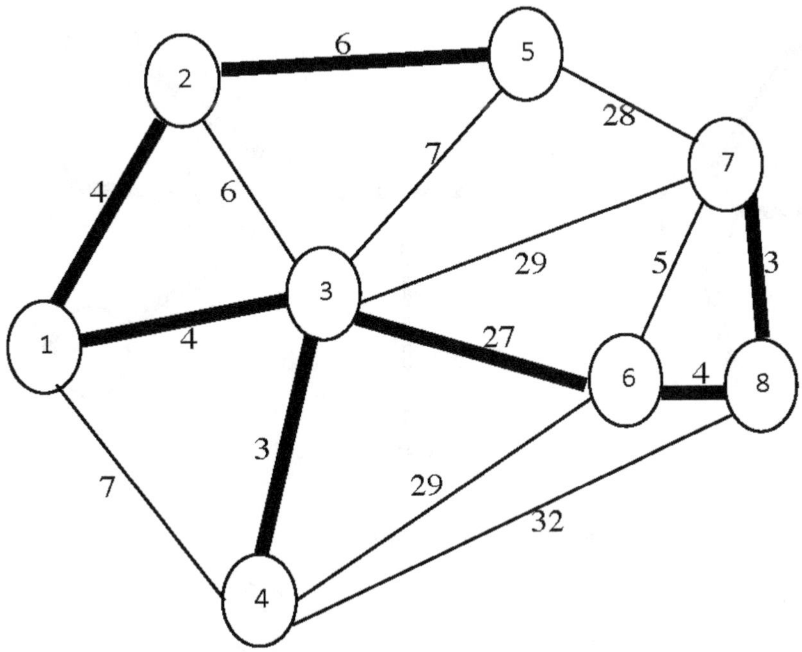

Also note that arm (3,6) is the **only** arc in the column or row of arcs with special features. The other arcs in this column or row are (3,7); (4,6); (4,8) & (5,7).

SUB-TOURS

Detecting Sub-Tours

A minimum spanning tree can be used to detect existence of sub-tours. In a spanning tree a sub-tour may exist if there is an arm (α) and fingers and the distance or length of the arm is significantly bigger than any of the fingers.

In Figure 10 the arc with arc length α is an arm and those with arc lengths $f_1, f_2, \ldots, f_r$ are fingers. The TSP may have a sub-tour if $\alpha > f_1, f_2, \ldots, f_r$. In other words we may have a sub-tour if α is significantly bigger than any of the fingers $f_1, f_2, \ldots, f_r$. We can use this important fact to detect sub-tours and once detected then they can be eliminated. An arm can form a sub-tour if the number of nodes in the finger is 3 or more. The arm is usually the only arc connected in the column or row of arcs with special features. Using a minimal spanning tree to detect sub-tours was used by Munapo (2020).

Generating Sub-Tour Elimination Constraints

The sub-tour elimination constraint is generated from the arm and arcs with special features.

Figure 10. Arm of a spanning tree

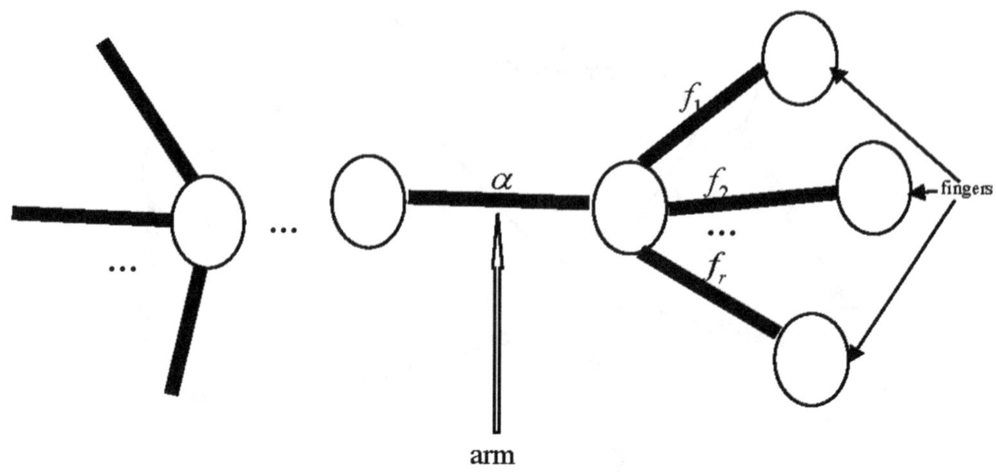

Figure 11. Generating sub-tour elimination constraint

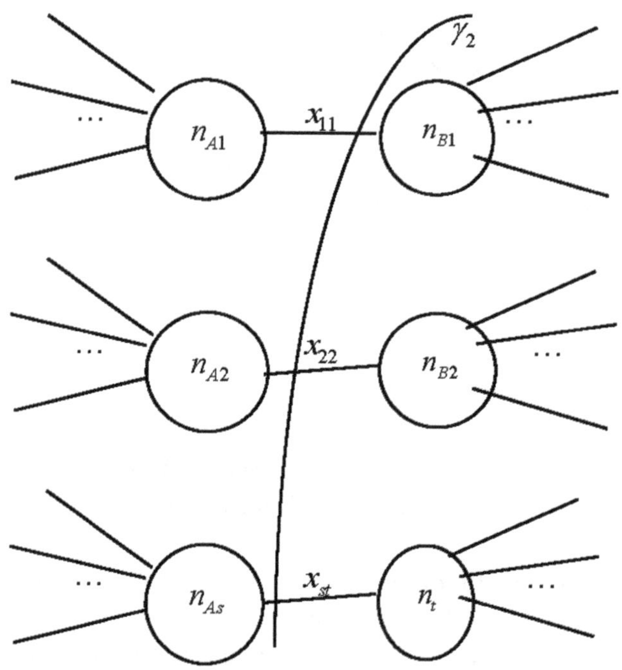

From Figure 11 the sub-tour is generated as shown in (9).

$$x_{11} + x_{22} + \ldots + x_{st} \geq 2. \tag{9}$$

From Figure 12 the sub-tour is generated as shown in (10).

$$x_{36} + x_{37} + x_{46} + x_{48} + x_{57} \geq 2. \tag{10}$$

Figure 12. Generation of a sub-tour elimination constraint

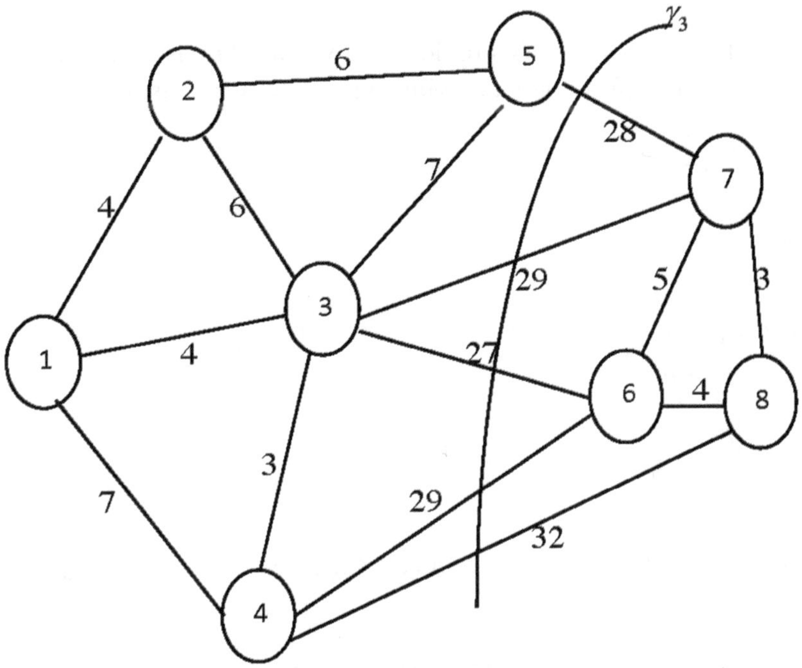

FROM LINEAR INTEGER TO QUADRATIC CONVEX PROGRAM (QP)

Formulated Linear Integer Form

Suppose the formulated linear integer model is given in (11)

$$\text{Min}\, c_{12}x_{12} + c_{13}x_{13} + \ldots + c_{in}x_{in}$$

Such that:

$$\left.\begin{array}{l} x_{12} + x_{12} + \ldots + x_{1r} = 2 \ \text{ standard constraints} \\ \ldots \\ x_{s1} + x_{s2} + \ldots + x_{st} \geq 2 \ \text{ sub-tour elimination constraints} \end{array}\right\} \quad (11)$$

Where $x_j \geq 0$ and integer $\forall j$.

In linear integer form there is no algorithm that can solve directly in polynomial. A way out is to change it to nonlinear form. Munapo (2016) was able to transform the linear integer form into a convex quadratic form and then applied interior point algorithms to obtain an optimal integer solution.

REQUIRED LINEAR BINARY FORM

Suppose the formulated TSP linear binary problem is given as (11). Note that this is in minimization form and we use $x_j=1-z_j$, to transform into a maximization form which is the required form.

Maximize $c_1z_1 + c_2z_2 + ... + c_nz_n$

Such that

$$\begin{aligned} &a_{11}z_1 + a_{12}z_n + ... + a_{1n}z_n = b_1 \\ &a_{21}z_1 + a_{22}z_n + ... + a_{2n}z_n = b_2 \\ &\dots \\ &a_{m1}z_1 + a_{m2}z_n + ... + a_{mn}z_n = b_m \end{aligned} \tag{12}$$

Where $c_j \geq 0$, a_{ij} and b_j are constants, z_j is also a binary variable, $i=1,2,\ldots,m$ and $j=1,2,\ldots,n$

CONVERTING TO CONVEX QUADRATIC FORM

Munapo (2016) presented a very interesting and important property of binary variables. The special property is given in (13).

$$z_j^2 + s_j^2 = 1. \tag{13}$$

Where s_j is a slack and when $z_j=1$ then $s_j=0$ and vice versa. The special feature given in (13) will force variables to assume binary values and is incorporated into a quadratic function. The linear binary problem in convex quadratic is given in (14).

$$\text{Maximize} \quad \begin{aligned} &c_1x_1 + c_2x_2 + ... + c_nx_n + \\ &h(x_1^2 + x_2^2 + ... + x_n^2 + s_1^2 + s_2^2 + ... + s_n^3). \end{aligned}$$

Such that

$$\begin{aligned} &a_{11}z_1 + a_{12}z_n + ... + a_{1n}z_n = b_1 \\ &a_{21}z_1 + a_{22}z_n + ... + a_{2n}z_n = b_2 \\ &\dots \\ &a_{m1}z_1 + a_{m2}z_n + ... + a_{mn}z_n = b_m \end{aligned} \tag{14}$$

$$z_1 + s_1 = 1 \;\; z_2 + s_2 = 1 \;\; \ldots \;\; z_n + s_n = 1$$

Where h>0 and is a very large constant relative to all coefficients c_j. Munapo (2016) showed that convex quadratic objective form in (16) can reduce to the linear form given in (15).

$$\begin{aligned} & c_1z_1 + c_2z_2 + \ldots + c_nz_n + h(z_1^2 + z_2^2 + \ldots + z_n^2 + s_1^2 + s_2^2 + \ldots + s_n^2) \\ & = c_1z_1 + c_2z_2 + \ldots + c_nz_n + h\{(z_1^2 + z_1^2) + (z_2^2 + s_2^2) + \ldots + (z_n^2 + s_n^2)\} \\ & = c_1z_1 + c_2z_2 + \ldots + c_nz_n + h\{(1) + (1) + \ldots + (1)\} \\ & = c_1z_1 + c_2z_2 + \ldots + c_nz_n + hn \\ & = c_1x_1 + c_2x_2 + \ldots + c_nx_n + \text{constant} \end{aligned} \tag{15}$$

POSITIVE DEFINITENESS

Let

$$f(Z) = c_1z_1 + c_2z_2 + \ldots + c_nz_n + h(z_1^2 + z_2^2 + \ldots + z_n^2 + s_1^2 + s_2^2 + \ldots + s^2).$$

Since $f(Z)$ has continuous second order partial derivatives, the Hessian matrix H exists and is given in (16).

$$H = \begin{bmatrix} 2h & 0 & \ldots & 0 \\ 0 & 2h & \ldots & 0 \\ \ldots & & & \ldots \\ 0 & 0 & \ldots & 2h \end{bmatrix}. \tag{16}$$

The matrix H is positive definite since $X^THX>0$, $\forall X \neq 0$. More on positive definiteness is given in Jensen and Bard (2003).

PROPOSED ALGORITHM

Algorithm

Step 1: Construct a minimal spanning tree of the TSP and use it to detect sub-tours. If there is no sign of sub-tours then go to Step 3 else go to Step 2.

Step 2: From the minimal spanning tree identify arcs with special features and use these to generate sub-tour elimination constraints.

Step 3: Formulate TSP as a convex quadratic problem and solve using interior point algorithms.

Figure 13. Proposed algorithm

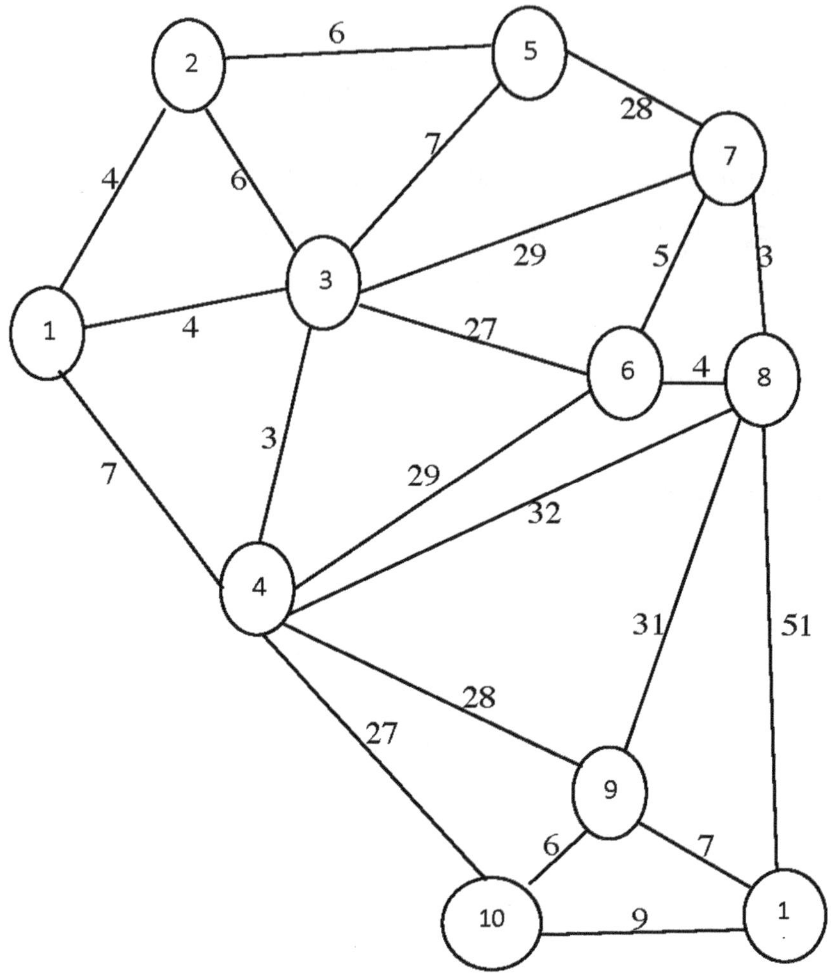

NUMERICAL ILLUSTRATION: PROPOSED ALGORITHM

Use a minimum spanning tree to detect sub-tours, generate sub-tour elimination constraint for the sub-tours and then solve to get an optimal solution.

SOLUTION BY PROPOSED ALGORITHM

Constructing the minimal spanning tree we have Figure 14. Two arms that can form sub-tours can be detected as shown in Figure 14. These arms are circled as shown in Figure 15 and are arm (3.6) and arm (4,10).

From these two arms it is clear that we need to generate two sub-tour elimination constraints. The two sets of arcs with special features can be identified as given in Figure 15.

Figure 14. Minimal spanning tree & two sets of arcs with special features

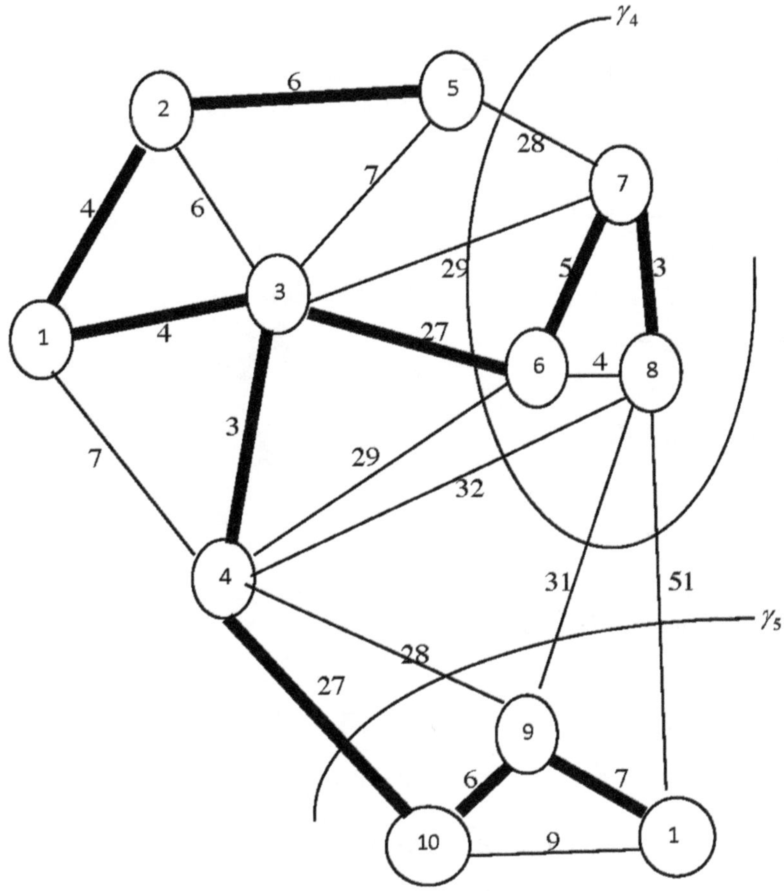

The linear integer formulation is given as (34). Note that the arm (3,6) is the only arc connected out of 7 arcs with special features. The other arcs are (3,6); (3,7); (4,6); (4,8); (8,9) and (8,11). As for the arm (4,19) it is the only arc in its column of 4 arcs with special features and these arcs are ((4,9); (8,9) and (8,11).

$$\text{Min} \begin{array}{l} 4x_{12}+4x_{13}+7x_{14}+6x_{23}+6x_{25}+3x_{34}+7x_{35}+27x_{36}+29x_{37}+29x_{46}+32x_{48}+28x_{49} \\ +27x_{4(10)}+28x_{57}+5x_{67}+4x_{68}+3x_{78}+31x_{89}+51x_{8(10)}+6x_{9(10)}+7x_{9(11)}+9x_{10(11)} \end{array}$$

Such that

Node 1: $x_{12}+x_{13}+x_{14}=2.$

Node 2: $x_{12}+x_{23}+x_{25}=2.$

Node 3: $x_{13}+x_{23}+x_{34}+x_{35}+x_{36}+x_{37}=2.$

Node 4: $x_{13}+x_{34}+x_{46}+x_{48}+x_{49}+x_{4(10)}=2.$

Node 5: $x_{26}+x_{35}+x_{57}=2.$

Node 6: $x_{36}+x_{46}+x_{67}+x_{68}=2.$ Standard constraints

Node 7: $x_{37}+x_{57}+x_{67}+x_{68}=2.$

Node 8: $x_{48}+x_{67}+x_{78}+x_{89}+x_{9(11)}=2.$

Node 9: $x_{49}+x_{89}+x_{9(10)}+x_{9(11)}=2.$

Node 10: $x_{4(10)}+x_{9(10)}+x_{10(11)}=2.$

Node 10: $x_{8(11)}+x_{9(11)}+x_{10(11)}=2.$ (17)

$$\left.\begin{aligned}&\gamma_1: x_{37}+x_{36}+x_{46}+x_{48}+x_{57}+x_{89}+x_{8(11)}\geq 2 \text{ sub-tour elimination}\\&\gamma_1: x_{49}+x_{4(10)}+x_{89}+x_{8(11)}\geq 2 \text{ constraints}\end{aligned}\right\}$$

$x_{12}+s_{12}=1, x_{13}+s_{13}=1,\ldots,x_{10(11)}+s_{10(11)}=1.$

Let x_j=1-z_j and the linear binary problem changes to (18).

$$\text{Max}\;\begin{aligned}&4z_{12}+4z_{13}+7z_{14}+6z_{23}+6z_{25}+3z_{34}+7z_{35}+27z_{36}+29x_{37}+29z_{46}+32z_{48}+28z_{49}\\&+27z_{4(10)}+28z_{57}+5z_{67}+4z_{68}+3z_{78}+31z_{89}+51z_{8(10)}+6z_{9(10)}+7z_{9(11)}+9z_{10(11)}\end{aligned}$$

Such that

Node 1: $z_{12}+z_{13}+z_{14}=1.$

Node 2: $z_{12}+z_{23}+z_{25}=1.$

Node 3: $z_{13}+z_{23}+z_{34}+z_{35}+z_{36}+z_{37}=4.$

Node 4: $z_{13} + z_{34} + z_{46} + z_{48} + z_{49} + z_{4(10)} = 4.$

Node 5: $z_{26} + z_{35} + z_{57} = 1.$

Node 6: $z_{36} + z_{46} + z_{67} + z_{68} = 2.$ Standard constraints

Node 7: $z_{37} + z_{57} + z_{67} + z_{68} = 2.$

Node 8: $z_{48} + z_{67} + z_{78} + z_{89} + z_{9(11)} = 3.$

Node 9: $z_{49} + z_{89} + z_{9(10)} + z_{9(11)} = 2.$

Node 10: $z_{4(10)} + z_{9(10)} + z_{10(11)} = 1.$

Node 10: $z_{8(11)} + z_{9(11)} + z_{10(11)} = 1.$ (18)

$$\left.\begin{array}{l}\gamma_1 : z_{37} + z_{36} + z_{46} + z_{48} + z_{57} + z_{89} + z_{8(11)} \leq 5 \text{ sub-tour elimination} \\ \gamma_1 : z_{49} + z_{4(10)} + z_{89} + z_{8(11)} \leq 2 \text{ constraints}\end{array}\right\}$$

$z_{12} + s_{12} = 1, z_{13} + s_{13} = 1, \ldots, z_{10(11)} + s_{10(11)} = 1.$

The linear integer problem be can transformed into a convex quadratic problem as given in (19).

Max

$$4z_{12} + 4z_{13} + 7z_{14} + 6z_{23} + 6z_{25} + 3z_{34} + 7z_{35} + 27z_{36} + 29z_{37} + 29z_{46} + 32z_{48} + 28z_{49} + 27z_{4(10)}$$
$$+28z_{57} + 5z_{67} + 4z_{68} + 3z_{78} + 31z_{89} + 51z_{8(10)} + 6z_{9(10)} + 7z_{9(11)} + 9z_{10(11)} + 353000\{(z_{12}^2 + s_{12}^2)$$
$$+(z_{13}^2 + s_{13}^2) + (z_{14}^2 + s_{14}^2) + (z_{23}^2 + s_{23}^2) + (z_{25}^2 + s_{25}^2) + (z_{34}^2 + s_{34}^2) + (z_{35}^2 + s_{35}^2) + (z_{36}^2 + s_{36}^2) + (z_{37}^2 + s_{37}^2)$$
$$+(z_{46}^2 + s_{46}^2) + (z_{48}^2 + s_{48}^2) + +(z_{49}^2 + s_{49}^2) + (z_{4(10)}^2 + s_{4(10)}^2) + (z_{57}^2 + s_{57}^2) + (z_{67}^2 + s_{67}^2) + (z_{68}^2 + s_{68}^2)$$
$$+(z_{78}^2 + s_{78}^2) + (z_{89}^2 + s_{89}^2) + (z_{8(10)}^2 + s_{8(10)}^2) + (z_{9(10)}^2 + s_{9(10)}^2) + (z_{9(11)}^2 + s_{9(11)}^2) + (z_{10(11)}^2 + s_{10(11)}^2)$$

Such that

Node 1: $z_{12} + z_{13} + z_{14} = 1.$

Node 2: $z_{12}+z_{23}+z_{25}=1.$

Node 3: $z_{13}+z_{23}+z_{34}+z_{35}+z_{36}+z_{37}=4.$

Node 4: $z_{13}+z_{34}+z_{46}+z_{48}+z_{49}+z_{4(10)}=4.$

Node 5: $z_{26}+z_{35}+z_{57}=1.$

Node 6: $z_{36}+z_{46}+z_{67}+z_{68}=2.$ Standard constraints

Node 7: $z_{37}+z_{57}+z_{67}+z_{68}=2.$

Node 8: $z_{48}+z_{67}+z_{78}+z_{89}+z_{9(11)}=3.$

Node 9: $z_{49}+z_{89}+z_{9(10)}+z_{9(11)}=2.$

Node 10: $z_{4(10)}+z_{9(10)}+z_{10(11)}=1.$

Node 10: $z_{8(11)}+z_{9(11)}+z_{10(11)}=1.$ (19)

$$\gamma_1: \left.\begin{array}{l}\gamma_1: z_{37}+z_{36}+z_{46}+z_{48}+z_{57}+z_{89}+z_{8(11)}\leq 5 \text{ sub-tour elimination}\\ \gamma_1: z_{49}+z_{4(10)}+z_{89}+z_{8(11)}\leq 2 \text{ constraints}\end{array}\right\}$$

$$z_{12}+s_{12}=1, z_{13}+s_{13}=1,\ldots,z_{10(11)}+s_{10(11)}=1.$$

Using interior point algorithm we obtain the optimal solution given in (20).

$$z_{12}=z_{13}=z_{25}=z_{34}=z_{4(10)}=z_{57}=z_{67}=z_{68}=z_{78}=z_{89}=z_{9(11)}=z_{10(11)}=0 \quad (20)$$

This gives the optimal solution to the original problem as (21).

$$x_{12}=x_{13}=x_{25}=x_{34}=x_{4(10)}=x_{57}=x_{67}=x_{68}=x_{78}=x_{89}=x_{9(11)}=x_{10(11)}=1 \quad (21)$$

Figure 15. Optimal solution in diagram form

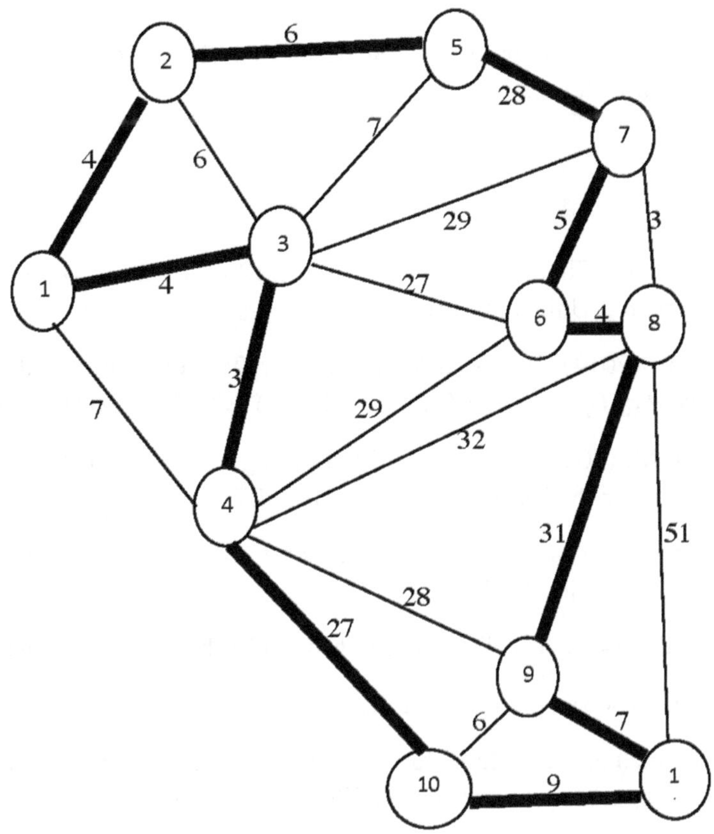

Limitation

There are no computational experiments to support the proposed approach.

CONCLUSION

The chapter presented the traveling salesman problem, its network properties, a way to detect sub-tours and using the detected sub-tours to generate sub-tour elimination constraints. The formulated linear integer model is then transformed into a convex quadratic problem and then solved in polynomial time by interior point algorithms. The proposed algorithm has a bright future in TSP related models. This algorithm will not explode unlike the branch and bound related approaches. Interior point algorithms can handle large sizes of the TSP problems. When generating sub-tour elimination constraints, only the necessary ones are generated. The other formulations require all sub-tour elimination constraints which is computationally expensive. With this approach we expect to solve any size of TSP model.

Area for Further Research

There is a need for computational experiments on standard TSP benchmark problems to support the efficiency of the proposed procedure.

REFERENCES

Applegate, D. L., Bixby, R. E., Chvatal, V., & Cook, W. L. (2006). *The traveling salesman problem: a computational study*. Princeton University Press.

Bektaş, T., & Gouveia, L. (2014). Requiem for the Miller–Tucker–Zemlin subtour elimination constraints? *European Journal of Operational Research, 236*(3), 820–832. doi:10.1016/j.ejor.2013.07.038

Berman, P., & Karpinski, M. (2006). 8/7 - approximation algorithm for *(1,2)* – TSP. *Proc. 17th ACM-SIAM SODA conference*, 641-648. 10.1145/1109557.1109627

Gondzio, J. (2012). Interior Point Methods: 25 Years Later. *European Journal of Operational Research, 218*(3), 587–601. doi:10.1016/j.ejor.2011.09.017

Gutin, G., & Punnen, A. P. (2006). The Traveling Salesman Problem and Its Variants. Springer.

Jensen, P. A., & Bard, J. F. (2003). *Operations Research Models and Methods*. John Wiley & Sons, Inc.

Mitchell, J. F. (2001). Branch and cut algorithms for integer programming. In A. F. Christodous & P. M. Pardalos (Eds.), *Encyclopedia of Optimization. Kluwer Academic Publishers, 2001*. doi:10.1007/0-306-48332-7_215

Munapo, E. (2016). Solving the Binary Linear Programming Model in Polynomial Time. *American Journal of Operations Research, 6*(01), 1–7. doi:10.4236/ajor.2016.61001

Munapo, E. (2020). Network Reconstruction – A New Approach to the Traveling Salesman Problem and Complexity. In P. Vasant, I. Zelinka, & G. W. Weber (Eds.), *Intelligent Computing and Optimization. ICO 2019. Advances in Intelligent Systems and Computing, 1072*. Cham: Springer. doi:10.1007/978-3-030-33585-4_26

Nadef, D. (2002). Polyhedral theory and branch and cut algorithms for the symmetric TSP. In The traveling salesman problem and its variations. Kluwer.

Padberg, M., & Rinaldi, G. (1991). A branch and cut algorithm for the resolution of large-scale symmetric traveling salesman problems. *SIAM Review, 33*(1), 60–100. doi:10.1137/1033004

Papadimitriou, C. H. (1977). The Euclidean traveling salesman problem is NP-complete. *Theoretical Computer Science, 4*(3), 237–244. doi:10.1016/0304-3975(77)90012-3

Winston, W. L. (2004). *Operations Research: Applications and Algorithms* (4th ed.). Thomson Brooks/Cole.

Wolsey, L. A. (1980). Heuristics analysis, linear programming and branch and bound. *Mathematical Programming Study, 13*, 121–134. doi:10.1007/BFb0120913

KEY TERMS AND DEFINITIONS

Algorithm: A procedure or set of rules to be followed in problem-solving operations, specifically by a computer.

Branch and Bound Method: This is a solution method which partitions the feasible solution space into smaller subsets of solutions.

Hessian Matrix: Is a square matrix of second-order partial derivatives which describes the local curvature of a function of many variables.

Heuristic: Is a mental shortcut that allows people to solve problems and make judgments quickly and efficiently.

MST: Given any n set of nodes, a minimal spanning tree (MST) is the network of arcs that connects all the n nodes.

Objective Function: Is a mathematical expression that minimizes or maximizes a linear or nonlinear problem.

Optimal Solution: A feasible solution at the point where the objective function reaches its maximum/minimum value.

Positive Definite: A square matrix H is positive definite if $Y^T HY>0, \forall Y \neq 0$.

Principal Minor: A principal submatrix of a square matrix A is the matrix obtained by deleting any k rows and the corresponding k columns.

Sub-Tour: Given any n nodes, any connection of nodes made of up k nodes where k is less than n, is called a sub-tour. The connection is such that when visiting the nodes each node is visited once and we return to the initial node.

Tour: Is a traversal of each node exactly once that returns to the initial node. The cost of a tour is the sum of the distance between each pair of nodes on the tour.

TSP: Is the traveling salesman problem (TSP) where we move from an origin node and return to it, traversing the nodes in such a way that every node is visited once and that the total distance travelled is minimized.

Chapter 7
Optimal Sizing of Hybrid Wind and Solar Renewable Energy System:
A Case Study of Ethiopia

Diriba Kajela Geleta
Department of Mathematics, Madda Walabu University, Oromia, Ethiopia

Mukhdeep Singh Manshahia
https://orcid.org/0000-0003-3342-8030
Punjabi University, Patiala, India

ABSTRACT

If properly designed and utilized, earth has rich potential of clean energy in satisfying the energy demand of the world. In this chapter, nature-inspired methodology was employed to optimize hybrids of renewable energy system in the case of Jeldu district of Ethiopia. The main goal of the researchers here is to minimize the total annual cost of the system, which can be designed by using appropriate numbers of components based on the pre-designed constraints to satisfy the load demand. MATLAB code was designed for the proposed methodology, and the results were discussed. It was seen from the result that the proposed approach has solved the optimum sizing of defined problem with high convergence. The results show that energy demand of the village can be optimally satisfied by the use of wind and solar hybrid system. Moreover, the application of this chapter is important for countries like Ethiopia to increase access to electricity.

INTRODUCTION

A lot of efforts have been made in the direction of improvement in order to curb the issue of climate change and global warming through cutting an inevitable production of CO2 which produce harmful emissions from the power generating schemes. Now days, based on the increment of population and

DOI: 10.4018/978-1-7998-3970-5.ch007

energy demand all over the world, different decentralized power production schemes have been raised. Till now, world is using fossil fuels as a central source of energy (Marc Anthony Mannah, 2017, Kharchenko, V., & Vasant, P., 2018, 2020, 2020). These conventional energy sources are subject to depletion and have multi directional effects, like the issue of environmental pollution, diminution of its sources and continuous increase in oil prices (Thomas, J. J., Karagoz, P., Ahamed, B. B., & Vasant, P., 2020). As a result, the global population without access to electricity was about 840 million in 2017 (Guardian Agencies Report, 2019). Since many efforts have been made in all developing countries of the world to achieve millennium development goal of 2030, the global electrification rate reached 89% in 2017 from 83% in 2010, which means still about 840 million people out of 7.7 billion current world population leaving without access (UNSD, WB, WHO, 2019).

This large number of populations which accounts about 10.9% of the total world population without power access needs global effort to bring the solution. Among this, significant number was the population settled in the rural areas of the world. In most cases, the citizens reside in the rural area of the world lives in scattered manner in very difficult geographical location to extend electricity from national grid by the help of government. Instead the government designs different off grid power generating mechanisms to satisfy the basic need of that society. These off-grid mechanism may include the use of diesel generator which may not affordable for most the society living developing countries (Bhandari, B. et al. 2016). All these challenges have pushed the world attention for the development and utilization of alternative renewable energy sources (Geleta & Manshahia, 2020; Luna, Trejo, Vargas, & Os-Moreno, 2012; Rubio, Perea, Vazquez, Os-Moreno., 2012; Vasant, Kose, & Watada, 2017). Encouraging the off-grid power generating mechanisms are most advantageous in terms economic aspects, sustainability of the power, reliability and environmental protections for the nations living far from national grids. As explained above, in addition to shortage of power there was unfair distribution among urban and rural settlers in all countries, especially sub-Saharan African countries and South Eastern Asia. The rural area of these regions uses kerosene for lighting, wood and animal dugs for cooking. Due to less communication and technologies, the life standard of these countries is still under expected in 21st century. Serves giving sectors like schools, health centers and Agri-processing centers are need electricity in order to properly functioning.

Here we employ nature inspired algorithm to optimize hybrids of wind and solar which may applicable at any remote areas of the world and relatively easier and cheaper to implement. This paper is presenting a case study of Kabi village of Jeldu district, Ethiopia to find the optimal size of hybrids of wind and solar renewable energy system. The main concern is to determine the numbers of wind turbine, solar system and batteries, so that the desired load can be satisfied with minimum possible annual cost.

BACKGROUND

The rural area of sub-Saharan African countries and South Eastern Asia uses kerosene for lighting, wood and animal dugs for cooking. Due to less communication and technologies, the life standard of these societies is still under expected in 21st century. Serves giving sectors like schools, health centers and Agri-processing centers are need electricity in order for proper functioning.

Sub-Saharan African countries, which hosts more than 950 million people, is the most electricity-poor region in the world. More than 600 million people lack access to electricity. This access, a smaller number of the region was less due to the only country South Africa has high coverage of electricity. Ethiopia, is one of the countries in east Africa with large amount of population (about 100 million) and located at 9.15^0 North, 40.49^0 East in the tropical zone laying between the Equator and the Tropic of Cancer. The total access to electricity of the country was about 45% and the number of households connected to this access is 15%. About 89.6% of electricity in Ethiopia consumed in urban area. Among these about 85% of the population lives in the rural with the access to electricity less than 15%. Even though a number of action actions was taken by different stakeholder to improve the energy sector, the problem of rural areas of the country didn't solved yet (MoWIE, 2019). There is high demand of electricity in the country. The need for electric power is highly increasing at every corner of the country. In addition to for lighting, cooking food and mailing, there is maximum interest for using electronic devices like mobile charging which is used communication technology.

Jeldu is one of the Oromia region districts located at central part of Ethiopia. It is the mountainous district located between about 9. 05° to 9.25° North latitude and 37 .40° to 38.11° East longitude. According to 2007 national census reports, total number of populations of this district was 202,716. It was part of the West Shewa Zone with an elevation of about 2360 meters above sea level. The district located to the west and about 110 kms from Addis Ababa, the capital city of Ethiopia. The particular place where this research was conducted is Kabi village of Jeldu district. The village contains 164 households with total population of 984, one primary school, one health center one protestant church and one small milk processing center. Since there is no eccentricity for this village at all, people of the village use kerosene for lighting during night time, woods and animal dugs for cooking their food and diesel generator at church, school and health center and for mill to grind their crops. The need for electricity was high for the village because they forced to pay additional transportation cost even to chare their mobiles. Committees are assigned and communicated with governmental officials at different levels to get the access of electricity in the past few decades. But till now, the village is in serious difficulty by lack electricity not only for day to day life but also for information communication and technology.

Even though the district has high potential of renewable energy sources, any project was not tried yet. So, this research work will be very important for remote areas of the world in general, and for Kabi village of Jeldu district in particular. It can be used as bench mark for the researchers to investigate further by applying their own techniques on simulation tools and hybrids of optimization techniques. In particular, the work will be very important to show the direction how public and private sectors can engage to improve the life standard of Jeldu society by supply the demand of energy which can solve plenty of problems in addition to the improving services like, health care, education quality, information access, using technologies etc. Large communities live in rural areas are far from the main grid extension never got electric access yet, due to economic and geographical constraints. This off grid electrification will be the best option for rural areas the country. These communities in the country use traditional ways of accessing energy. So, expanding renewable energy system in optimum manner is much important for the country.

MOTIVATION OF RESEARCH

In recent years, the search for environmentally clean, inexhaustible, and friendly use energy sources has been undertaking the concerns of public policy. The limitations of traditional energy sources like shortage of its source, environmental degradation and continuously raising up of oil price have mainly motivated all the public and private sectors to give attention in the development and usage of alternative energy sources.

In order to decrease the problems associated with these traditional energy sources, a lot of actions have been taken by public, private sectors and professional efforts have made by scholars of the field. As a result, an alternative energy resources which are abundant, has less effect on the environment, getting much attention and rapidly growing up nowadays (Geleta & Manshahia, 2017; Luna, et al., 2012). Various studies have shown that there is a direct relationship between electricity access and the socio-economic development of a country. Nowadays, electricity has become an indispensable prerequisite for enhancing economic activity and improving human quality of life. Agricultural and industrial production processes are made more efficient through the use of electricity which can directly affect the economic activity of the country.

Renewable energy sources are going to play an important role in supplying clean energy to the world for the future by diversifying and maintaining the energy supply market exist now (Kaabeche, et al. 2010). Nowadays, the electrification of rural villages depends upon extension of main grids, installation of diesel generators, woods and animal products as an option. In reality, grid expansion to all the places of such areas is very difficult to satisfy the power demand because of, either financial constraint or infeasibility of the places. Most of such societies are living at difficult geographical location with most probably low densely populated and a very low power demand. Thus, to improve the power utility coverage of the world, applying standalone renewable energy sources will make such society more beneficiary (Geleta & Manshahia, 2018; Zong, 2012). The most commonly known renewable energy sources are hydro power, wind, solar, hydro, biomass, ocean wave, geothermal and tides which are naturally replaceable. Even though, these energy sources abundant and endless, all of them have basic limitations in generating maximum power output due to their whether dependency. This drawback of renewable energy technologies can be managed by applying hybrids renewable energy technology (Vasant, P.,2018; Geleta & Manshahia, 2017).

RELATED WORK

Different scholars have been applied various methods to optimize the hybrid renewable energy systems for different problems. Optimization problems can be solved either by conventional methods which involves the concept of Gradient and Hessian matrix for continuously differentiable objective functions or non-conventional/ stochastic methods by which hard problems are treated. Both type of methods has their own advantages and limitations when employed to solve a particular problem. Nowadays, since multi-dimensional and stochastic optimization problems are getting much attention, the non-conventional algorithms getting upper hand to be implemented.

For optimizing the total annual cost of the hybrid renewable energy systems, many scholars have implemented a lot of algorithms on the literature. Here, some literatures are reviewed by the researchers.

Berihun G. (2013) has conducted a case study on rural area of Ethiopia by using HOMER software. He has modelled micro hybrid of hydro-wind power generating for the rural areas of the country. His objective was to develop a hybrid system cost competitive to supply energy for remote area. He has got result that, the COE most favorable Wind/Micro hydro hybrid system is \$0.112/kWh. Moreover, COE of standalone Micro-hydro system is \$0.035/kWh and concluded micro hydro system is the most economical, technical and can only satisfy the energy demand of the village and technically feasible option.

Bekele, G., & Palm, B. (2010) have employed an alternative methodology of producing electric power from hybrids of solar-wind renewable system to a remote community of Ethiopia through HOMER software. They have designed the hybrid system to optimize the power output. Their results were compared from the list of feasible renewable power sources. Based on the net present cost they found from simulation, they have shown that hybrid solar and wind system is only the promising technology to satisfy the power demand of these communities. On their other work, Bekele, G., & Palm, B. (2012) they have design of a hybrid electric power generation system by using wind and solar energy in order to supply power for the community living in Ethiopian remote area. In the later work they have begun their work by collecting all primary and secondary data of the selected site and simulate by the help of HOMER software. Their result shows there is a huge potential and usable solar and wind sources of the site. Additionally, by changing net present cost of the system it is possible to select the best optimal solution of the system among the given alternatives of simulation results.

A research conducted by Weldemariam, L. E. (2010) on rural areas of Ethiopia has shown there is a promising future of hybrid renewable energy to satisfy the load demand of the rural communities. His study intended to enhance reliable, an efficient and cost competitive system configuration of hybrid of wind and solar power generating system to improve the life of the rural societies which are not connected to the central grid. He concluded that renewable energy sources and/or their hybrid configuration are promising and cost competitive for the rural settlers when the huge initial capital investment of the renewable energy resources is supported by the concerned bodies. Golma, B. (2011) has developed Design of a Photovoltaic/ Wind hybrid Power generation system for Ethiopian remote area. His aim was to design and model a stand-alone PV-Wind hybrid power generation system to optimize the total net present cost of the system. After simulation, he has proposed hybrid system is good solution to satisfy the energy demand of the rural community and the result obtained affected by the renewable energy access is the promising to implement the project for the rural area. After simulation was made, the possible cost of energy for the proposed hybrid set up configuration is range between 30 cents to 40 cents per kilowatt hour.

In the paper presented by Benatiallah, A., et al. 2010, the methodology GA was applied for calculation the optimum size of a Wind system to satisfy the load demand of a typical house in south of Algeria (desert area). For a given load and a mixed multiple-criteria integer programming problem, the types and sizes of wind turbine generators (WTG) was calculated based on the minimum cost of system. They have employed the nature inspired algorithms, genetic algorithm (GA) for optimally sizing a wind power system. The authors have defined that the objective function is the total cost, where the total cost is the sum of initial cost, an operation cost, and a maintenance cost and have determined an optimal configuration of wind generating systems, where total cost is more optimal using GA. In order to match the load of the site, a computer program has been developed to size system components. A cost of electricity, an overall system cost is also calculated for each configuration. The study was performed using a graphical user interface programmed in MATLAB.

Kellogg, W. G., et al.1998 have used the deterministic Optimization method called an iterative optimization method to select the wind turbine size and PV module number needed to make the power difference is zero. In this paper was to determine the optimum generation capacity and storage needed for a stand-alone, wind, PV, and hybrid wind/PV system at an experimental site in a remote area in Montana with a typical residential load. The total annual cost for each configuration is then calculated and the combination with the lowest cost is selected to represent the optimal mixture, an economic analysis has been performed for the above three scenarios and is used to justify the use of renewable energy versus constructing a line extension from the nearest existing power line to supply the load with conventional power. Annual average data for hourly values for load, wind speed, and insolation have been used in order to determine the optimal value. The Iteration method applied by (Geleta & Manshahia, 2018) was discussed the optimal value of the hybrid of wind and solar renewable energy system by taking the load balance in to consideration. The main objected they set up was optimizing the total annual cost of hybrid system which can limit the numbers of wind turbine, solar panels and batteries which can satisfy the desired load through high reliability. For different iteration they consider numbers of the components at which reliability of the system become zero. Their result shown iteration method can optimize hybrid energy system which can satisfy the desired load.

Fulzele, J. B. et al., 2018, have systematically designed hybrids of renewable energy system for the development of highly reliable, cost effective and continuously supply of energy for the rural area. They have used hourly electrical load profile, monthly solar radiations and wind speeds to minimize the total net present cost of the system. They have employed HOMER to optimize the system. Their result shows the hybrid system gives the optimal solution of power generation through HOMER for the rural area. Mohamed, M. A., et al. 2016 have employed a method called PSO, for an optimal sizing algorithm for a hybrid renewable energy system using smart grid load management application based on the available generation. Their main goal was to optimize the system cost by which the load demand satisfied under a set of optimization constraints. PSO has been presented to determine the optimal size of the system components under pre designed set of conditions. The simulation results affirmed that PSO is the promising optimization techniques due to its ability to reach the global optimum with relative simplicity and computational proficiency contrasted with the customary optimization techniques. Finally, parallel implementation of PSO has been utilized to speed up the optimization process, and the simulation results confirmed that it can save more time during the optimization process compared to the serial implementation of PSO.

The trade-off method by (Gavanidou, Bakirtzis, & Dokopoulos, 1993; Elhadidy & Shaahid, 1999) and the least square method used by (Borowy & Salameh, 1996) have presented a methodology for optimization of a PV/wind system based on deficiency of power supply probability (DPSP), relative excess power generated (REPG), unutilized energy probability (UEP). The Artificial Bee Colony (ABC) employed by (Geleta & Manshahia, 2020) has designed the total system hybrids of wind and solar renewable energy system including some technical aspects to be considered. The onjective of their research was to optimize the total annual cost of hybrid of wind and solar renewable energy system by identifying the number of individual components participating in the system to satsfy the desired load. The MATLAB code designed for the method was used for simulation for the particular data. The Optimal result optained from simulation was compared with some results on literature. It is shown that the employed method ABC was supperior to the others with result and convergence rate good optimization result.

The technical approach also called loss of power supply probability (LPSP) by (Diaf, Belhamel, Haddadi, & Louche, 2007) have optimized a hybrid system size based on loss of power supply probability (LPSP) and the levelized cost of energy (LCE). (Singh & Kaushik, 2016; Alireza, 2013) have presented artificial bee colony algorithm for optimal sizing of grid integrated hybrid PV-biomass energy system by the least levelized cost of energy while minimizing annualized cost of the system to find optimum hybrid system configuration. It has been observed from the results that a grid connected hybrid PV- biomass system is cost effective, reliable choice and the proposed algorithm provides better results as compared with other deterministic method and HOMER (Koutroulis, Dionissia, Potirakis, & Kostas, 2006; Ashok, 2007; Wang & Singh, 2009). Lotfi, S., et al. (2013) have introduced a deterministic approach technique to optimal unit sizing for hybrid power generation systems utilizing photovoltaic and wind energy. The optimization problem is considered as a multi objective one, and a discrete set of Pareto optimal solutions is derived numerically by using the weighting method. To test the validity and effectiveness of the method and to evaluate the system performance, a numerical study on an on-site system has been carried out by using data on natural energy obtained through real measurement. They conclude from these as relationships between economy and energy savings or environmental protection has been clarified, as it is use full for large investment due to its economic capacity demand of electricity satisfied.

Scholars like (Satish, K. R. et al.,2014; Nafeh, A. E., 2011) have presented their paper on a new approach of optimum design for a hybrid PV/Wind energy system in order to optimize the total net present cost of the system in order to assist the designers to take in to consideration both the economic and ecological aspects when designing the system .They have used photovoltaic panels, wind turbine s and storage batteries to get the hybrid system and employ, the Optimization Technique Genetic Algorithm to minimize the formulated objective function *i.e.* the total cost of the system and ensure that the load is served reliably. In their work they design a computer program by using MATLAB code to formulate the optimization problem by computing the coefficients of the objective function and arrive at the conclusion that GA converges very well and feasible for sizing the hybrid system. Additionally, they got a number of potential optimal solutions under this technique and PV-wind hybrid energy systems are the most economical and reliable solution for electrifying remote area loads. (Bhandari, B., et al., 2016) have applied the conventional optimization method, called linear programming for minimizing the total operation cost of the system by using maximum RES potentials. They have considered the monthly power fluctuation and electric demand of the particular places. They have tested the robustness of the power generated for the defined two particular areas. The optimization results indicate that the optimized hybrid system can help to completely shifts from current highly fossil fuel dependence power system to environmentally clean renewable energy-based power system in wide geography.

Most of the abovementioned methods are deterministic approach and/or apply HOMER software, to optimize hybrids of renewable energy system by considering their own fitness functions and constraints. But non off them did show the parameters opportunities of their method when it was applied to the particular area. Some nature inspired algorithms like Artificial Bee Colony algorithm by (Geleta & Manshahia, 2020 and Singh & Kaushik, 2016; Alireza, 2013), Genetic Algorithm by (Benatiallah, A., et al. 2010; Satish, K. R. et al.,2014 and Nafeh, A. E., 2011) have used to optimize the hybrid system.

But the researchers identified some gaps like:-

- Intensive collected data of the particular place was not implemented
- Initial capital cost of the project was not identfied with possible suggetion of its improvement.
- Comparision with the results in the literature were not got attention.

In this paper the researchers presents a case study for solving the economic aspect meaning, minimizing the total annual cost of the defined problem which can satisfy the desired power of one particular place Kabi, in Jeldu district, Oromia, Ethiopia. In this research some technical aspects of the system such as reliability, energy not supplied by the system and sensitivity analysis of some impute variables are also conspired. Other further applications and comparisons are left for future deal. Researchers hopes that the output of this paper become good impute for the beneficiaries at local level and researchers to follow the way by applying their own methods.

By this research, data of particular rural area was intensively collected to optimize the total annual cost of wind and solar renewable energy system. As a result, it was novel by: -

- Identifying optimal total annual cost needed for the project
- Recommending type of renewable energy was affordable for a particular area especially for rural settlers.

Researchers hopes that the output of this paper become good impute for the beneficiaries at local level and researchers to follow the way by applying their own methods.

ENERGY SCENARIO IN ETHIOPIA

The exploitation, distribution and usage of energy is different from continent to continent and within the continent too. Most the exploited energies were confined in developed countries and others like central and south eastern Asian countries and most Sub-Saharan African countries are high electric access deficit regions. The global population access to electricity was 89% in 2017 with an amount annual electrification growth rate 0.8% (UNSD, WB, WHO, 2019). The amount of lack of access with the defined growth rate could not solved in the next few years. So, it needs argent progress to give light for this significant number of world population their basic needs.

Most African countries especially Sub Saharan African countries which is home to more than 950 million people has about 600 million people lack access to electricity. As seen from the figure 1 the globally population including central and southern Asia was about 840 million people lack access to electricity.

Figure 1. Major world electric deficit (UNSD, WB, WHO, 2019)

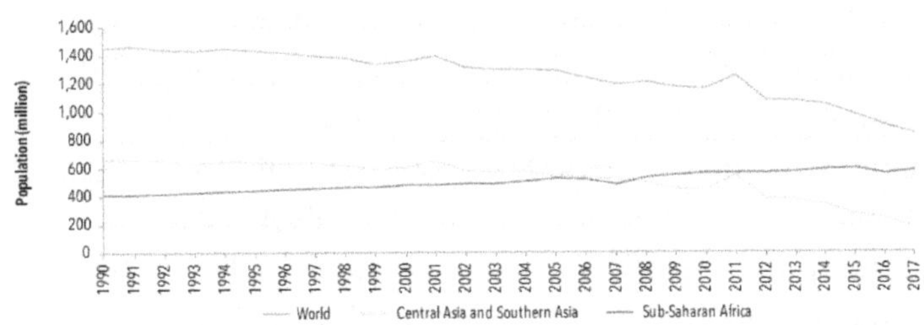

Source: World Bank.

In order to decrease electricity access deficit, the world is making progress towards achieving Sustainable Development Goal (SDG) to meet the targets set for 2030. These targets are: -

- To ensure universal access to affordable, reliable, and modern energy services that focuses on the proportion of the population with access to electricity, relying primarily on clean fuels and technologies for cooking,
- To increase substantially the share of renewable energy in the global energy mix and
- To double the global rate of improvement in energy efficiency [Guardian Agencies Report, 2019].

In recent years, encouraging progress was made in expanding access to electricity in several countries of the world, notably India, Bangladesh, and Kenya. As a result of these countries effort, the global population without access to electricity decreased to about 840 million in 2017 from 1.2 billion in 2010 (UNSD, WB, WHO, 2019).

Globally, since many efforts have been made in developing world, the electrification rate reached 89% in 2017 from that of 83% in 2010, still leaving about 840 million people without access. The progress amounts to an average annual electrification rate of 0.8 percentage points, and newly gained access for more than 920 million people since 2010.

Large amount of the population lacking access to electricity are increasingly concentrated in Sub-Saharan African countries as shown in the figure 2.

Figure 2. Countries with low access to electricity from 2010-2017(UNSD, WB, WHO, 2019)

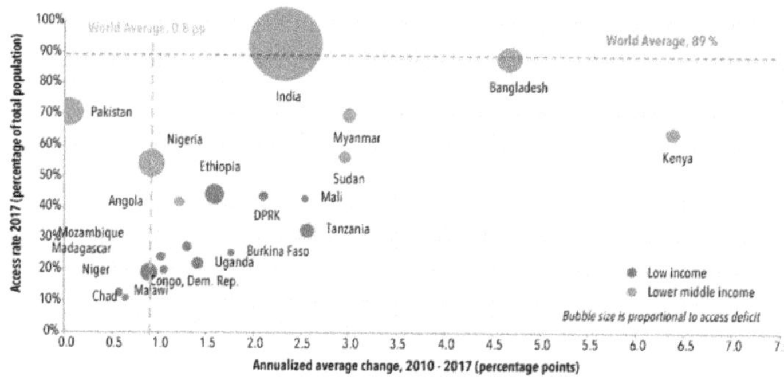

Africa has made an improvement over the past years, in terms of poverty reduction and economic growth, even though some of the challenges are continue in many countries of the region. Among the challenges boldly seen in the region, lack of access to energy was the main obstacle and negatively holding the expansion of infra structure which have direct relationship with advancement of civilization and technology. As shown in figure 3, Expanding access to modern renewable energy is still under challenging in Sub Saharan African countries. It was substantially lower than what it could be to force the economic growth of the region. This lack of access to electricity imposes significant impacts on modern economic activities, provision of public services, and quality of life, as well as on adoption of new technologies in various sectors such as education, health, agriculture, and finance (Blimpo, M. P., & Cosgrove-Davies, M., 2019).

Ethiopia is a large country located at eastern horn of Africa, between 3° to 14.8^o north, latitude and 33° to 48° east, longitude with the total are of 1.1million km^2. The total population of Ethiopia was about 100 million with about 85 different ethnics of nations and nationalities. Since its location was in moderate template zone, the country was endowed with suitable climate throughout the year. As a result, it is the home of different precious animals and birds found only in Ethiopia and becoming high tourist attraction by now. Geographical location, ethnic differences and capture of the society are among the beauty of Ethiopia. Even though, a lot of efforts have been made in the past three decades, due to its political instability the development of the country was not as much said and planned. The Growth and Transformation Plans (GTP I, GTP II) launched by Ethiopian government from 2011-2015 and 2016-2020 respectively are good, ambitious and intensive plan document to bring the country from low income to

Figure 3. Sub-Saharan countries electric access and without access (Mante, O. D., 2018)

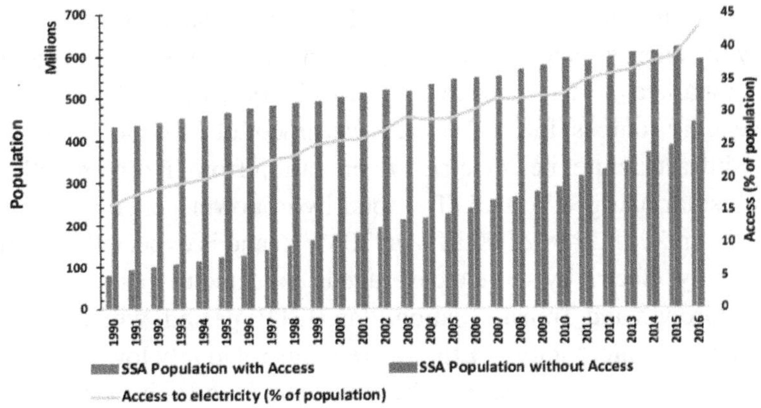

middle income countries by the end of the plan date. In these documents all sectors are accessed and planned to their maximum possible values.

Among these sectors, Ethiopian Electric Power Corporation (EEPCo) is the one the main attention given sector. As seen from the documents this sector, starting from current base line in terms of electricity generation, expansion and access to electricity, its importance and impact on other sectors, all potential sources, opportunities and treats are identified. These documents clearly put the importance of energy and the role it has in all the other sectors to enhance the countries development by two digits to achieve the millennium development goal and become middle income country (Beyene, G. E., el al. (2018).

These governments valuable and ambitious plan, which can really transform the country was not implemented as said at the end of the date. of course, there are a lot of reasons which directly affect the implementation of this plan. Among these, budget constraint, highly skilled man power to all the sectors are some. In addition to these, the political instability of the country affects the implementation of the plan. The Oromo people protests to ask their self-determination since 1991 was the main challenges of this country.

Ethiopia has high potential of renewable power generation sources. Among these, from hydro power 45,000MW, from geothermal 7,000MW, from wind 1035GW and from solar 5.2kwh/m^2 which can enhance the development of the country in general and electricity access of the society in particular. But electricity generation capacity of Ethiopia is about 4228MW of which 90% was only from hydro power (Beyene, G., 2018).

The electricity access of the countries is also one of poorest among the Sub Saharan region. It was the lowest when compared with global access as shown in fig 2. In Ethiopia only 25% of the population has access to electricity. The per capita electric consumption is also one of the lowest according to the standard published by International Renewable Energy Agency (IREA, 2011), which defines electric access as the annual consumption of at least 250kwh electricity for the rural areas and 500kwh for the urban area households. By now, the electric access in Ethiopia was about 100kwh/ year which is less than the Sub-Sahara average, that is 510kWh (MoWIE, 2015; Beyene, G., 2018). However, the country has utilized insignificant amount of its energy resource potential from all the sources. As a result, out of 69% sub-Saharan African countries which has no access to electricity Ethiopia has high share about 63% (Gatamesay B. et al. 2015).

In Ethiopia, the lack of access to modern energy services that are clean, efficient and environmentally sustainable is a critical limitation on economic growth and sustainable development. Recognizing the critical role played by the energy sector in the economic growth and development process, the Government of Ethiopia has embarked on large scale hydroelectricity projects, with a view to developing renewable and sustainable energy sources. In GDP II the government of Ethiopia has injected a huge amount of money into energy infrastructure, i.e., electricity generation from hydro and other renewable energy sources such as wind, solar and geothermal. The total hydropower generation capacity country-wide increased from 714 MW in 2011/12 to 2,000 MW in 2015/16 and is expected to increase to over 10,000 MW by 2020. Accordingly, electricity service coverage is expected to increase from 41% in 2014/15 to 75% by the end of the Growth and Transformation Plan (GTP) period, which is 2019/2020. Currently, the per-capita consumption of electricity in Ethiopia remains relatively low at about 200 kWh per year. The national energy balance is dominated by a heavy reliance on traditional biomass energy sources such as wood fuels, crop residues, and animal dung (Fantu G. et al. 2015)

The Government of Ethiopia, under its latest Growth and Transformation Plan (GTP), set the mission of transforming from a developing country to a middle-income country by 2025. Ethiopia's ability to achieve this ambitious goal in such key sectors as agriculture and industry is significantly constrained by current challenges in the power sector. Although Ethiopia is endowed with abundant renewable energy resources and has a potential to generate over 60,000 megawatts (MW) of electric power from hydroelectric, wind, solar and geothermal sources, currently it only has approximately 2,300 MW of installed generation capacity to serve a population of over 95 million people. The targets for increasing generation capacity to 10,000 MW established under the first iteration of the GTP will be met by completion of two major hydro power plants in2017 and 2018. The current GTP has a new target to increase generation capacity to over 17,000 MW by 2020, with an overall potential of 35,000 MW by 2037, which would help sustain Ethiopia's continued economic growth and enable it to become a regional renewable energy hub in East Africa (Power Africa in Ethiopia, 2018).

Some of the broad targets set in (GTP II, 2016) to enhance the total activity of the country in order to achieve the desired goal are:

- To increase the electricity generation capacity from multi sources to 17,000MW from the base 4180MW in 2015.
- To make the household electricity connection more than double meaning to 7 million from 2.3 million at the end of GTP 1.
- To increase the electricity grid network coverage to 90% from 60% in 2015.
- To increase the total length of power transmission lines from 16,018km in 2014/15 to
- 21,728 km in 2020

Resent reports show that, similar to GTP I, the targets set to accomplish at the end of GTP II (2020) is also not on the normal track. The actual visible situation is also didn't indicate to the accomplishment of the goals. Ethiopian Ministry of Water, Irrigation, and Energy has also declared in National Electrification Program 2, Integrated Planning for Universal Access (2019) as the country will achieve millennium development goal by launching mega hydroelectric projects to double the power generation capacity of Ethiopia at 2025. The following table was taken from the Ministry plan National Electrification Program 2, Integrated Planning for Universal Access (2019) shows the present electric access of the country with the future plan when it was achieved.

As stated in the National Electrification Program 2.0 set by Ministry of Water, Irrigation and Energy of Ethiopia (MoWIE,2019), the total access rate will be reached 100% by the 2025, from it' 47% by 2019. It was ambitious plan unless effort of highly committed administration of the sector will be supported by adequate finance. These national planed programs need multi direction supports to accelerate toward its accomplishment. In addition to the devotion made by concerned bodies, all citizens of the country, especially researchers should support by designing different alternatives how access to electricity will improve. The researchers have analyzed and committed to give valuable professional support to balance the high electric shortage and ambitious plan developed to avoid the problem with in the next five, six years.

Energy Demand of the Study Area

Profile of Jeldu District

As discussed under introduction Jeldu was located in West Showa zone, Oromia reginal state of Ethiopia. It was high land district with moderate temperature and rain fall. Agricultural activity is the main income scheme of the society.

The rural village kabi, is one of the kebeles (villages) in Jeldu district and located to the west and around 15km away from the district's capital Gojo. Kabi village was selected for the application of this case study due to its centeredness of the rest villages of the district. Figure 4 shows the location of the district in Oromia regional state, Ethiopia.

Figure 4. Map of Jeldu District (Dandessa, C. et al.; 2018)

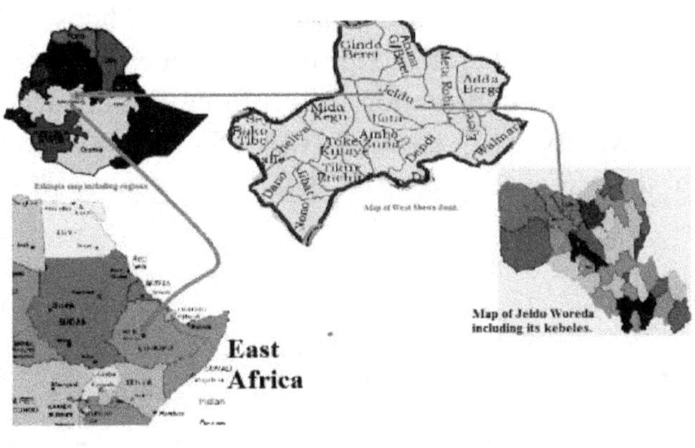

Less amount of the area under consideration is hilly with the rest flat plateau. The village has no basic infrastructures like clean water, electricity, road and internet access. One primary school is available, students of the village has to went 3hrs/day to get high school education by foot. Medical care of the society is also very low from the standard. As a result, researchers select the village to conduct this case study and recommend the concerned government body or non-governmental organizations to remind about that village. Data of the village was collected under the consultancy of Jeldu district electric office, Jeldu district administration office, local administration of the village and elders of the village.

Primary and Secondary Data of the Village

1. **Primary data:** Primary data are those data collected by the researchers through intensive field survey, interview with some peoples of the village. During the field survey, the primary data necessary for this study like topography of the village, types of the society, numbers of institutions or religious institutions and type of community service giving institutions such as, school, health center are identified. The following table 1 shows the survey result of the village.

Table 1. Primary data of the Village

Data Collected	Value	Data Source
States of wind and solar radiation	Rich	Jeldu district electricity office
Category of population	2	Leader of the village
Number of Primary schools	1	Jeldu district Education Offices
Number of Health center	1	Jeldu district Health Offices
Number of Institutions	3	Jeldu district Administration Office
Grinding mill	1	Field survey
Milk processing center	2	Field survey

2. **Secondary data:** This is the data that is not directly collected by the involvement of the researcher for his purpose; others organize and document it for different purpose. Secondary data which more appropriate for this case study are population size & number of households, power demand of the village, solar radiation, wind speed profile and some equipment cost related to the case study are organized from Jeldu district Administration office, From Ethiopian Ministry of Energy, Water and Irrigation, Ethiopian Metrology Agencies & NASA Surface metrology web site. The following table 2 shows some of the organized secondary data.

Table 2. Secondary data of the village

Secondary Data Collected	Value	Data Source
Number of populations	984	Jeldu district administration
Number of households	164	Jeldu district administration
Power demand	589.64KWh/day	Researcher
Solar radiation	8.29KWh/m^2/day	NASA web site
Wind speed	1.77m/sec	NASA web site

Energy Demand Assessment and Load Scheduling of the Village

Energy is the back bone of the economy in order to enhance the life standard of the society. Without at least for basic consumption life is meaningless for one society. So, it should be national issue to improve the life of the society by applying multi dimension energy sources to the rural areas in order to fulfil basic energy demand. Home consumption includes lighting, cooking, TV, refrigerator and charging mobiles also needs electricity. In the village under consideration, energy is required for lighting, charging electronic equipment's, and clinical equipment's, water pumping and agricultural service includes lighting & pumps for small irrigation activities and some milk processing.

Even though the primary consumption of the village is limited to the above-mentioned activities, by considering future expansion of the village with the potential activities the village can perform, we also include the load demand for small-scale industrial loads, like milk processing. The total load of the village is composed of the household devices such as bulbs, stoves, water heater, TVs and radios and refrigerators. It is assumed that the peoples of the village are categories in to small and medium families based on their income and living standard. The possible power consumption for the categories of the village is shown in next tables. This table shows estimation of power consumption by each house hold, school, health center, churches, agricultural and in a single day.

As information gathered from village administration and Jeldu district social affairs office, based on their income and living standard, there are 68 medium income households and 96 low income households out of the total 164 households of the village.

Scheduling of Electrical load

- Based on the status of the people and their need to use electricity if available, we schedule the consumption of power for the two categories as follows.
- For low income families, the estimated load is three 11W (CFL) Compact Fluorescent Lamps one for outdoor and two for indoor lighting, a 15W radio receiver operated for 6 hours per day, 70W rated TV working for 8 hours per day and 40W for other purposes for an average of 4 hours per day.
- For medium income households, four 11W Compact Fluorescent Lamps for each family, one for outdoor and two for indoor lighting, 30W radio receiver, 100W TV and 200W for injera baking stove was also considered for this category.
- Other service giving institutions: - primary school has used eight 15W Compact Fluorescent Lamps for lighting, 100W for electronic usage and 50W for heater. Similarly, for the health center four 15W lamps are considered for lighting, 60W rated vaccine refrigerator, 200W normal refrigerator and 1000W water heater for sterilizing medical equipment, 15W microscope is included and 100W for electronic usage. The church uses ten 11W CFL for lighting, and four 15W for megaphone purpose.
- Two milk processing machine rated capacity of 4KW, one flour mills with rated capacity of 7.5kW with milling capacity and three water pumps of rated capacity of 2kW assumed.

Since all of the village population is farmers, they spent much of their time on agricultural activities like farming and cattle protection. As a result, energy consumption of the village is not constant. Service giving institution is also varying their use based on calendar. The electrical load demands of the village are summarized in the following table 3 to 8.

Table 3. Daily Load Demand of low-income families

No	Used Device	Unit Power(W)	Qty	Total Load (KW)	Total Operating time	Total KWh/Day
1	CFL	30	3	3.2	8	19.2
2	TV	70	1	6.7	8	53.6
3	Radio	30	1	2.9	6	17.4
4	Others	40	1	3.8	4	15.2
Sub Total				16.6		105.4

Table 4. Daily Load Demand of middle-income families

No	Used Device	Unit Power(W)	Qty	Total Load (KW)	Total Operating time	Total KWh/Day
1	CFL	11	4	3.0	6	18.0
2	TV	100	1	6.8	4	27.2
3	Radio	30	1	2.0	6	12.0
4	Stove for cooking	1000	1	68.0	2	136.0
5	Stove for Injera	3000	1	204.0	1	204.0
6	Others	60	1	3.8	4	15.2
Sub Total				279.7		412.4

Table 5. Daily Load Demand for Health Centre

No.	Used Device	Unit Power(W)	Qty	Total Load (KW)	Total Operating time	Total KWh/Day
1	CFL	15	4	0.06	6	0.36
2	Electronics and/or TV	100	1	0.1	4	0.4
3	Refrigerator type 1.	60	1	0.06	24	1.44
4	Refrigerator type 2.	80	1	0.08	12	0.96
5	Heater	200	1	0.2	18	3.6
6	Microscope	15	2	0.03	8	0.24
6	Others	40	1	0.04	8	0.32
Sub Total				0.57		7.32

Table 6. Daily Load Demand of School

No.	Used Device	Unit Power(W)	Qty	Total Load (KW)	Total Operating time	Total KWh/Day
1	CFL	15	8	0.12	8	0.96
2	Electronic usage (TV)	100	1	0.1	8	0.8
3	Radio	30	1	0.03	6	0.18
4	Heater	50	1	0.05	4	0.2
5	Others	40	1	0.04	4	0.16
Sub Total				0.34		2.3

Table 7. Daily Load Demand of Church

No.	Used Device	Unit Power(W)	Qty	Total Load (KW)	Total Operating time (Hrs)	Total KWh/Day
1	CFL	11	10	0.11	8	0.88
2	Megaphone	15	4	0.06	3	0.18
5	Others	20	1	0.02	2	0.04
Sub Total				0.19		1.1

Table 8. Daily Load Demand of small Industry

No.	Used Device	Unit Power(W)	Qty	Total Load (KW)	Total Operating time	Total KWh/Day
1	Milk processing Machine	4000	2	8	4	32
2	Flour mill	7500	1	7.5	8	60
3	Water Pumping	2000	4	8	8	64
4	Others	50	1	0.05	4	0.02
Sub Total				23.55		156.02
Grand Total				338.95		589.64

Future Power Projection

Activity of the society is not static. In addition to the government plan to increase the life standard of the society, peoples are also straggling with nature to win and improve their life. The same thing is true in Kabi Village of Jaldu district. Some people are opening shop, some are trying to invest on cafeteria, animal fattening, poultry production. Construction of a private mill is also under process. Based on this activity the researcher forced to predict the load demand of the village. Prediction of the future condition so called forecasting is very important in decision making. Considering futurity of the village and governments policy is essential for supplying power demand of the society for future.

For this study, Load, projection for the village is considered from the country average load profile and factual of the village. The electricity demand forecast depends on both the peak demand and average demand throughout the year under consideration. According to EEPCO the Target Scenario electricity, demand of the country will be expected to grow by 38% for the period 2020(EEPCO). From this Forecasting Estimation we assume the village Energy Demand also will increases by 30%. Fewer amounts is assumed due to the national estimation contains both urban and rural areas and the fact that the need for rural area is less than the urban areas. As a result, the daily average power dements, 766.6KWh/day is taken as the power demand of the village under this research work. The monthly power demand of the village becomes 23,298KWh/month.

MODELLING OF HYBRID OF WIND AND SOLAR ENERGY SYSTEM

In order to get the maximum desired load from the hybrid source under consideration for certain particular area, all the power producing sources should be designed to its maximum power output. This systematic mathematical manipulation is known as modelling. It is a physical representation that shows what the system looks like or how it works. Once the system is properly modelled by incorporating all the necessary variables and parameters that really shows the physical representation of the system, far distance of optimization of Hybrids of Renewable Energy System (HRES) has been done. Here the system consists of wind turbine, solar panel, battery bank, bidirectional converter. The primary power sources are wind turbine and solar panel through the bidirectional converter and battery is used as storage to balance the produced power with the demand.

The following diagram shows the model of hybrids of wind, solar and battery with bidirectional converters connected to AC and DC BUS

Figure 5. System component of Hybrid of Wind and Solar Renewable Energy System

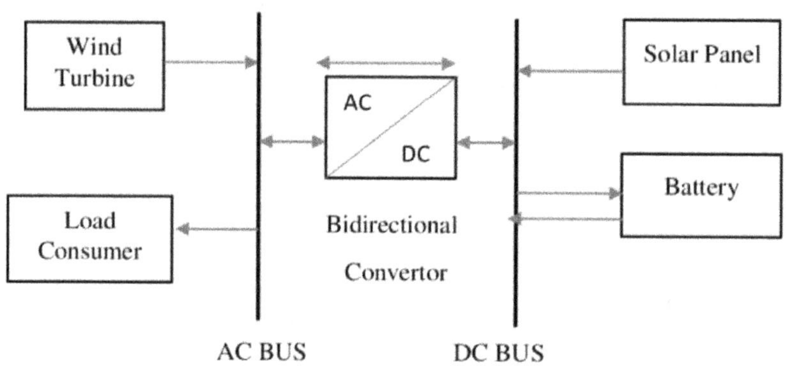

There are a number of techniques in literature to model Hybrids of Wind and Solar Renewable Energy System components. For this case study, the researchers have included three principal power generator subsystems: - these are the solar panel generator, the wind turbine and the battery storage as shown in figure 5. (Eltamaly & Mohamed, 2014).

Based on the power generating potential of Jeldu district, it was assumed that the power generated from each renewable source is constant during one-hour duration when the model was done. The mathematical modelling of each component of the proposed hybrid wind and solar energy with battery system in order to analyze the system performance is discussed in this section.

Modelling Wind Generator

Wind is more or less constant for Jeldu district due to its geographical location. It was high potential throughout the year. The electric power output from wind turbine at a particular place under study can be affected by air density of the place, wind speed at hub height and radius of the rotor of the turbine.

Adjusting the height of the wind turbine has a valuable effect on output energy of the whole system, the adjustment of the wind profile for height at the time of installation can be taken into account by using a height adjustment equation. Wind speed at certain height can be calculated by using power law, as shown in equation below (Ilinca, McCarthy, Chaumel, & Retiveau, 2003).

$$V = V_i \left(\frac{h}{h_i} \right)^{\alpha} \tag{1}$$

Where, V is the wind speed at hub height h, V_i is the wind speed measured at the reference height h_i and α is the power law exponent who varies with the elevation, the time of day, the season, the wind speed and the temperature. the typical value 1/7 corresponding to low roughness surfaces and well exposed sites, is used in this study.

After height adjustment, the Modell of power output information from the wind turbine in order to maximize as possible. Different scholars use different types of their own models to maximize the desired energy output of different components. Some of them are (Chedid, Akiki, & Rahman, 1998; Eftichios, Dionissia, Antonis, & Kostas, 2006; Lysen, 1983) assume that the turbine power curve has a linear, quadratic or cubic form. (Borowy & Salameh, 1996; Borowy & Salameh, 1996) are model the component by taking Weibull parameters in to account. Others like; (Bueno & Carta, 2005) approximate the power curve with a piecewise linear function with a few nodes.

The power output obtained from each wind turbine is predicted by using the following equation in terms of the wind speed (Javadi, Mazlumi, & Jalilvand, 2011).

$$P = \frac{\pi}{2} \mu \rho r^2 v^3 \tag{2}$$

Where, μ is the efficiency factor which tells part of the wind blowing through the area spanned by the rotor blades is converted into electric energy. ρ depends on air pressure, temperature and humidity. 1.2 is a good average value at sea level. r is radius of a rotor blade and v is speed of the wind at a given time.

Modelling Solar Generator

Solar radiation is also the richest potential for Jeldu district among some possible sources of renewable energy. Thus, proper modelling for operation and the performance of a PV generator to its maximum power output plays great role to supply the desired power of the village. The output power from PV energy subsystem is depend upon the manufacturers data of the PV modules the solar radiation on tilted surface, orientation of the PV array against the movement of the sun, and the ambient temperature at the given time as follows (Etamaly, Mohamed, & Alo-lah, 2015).

$$P = RA\sigma(t) \tag{3}$$

Where, R is the solar radiation on the tilted surface, A is the total cell area and $\sigma(t)$ is the solar cell efficiency at a given time t.

Modelling of Battery Storage

Modelling the capability of battery in the hybrid system is very essential to continuously balance the demand and supply power by charging and discharging. It is used as temporary power storage when there is excess power and as back up when there is a shortage of power. Since the energy output from the PV cells and the wind turbines are vary and time dependent, proper battery sizing to maintain the load demand is important. The maximum efficiency of the state of charge of battery at any time, t is depending up on the past state of charge and to the energy production and load demand situation of the system during the time t-1. The case of over charge of the battery is occur when excess amount of power is generated by the hybrid system or demand of the load is low at a time. At the time of the state of charge of the battery reaches its maximum value $C_{(batt,\ max)}$, the control system intervenes and stop the charging process where as when it reaches its minimum level $C_{(batt,\ min)}$ the control system disconnects the load to prevent batteries against shortening their life or being distracted (Diaf et al.,2007; Borowy & Salameh, 1996).

At the time of the total output power of the hybrid system is greater than that of the desired load power, the battery is in the state of charging, and the charged quantity of the battery at the moment of t is expressed by (Diaf et al.,2007; Borowy & Salameh, 1996).

$$C_{batt}(t)=C_{batt}(t-1)(1-\sigma)+\left(E_{PV}(t)+E_{WT}(t)-\frac{E_L(t)}{\eta_{inv}}\right) \tag{4}$$

On the other hand, when the desired load is greater than the total energy generated, the battery bank is in discharging state to satisfy the demand. Therefore, the available battery bank capacity at hour t can be expressed as

$$C_{batt}(t)=C_{batt}(t-1)(1-\sigma)+\left(\frac{E_L(t)}{\eta_{inv}}-\ E_{PV}(t)+E_{WT}(t)\right) \tag{5}$$

where, $C_{batt}(t)$ and $C_{batt}(t\text{-}1)$ are the available battery bank capacity in watt hour at hour t and t-1, respectively; $\eta_{bat\,t}$ is the battery efficiency (during discharging process, the battery discharging efficiency was set equal to 1 and during charging, the efficiency is 0.65 - 0.85 depending on the charging current) (Bin et al., 2003). σ is the self-discharge rate of the battery bank. $E_{PV}(t)$ and $E_{WT}(t)$ are the energy generated by PV and wind generators, respectively; $E_L(t)$ is the load demand at hour t and η_{inv} is the inverter efficiency which may considered as constant.

Modelling of System Reliability

In modelling the hybrid system, along with its economic aspects, various technical analysis approaches are used in literature to optimizing hybrid renewable energy systems. Among these employed techniques, the trade-off method employed by (Elhadidy & Shaahid, 1999), the least square method used by (Kellog et al., 1998; Borowy & Salameh, 1996) loss of power supply probability (LPSP) applied by (Lu, Yang, & Burnett, 2002; Kellog et al., 1998). Here, the researchers apply the technical sizing model for the hybrid of wind and solar renewable energy system is developed according to the concept of LPSP to evaluate the reliability of the system. The methodology used can be summarized in the following steps (Diaf et al., 2007).

The total power, $P_{tot}(t)$, generated by the wind turbine and PV generator at hour t is calculated as follows.

$$P_{tot}(t) = P_{PV}(t) + P_{WT}(t) \tag{6}$$

Then, the inverter total input power, $P_{inv}(t)$, is calculated using the corresponding load power requirement as follows.

$$P_{inv}(t) = \frac{P_{load}(t)}{\eta_{inv}} \tag{7}$$

where, $P_{load}(t)$ is the power consumed by the load at hour t, η_{inv} is efficiency of the inverter. It is to be understood that, the desired load demand at any time t $P_{load}(t)$, may or may not be satisfied according to the corresponding values of the total generated power $P_{tot}(t)$ and $SOC(t)$ at that hour. Proper management of the load is also important. some energy management conditions of the PV/Wind hybrid system are summarized as follows:

- If [$P_{tot}(t) \geq P_{inv}(t)$] and [$SOC(t-1) \leq SOC_{max}$], then, charge the battery with excess power $P_{Batt} = (P_{tot}(t) - P_{inv}(t)) * \eta_{Batt}$. Next, check if $SOC(t) > SOC_{max}$. Then, stop the battery charging.
- If [$P_{tot}(t) > P_{inv}(t)$] and [$SOC(t-1) > SOC_{max}$], then satisfy the load, Stop charging the battery and Dump the surplus power $P_{dump} = (P_{tot}(t) - P_{inv}(t))$
- If [$P_{tot}(t) = P_{inv}(t)$], then satisfy the load only. No extra process of charging /discharging or dumping power is here.
- If $P_{tot}(t) < P_{inv}(t)$] and $DOP(t-1) < DOP_{max}$, then satisfy the load by dis- charging the battery to overcome the load deficit and check if $DOP(t) > DOP_{max}$, then stop battery discharge and set $DOP(t) = DOP_{max}(t)$. Here may be less amount of subordinate power generating is needed.
- If $P_{tot}(t) < P_{inv}(t)$] and $DOP(t-1) > DOP_{max}$, then stop the battery discharge and Set deficit $P_{Def} = P_{tot}(t) < P_{inv}(t)$. Here more power generating support will be needed.

In addition to the output power management, knowing the load that is not supplied by the system and not used by the system are essential to either increase efficiency of the system to maximize the power output and/or to design efficiently usage of the power.

The wasted energy (WE), defined as the energy produced and not used by the system, for hour t is calculated as follows (Diaf et al., 2007).

$$WE(t) = P_{tot}(t)\sigma t - \left(\frac{P_{Load}(t)}{\eta_{inv}} \Delta t + \left(\frac{C_{batt,\max} - C_{batt}(t-1)}{\eta_{cha}} \right) \right) \tag{8}$$

The loss of power supply probability (LPSP), for a given period of time T is defined as the ratio of all the (LPS(t)) values over the total load required during that period. It was considered as technical implemented criteria for sizing a hybrid PV/wind system employing a battery bank. Here, some reliability indices are applied to evaluate capacity of the hybrid system sizing developed to supply the load demand. These are the LPSP technique, which is defined as the probability that an insufficient energy results when the hybrid system is unable to supply the load, meaning the system is reliable when it is able to supply enough power to the electrical load during a certain period. The energy not supplied (ENS) to the system which is considered when generated power is less than the demanded power and the reliability of power supply (RPS) when there is a power loss. These technical analyses can be obtained by using the following formulas (Diaf et al, 2007; Bin et al., 2003).

$$LPSP = \frac{\sum_{t=1}^{T} LPS(t)}{\sum_{t=1}^{T} P_{Load}(t)\Delta t} \tag{9}$$

Where, T is the operation time (in this study, T= 1 year).

When there is energy deficit in the system, the Loss of power supply (LPS) of the hybrid system should take in to consideration to satisfy the desired load and can be calculated as:

$$LPS(t) = P_{Dem}(t) - P_{Gen}(t) \tag{10}$$

Another important issue when power generation and management are taken in to consideration is that, the energy not supplied (ENS) by the hybrid system. This energy can be calculated from power difference as given below:

$$ENS = \sum \left(P_{Dem} - P_{Gen} \right) \tag{11}$$

Reliability is related to the total amount of power that a renewable energy source can produce. The Reliability of power supply (RPS) certain hybrid system can be found in terms of Loss of power supply probability (LPSP) as follows

$$RPS = 1 - LPSP \tag{12}$$

OPTIMIZATION FORMULATION

Optimization is the process of finding the best value among the given alternatives. This value can be global optimal or the value which better for that time. Getting global optimal value by using nature inspired algorithms needs experience of the researcher in designing the system, controlling all the variables and parameters that can disturb the value. After designing, developing the fitness function to be optimized is also important.

Here, in order to optimal sizing the hybrid of renewable energy system researchers develop the proper fitness function which includes all the components and possible assumptions which affect the desired result directly or indirectly. The main Objective of the project here is to minimize the total annual cost

(f_{TAC}) of the system which satisfy the load demand through predetermined constraints. For this problem the total annual cost is taken as the sum of initial capital cost and annual maintenance cost. Thus, the problem to be minimized will be stated as follows: (C_{mnt}) (Geleta & Manshahia, 2018; Zong, 2012; Yang, Zhou, & Lou, 2008).

$$\min f_{TAC} = C_{ICC} + C_{mnt} \tag{13}$$

Maintenance cost C_{mnt} of the system occurs during the whole project life time while capital cost C_{ICC} occurs at the beginning of the project time. In order to compare these two different costs which have different time of implementation, the initial capital cost has to converted to annual capital cost by the capital recovery factor (*CRF*) can be defined as (Hadidian, Arabi, & Bigdeli, 2016).

$$CRF = \frac{i(i+1)^n}{(i+1)^n - 1} \tag{14}$$

Where, i the interest rate occurs during the project life time and n denote the life span of the system.

The fitness function defined above involves many components and variables. Thus, the initial capital cost and maintenance cost have contained these all components and redefined. Now the considered total initial capital cost of the system can be broken in the annual costs of the wind turbine, solar panel, batteries and backup generator involved in the system and will be given as follows:

$$C_{ICC} = \frac{i(i+1)^n}{(i+1)^n - 1}\left[N_{pv}C_{pv} + N_{WT}C_{WT} + \left(\frac{n}{LS_{Batt}}\right)N_{Batt}C_{Batt} + C_{Backup}\right] \tag{15}$$

Where, LS_{Batt} is batteries life span, N_{pv},N_{WT} and N_{Batt} are numbers of PV panels, wind turbine and batteries respectively, C_{pv},C_{WT},C_{Batt} and C_{Backup} are unit costs of each components respectively The unit costs mentioned above are taken as unit costs of the material and its installation fee. This cost is calculated for both solar panel and wind turbine as shown on equation (16) next.

$$C_{pv} = C_{pv,unit} + C_{inst,unit} \qquad C_{WT} = C_{WT,unit} + C_{inst,unit} \tag{16}$$

Batteries are important component to store the generated power. The life span of battery is relatively short when compared with the life span of the project. The number of batteries N_{Batt} needed for the project, depends on the number of photovoltaic panel and number of wind turbines and determined by the following function

$$N_{Batt}\left(N_{pv}, N_{WT}\right) = Roundup\left(\frac{S_{Req}}{\eta S_{Batt}}\right) \tag{17}$$

Where, Roundup (.) is a function returns rounded an integral output; S_{Req} is the amount of needed storage capacity; η is the rated capacity usage in % and S_{Batt} is rated capacity of each battery.

Similar to the number of batteries, the required storage capacity S_{Req} which is defined as the number of solar panels and wind turbines in the hybrid system can be obtained by using energy curve ΔW defined as:

$$\Delta W = W_{Gen} - W_{Dem} = \int \Delta P dt = \int \left(P_{Gen} - P_{Dem} \right) dt \tag{18}$$

Where, W_{Gen} and P_{Gen} are the total energy and power generated respectively and W_{Dem} and P_{Dem} are their respective demand values. Thus, the required storage capacity S_{Req} defined as the number of solar panels and wind turbines is given by:

$$S_{\text{Req}}\left(N_{pv}, N_{WT}\right) = \sum_{t=1}^{Max\ t}\left(P_{PV}^{t} + P_{WT}^{t} - P_{Dem}^{t}\right)\Delta t - \sum_{t=1}^{Min\ t}\left(P_{PV}^{t} + P_{WT}^{t} - P_{Dem}^{t}\right)\Delta t \tag{19}$$

Where, Max t is the maximum time when total energy (kwh) is highest; Min t is the minimum time when total energy (kwh) is lowest. Δt is unit time under consideration (1hr) here. P_{PV}^{t} and P_{WT}^{t} are the powers generated by solar panel and wind turbine at time t respectively; and P_{Dem}^{t} the total power demand at time t. The total power generated by the components at time t is given by:

$$P_{PV}^{t} = N_{PV} \times P_{PV,Each\ unit}^{t} \qquad P_{WT}^{t} = N_{WT} \times P_{WT,Each\ unit}^{t} \tag{20}$$

Where, P_{PV}^{t} and P_{WT}^{t} are the total powers generated by the wind turbines and solar panels, whereas $P_{PV,Each\ unit}^{t}$ and $P_{WT,Each\ unit}^{t}$ are the power generated from each respective component at a time t.

The annual maintenance cost of the system is also constituted from the operation and maintenance costs of each of the components. This cost is active throughout the project and can be calculated by the following equation.

$$C_{Maint} = \left(C_{PV,Maint} \times \sum_{t=1}^{24} P_{PV}^{t} \Delta t + C_{WT,Maint} \times \sum_{t=1}^{24} P_{WT}^{t} \Delta t \right) \times 365 \tag{21}$$

Constraints

The fitness function defined by equations (13-21) for the optimization of hybrid system of wind and solar renewable energy by unit sizing the system components. In order to meet the desired load by possible minimum total annual cost, there should be conditions and assumptions taken in to consideration. Here the following categories of conditions have to be fulfilled when the process of optimization is done.

1. **Sign constraints**

For installation of the project, there should be some amount of the components which is not greater than the maximum component. This was given by the following equations.

$$N_{PV} \in \mathbb{Z}, N_{PV} \geq 0 \text{ and } N_{PV} \leq N_{PV,max}$$

$$N_{WT} \in \mathbb{Z}, N_{WT} \geq 0 \text{ and } N_{WT} \leq N_{WT,max} \tag{22}$$

2. Power Generated Constraint

The total transferred power from PV and WT to the battery bank is calculated using the following Equation

$$P_{Total}^{k}(t) = N_{PV} P_{PV}^{k}(t) + N_{WG} P_{WG}^{k}(t) \tag{23}$$

$1 \leq k \leq 365$, $1 \leq t \leq 24$

3. Battery Constraint

When the activity of operation and performance of hybrid system is made, proper attention must be given to the life span of battery to satisfy the energy demand and supply of the whole system. In order to protect it from damage, battery should not be over discharged or overcharged. This means that the battery SOC at any hour t must be subject to the following constraint:

$$(1 - POD_{\max}) \leq SOC \leq SOC_{\max} \tag{24}$$

Where, $POD_{\max}$ and $SOC_{\max}$ are the battery maximum permissible depth of discharge and SOC, respectively.

4. **System Reliability Constraint**

As far as the reliability of the system is considered as constraint, the loss of power supply probability (LPSP) should be less than the maximum designed LPSP by the user.

$$LPSP \leq LPSP_{\max} \tag{25}$$

NATURE INSPIRED ALGORITHMS

All activities of human being were in ocean of nature. Due to its wonderful ability, nature plays great role in showing natural way how peoples solve the problems they face. Nature gives certain logical and effective ways to find solution to these real-world problems. Since the algorithms are taken by following the principles in nature, they are named as "Nature Inspired Algorithms" in solving optimization problems (Vasant, Zelinka & Weber, 2018, 2019).

There are different categories of algorithms which mimics the principles in nature. The most predominant and successful classes of nature inspired algorithms are Evolutionary Algorithms, Swarm based Algorithms and Ecological based algorithms. Their inspirations were from natural evolution, collective behavior in animals and the phenomena observed in the environment respectively (Geleta & Manshahia, 2017). These categories of nature inspired algorithms are classified in to different sub groups to give timely good solutions of different optimization problems (Ganesan, Vasant & Elamvazuthi, 2016; Vasant, Marmolejo, Litvinchev & Aguilar, 2019; Zelinka, Tomaszek, Vasant, Dao, & Hoang, 2018)

Evolutionary algorithm is a part of Evolutionary computation which is a sub field of

nature inspired algorithms, which mimics the biological evolution of Charles Darwin theory of Natural Evolution. Over many stages of life, biological organisms develop according to the principles of natural selection like "Survival of The Fittest" to attain some outstanding accomplishments. In such algorithm only the fittest candidates are selected. The family groups of Evolutionary algorithms are Genetic Algorithm (GA), Genetic Programming (GP), Differential Evolution (DE) and evolutionary strategy (ES).

Swarm intelligence (SI) is the characteristics of collective, emerging behavior of multiple, interacting agents who follow some simple rules which can accomplish very trajectory work in group. While each agent may be considered as unintelligent which cannot do nothing, the whole system of multiple agents may show some self-organization behavior and thus can behave like some sort of collective intelligence. Some sub groups of this category are Ant colony optimization (ACO), Grey Wolf Optimization (GWO), Firefly Algorithm (FA), Particle Swarm Optimization (PSO), Bat Algorithms (BA), Artificial Bee Colony Optimization (ABC), Fish swarming Algorithm (FSA), and Cuckoo Search (CS) can be mentioned (Geleta & Manshahia, 2017).

Ecology Based Computation algorithms are taken its inspiration from the processes of natural ecological phenomena. These group of algorithms imitate different principles of natural phenomena occurred in our environment to solve the problem in its better way. Some families of this category are Water Wave Optimization (WWO), Gravitational Search Algorithm (GSA), Flower Pollination Algorithm (FPA) and Teaching- Learning Based Algorithm (TLBO) (Geleta & Manshahia, 2017).

In this study, authors have implemented Grey wolf optimization on Ethiopian data for optimal sizing of hybrid renewable systems. The concept of Grey wolf optimization (GWO) first introduced by (Mirjalili, Seyed M, & Andrew, 2014) in 2014, which is recently developed population based meta-heuristics technique that takes its inspiration from the hunting behaviour and social hierarchy in leadership mechanism of grey wolves in nature. Nowadays it was becoming popular optimizing different real-world problems.

There are two interesting phenomena in the behaviour of grey wolves: Their social hierarchy and hunting mechanism. Mostly, grey wolves prefer to live in a pack of average 5-12 wolves with democratic and very strict social hierarchy. The most powerful leader of the pack is called alphas wolf. Since all the commands are given by alpha wolf, it is very predominant wolf in the pack on making decisions about hunting, sleeping place, sleeping and wakeup time, feeding, guarding and so on (Eid, M., & Grosan., 2018; Hadidian, Arabi, & Bigdeli, 2016). Very interesting nature gifted behaviour of grey wolves is that the alpha is not necessarily the strongest member of the pack but the best in terms of managing the pack. This shows that then organization and discipline of a pack is much more important than its strength (Geleta & Manshahia, 2017).

The second social hierarchy of wolves is called betas and they are assistant wolves as they help the alpha in making decisions and other respective pack activities. It was probably the best candidate of the group to be the alpha in case one of the alpha wolves passes away. The beta wolf should respect the alphas decision, reinforce commands to the other lower-level and mediate the pack with the leader.

(Sharmistha, Bhattacharje, & Bhattacharya, 2016; Eid, M., & Grosan., 2018). The lowest ranking grey wolves of the hierarchy are Omegas and they have to submit to all other dominant wolves. Commonly they had the role of scapegoat and the last wolves that allowed eating the prey (Mirjalili et al., 2014).

Delta wolves come in the hierarchy next to the alphas and betas but they lead the omega nearby. They submitted themselves for alphas and betas and highly dominated the omega wolves in the pack. There was responsible keeping their boundary and warning the pack in case of any danger (Mirjalili et al., 2014; Hadidian, Arabi, & Bigdeli, 2016). The second very important in the grey wolves is that, their group hunting mechanism which includes tracking, chasing, encircling and harassing the prey until they stop moving.

INPUT DATASETS

The total annual cost of power demand for Kabi village of Jeldu district was going to be optimized under the given set of assumptions. The developed fitness function given in (10-20) above has to evaluated by the employed algorithm GWO through MATLAB code. The code was developed for hybrid solar and wind renewable energy system and evaluated by the algorithm in order to get the optimal value

Based on the assessment made by the researcher, the energy demand of the village is not constant throughout the year. There are seasons in which power is much needed, like January to April, due to most weddings culturally preferred by these months. This time is after crops harvesting and farmers have time to relax. Schools and religious institutions are also active at this period. On the other hand, months from June to October are difficult period for that society. At these months there is heavy rain, pick agricultural activities and no school time. As a result, less power is consumed.

The researchers estimate the monthly load demand of the village from calculated monthly demand by taking $\pm 1\%$ on each successive month based on the real situation observed village.

Kabi village of Jeldu district is with latitude 9.1° North, longitude 37.4° East at an elevation of 2360 meters above sea level. Based on these location data the average monthly wind speed (m/s) at the height of 10 meters and solar radiation (KWh/ m^2/day) was taken from National Aeronautics and Space Administration (NASA) web site. As shown on the table 8., there are high potential of wind and solar radiation in the district as general and in the village of Kabi in particular. These shows that there is a promising to satisfy the load demand of the village by the off-grid power source

Moreover table 7. Shows come materials to be purchased for installation of the project and total fees are also included. As we have (Geleta & Manshahia, 2018) considered the data for optimization of hybrids of wind and solar renewable energy system in the previous study, here also taken for this study. The data of table 7 was improved by factors of inflation rate 3.31%, of 2018. So, all the numerical data in this paper is similar to the undated data on (Geleta & Manshahia, 2018).

Table 9 shows the valves of the updated decision variables for the test of the system and initialize the employed nature inspired algorith. The estimated load demand, wind speed and solar radiation of the village was summarized in table 8. Because of life span of each battery is taken as 4 years, 5 times installations are needed during the whole life span of the system.

The data shown in table 10 was organized by the researchers. The need demand of the village was organized as shown on tables 1-8 above. Regarding wind power and solar power, first the researches take monthly average of wind speed and solar radiation of the place at about 9. 05° to 9.25° North latitude and 37 .40° to 38.11° East longitude from NASA website and convert it to electrical power by

Table 9. Design variables used for Solar and Wind Hybrid System

Variables	values
Annual interest (i)	6%
Life span of the system(n)	20 years
Solar panel price	$350/panel
Solar panel installation fee	50% of the price
Wind turbine price	$20000/Turbine
Wind turbine installation fee	25% of the price
Unit cost of the battery	$170
Cost of backup generator	$2000
Usage % of battery rated capacity (η)	80%
Batteries rated capacity	2.1Kwh
Batteries life span	4 years
Unit time (Δt)	1hr
Maintenance cost of PV array	0.5cents/Kwh
Maintenance cost of wind turbine	2cents/Kwh

using equations 2 and 3 stated in modelling part. The calculation was done by the help of excel sheet by controlling constants mentioned in the equations. Here the data specified on table 11. is used for the optimization of the hybrid system as monthly power demand and the power we can monthly generate from wind and solar of Kabi village.

Table 9 shows the valves of the updated decision variables for the test of the system and initialize the employed nature inspired algorithm. The estimated load demand, wind speed and solar radiation of the village was summarized in table 8. Because of life span of each battery is taken as 4 years, 5 times installations are needed during the whole life span of the system.

The data shown in table 10 was organized by the researchers. The need demand of the village was organized as shown on tables 1-8 above. Regarding wind power and solar power, first the researches take monthly average of wind speed and solar radiation of the place at about 9. 05° to 9.25° North latitude and 37 .40° to 38.11° East longitude from NASA website and convert it to electrical power by using equations 2 and 3 stated in modelling part. The calculation was done by the help of excel sheet by controlling constants mentioned in the equations. Here the data specified on table 11. is used for the optimization of the hybrid system as monthly power demand and the power we can monthly generate from wind and solar of Kabi village. Figure 6 and Figure 7 show Monthly Wind Potential of Jeldu district and Solar Radiation Potential of Jeldu district respectively.

From figure 8, one can easily identify that the pick power demand of the village is at the month April. This is due to most of ceremonies of certain cultural and religious festivals of the district in general are held in this month. On the contrary, months from August to mid of November are the time that some peoples of the village lack food. Crop harvest of the place was starts from mid of November, which in turn increase the usage of power.

Table 10. Wind speed and Solar radiation Data from NASA

No.	Month	Estimated Demand (KWh/ Month)	Average wind speed (m/sec)	Solar Radiation (KWh /m2/day)
1	January	23,298	1.65	7.57
2	February	23,530	1.90	8.25
3	March	23,764	2.17	8.24
4	April	23,996	1.81	8.67
5	May	23,056	1.72	8.67
6	June	22,832	1.59	8.72
7	July	22,599	1.84	8.69
8	August	22,366	1.85	8.68
9	September	22,832	1.61	8.24
10	October	22,065	1.77	8.13
11	November	23,764	1.82	7.92
12	December	23,530	1.55	7.66
Monthly Average		23136	1.77	8.29

Table 11. Average of monthly energy demand and powers of Kabi village

No.	Months	Power Demand (KWh/ Month)	Wind power (KW/ Month)	Solar Power (KW/Month)
1	January	23,298	0.34	0.77
2	February	23,530	0.52	0.84
3	March	23,764	0.77	0.84
4	April	23,996	0.45	0.88
5	May	23,056	0.38	0.88
6	June	22,832	0.30	0.89
7	July	22,599	0.47	0.89
8	August	22,366	0.48	0.89
9	September	22,832	0.32	0.84
10	October	22,065	0.42	0.83
11	November	23,764	0.46	0.81
12	December	23,530	0.23	0.78

The monthly wind power of the village organized in table 9. is illustrated on figure 9. As this figure shows the wind power of the village in not constant through the year to produce electricity for the village. Months like January, February and March are dry and windy season for Jeldu district. As a result, maximum power can be extracted with in these months. In the rest months, since the season is rainy, the wind is relatively less.

Figure 6. Monthly Wind Potential of Jeldu district

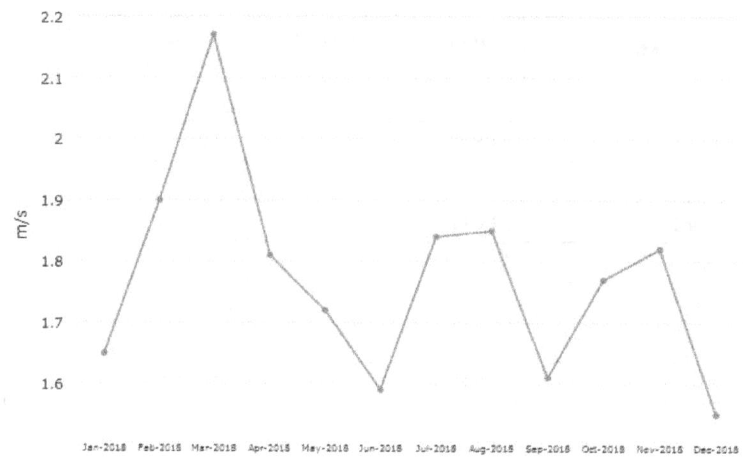

Figure 7. Solar Radiation Potential of Jeldu district

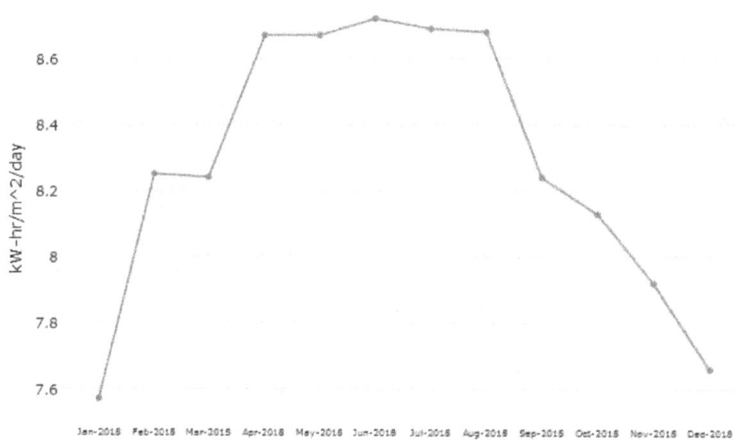

Figure 8. Monthly Power Demand of Kabi Village

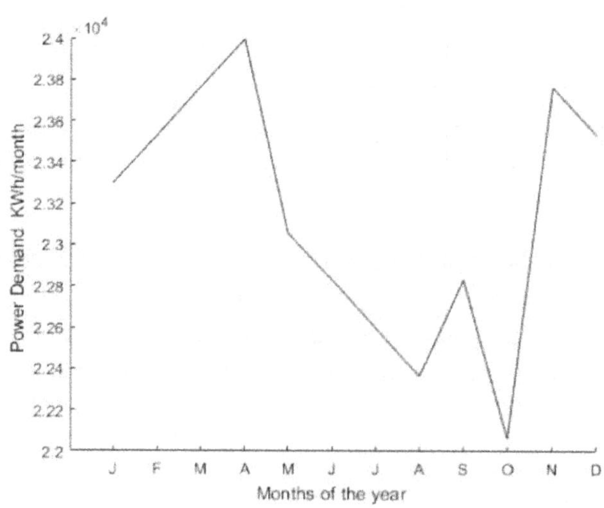

Figure 9. Monthly Wind Power of Kabi Village

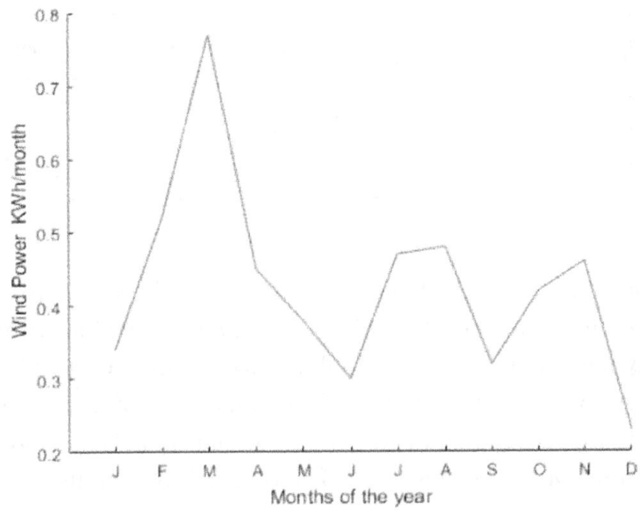

Figure 10. Monthly Solar Power of Kabi Village

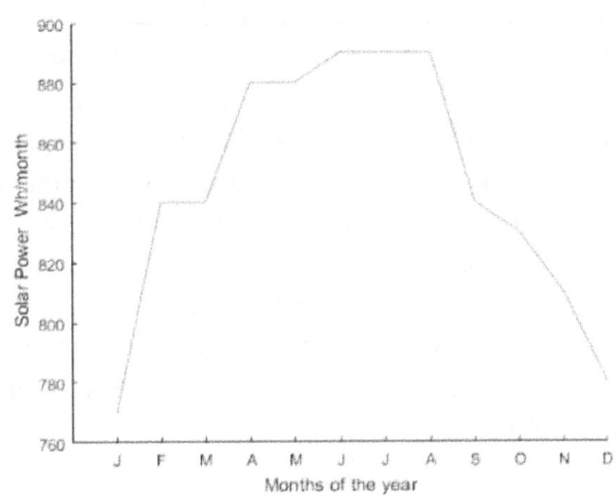

The monthly solar power enegy which can be eaxtracted was illustrated on figure 10. This show as if properly designed, the power demand of the village is satisfyed either from the wind alon, or solar alon or the combinations of the two power generating sources.

DISCUSSION

To minimize the total annual cost of Hybrid Wind and Solar Renewable Energy System under the determined constraints, the algorithm was run for the fitness function by the input data give in the tables 9 and 11. After simulation was done the results are shown in tables 12. Optimization of the proposed system is done with the objective of minimizing the annual cost of the system which can balance the desired load by satisfying all the constraint from equations 13-25 specified under section 4.

Table 12. Optimization Results of wind-solar-battery system

No of panels	No of Turbines	No of Batteries	LPSP	Reliability	Total Cost ($)	annual
161	0	16	0	0.998	8936.4	
74	1	11	0	1.002	6827.9	
0	2	9	0	1.083	5572.0	

From the results given in table 12, the first row of the result is when only solar panel was used to satisfy the load demand of the village. For this solar panel and battery alone project 161 solar panels and 16 batteries are needed to generate power for the village with annual cost of $8936.4 as optimal value.

At the second row the algorithm selects the combinations of solar panels and wind turbine to generate the load to fulfill the Kabi village's demand. Here 74 solar panels 1 wind turbine and 11 batteries are needed for the generations of the desired load. The optimal total annual cost for this hybrid system is $6827.9

The result of third row shown only when wind turbine with battery was employed to satisfy the desired load of the particular place. For this combination 2 wind turbines and 9 batteries have to employed to accomplish the aim of the project with total annual cost of $5572.0

As shown on the table reliability of the system and loss of power supply probability of the system was tested. The value of LPSP and reliability of the system in all the three rows are zero and almost one respectively. This shows based on the interest of the society and the capital needed for the project, all the three combinations can generate the desired load to satisfy their demand.

The optimal solution of the hybrid system is found when N_{PV}=0, N_{WT}=2, and N_{Batt}=9 with the optimal cost of $5572.0 meaning when only wind panel was used as power generating mechanism. But, as shown on Fig 6. the nature of wind is not constant throughout the year. Thus, applying wind turbine only may disturb and couldn't be controlled by backup battery only. Therefore, the solution for such problem is using the hybrid of wind along with solar to overcome the drawback comes by weather change and in order to make the power output stable.

LIMITATION OF RESEARCH

Generally, when the total generated powers through the renewable energy systems are more than the required demand, it can be stored in the batteries. In other words, when the system in standalone or hybrid system cannot satisfy the demand, the battery energy storage systems is needed to compensate the load deficit and improve the supply reliability of the system.

ADVANTAGES AND DISADVANTAGES

Optimization is the process of getting best value from different alternatives of candidates under a set of conditions. Nowadays, by mimicking the process in nature, different algorithms are developed by scholars of the field. The application of these algorithms is promising for the future to handle real world problems including hard optimization problems. Energy was becoming the main engine for the socio-economic development of the society, employing this nature inspired algorithm to optimize renewable energy system is significantly important (Krishnamoorthy, M., Suresh, S., & Alagappan, S., 2020). The employed approach was effectively solved the designed problem. Even though, all the three candidates have to converge to the same solution in algorithm, since the number of parameters of this problem was less, it has shown high ability to get the optimal value. Advantages and disadvantages of proposed approach are given in Table 13.

Table 13. Advantages and disadvantages of proposed approach

Advantages	Disadvantages
• Can be applied for wide search space • It has relatively fewer parameters to be adjusted • The obtained result was good for the time given • The parameters and variables can fit to many real-world Optimization problems • It's ability to handle a large number • of variables and escaping to local solutions	• Low capability to handle the difficulties of a multi-modal search landscape, as it seems that all three alpha, beta and gamma wolves tend to converge to the same solution. • It is costly to apply for narrow search space. • Relatively less convergence rate. • It needs researcher's ability and experience to set the parameters. • Less power full for multi- objective function

RECOMMENDATIONS

The researchers defined the fitness function by using appropriate variables and parameters. Numerical data of the particular Kabi village of Jeldu district was also taken from NASA web site to optimize the system and illustrate the effectiveness of the methodology.

Based on the survey made at Jeldu district, assessment made in energy sector of Ethiopia from Jeldu district administration, Ethiopia Electric Corporation and NASA web site and the result form simulation for the data organized from different sources, the researchers recommend the following.

- Even though there is an extended plan and encouraging commitments are there in the Ministry of Water, Irrigation and Energy of Ethiopia to scale the electricity access of the nation to 100% by the year of 2025, the actual base of year 2019 47% is not on the right trac. So, it needs great attention.
- Ethiopian government has also shown great commitment by launching mega hydro projects to transform the energy sector in particular in the growth and transformation plan (GTP-II). Since the actual assessment showns implementation of GTP-I was much less than expected, the government should increase support and control of the ministry.
- Administration of Jeldu district should plan and made effort to give electric access to the village.
- Ethiopian government should look other power generating sources to increase the electric access of the rural areas.
- For Jeldu district, specially Kabi village renewable energy source like wind and solar is the best solution to supply electric demand.

FUTURE RESEARCH DIRECTIONS

Hopefully, the way of organizing data to optimize hybrid wind and solar renewable energy system in the case of Kabi village of Jeldu district would be very beneficial and become a good benchmark for us and other researchers to apply other nature inspired meta heuristic algorithms to their own methods and compare their result with other nature inspired algorithms. Thus, the following should be our future focus

- Applying other meta heuristic algorithm to this particular data and compare the results.
- Improving the optimal results of proposed methodology by applying certain technical tests such as sensitivity analysis and uncertainty analysis.
- Applying hybrids of nature inspired algorithms and compare the results with proposed approach.

CONCLUSION

In this paper, the researcher presented a case study to optimize the total annual cost of a hybrid Wind and Solar renewable energy system with standalone monthly average of about 23136 KWh/month power generation by considering certain predesigned constraints to satisfy the load demand of Kabi village of Jeldu district, Oromia reginal state, Ethiopia. The main objective of this research is minimizing the total annual cost of system by sizing the number of individual components while the load demand of the village is satisfied. This research can be used as benchmark for the government of Ethiopia in particular and sub-Sahara region in general to decrease the problem related to large number of lacks of electricity access. Since the application of nature inspired algorithm was tested for particular rural area, by gathering data of the place from primary and secondary sources, it will useful for the researchers who wants to apply their own algorithms.

MATLAB code was designed for the fitness function, and the employed algorithm. After simulation as shown on table 12, the system was optimized and the optimal value $5572.0 was obtained to satisfy the load demand of the village. Based on its smaller number of parameters and high ability to escape to optimal value, proposed approach has effectively solved the problem. The optimal number of wind turbine, PV panels, and batteries are determined to supply the desired load. Generally, the researchers conclude that applying nature inspired algorithms to solve hybrids of renewable energy system is encouraging due to their ability of reaching optimal solutions in complex search space.

ACKNOWLEDGMENT

The authors wish to thank all the research scholars cited in this paper.

REFERENCES

Al-Shamma, A. A., & Khaled, E. (2012). Optimum sizing of hybrid PV/wind/battery/diesel system considering wind turbine parameters using Genetic Algorithm. *Power and Energy (PECon), IEEE International Conference*, 121-126.

Alireza, A. (2013). Developing a discrete harmony search algorithm for size opti- mization of wind–photovoltaic hybrid energy system. *Solar Energy*, *98*, 190–195. doi:10.1016/j.solener.2013.10.008

Ashok, S. (2007). Optimised model for community-based hybrid energy system. *Renewable Energy*, *32*(7), 1155–1164. doi:10.1016/j.renene.2006.04.008

Bekele, G., & Palm, B. (2010). Feasibility study for a standalone solar–wind-based hybrid energy system for application in Ethiopia. *Applied Energy*, *87*(2), 487–495. doi:10.1016/j.apenergy.2009.06.006

Belfkira, R., Zhang, L., & Barakat, G. (2011). Optimal sizing study of hybrid wind/PV/diesel power generation unit. *Solar Energy*, *85*(1), 100–110. doi:10.1016/j.solener.2010.10.018

Benatiallah, A., Kadia, L., & Dakyob, B. (2010). Modelling and optimization of wind energy systems. *JJMIE*, *4*(1).

Berihun, G. (2013). *Modeling and Simulating of a micro hydro wind hybrid power generation system for rural area of Ethiopia* (Mater thesis). Addis Ababa University.

Beyene, G. (2018). *The Challenges and Prospects of Electricity Access in Ethiopia* (Doctoral dissertation). Addis Ababa University.

Beyene, G. E., Kumie, A., Edwards, R., & Troncoso, K. (2018). *Opportunities for transition to clean household energy in Ethiopia: application of the household energy assessment rapid tool (HEART)*. Academic Press.

Bhandari, B., Ahn, S. H., & Ahn, T. B. (2016). Optimization of hybrid renewable energy power system for remote installations: Case studies for mountain and island. *International Journal of Precision Engineering and Manufacturing*, *17*(6), 815–822. doi:10.100712541-016-0100-2

Binitha, S., & Sathya, S. S. (2012). A survey of bio inspired optimization algo-rithms. *International Journal of Soft Comput-ing and Engineering*, *2*(2), 137–151.

Blimpo, M. P., & Cosgrove-Davies, M. (2019). *Electricity Access in Sub-Saharan Africa: Uptake, Reliability, and Complementary Factors for Economic Impact*. World Bank Publications. doi:10.1596/978-1-4648-1361-0

Borowy, B. S., & Salameh, Z. M. (1996). Methodology for optimally sizing the combi-nation of a battery bank and PV array in a wind/PV hybrid system. *IEEE Transactions on Energy Conversion*, *11*(2), 367–375. doi:10.1109/60.507648

Bueno, C., & Carta, J. A. (2005). Technical–economic analysis of wind-powered pumped hydrostorage systems. Part II: Model application to the island of El Hierro. *Solar Energy*, *78*(3), 396–405. doi:10.1016/j.solener.2004.08.007

Chedid, R., Akiki, H., & Rahman, S. (1998). A decision support technique for the design of hybrid solar-wind power systems. *IEEE Transactions on Energy Conversion*, *13*(1), 76–83. doi:10.1109/60.658207

Dandessa, C., Negisho, T., & Bekele, G. (2018). *Assessment of Traditional Medicinal Plants Used for Treatment of Human Diseases in Oromia Region, West Shewa Zone*. Jeldu Woreda, Ethiopia: Asian Plant Research Journal.

Diaf, S., Belhamel, M., Haddadi, M., & Louche, A. (2007). A methodology for op-timal sizing of autonomous hybrid PV/wind system. *Energy Policy*, *35*(11), 5708–5718. doi:10.1016/j.enpol.2007.06.020

Eftichios, K., Dionissia, K., Antonis, P., & Kostas, K. (2006). Methodology for optimal sizing of stand alone photovoltaic/wind generator systems using genetic algorithms. *Solar Energy*, *80*(9), 1072–1088. doi:10.1016/j.solener.2005.11.002

Eldos, T., & Al Qasim, R. (2013). *On the performance of the Gravitational Search Algorithm*. West Yorkshire, UK: International Journal of Advanced Computer Science Applications. doi:10.14569/IJACSA.2013.040811

Elhadidy, M. A., & Shaahid, S. M. (1999). Optimal sizing of battery storage for hybrid (wind+ diesel) power systems. *Renewable Energy*, *18*(1), 77–86. doi:10.1016/S0960-1481(98)00796-4

Eltamaly, A. M., & Mohamed, M. A. (2014). A novel design and optimization software for autonomous PV/wind/battery hybrid power systems. *Mathematical Problems in Engineering*, *2014*, 2014. doi:10.1155/2014/637174

Etamaly, A. M., Mohamed, M. A., & Alo-lah, A. I. (2015). A smart technique for op-timization and simulation of hybrid photo-voltaic/wind/diesel/battery energy systems. *Smart Energy Grid Engineering (SEGE), 2015 IEEE International Conference on*, 1-6.

Fulzele, J. B., Daigawane, M. B., & Daigawane, P. M. (2018). Design of hybrid PV-wind stand-alone renewable energy system: Case study. *Journal of Information and Optimization Sciences*, *39*(1), 345–355. doi:10.1080/02522667.2017.1374744

Ganesan, T., Vasant, P., & Elamvazuthi, I. (2016). *Advances in Metaheuristics: Applications in Engineering Systems*. CRC Press.

Gavanidou, E. S., Bakirtzis, A. G., & Dokopoulos, P. S. (1993). A probabilistic method for the evaluation of the performance and the reliability of wind-diesel energy systems. *IEEE Transactions on Energy Conversion*, *8*(2), 197–206. doi:10.1109/60.222705

Geleta, D., & Manshahia, M. S. (2017). Optimization of Renewable Energy Systems: A Review. *International Journal of Scientific Re-search in Science and Technology, 8*(3), 769-795.

Geleta, D., & Manshahia, M. S. (2018). Opti-mization of Hybrid Wind and Solar Renewable Energy System by Iteration Method. *International Conference on Intelligent Computing and Optimization Springer*, 98-107.

Geleta, D. K., & Manshahia, M. S. (2017). Nature Inspired Computational Intelligence: A Survey. *International Journal of Engineering. Science and Mathematics*, *6*(7), 769–795.

Geleta, D. K., & Manshahia, M. S. (2020). Artificial Bee Colony-Based Optimization of Hybrid Wind and Solar Renewable Energy System. In *Handbook of Research on Energy-Saving Technologies for Environmentally-Friendly Agricultural Development* (pp. 429–453). IGI Global. doi:10.4018/978-1-5225-9420-8.ch017

Hadidian, M. J., Arabi, N. S., & Bigdeli. (2016). Optimal sizing of a stand-alone hybrid photovoltaic/wind system using new grey wolf optimizer considering re-liability. *Journal of Renewable and Sustainable Energy, 8*(3). https://earthobservatory.nasa.gov/features/EnergyBalance

Ilinca, A., McCarthy, E., Chaumel, J. L., & Retiveau, J. L. (2003). Wind potential assessment of Quebec Province. *Renewable Energy, 28*(12), 1881–1897. doi:10.1016/S0960-1481(03)00072-7

Javadi, M. R., Mazlumi, K., & Jalilvand, A. (2011). Application of GA, PSO and ABC in optimal design of a stand-alone hybrid system for north-west of Iran. In *Electrical and Elec-tronics Engineering* (pp. 1–203). ELECO.

Kaabeche, A., Belhamel, M., & Ibtiouen, R. (2010). Optimal sizing method for stand-alone hybrid PV/wind power genera-tion system. *Revue des Energies Renouvelables (SMEE'10) Bou Ismail Tipaza*, 205- 213.

Karaboga, D., & Basturk, B. (2007). A powerful and e cient algorithm for numeri-cal function optimization: Arti cial bee colony (ABC) algorithm. *Journal of Global Optimization, 39*(3), 459–471. doi:10.100710898-007-9149-x

Karaboga, D., Gorkemli, B., Ozturk, C., & Karaboga, N. (2014). A comprehensive survey: Artificial bee colony (ABC) algorithm and ap-plications. *Artificial Intelligence Review, 42*(1), 21–57. doi:10.100710462-012-9328-0

Kefayat, M. A., Lashkar, A., & Nabavi, N. (2015). A hybrid of ant colony optimiza-tion and arti cial bee colony algorithm for probabilistic optimal placement and sizing of distributed energy resources. *Energy Conversion and Management, 92*, 149–161. doi:10.1016/j.enconman.2014.12.037

Kellog, W., Nehrir, M., Venkataramanan, G., & Gerez, V. (1998). Generation Unit Sizing and Cost Analysis for Stand -alone Wind, Phovoltaic, and Hybrid Wind/PV Systems. *IEEE Transactions on Energy Conversion, 13*(1), 70–75. doi:10.1109/60.658206

Kharchenko, V., & Vasant, P. (2018). *Handbook of Research on Renewable Energy and Electric Resources for Sustainable Rural Development*. Hershey, PA: IGI Global. doi:10.4018/978-1-5225-3867-7

Kharchenko, V., & Vasant, P. (2020). *Handbook of Research on Smart Computing for Renewable Energy and Agro-Engineering*. Hershey, PA: IGI Global. doi:10.4018/978-1-7998-1216-6

Kharchenko, V., & Vasant, P. (2020). *Handbook of Research on Energy-Saving Technologies for Environmentally-Friendly Agricultural Development*. Hershey, PA: IGI Global. doi:10.4018/978-1-5225-9420-8

Kosmadakis, G., Sotirios, K., & Emmanuel, K. (2013). Renewable and conventional elec-tricity generation systems. *Technologies and diversity of energy systems*. In *Renewable En-ergy Governance* (pp. 9–3). Springer London. doi:10.1007/978-1-4471-5595-9_2

Koutroulis, E., Dionissia, K., Potirakis, A., & Kostas, K. (2006). Methodol-ogy for optimal sizing of stand-aloMethodol-ogy for optimal sizing of stand-alone photovoltaic/wind-generator systems us-ing genetic algorithms. *Solar Energy, 80*(9), 1072–1088. doi:10.1016/j.solener.2005.11.002

Lu, L., Yang, H., & Burnett, J. (2002). Investigation on wind power potential on Hong Kong islandsan analysis of wind power and wind turbine characteristics. *Renewable Energy, 27*(1), 1–12. doi:10.1016/S0960-1481(01)00164-1

Luna, R. R., Trejo, P. M., Vargas, D., & Os-Moreno, G. (2012). Op-timal sizing of renewable hybrids energy sys-tems: A review of methodologies. *Solar Energy*, *88*(4), 1077–1088. doi:10.1016/j.solener.2011.10.016

Lysen, H. (1983). Introduction to wind energy. Consultancy Services. *Wind Energy (Chichester, England)*, 82–91.

Manish, D., Upadhyay, N., & Silakari, S. (2015). An exhaustive survey on nature An exhaustive survey on nature inspired optimization algorithms. *International Journal of Software Engineering and Its Applications*, *9*(4), 91–104.

Mante, O. D. (2018). *Sub-Saharan Africa Is Lighting Up: Uneven Progress on Electrification*. Academic Press.

Mohamed, M. A., Eltamaly, A. M., & Alolah, A. I. (2016). PSO-based smart grid application for sizing and optimization of hybrid renewable energy systems. *PLoS One*, *11*(8), e0159702. doi:10.1371/journal.pone.0159702 PMID:27513000

MoWIE. (2019). *National electrification program 2.0; Intgrated planning for universal access*. Academic Press.

Nafeh, A. E. S. A. (2011). Optimal economical sizing of a PV-wind hybrid energy system using genetic algorithm. *International Journal of Green Energy*, *8*(1), 25–43. doi:10.1080/15435075.2010.529407

Ramoji, S. K., Rath, B. B., & Kumar, D. V. (2014). Optimization of hybrid PV/wind energy system using genetic algorithm (GA). *Journal of Engineering Research and Applications*, *4*, 29–37.

Rashedi, E., Nezamabadi-Pour, H., & Saryazdi, S. (2009). A gravitational search algorithm. *Information Sciences*, *179*(13), 2232–2248. doi:10.1016/j.ins.2009.03.004

Rubio, R., Perea, M. T., Vazquez, D. V., & Os-Moreno, G. J. (2012). Op-timal sizing of renewable hybrids energy sys-tems: A review of methodologies. *Solar Energy*, *86*(4), 1077–1088. doi:10.1016/j.solener.2011.10.016

Sabri, N. M., Puteh, M., & Mahmood, M. R. (2013). A review of gravitational search algorithm. *Int. J. Advance. Soft Comput.*, *5*(3), 1–39.

Singh, S., & Kaushik, S. C. (2016). Op-timal sizing of grid integrated hybrid PV-biomass energy system using arti cial bee colony algorithm. *IET Renewable Power Generation*, *10*(5), 642–650. doi:10.1049/iet-rpg.2015.0298

Sun, G., & Zhang, A. (2013). A hybrid genetic algorithm and gravitational search algorithm for image segmentation using multilevel thresholding. *Iberian Conference on Pattern Recognition and Image Analysis*, 707-714. 10.1007/978-3-642-38628-2_84

Tesfaye, Y. (2014). *Application of Micro-Hydro Pv/Battery off-Grid Hybrid Energy System for Ethiopian Rural Area* (Masters Thesis). Addis Ababa University.

Thomas, J. J., Karagoz, P., Ahamed, B. B., & Vasant, P. (2020). *Deep Learning Techniques and Optimization Strategies in Big Data Analytics*. Hershey, PA: IGI Global. doi:10.4018/978-1-7998-1192-3

Tong, C. (2014). Gravitational search algorithm based on simulated annealing. *Journal of Convergence Information Technology*, *9*(2), 231.

UNSD & WHO. (2019). Tracking SDG7: The energy progress report 2019. Authors.

Vasant, P., Kose, U., & Watada, J. (1703). Metaheuristic (2017). Techniques in Enhancing the Efficiency and Performance of Thermo-Electric Cooling Devices. *Energies*, 10.

Vasant, P., Marmolejo, J. A., Litvinchev, I., & Aguilar, R. R. (2019). Nature-inspired meta-heuristics approaches for charging plug-in hybrid electric vehicle. *Wireless Networks*. doi:10.100711276-019-01993-w

Vasant, P., Zelinka, I., & Weber, G. W. (2019). Intelligent Computing and Optimization. In *Proceedings of the 2nd International Conference on Intelligent Computing and Optimization, (ICO) 2019*. Springer International Publishing. 10.1007/978-3-030-00979-3

Wang, J. S., & Song, J. D. (2017). A Hybrid Algorithm Based on Gravitational Search and Particle Swarm Optimization Algorithm to Solve Function Optimization Problems. *Engineering Letters, 25*(1).

Wang, L., & Singh, C. (2009). Multicriteria design of hybrid power generation systems based on a modi ed particle swarm optimization algorithm. *IEEE Transactions on Energy Conversion*, *24*(1), 163–172. doi:10.1109/TEC.2008.2005280

Weldemariam, L. E. (2010). *Genset-solar-wind hybrid power system of off-grid power station for rural applications*. Academic Press.

Yangg, H., Zhou, W., & Lou, C. (2008). Optimal siz-ing method for stand-alone hybrid solar{wind system with LPSP technology by using genetic algorithm. *Solar Energy*, *82*(4), 354–367. doi:10.1016/j.solener.2007.08.005

Zelinka, I., Tomaszek, L., Vasant, P., & Duy, V. H. (2018). A novel approach on evolutionary dynamics analysis – A progress report. *Journal of Computational Science*, *25*, 437-445.

Zong, W. G. (2012). Size optimiza-tion for a hybrid photovoltaic wind energy system. *Electrical Power and Energy Systems*, *42*(1), 448–451. doi:10.1016/j.ijepes.2012.04.051

KEY TERMS AND DEFINITIONS

Constraints: Are the conditions set by the researcher to control the boundary optimal solution and can also affect solutions of the given Optimization problem.

Electricity Access: Getting the chance of using electricity.

Fitness Function: It is a kind of objective function that developed for a particular problem by containing all its parameters in order to find the desired solution.

Hybrid Energy: Is the newly emerging technology by combining two or more energy generating sources to maximize the efficiency of the system.

Loss of Supply Probability: The ratio of power deficit to the total power that the power generating system can produce.

Nature-Inspired Algorithm: Is the algorithm which takes its inspiration from nature and used to solve hard real-world Optimization problems where conventional Optimization methods were failed to solve it (Geleta & Manshahia, 2020).

Optimization: Is the process of searching timely best optimal value among all the given alternatives.

Reliability: Is the total amount of power that a renewable energy source can produce.

Chapter 8
Prospects for Energy Supply of the Arctic Zone Objects of Russia Using Frost-Resistant Solar Modules

Vladimir Panchenko
https://orcid.org/0000-0002-4689-843X
Russian University of Transport, Russia

ABSTRACT

The scientific work is devoted to the prospect of using frost-resistant solar modules with extended service life of various designs for energy supply of infrastructure facilities of the Arctic zone of Russia. The general characteristic of the region under consideration is given, and its energy specifics, directions of energy development based on renewable energy sources are considered. In the work, frost-resistant planar photovoltaic modules and solar roofing panels with an extended service life for power supply of objects are proposed. For simultaneous heat and electrical generation, frost-resistant planar photovoltaic thermal roofing panels and concentrator solar installation with high-voltage matrix solar modules with a voltage of 1000 V and an electrical efficiency of up to 28% are proposed. The considered solar modules have an extended rated power period due to the use of the technology of sealing solar cells with a two-component polysiloxane compound and are able to work effectively at large negative ambient temperatures and large ranges of its fluctuations.

INTRODUCTION

Currently, the economy faces a technological challenge – the transition to a new technological structure in production, services, and management. The resource-saving approach suggests that the pilot should consider those sectors and regions where the successful implementation of advanced technologies is vital. Given the difficult natural and climatic conditions, the Arctic is a favorable testing ground for technological innovation. In addition, the Arctic is becoming the center of attention of the country's leadership. The

DOI: 10.4018/978-1-7998-3970-5.ch008

President of Russian Federation instructed to restore and expand the degree of development and control of the territories of the Arctic zone (we are talking about transport and logistics infrastructure, places of extraction and processing of raw materials in the Arctic). All this must be done taking into account the use of advanced technologies (often having a dual purpose) (Degtyarev, Panchenko & Mayorov, 2018).

ARCTIC ZONES OF THE RUSSIAN FEDERATION AND PREREQUISITES FOR THE DEVELOPMENT OF ENERGY BASED ON RENEWABLE ENERGY SOURCES

The Arctic is defined as the area around the North Pole, including the Arctic Ocean and surrounding areas. The southern border is conditional, there is no clear concept of the territory of the Arctic, it can be drawn along the southern border of the Arctic climatic zone (the dominance zone of the Arctic air masses), the zone of Arctic deserts or the tundra zone, or, for example, "mechanically" – along the Arctic Circle (Figure 1) (ael-msu.org). The Arctic zone of the Russian Federation (AZRF) is a heterogeneous region that requires a differentiated approach to the development of energy from renewable energy sources (RES).

Figure 1. Arctic zone of the Russian Federation and soil temperature in the summer in a depth of 40 centimeters in this area

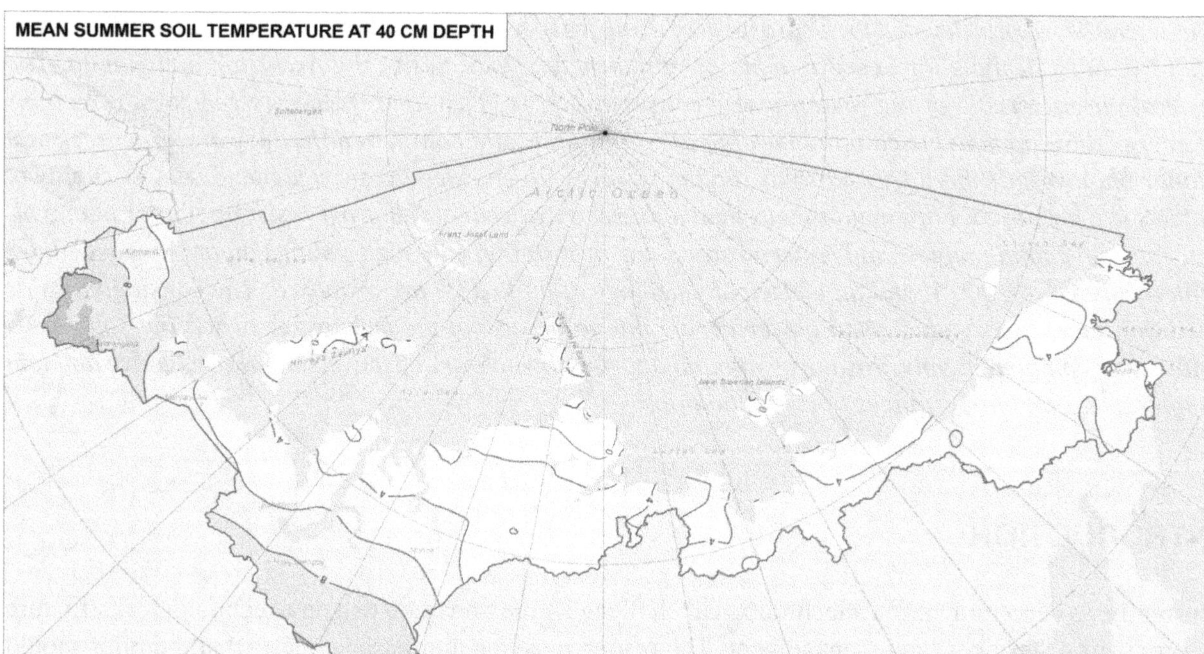

With a certain degree of conventionality and administratively, the concept of the Arctic can include the Murmansk region, the coastal regions of the Arkhangelsk region, the territories of the Nenets and Yamalo-Nenets autonomous regions, the polar regions of the Krasnoyarsk Territory and Yakutia and Chukotka, and also the islands of the Arctic Ocean of the Russian sector of the Arctic. In general, the Russian Arctic is about 4 million km^2 of territory and more than 1,5 million people.

In turn, the Russian Arctic can also, with a certain degree of conventionality, be divided into the "Near Arctic" and "Far Arctic", the border between which can be drawn along the lower reaches of the Yenisei. Murmansk region (Kola Peninsula) enters the Near Arctic zone, north of the Arkhangelsk region and Nenets autonomous region, Yamal. These territories, in turn, have their own isolation and specificity.

The Near Arctic as a whole – it is more than 90% of the population, industry and, in general, the economy, the comparative accessibility of the territory - not only by sea, but also by rail, milder natural conditions. It is in the Near Arctic that the currently developed reserves of fossil hydrocarbons are concentrated.

The natural resource potential of renewable energy in the Near Arctic is large, highly diversified and includes: bioresources: peat reserves, locally (Arkhangelsk region) – waste from woodworking industries; hydropower: hydropower of small rivers (Kola Peninsula, Subpolar and Polar Urals); water resources: tidal energy (bays of the White and Barents Seas); wind energy resources; solar resources (in summer); geothermal resources (probably most associated with the Kola Peninsula).

In addition, the Near Arctic is a territory where renewable energy development projects (mainly related to the Kola Peninsula) were carried out in Soviet times, in particular, Kislogubskaya tidal power plant and small hydropower plants. In the post-Soviet era, a number of wind energy projects of various sizes are also being implemented in the Murmansk region.

At the same time, "traditional" energy and network energy supply are developed in the Near Arctic. The largest energy facility is the Kola nuclear power plant; a number of thermal power plants and hydroelectric power stations are operating; also, the largest hydrocarbon reserves in Russia (the Yamalo-Nenets Autonomous region the Nenets region) are available and developed in the Near Arctic.

The Near Arctic has powerful prerequisites for the development of energy on renewable energy sources, both of a physical-geographical and economic-geographical nature. A relatively large absolute amount of renewable energy capacities can be installed there. But renewable energy in the foreseeable future can only have an auxiliary character, to some extent supplementing the "traditional" hydrocarbon and nuclear energy supply, which will remain the main one in the region.

The Far Arctic can also be divided into three parts with a certain degree of conventionality: Taimyr; north of Yakutia; Chukotka. The Far Arctic probably has a lower total natural potential of renewable energy per unit area. In addition, the development is hindered by the remoteness of the territory, extreme environmental conditions and a small number of consumers.

On the other hand, small autonomous energy based on renewable energy sources may be more in demand there due to the same unfavorable circumstances, since they also determine the too high cost of hydrocarbon energy carriers that have to be imported into the territory.

In this regard, the Far Arctic may become a territory where interesting and innovative solutions in the field of renewable energy are being implemented, and the renewable energy sources themselves will occupy a significant share in the energy balance, despite the fact that in absolute terms the level of development of renewable energy will be lower than in the Near Arctic.

The Far Arctic has its own interesting features, suggesting the possibility of certain local solutions. In particular, due to the greater number of hours with sunshine per year, it has a higher potential for solar energy. In addition, locally (Putorana Plateau, the mountainous regions of Chukotka), the potential of small river hydropower is high. In addition, the potential of tidal energy (also in Chukotka) and geothermal energy (also the mountainous regions of Taimyr and Chukotka) should be considered.

As for autonomous solar energy, the most promising inland areas of Yakutia and the north of the Krasnoyarsk region. This is due to both physical and geographical, economic and geographical features. The prevalence of anticyclone atmospheric circulation in this territory determines the prevalence of clear weather for most of the year, including the summer months. At the same time, it is the most sparsely populated part of Russia (population density is about 0,02 people/km^2, or 1 person per 50 km^2). The population is dispersed over a number of remote villages with a distance between settlements of the order of hundreds of kilometers that do not have a power supply network and without a ground transportation network. In addition, a significant part of the population is a traditional nomadic economy. It is also a zone of sharply continental climate with sharp daily and seasonal temperature differences. In summer, they can reach 30°C, in winter – decrease to -50°C and lower (absolute records belong to the northeast of Yakutia (Oymyakon, Verkhoyansk) - to -68°C and -71°C).

It should be noted that in the north and in other regions of Yakutia, under the management of Sakhaenergo (structural division of RusHydro), about 15 projects of autonomous solar stations (in combination with fuel generators) have already been implemented to power remote villages. Since power plants have been built recently (in the last 3-5 years), it is still difficult to judge the effectiveness of their work. In this case, the use of solar panels with resistance to severe frosts and sudden changes in temperature looks promising (Poulek, Strebkov, Persic & Libra, 2012; Strebkov, Persits & Panchenko, 2014; Panchenko, Strebkov & Persits, 2015).

Table 1 shows the zoning scheme of the AZRF in relation to the prospects and possible directions of energy development based on renewable energy sources.

Table 1. Arctic zones of the Russian Federation and prerequisites for the development of energy based on renewable energy sources

Region	General characteristics	Energy specifics, directions of energy development in renewable energy sources and the place of renewable energy in the fuel and energy complex
1. The Near Arctic (Murmansk region, north of the Arkhangelsk region, Nenets region, Yamalo-Nenets region)	Comparative proximity to the economic center of the country, more than 90% of the population of the entire Russian Arctic (more than 1,5 million people), a relatively developed transport network and economy.	Large and diversified renewable energy potential; at the same time, a powerful fuel and energy complex based on fossil energy sources (hydrocarbon and nuclear energy). A wide range of renewable energy sources; supporting role regarding "traditional" energy supply; a number of renewable energy projects have been implemented.
1.1. Murmansk region (Kola Peninsula)	The most "closest", populated and economically developed part of the Arctic with a developed transport network; population – more than 700 thousand people. Proximity to Western Europe, border area.	Kola nuclear power plant, a number of hydroelectric power stations; renewable energy potential: small hydropower, wind power, tidal energy, solar energy in the summer. Kislogubskaya tidal power station was built; there are projects for the construction of other tidal power stations; separate small wind farms were built. Surveys in geothermal energy may be promising. The sparsely populated and remote central and eastern regions of the Kola Peninsula and nomadic farms are more promising for small autonomous energy.

continues on following page

Table 1. Continued

Region	General characteristics	Energy specifics, directions of energy development in renewable energy sources and the place of renewable energy in the fuel and energy complex
1.2. North of the Arkhangelsk region and Nenets region	Center – Arkhangelsk. Outside it is a more remote territory with a rarer population and less developed transport network (compared to the Murmansk region).	Mostly hydrocarbon energy on imported energy. At the same time, rich and currently developed hydrocarbon deposits (Nenets region). The renewable energy potential is highly associated with bioenergy (peat and forest waste). In addition, there is the potential of tidal energy, wind energy, solar energy in the summer, and locally small hydropower. Sparsely populated and remote areas and populated in the center and in the east, not associated with large energy nodes, and nomadic farms, are most promising for small autonomous energy.
1.3. Yamalo-Nenets region	The specifics of the territory are determined by the presence and development of the largest gas and oil fields in Russia. At the same time, remote settlements and nomadic farms (the largest number of nomadic farms among Russian regions) are promising for the development of small autonomous energy.	Promising areas of renewable energy: locally – small hydropower, locally (coastal areas) – wind energy; solar power in the summer.
2. The Far Arctic (North of Krasnoyarsk region (Taimyr and adjacent territories), north of Yakutia, Chukotka)	Extreme natural conditions. Remoteness from the economic centers of the country, lack of railway communication. The economy is represented by individual foci. Population – 100-200 thousand people; on average – 1 person/km^2, in places – less than 1 person/50 km^2. The most sparsely populated and sparsely populated part of Russia.	With some exceptions, it is almost completely dependent on imported energy sources (coal, fuel oil) or the use of local archaic fuel (firewood, peat). The specificity of renewable energy varies by district. In general, the natural and natural-economic potential of renewable energy is lower than in the Near Arctic per unit area, but higher per capita. Small autonomous stations based on renewable energy sources can occupy a significant place in the energy balance of the territory.
2.1. Taimyr	Outside Norilsk, it is the most sparsely populated and little explored part of Russia. Separate households; nomadic farms; separate coastal settlements (Dickson).	Lack of own modern energy generation. Promising areas of renewable energy: wind energy in coastal settlements, solar energy in the summer; locally (area of the Putorana plateau, Byrranga mountains) – hydropower of small rivers. Emphasis on small autonomous energy.
2.2. North of Yakutia	Also sparsely populated territory with separate settlements, nomadic farms, separate coastal settlements (Tiksi).	Lack of own modern energy generation. Promising areas of renewable energy: wind energy in coastal settlements, solar energy in the summer. High potential for solar energy in the Arctic; a number of autonomous solar energy projects have been implemented.
2.3. Chukotka	Relative to other regions of the Far Arctic, it differs in the specificity of its geographical position at the junction of the basins of the Arctic and Pacific oceans; better accessibility by sea. In the presence of a relatively large center (Anadyr), the majority of the population is dispersed over small remote settlements and nomadic farms. The specific nature of the environment determines the relatively high potential of renewable energy.	Own energy generation is connected with the Bilibino nuclear power plant and Anadyr thermal power plant (it worked on local coal, now it is switching to gas). At the same time, most of the territory and population is outside the grid. Promising areas of renewable energy: tidal energy, wind energy, hydropower of small rivers, geothermal energy, solar energy in the summer. Along with the Kola Peninsula, the highest gross potential of renewable energy, represented by the widest spectrum.

At present, a new stage in the development of the AZRF is being announced, which, by its natural-economic, demographic and other conditions, differs significantly from other regions of the Russian Federation and has its own distinctive features: extreme climatic conditions, including permanent ice cover or drifting ice in the Arctic seas; focal nature of industrial and economic development of territories and low population density (1-2 people per 10 km^2); remoteness from the main industrial centers, high resource consumption and the dependence of economic activity and livelihoods of the population on the supply of fuel, food and essential goods from other regions of Russia; vulnerability of nature from industrial emergencies and human production activities.

The development of the AZRF requires significant energy consumption both to ensure technological processes, and to create comfortable living conditions. The infrastructure of the new Arctic will consist of several circuits: transport and logistics, industrial, resettlement, tourism, defense. Each circuit will, among other things, contain a system for monitoring the status of the main nodes. The Northern Sea Route will act as a unifying framework for the development infrastructure.

Also, do not forget about the need for assistance in organizing energy supply for the resettlement places of indigenous peoples of the North, Siberia and the Far East of the Russian Federation. The principles of sustainable development indicate the need to provide residents with electricity, communications, and medical care. The whole range of measures to ensure a quality standard of living requires a constant supply of electricity. Therefore, the task is to create and implement autonomous, distributed energy supply systems. The resources of local renewable energy are many times higher than the current and future energy consumption in the Arctic. If we talk about the economic feasibility of introducing distributed, adaptive energy systems based on renewable energy sources, there is considerable potential for energy consumers in the Russian Arctic, created by the tariff for electricity produced by diesel generators operating under the northern delivery program, for which the cost of electricity reaches 120 rubles per kW·hour. In addition, the use of autonomous energy supply systems based on renewable energy sources reduces the dependence of settlements and objects on external supplies and increases the reliability of energy supply.

Given the geographical features of the Arctic, namely the polar day and night season, it is advisable to talk about distributed energy supply systems, including wind generation, solar generation and thermal sources. A separate engineering task is the development of energy storage systems and the reduction of energy consumption of key technological processes, while maintaining their quality and productivity. In some areas of coastal Arctic zones, wind speeds exceed 5-7 m/s. If we consider the Arctic zone, then a number of mainland territories have a rather high degree of insolation (Figure 2 on the left) (agroatlas.ru/ru/content/Climatic_maps/). At the same time, the use of this resource will require the improvement of technologies for collecting and converting solar radiation, where a special temperature regime should be taken into account (Figure 2 on the right), which sets the material science requirements for the components of the power system elements.

The intensity of solar radiation in several regions of Siberia is close to the intensity of solar radiation in Spain, and low temperatures increase the efficiency of silicon photovoltaic converters, which increases the generation of electricity. However, it is necessary to take into account the ability of solar photovoltaic modules to operate at the nominal power level over a wide range of operating temperatures, negative values of which can reach -70 °C. Modern planar photovoltaic solar modules manufactured using the technology of laminating ethylene vinyl acetate films of silicon photovoltaic cells are not able to operate at such negative temperatures and large ranges of operating temperatures.

Figure 2. Distribution of total solar radiation (on the left) and minimum air temperature during the year (on the right)

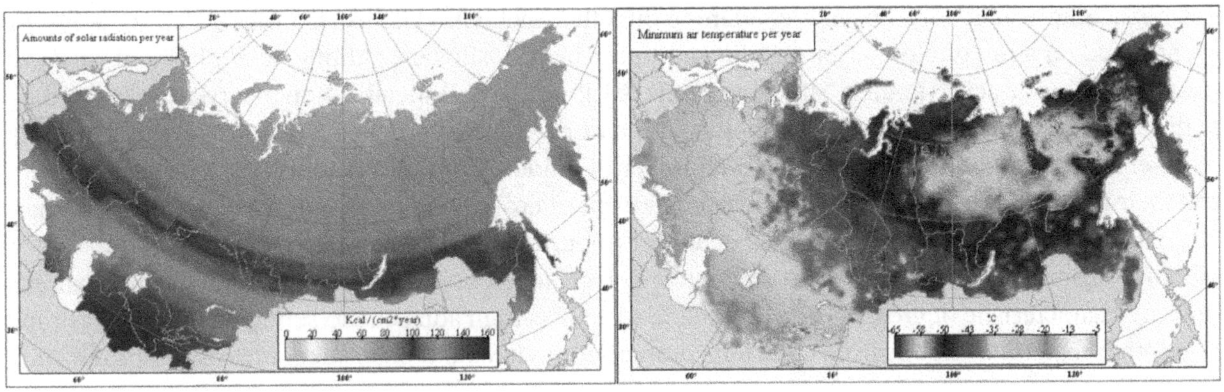

BACKGROUND

Almost all companies in the world produce solar modules with planar solar photovoltaic silicon cells (Poulek, Strebkov, Persic & Libra, 2012; Kharchenko, Nikitin, Tikhonov, Panchenko & Vasant, 2019; Wohlgemuth, 2003; Reda, 2007; Daliento & Lancellotti, 2010; Parretta, Bombace, Graditi & Schioppo, 2005; Ketola, McIntosh, Norris & Tomalia, 2008). The most common manufacturing technology is lamination technology using films based on a copolymer of ethylene with vinyl acetate (EVA). The main disadvantages of this technology are the high energy intensity of the process, the limited term of the rated power of the module (20 - 25 years), due to the low light, heat and weather resistance, high corrosion activity of EVA. Such modules cannot be used in installations with solar concentrators.

The main reasons for the degradation of solar photovoltaic modules are corrosion of the contact grid of photoconverters and an increase in optical loss in the laminating layers. When using a solar concentrator, the darkening process of the laminating material is significantly accelerated. Other disadvantages of lamination technology include the release of toxic volatile substances in the lamination process, as well as the significant energy intensity of the lamination process.

The mechanism of cracking on photoelectric converters, like most of the causes of solar module failures, is associated with the properties of the encapsulating material – a film based on a copolymer of ethylene with vinyl acetate. The mechanical properties of this material, in particular, the elastic modulus is highly dependent on temperature - with decreasing temperature from +20 °C to -40 °C, the values of the elastic modulus increase by several orders of magnitude (Poulek, Strebkov, Persic & Libra, 2012). Low temperature peeling is a big problem with solar modules made using the lamination process. The thermal coefficient of expansion of an ethylene vinyl acetate film is 20 times higher than that of glass. At a low temperature, ethylene vinyl acetate is not elastic – it shrinks much more than glass, which leads to the separation of films and glass and leads to a rapid failure of the solar module.

Recently, the technology for the production of photovoltaic modules has been actively improved, both in terms of replacing the ethylene vinyl acetate film with film materials with a different polymer base, and in terms of technological processes replacing the lamination stage (Poulek, Strebkov, Persic & Libra, 2012; Kharchenko, Nikitin, Tikhonov, Panchenko & Vasant, 2019; Wohlgemuth, 2003; Reda, 2007; Daliento & Lancellotti, 2010; Parretta, Bombace, Graditi & Schioppo, 2005; Ketola, McIntosh, Norris

& Tomalia, 2008). To seal microcircuits and semiconductor devices, polysiloxane gels are used, which are a rare crosslinked structure formed in the process of hydrosilicon – the reaction of the interaction of low molecular weight polysiloxanes containing dimethyl-methylvinylsiloxane units with a crosslinking agent based on a mixture of various cyclic and linear hydridosiloxane sulfonic acid platinum catalysts (platinum chloride acid) (Poulek, Strebkov, Persic & Libra, 2012). Vulcanization is carried out according to the "polymer-polymer" scheme without isolation of reaction by-products with the formation of long transverse bridges giving the vulcanizate a number of unique properties: high dielectric characteristics and their preservation at low temperatures; adjustment of crosslinking frequency and viscoelastic characteristics; high degree of purity in the content of impurities; lack of internal mechanical stress; good vibration absorption (damping); the correction of defects inherent in liquids, along with the dimensional stability characteristic of crosslinked elastomers, as well as high adhesion to semiconductors, glass and most other materials; high resistance to temperature, ultraviolet and ozone degradation; environmental safety of use.

STATEMENT OF THE OBJECTIVE AND RESEARCH METHODS

The objective of the study is the creation of solar energy converters for heat and electricity supply to consumers, the design of which allows for a long time to work at the level of rated power at low negative ambient temperatures.

To create the design documentation for the manufactured modules, the Kompas 3D (kompas.ru) computer-aided design system was used. To simulate and visualize the thermal state of photovoltaic thermal modules, the ANSYS (ansys.com) finite element analysis system was used. For the manufacture of solar modules, encapsulation technology was used with a two-component polysiloxane compound. For laboratory measurements of the current-voltage characteristics of solar modules, the PICOSOLAR solar radiation simulator was used. For accelerated testing of solar modules, various BINDER climate chambers were used.

PLANAR FROST-RESISTANT PHOTOVOLTAIC MODULES WITH EXTENDED LIFE AT RATED POWER

As a result of work carried out in cooperation with Poulek Solar Ltd, Czech Republic, it was shown that a significant increase in the rated power period of solar modules, their stable operation with concentrators, as well as a reduction in production energy costs, is ensured by using a two-component polysiloxane as a filler material a compound cured in the presence of a platinum catalyst to a state of low modulus gel. A technology has been developed for sealing solar PV modules with an extended rated power life compared to standard laminated solar modules. The technology of manufacturing a vacuum double-glazed window with a thermoplastic spacer around the perimeter, where the previously evacuated cavity is filled with a two-component liquid polysiloxane compound, structured at low temperature into a low-modulus gel, is adopted as the basis. To implement this process, an automatic mixing and batching system for a two-component polysiloxane compound was developed, which is shown in the Figure 3. Using the developed installation, serial production of solar modules with various capacities was established. A solar module

Figure 3. Installation of automatic mixing of a two-component polysiloxane compound and solar modules made using it

with a rated power of 160 W, manufactured using the developed encapsulation technology with a two-component polysiloxane compound, is shown in the Figure 3 on the right.

Thanks to the developed manufacturing technology, new designs of solar modules have been created, which include a two-component polysiloxane compound – a protective material that completely envelops and isolates the photoelectric converters from moisture, thermal and mechanical influences, as well as providing optical contact between the light-reflecting surface of the photoelectric converters and the protective outer coating (Strebkov, Persits & Panchenko, 2014; Panchenko, Strebkov & Persits, 2015). According to the results of testing solar modules, it is possible to identify positive differences between the sealing technology with polysiloxane compound and the standard technology for laminating solar cells (Table 2).

Table 2. Comparative characteristics of the technological processes of lamination (EVA) and sealing (polysiloxane compound)

	EVA (ethylene vinyl acetate)	Polysiloxane compound
Operating temperature	- 30 ÷ + 60 °C	- 70 ÷ + 110 °C
UV resistance	low	high
Rated power period	20 - 25 years	40 - 50 years
Electricity consumption	40 kW·h	5 kW·h
Refractive index	1,482	1,406
Transparency for solar radiation with different wavelengths	8% (λ = 360 nm)	90% (λ = 360 nm)
	62% (λ = 400 nm)	92% (λ = 400 nm)
	91% (λ = 600 ÷ 1000 nm)	93% (λ = 600 ÷ 1000 nm)
Corrosive agent in the manufacture	acetic acid	no
Corrosive agent for aging	acetic acid	no
Mechanical stress - manufacturing	yes	no
- aging	yes	no
Elastic modulus	10,0 N/mm^2	0,006 N/mm^2
Linear coefficient of thermal expansion	4,0 × 10^{-4} K^{-1}	2,5 × 10^{-4} K^{-1}

Solar modules manufactured by the developed sealing technology with a two-component polysiloxane compound retain a higher level of energy production for a period twice the life of a standard laminated module; they lack internal mechanical stresses; high resistance to temperature, ultraviolet and ozone degradation remains, and the technology itself is environmentally friendly.

PLANAR FROST-RESISTANT PHOTOVOLTAIC ROOFING PANELS IN THE FORM OF TILES

One of the architectural solutions for the power supply of facilities is solar modules built into the roofing itself, that is, the so-called "solar roofing tiles", since a well-known drawback of solar modules, which are now widely used, is the need to install roofing under the solar module to protect buildings from external influences, which increases the cost of the buildings themselves (Strebkov, Panchenko, Irodionov & Kirsanov, 2015; Strebkov, Kirsanov & Panchenko, 2017).

Solar tiles are used as roofing material in the construction of buildings with simultaneous electrical generation from solar radiation. When using solar tiles, both architectural and construction tasks are solved, and in the manufacture of tiles, secondary plastic raw materials are used, as well as autonomous or parallel power supply to the consumer. The roofing solar panel is a tile of a standard shape made from recycled materials (polyethylene bottles or stretch film and adhesive components), which reduces the cost of production and favorably affects the environment (Figure 4). The composition of the solar tile also includes solar cells in the polysiloxane compound, which increases the duration of their rated power (Poulek, Strebkov, Persic & Libra, 2012; Strebkov, Persits & Panchenko, 2014; Panchenko, Strebkov & Persits, 2015).

Figure 4. Planar (on the left) and concentrator (on the right) solar tiles with polysiloxane compound

Switching between the solar roofing panels is carried out sequentially in order to obtain a large voltage at the output, where the voltage of each panel is 1 - 1.2 V. A Schottky diode is provided in the junction box, and a pair of sealed plastic MC 4 plugs is installed at the output.

At a detailed examination, the roofing solar panel includes a supporting base with a solar battery placed on it based on semiconductor photoelectric converters with an electric current collection cable. The solar battery is placed on a flat platform recessed relative to the upper surface of the base and protected by a two-component polysiloxane compound. The solar panel, which is part of the tile, contains built-in commutated silicon solar cells measuring 156 × 156 mm or 125 × 125 mm, has a protective coating of tempered glass, different voltage and electric power depending on the number of connected solar cells. Roofing concentrator solar panels are installed on the southern slope of the roof of the building at two possible angles – for maximum production in the summer months or for maximum production in the winter months of the year.

Table 3 presents the technical characteristics after full-scale tests of the developed solar roofing panels of planar design, manufactured using the processes of lamination and encapsulation of solar cells.

Table 3. Technical characteristics of the developed planar solar roofing panels (encapsulated and laminated)

Indicator	Unit of measurement	Value (when encapsulation; when laminating)
Open circuit voltage	V	1,11; 1,08
Operating voltage	V	0,85; 0,83
Short circuit current	A	3,5; 3,32
Operating current	A	3,06; 3,01
Electric power	W	2,6; 2,5
Volt-ampere characteristics fill factor	-	0,67; 0,64
The temperature of the faces and rear parties	°C	40 and 32; 42 and 34
Module dimensions	mm	420 × 310 × 50
The term of the rated power	years	40 - 50; 20 - 25
Module weight	kg	2,3; 2,1

At an ambient temperature of about 15 °C, the temperature of the front surface of a planar laminated roofing solar panel during field tests was 42 °C, and the rear 34 ° C. The temperature of the front surface of a planar encapsulated roofing solar panel during field tests was 40 °C, and the rear 32 °C, which indicates a more favorable thermal regime of solar cells in an encapsulated solar module.

According to the developed models of the solar roofing panel, a mold and composite foundations of tiles were made, which allows for small-scale production. The technology of encapsulating solar cells with a two-component polysiloxane compound allows increasing the period of the rated power of the photovoltaic part up to 40-50 years. Designed roofing solar panels are ready for industrial production. The localization of the production of the roofing solar panel is more than 80% due to the lack of complex foreign microelectronics and the use of domestic components to ensure import substitution of civilian, military and agricultural products. Small-scale batches of roofing solar modules are currently being produced.

STATIONARY FROST-RESISTANT PHOTOVOLTAIC THERMAL MODULES FOR AUTONOMOUS AND PARALLEL POWER GENERATION

Cooling of solar cells with a coolant will increase the operating voltage of the cells and, accordingly, the output electric power, as well as the electrical efficiency (Ibrahim, Othman, Ruslan, Mat & Sopian, 2011; Hosseini, Hosseini & Khorasanizadeh, 2011; Dubey & Tay, 2012; Buonomano, Calise & Vicidimini, 2016; Sandnes & Rekstad, 2002; Kalogirou, 2001; Ji, Lu, Chow, He & Pei, 2007). The main reasons for reducing the efficiency of photovoltaic converters is the deposition of dust on their surface and heating of the material under the influence of solar radiation. As a result, an increase in temperature reduces the amount of electricity generated. For example, an increase in the temperature of silicon materials by 1 °C reduces the conversion coefficient by 0.4 ... 0.5% (Kharchenko, Nikitin, Tikhonov, Panchenko & Vasant, 2019). The search for ways to overcome the above negative factors led to the integration of photovoltaic converters with flat solar collectors and the creation on their basis of a new type of installations, the so-called photovoltaic thermal modules (PVTM) (Kharchenko, Panchenko, Tikhonov & Vasant, 2018; Panchenko, Kharchenko & Vasant, 2019). In such an installation, two tasks are simultaneously solved: obtaining thermal and electric energy with a significant reduction in the area occupied by the installation compared to the separate placement of solar modules and solar collectors. Cooling with a coolant of the solar collector allows increasing the operating voltage of the elements and, accordingly, the output electric power, as well as the electrical efficiency (Panchenko, Chirskiy & Kharchenko, 2019; Panchenko & Kharchenko, 2019).

PLANAR FROST-RESISTANT PHOTOVOLTAIC THERMAL ROOFING PANELS IN THE FORM OF TILES

Based on the thermal operation of solar photovoltaic roofing panels, it can be concluded that it is advisable to use cooling of solar cells in view of the decrease in their efficiency with increasing temperature. Thus, along with the planar and concentrator photovoltaic solar roofing panels, photovoltaic thermal roofing panels are of great interest in view of the increase in the solar cells electrical efficiency and the generation of warm water at the outlet, which also increases the overall module efficiency (Figure 5) (Panchenko, 2018).

Figure 5. Drawing of a planar photovoltaic thermal module in the form of a roofing panel (left) and a fabricated sample (right)

Table 4 presents the weight, physical, structural and cost parameters of the manufactured sample of a stationary photovoltaic thermal module in the form of a roofing panel.

Roofing solar modules can provide independent or parallel with the network power supply of various stationary facilities. By combining and optimizing the composition of a solar power plant, consisting of various types of solar roofing modules, it is possible to achieve the necessary generation of electricity and warm water (Babaev, Kharchenko, Panchenko & Vasant, 2019; Babaev, Kharchenko & Panchenko, 2019).

Table 4. The main parameters of the photovoltaic thermal roofing panel

Length of roofing panel, mm	434
Width of roofing panel, mm	312
Thickness of roofing panel, mm	26
Mass of roofing panel, kg	2,5
Period of rated power, years	40 - 50
Type of solar cells	Monocrystalline silicon
Number of solar cells, pcs	6
Switching	Serial
Size of the solar cell, mm	125 × 125
Area of solar cells, m^2	0,09
Photodetector area, m^2	0,1
Absorber material	Aluminum (Copper)
Short circuit current, A	4,7 (6,2)*
Current at the operating point, A	4,6 (5,8)*
Open circuit voltage, V	3,2 (4,1)*
Voltage at the operating point, V	2,5 (3,4)*
Power, W	11 (20)*
Volt-ampere characteristics fill factor, %	60 (75)*
Photoconversion efficiency,%	13 (20)*
Substrate material	Plastic (recycled plastic)
Filler-sealant	Organosilicon two-component polysiloxane compound
Panel operating temperature, °C	- 70 ... + 110
Fastening the panel	2 screws in a wooden beam
Thermal insulator front	Air (inert gases)
Heat insulator rear	Air (polystyrene, mineral wool)
Protective front material	Tempered glass with a low content of iron oxide (optiwhite) 3 - 4 mm thick
Enclosure class	IP 65
Connectors	MC4 sealed
Cable length, mm	40
Cable section, mm^2	4 (6)
Heat carrier	Water (air, freon, antifreeze)

FROST-RESISTANT HIGH EFFICIENCY MATRIX

Number of coolant holes	1 (2) inlet, 1 (2) outlet
Fluid line connection	Hydraulic fittings
Thermal line of panels	Thermal insulation pipelines
Heat carrier volume in the panel radiator, l	0,3
The flow rate of the coolant, l/min	… 0,1 - 1 … (adjustable)
The temperature of the coolant at the inlet, °C	10 - 30
The temperature of the coolant at the outlet, °C	35 - 55
The cost of the panel, dol.	10 - 15

* Electrical values are shown in brackets when using solar cells with a one-sided contact grid and an electrical efficiency of about 20%, which are planned to be used in the mass production of stationary photovoltaic thermal modules in the form of roofing panels

HIGH VOLTAGE SOLAR MODULES

Matrix solar modules have a two-sided working surface, where due to the reflection of solar radiation on the back surface, the electric power and electrical efficiency of the module increases. Also, matrix solar modules have 2 times longer service life (40-50 years) (in comparison with foreign planar solar modules), the electrical efficiency is 22-28% at 50-200 times the concentration, and this value remains the same with increasing temperature to 60 °C, which simplifies the cooling system of the modules, and the module current increases in proportion to the concentration.

Along with planar photovoltaic thermal solar modules, concentrator photovoltaic thermal modules with the matrix solar cells are of great interest in view of the economy of silicon with a high degree of purification and the possibility of heating the coolant to higher temperatures. The solar concentrator photovoltaic thermal module (Kharchenko, Panchenko, Tikhonov & Vasant, 2018; Sinitsyn, Panchenko, Kharchenko & Vasant, 2020) consists of a paraboloid type concentrator that increases the concentration of solar radiation in the focal region on the cylindrical photoelectric part and the thermal part of the cylindrical radiator. Solar radiation entering the surface of the solar concentrator is reflected at tilt angles oriented in their zones in such a way that a sufficiently uniform illumination concentration is ensured on the photoelectric detector in the form of a cylinder from commutated high-voltage matrix solar modules and on the front end heat receiver of a cylindrical radiator for reheating running water. Concentrator solar photovoltaic thermal module, representing a solar module with a paraboloid type concentrator, is used in conditions where it is necessary to generate hot water along with generating electricity. In this module, due to the use of matrix high voltage solar cells, it was possible to achieve an increase in electrical efficiency along with an increase in the term of the rated power of solar cells.

The photoelectric characteristics of matrix solar modules were measured to determine the electrical efficiency at various levels of illumination at radiation intensities from 5 W/cm^2 to 25 W/cm^2 (250 times the concentration of solar radiation). The maximum values of electrical efficiency were obtained at a concentration of 50 times (Table 5). Table 5 presents the electrical characteristics of a matrix solar module with an area of 3 cm^2 at various degrees of exposure to light radiation.

At a radiation concentration of about 50 W/m^2, the electrical efficiency of a 3 cm^2 matrix high-voltage solar module was 28%, which significantly exceeds the electrical efficiency of standard planar solar modules (15 - 19%).

Table 5. Electrical characteristics of the matrix solar module at various degrees of exposure to light

Concentration	Electrical efficiency, %	Short-circuit current, mA	Open circuit voltage, V	Electric power, W	Fill factor current-voltage characteristics
51	28	233,7	24,4	4,3	0,75
99	25,2	415,9	24,4	7,5	0,74
142	24,4	577,6	24,9	10,4	0,73
192	22,3	711,6	24,9	12,9	0,73
221	22,2	825,5	25,2	14,7	0,71

The technology for the production of matrix solar modules is adapted to the conditions of industrial production, it does not use multi-stage diffusion, photolithography, network printing, vacuum metallization, etc., and the use of silver is excluded. The cost of manufacturing matrix solar modules is commensurate with the cost of planar modules per unit area. The development of the technology of third-generation matrix solar modules based on monocrystalline silicon will allow the creation of solar power plants with concentrators with lower unit costs per 1 kW of installed capacity and higher electricity production efficiency compared to coal-fired thermal power plants.

MATRIX HIGH-VOLTAGE FROST-RESISTANT SOLAR MODULE WITH INCREASED EFFICIENCY AND VOLTAGE OF MORE THAN 1000 V

A two-sided matrix solar module with dimensions of 700 × 100 mm (Figure 6) is designed to create high voltage direct current solar power plants (more than 1000 V) (Panchenko, Strebkov, Polyakov & Arbuzov, 2015). Such a high voltage allows using modules with transformerless inverters and connecting them to high-voltage DC lines with a voltage of 110-500 kV without converter substations. The development efficiency is noticeable when using a matrix module with concentrators in comparison with a planar module (of the same power). The matrix solar module with a length of 700 mm has an open circuit voltage of 1059 V and an operating voltage of 900 V. The cost of conversion substations is up to 30% of the cost of solar power plants, and to obtain an operating voltage of 900 V using traditional planar solar modules, more than 1500 planar need to be connected in series solar cells with dimensions of 156 × 156 mm each.

For the effective use of high-voltage matrix solar modules with solar concentrators, it is necessary to ensure uniform illumination of the solar module in the focal region and its cooling under the influence of concentrated solar radiation. The simplest way to do this is to use linear concentrators based on parabolic cylinders, linear lenses and Fresnel mirrors.

A comparison of the characteristics of planar and high-voltage matrix solar modules from single-crystal photoelectric solar cells is presented in the Table 6.

Modern processes of semiconductor electronics and nanotechnology will make it possible in the coming years to increase the electrical efficiency of converting concentrated solar radiation using matrix solar modules in industrial production to 30% and the limiting electric power to 50 W/cm^2 when converting concentrated solar radiation.

Figure 6. High-voltage matrix solar module with dimensions 700 × 100 mm

Table 6. Comparison of the characteristics of planar and high-voltage solar modules

Parameter	High-voltage solar module (compound filling technology)	Planar solar module (lamination technology)
Voltage, V	1000	12; 24
Warranty period of rated power, years	40 - 50	20 - 25
The average electrical efficiency with solar radiation of 1 kW/m^2, the spectrum of AM 1.5 and a temperature of 25 °C, %	12 - 14	18
Electrical efficiency with concentrated solar radiation of 100 kW/m^2, the spectrum of AM 1.5 and a temperature of 25 °C, %	22 - 28	1
Operating temperature	- 70 … + 110 °C	- 30 … + 60 °C
UV resistance	high	low
Corrosive agent in the manufacture	no	acetic acid
Corrosive agent for aging	no	acetic acid
Mechanical stress - manufacturing	no	yes
- aging	no	yes
Transparency for solar radiation with different wavelengths	90% (λ = 360 nm)	8% (λ = 360 nm)
	92% (λ = 400 nm	62% (λ = 400 nm)
	93% (λ = 600 … 1000 nm)	91% (λ = 600 … 1000 nm)

The production cost of high-voltage matrix solar modules is comparable with the cost of planar silicon solar modules per unit area and is 1000 times lower than the cost of cascade heterostructured solar modules based on $A_{III}B_V$ compounds with the same efficiency.

CONCLUSION

The considered frost-resistant solar modules of various designs find their application in solving the issues of autonomous energy supply to stationary consumers. The Arctic zone of the Russian Federation is an extremely promising area for the successful implementation and use of solar modules for energy supply of facilities. Stationary frost-resistant solar modules with an extended service life at the level of rated power allow receiving electricity both to an autonomous consumer and to an electric network, that is, to work with it in parallel. For an extended service life, the solar station will generate significantly more electrical energy, which increases the economic attractiveness of using such modules. When using frost-resistant solar photovoltaic roofing panels in the form of tiles, both construction and protective tasks are solved, and it becomes possible to receive electricity from solar radiation. The use of recycled plastic as a part of the roofing roof panel positively affects the environment, and the use of polysiloxane compound for sealing solar cells increases the duration of their rated power.

Unlike photovoltaic roofing panels, planar solar frost-resistant photovoltaic thermal roofing panels in the form of tiles, in addition to photovoltaic roofing panels, can also generate warm water, preserving their protective and construction functions, generating more electricity compared to standard laminated solar photovoltaic modules and having a competitive cost in terms of use secondary plastic in its composition. Concentrator frost-resistant solar photovoltaic thermal module, representing a solar module with a paraboloid type concentrator, is used in conditions where it is necessary to generate hot water along with generating electricity. In this module, due to the use of matrix high-voltage solar cells, it was possible to achieve an increase in electrical efficiency along with an increase in the term of the rated power of solar cells. The high-voltage matrix solar modules themselves have a great potential for implementation in concentrator solar modules, since an electrical efficiency of about 28% has already been obtained, which will reduce the cost of generated electricity and reduce the cost of installed capacity of the solar module. Such solar installations will be able to produce hot water, and their service life will be increased through the use of a two-component polysiloxane compound.

ACKNOWLEDGMENT

The researches were carried out on the basis of financing the state assignments of the All-Russian Institute for Electrification of Agriculture and the Federal Scientific Agroengineering Center VIM; on the basis of funding of the grant "Young lecturer of RUT" of the Russian University of Transport; on the basis of funding of the Scholarship of the President of the Russian Federation for young scientists and graduate students engaged in advanced research and development in priority areas of modernization of the Russian economy, direction of modernization: Energy efficiency and energy saving, including the development of new fuels, subject of scientific research: Development and research of solar photovoltaic thermal modules of planar and concentrator structures for stationary and mobile power generation.

REFERENCES

Agroecological atlas of Russia and neighboring countries. (n.d.). http://www.agroatlas.ru/ru/content/Climatic_maps/

ANSYS. (n.d.). https://www.ansys.com/

Babaev, B.D., Kharchenko, V., Panchenko, V. & Vasant, P. (2019). Materials and Methods of Thermal Energy Storage in Power Supply Systems. *Renewable Energy and Power Supply Challenges for Rural Regions*, 115-135. DOI: . doi:10.4018/978-1-5225-9179-5.ch005

Babaev, B. D., Kharchenko, V. V., & Panchenko, V. (2019). Development and Research of Phase-Transition and Thermochemical Materials for Heat Accumulation. Handbook of Research on Smart Computing for Renewable Energy and Agro-Engineering, 1-26. Doi:10.4018/978-1-7998-1216-6.ch001

Buonomano, A., Calise, F., & Vicidimini, M. (2016). Design, Simulation and Experimental Investigation of a Solar System Based on PV Panels and PVT Collectors. *Energies*, *9*(7), 497. doi:10.3390/en9070497

Daliento, S., & Lancellotti, L. (2010). 3D Analysis of the performances degradation caused by series resistance in concentrator solar cells. *Solar Energy*, *84*(1), 44–50. doi:10.1016/j.solener.2009.08.014

Degtyarev, K.S., Panchenko, V.A. & Mayorov, S.V. (2018). Perspektivy energosnabzheniya infrastrukturnyh ob"ektov na osnove vozobnovlyaemyh istochnikov energii v rossijskoj Arktike [Prospects for energy supply of infrastructure facilities based on renewable energy sources in the Russian Arctic]. *Sovremennye problemy sovershenstvovaniya raboty zheleznodorozhnogo transporta*, 79-91.

Dubey, S., & Tay, A. A. O. (2012). Experimental Study of the Performance of Two Different Types of Photovoltaic Thermal (PVT) Modules under Singapore Climatic Conditions. *Journal of Fundamentals of Renewable Energy and Applications*, *2*, 1–6. doi:10.4303/jfrea/R120313

Hosseini, R., Hosseini, N., & Khorasanizadeh, H. (2011). An Experimental study of combining a Photovoltaic System with a heating System. In *Proceedings of the World Renewable Energy Congress* (pp. 2993-3000). 10.3384/ecp110572993

Ibrahim, A., Othman, M. Y., Ruslan, M. H., Mat, S., & Sopian, K. (2011). Recent advances in flat plate photovoltaic/thermal (PV/T) solar collectors. *Renewable & Sustainable Energy Reviews*, *15*(1), 352–365. doi:10.1016/j.rser.2010.09.024

Ji, J., Lu, J., Chow, T., He, W., & Pei, G. (2007). A sensitivity study of a hybrid photovoltaic/thermal water-heating system with natural circulation. *Applied Energy*, *84*(2), 222–237. doi:10.1016/j.apenergy.2006.04.009

Kalogirou, S. A. (2001). Use of TRYNSYS for modeling and simulation of a hybrid PV-thermal solar system for Cyprus. *Renewable Energy*, *23*(2), 247–260. doi:10.1016/S0960-1481(00)00176-2

Ketola, B., McIntosh, K. R., Norris, A., & Tomalia, M. K. (2008). Silicones for photovoltaic encapsulation. *23rd European Photovoltaic Solar Energy Conference*, 2969-2973.

Kharchenko, V., Nikitin, B., Tikhonov, P., Panchenko, V., & Vasant, P. (2019). Evaluation of the Silicon Solar Cell Modules. *Intelligent Computing & Optimization. Advances in Intelligent Systems and Computing*, *866*, 328–336. doi:10.1007/978-3-030-00979-3_34

Kharchenko, V., Panchenko, V., Tikhonov, P., & Vasant, P. (2018). Cogenerative PV Thermal Modules of Different Design for Autonomous Heat and Electricity Supply. Handbook of Research on Renewable Energy and Electric Resources for Sustainable Rural Development, 86-119.

Laboratory of the Integrated Ecological and Geographical Research of the Arctic. (n.d.). http://www.ael-msu.org/

Panchenko, V., Chirskiy, S., & Kharchenko, V. (2019). Application of the Software System of Finite Element Analysis for the Simulation and Design Optimization of Solar Photovoltaic Thermal Modules. Handbook of Research on Smart Computing for Renewable Energy and Agro-Engineering, 106-131. DOI: 10.4018/978-1-7998-1216-6.ch005

Panchenko, V., & Kharchenko, V. (2019). Development and Research of PVT Modules in Computer-Aided Design and Finite Element Analysis Systems. Handbook of Research on Energy-Saving Technologies for Environmentally-Friendly Agricultural Development, 314-342. Doi:10.4018/978-1-5225-9420-8.ch013

Panchenko, V., Kharchenko, V., & Vasant, P. (2019). Modeling of Solar Photovoltaic Thermal Modules. *Intelligent Computing & Optimization. Advances in Intelligent Systems and Computing*, *866*, 108–116. doi:10.1007/978-3-030-00979-3_11

Panchenko, V. A. (2018). Solar Roof Panels for Electric and Thermal Generation. Applied Solar Energy, 54(5), 350-353. DOI: 10.3103/S0003701X18050146

Panchenko, V.A., Strebkov, D.S. & Persits, I.S. (2015). Razrabotka solnechnyh modulej s uvelichennym srokom nominal'noj raboty [Development of solar modules with extended nominal life]. *Nanostrukturirovannye materialy i preobrazovatel'nye ustrojstva dlya solnechnoj energetiki*, 91-94.

Panchenko, V.A., Strebkov, D.S., Polyakov, V.I. & Arbuzov, Yu.D. (2015). Vysokovol'tnye solnechnye moduli s napryazheniem 1000 V [High-voltage solar modules with a voltage of 1000 V]. *Al'ternativnaya energetika i ekologiya, 19*(183), 76-81.

Parretta, A., Bombace, M., Graditi, G., & Schioppo, R. (2005). Optical degradation of long-term, field-aged c-Si Photovoltaic modules. *Solar Energy Materials and Solar Cells*, *86*(3), 349–364. doi:10.1016/j.solmat.2004.08.006

Poulek, V., Strebkov, D. S., Persic, I. S., & Libra, M. (2012). Towards 50 years lifetime of PV panels laminated with silicone gel technology. *Solar Energy*, *86*(10), 3103–3108. doi:10.1016/j.solener.2012.07.013

Reda, S. M. (2007). Stability and photodegradation of phthalocyanines and hematoporphyrin doped PMMA as solar concentrators. *Solar Energy*, *81*(6), 755–760. doi:10.1016/j.solener.2006.10.004

Sandnes, B., & Rekstad, J. (2002). A photovoltaic/thermal (PV/T) collector with a polymer absorber plate: Experimental study and analytic model. *Solar Energy*, *72*(1), 63–73. doi:10.1016/S0038-092X(01)00091-3

Sinitsyn, S., Panchenko, V., Kharchenko, V., & Vasant, P. (2020). Optimization of Parquetting of the Concentrator of Photovoltaic Thermal Module. *Intelligent Computing & Optimization. Advances in Intelligent Systems and Computing*, *1072*, 160–169. doi:10.1007/978-3-030-33585-4_16

Strebkov, D., Panchenko, V., Irodionov, A., & Kirsanov, A. (2015). The development of roof solar panels. *Research in Agricultural Electric Engineering*, *3*(4), 123–127.

Strebkov, D. S., Kirsanov, A. I., & Panchenko, V. A. (2017). Solnechnye krovel'nye paneli dlya programmy "Odin million solnechnyh krysh v Rossii" [Solar roofing panels for the program "One Million Solar Roofs in Russia"]. *Upravlenie innovacionnym razvitiem arkticheskoj zony Rossijskoj Federacii. Sbornik izbrannyh trudov po materialam Vserossijskoj nauchno-prakticheskoj konferencii s mezhdunarodnym uchastiem*, 393-397.

Strebkov, D.S., Persits, I.S. & Panchenko, V.A. (2014). Solnechnye moduli s uvelichennym srokom sluzhby [Long life solar modules]. *Innovacii v sel'skom hozyajstve*, *3*(8), 154-158.

Three-dimensional modeling system. (n.d.). https://www.kompas.ru/

Wohlgemuth, J. H. (2003). Long Term Photovoltaic Module Reliability. *NCPV and Solar Program Review Meeting*, NREL/CD-520-33586.

KEY TERMS AND DEFINITIONS

Arctic Zone of the Russian Federation: Part of the territory of Russia, including the polar basin and the Arctic belt, which also includes the adjacent shelf with islands of mainland origin.

Computer-Aided Design System: An automated system that implements an information technology for performing design functions, is an organizational and technical system designed to automate the design process, consisting of personnel and a set of technical, software and other automation tools for its activities.

Cooling Radiator: A technical device used for removing heat for a liquid or gaseous coolant from a cooled object in order to cool the object and heat the coolant.

Finite Element Analysis System: A software package based on a numerical method for solving partial differential equations, as well as integral equations arising in solving problems of applied physics, which is widely used to solve problems of deformable solid mechanics, heat exchange, hydrodynamics and electrodynamics.

Heat Accumulator: Device for heat accumulation for the purpose of its further use.

Modeling: The study of objects on their models; building and studying models of real-life objects, processes or phenomena in order to obtain explanations of these phenomena, as well as to predict phenomena that interest the researcher.

Optimization: The process of maximizing profitable characteristics, ratios and minimizing costs.

Photovoltaic Thermal Module: A solar module, the design of which consists of photovoltaic solar cells for electrical conversion of solar radiation and heat absorber for their cooling and heat transfer to the coolant.

Receiving Surface: The surface of the photoelectric device/photovoltaic part of the device, which receives solar radiation.

Solar Battery: A combination of photoelectric converters (photocells) – semiconductor devices that directly convert solar energy into direct electric current, in contrast to solar collectors that which heat the coolant.

Solar Concentrator: A technical device designed to focus solar radiation into a focal spot with an increase in the concentration of solar radiation.

Solar Radiation Intensity: The density of solar radiation (energy illumination), coming per unit area of the photoelectric module.

Standard Conditions for Testing the Solar Cell: Test conditions, regulated by the density of the solar energy flux of 1000 W/m^2 and the temperature of photovoltaic solar cells of 25 °C.

Chapter 9
Development of Self-Organized Group Method of Data Handling (GMDH) Algorithm to Increase Permeate Flux (%) of Helical-Shaped Membrane

Anirban Banik
https://orcid.org/0000-0002-2987-2713
National Institute of Technology, Agartala, India

Sushant Kumar Biswal
https://orcid.org/0000-0002-1413-7206
National Institute of Technology, Agartala, India

Mrinmoy Majumder
National Institute of Technology, Agartala, India

Tarun Kanti Bandyopadhyay
National Institute of Technology, Agartala, India

ABSTRACT

The chapter focuses on enhancing the permeate flux of helical shaped membrane using group method of data handling (GMDH) algorithm. The variables such as operating pressure, pore size, and feed velocity were selected as input parameters, and permeate flux as model output. The uncertainty analysis evaluates the acceptability of the model, and it was found that values of Nash-Sutcliffe efficiency (NSE), the ratio of the root mean squared error to the standard deviation (RSR), percent bias (PBIAS) were close to the best value which shows the model acceptability. The effect of input parameters on model output is calibrated using sensitivity analysis. It shows that pore size is the most sensitive parameter followed by feed velocity. The optimum values of pore size, operating pressure, and feed velocity were calibrated and found to be 2.21μm, 1.31×10-03KPa, and 0.37m/sec, respectively. The errors in GMDH model were compared with multi linear regression (MLR) model. It shows that GMDH predicts results with minimum error. The predicted variable follows the actual variables with good accuracy.

DOI: 10.4018/978-1-7998-3970-5.ch009

INTRODUCTION

Membrane separation process consist of a semi-permeable barriers generally used to separate particle of interest from feed stream. The process separates the feed stream in two parts known as permeate and reject. Portion of effluent that passes through the membrane is known as permeate and the portion that cannot passes through the membrane is known as reject or concentrate. The process is widely used in treating effluent due to its ability to produce high quality of permeate flux. The process is considered as green technology as it doesn't include any addition of chemicals. De Souza et al studied the performance of two nano-filtration membranes in removing norfloxacin from synthetic pharmaceutical wastewater. Different parameters like solution concentration, pH and trans-membrane pressure are considered for evaluation. The study shows high norfloxacin rejection rate which was reported between 87 to 99.5%. The study also highlights the effect of pH on membrane selectivity and permeability (De Souza et al., 2018). Banik et al developed computational fluid dynamics model of disc membrane to evaluate the effect of membrane for improving the rubber industrial effluent. The study compares the ability of stationary and rotational membranes to enhance permeate flux. The Study shows superior ability of rotational membrane over stationary membrane (Banik, Bandyopadhyay, & Biswal, 2017). The chapter focuses on enhancing the permeate flux of helical shaped membrane implemented to treat the effluent generated from rubber industry.

Motivation, Background and Discussion

The membrane was known since eighteen century and was used to treat water. The separation ability of membrane depends on pore size. Based on pore size membrane can be divided into following categories like micro-filtration, ultra-filtration, nano-filtration and reverse osmosis (KUO & CHERYAN, 1983). Moreover membrane can also be classified based on membrane thickness, particle of transport, and charge of the membrane etc. Due to its ability to produce good quality of permeate flux, membrane is widely used in rubber industry, food industry, water and wastewater treatment, and paper industry(Mokhtar, Lau, Ismail, & Veerasamy, 2015)(Zhou, Zhao, Bai, Zhang, & Tang, 2012). Membrane separation still finds limited application due to its rapid fouling tendency. The deposition and accumulation of solutes and other constituents present in the feed stream on membrane bed is termed as fouling. Fouling of the membrane is due to partial and complete pore blocking. The fouling the membrane also leads to lowers the permeate flux thus reducing the membrane acceptability(Guo et al., 2019). So, GMDH algorithm was implemented to enhance the permeate flux and antifouling property of the membrane.

Choudhury et al prepared ceramic ultra-filtration membrane implementing CuO nano particles to remove chromium (VI). The study also used response surface methodology (RSM) to optimize the membrane process. The ceramic membrane illustrates high chromium (VI) ions rejection of 88.08% (Choudhury, Mondal, Majumdar, Saha, & Sahoo, 2018). Li et al used PAFSSB as pre treatment technique for RO membrane to treat pulp and paper wastewater. The paper illustrates PAFSSB increase the membrane efficiency and COD removal rate(Li & Zhang, 2011). Ejraei et al implemented a hybrid system combined of adsorption, photocatalytic degradation, and membrane separation process to treat wastewater generated from paper and pulp industry of Iran. The study highlights the best performance of membrane separation process among other tested method but superior separation performance is achieved when hybrid system combined of photocatalytic degradation, adsorption, and membrane separation process used in series(Ejraei, Aroon, & Ziarati Saravani, 2019). Zhao et al used flat sheet ceramic membrane for direct membrane filtration process for improving the quality of domestic sewage. The coagulant was used before membrane filtration process to remove pollutant and to mitigate fouling. The study shows that membrane removes 90% and 99% of chemical oxygen demand (COD) and phosphorus respectively. The membrane also exhibit long term operation and maintain stable flux throughout its operation(Zhao, Li, Li, & Li, 2019). Banik et al implemented response surface methodology to enhance the permeate flux of disc membrane to treat the rubber industrial effluent. The independent parameters like Pore size, inlet velocity and operating pressure were used to develop the RSM model. The proposed membrane exhibit permeate flux of 79.77% when membrane is allowed to operated under optimum conditions (Banik, Dutta, Bandyopadhyay, & Biswal, 2018). For achieving the goal of zero discharge Basu etal used microfiltration to treat molasses based alcohol distillery wastewater. Study illustrates that electro-coagulation followed by micro filtration is an effective method of treating the effluent generated from molasses based alcohol distilleries(Basu, Mukherjee, Kaushik, Batra, & Balakrishnan, 2015). Soleimani et al used polyacrylonitrile ultra-filtration membrane to improve the quality of oily wastewater. It implements artificial neural network (ANN) to predict the permeate flux and fouling resistance and then uses genetic algorithm to optimize the operating condition of the membrane. The study shows the ability of GA to predict maximum permeate flux and minimum fouling resistance (Soleimani, Shoushtari, Mirza, & Salahi, 2013). Chew et al combines the Darcy's law on cake filtration and artificial neural network to illustrate the dead end ultra filtration process. To develop the model the parameters such as turbidity, filtration time, and trans-membrane pressure were selected as model input. The method shows promising results and provides an alternative approach besides conventional laboratory approach. Predict(Chew, Aroua, & Hussain, 2017). Nourbakhsh et al designed artificial neural network combined with response surface methodology to predict permeate flux of red plum juice during membrane process. The input parameters such as TMP, temperature, Pore size, cross-flow velocity and processing time were used to develop the model. The artificial neural network model shows satisfactory non-linear dynamic behaviour of permeate flux of red plum juice at different operating conditions during membrane processing (Nourbakhsh, Emam-djomeh, Omid, & Mirsaeedghazi, 2014).

From the Study, It was found that membrane were limited in its use due to its fouling problem. To address the issue of fouling, many studies implements experimental investigation to enhance antifouling property and permeate flux. The major limitation of the present studies were time consuming, and expensive to perform. To overcome the problems, GMDH is used to predict the optimum parameters and to enhance permeate flux. GMDH is used for its adaptive quality and its ability to avoid any misleading assumption from the developer. It also has the ability to predict results with high accuracy. Some of the disadvantages of the GMDH algorithm are: algorithm is less responsive towards small dataset and gathering knowledge regarding the internal processing of the algorithm is also difficult. To accumulate information regarding the internal working of the algorithm, sensitivity analysis is implemented.

Novelty and Objective of the Study

GMDH algorithm was used to predict the optimum condition of helical shaped membrane to enhance permeate flux which has not been reported earlier. Moreover, the study also report about the effect of the input parameters on model output.

The Objective of the Chapter is to enhance the permeate flux of helical shaped membrane. Group method of data handling (GMDH) algorithm is proposed to predict the best operating variables. The GMDH model is developed by using parameters like operating pressure, inlet velocity and pore size as input parameters and permeate flux as model output. Uncertainty analysis used to evaluate the acceptability of the membrane. The individual effect of input parameters on model output is calculated using sensitivity analysis. The study also compares the average error of GMDH model with the error present in Multi linear regression model to justify the use of GMDH algorithm for prediction purpose.

METHODOLOGY

General Overview of GMDH

The general Group method of data handling (GMDH) is an adaptive algorithm used for developing a higher order polynomial form using equation 1.

$$Y = D_0 + \sum_{i=1}^{m} D_1 X_i + \sum_{i=1}^{m}\sum_{j=1}^{m} D_2 X_i X_j + \sum_{i=1}^{m}\sum_{j=1}^{m}\sum_{k=1}^{m} D_3 X_i X_j X_k + \ldots\ldots \quad (1)$$

Here m illustrates the number of input variables from X_1 to X_m and Y denotes the output of the model. In spite of the fact that equation 1 is similar to higher order polynomial regression, the manner in which the equation was developed but it differs from the techniques of regression analysis. The procedure of GMDH algorithm said to be similar to the way of natural selection.

GMDH Modelling and Data Normalization

Group method of data handling (GMDH) is an adaptive algorithm widely used for prediction, optimization and data mining etc (Ivakhnenko, 1971). Algorithm ignores any hazy assumptions of the developer and develop model with minimum error using the existing relation in the feed dataset. The algorithm consists of input, hidden and output layer where each layer consists of two input neuron and one output neuron. It implements polynomial function to predict the optimum parameters. The three independent parameters like operating pressure (KPa), pore size (μm), and feed velocity (m/sec) were used as input variables for model development. Before importing the feed dataset to the GMDH algorithm, all datasets were normalized in a scale of 0 to 1 using equation 2.

$$X_{normalization} = \left(\frac{X - X_{low}}{X_{high} - X_{low}} \right) \quad (2)$$

Here, X represents the data points, X_{high} and X_{low} are the maximum and minimum value of the feed data set respectively.

Problem Formulation

Optimization is a mathematical technique of finding optimum solution within the predefined boundary (Piryonesi & Tavakolan, 2017)(Radhakrishnan & Vijayarajan V., 2019)(Vaskov, Tyagunov, Shestopalova,

Deryugina, & Ishchenko, 2018). In the present study, the parameters such as operating pressure, inlet velocity and pore size were considered as inputs for the developed model. In the developed model, inputs were varied with a solo objective of maximizing permeate flux of the membrane. Numerically the problem of enhancing permeates flux can be defined by equation 3:

$$Y_{max}(permeateflux) = f(X_1;X_2;X_3) \quad (3)$$

Subjected to:

$0.2\mu m \leq X_1$(Pore size) $\leq 3\mu m$

200 KPa $\leq X_2$ (Operating pressure) $\leq 2.3x10^{-03}$ KPa

17.9 cm/sec $\leq X_3$ (Feed velocity) $\leq$ 51.26 cm/sec

Where X_1 denotes pore size, X_2 denotes operating pressure, and X_3 illustrates inlet velocity. Y is the output of the model and it represents permeate flux (%) of the membrane.

Uncertainty Analysis

The accuracy and acceptability of GMDH model were estimated by implementing different statistical techniques like percent bias (PBIAS), ratio of the root mean squared error to the standard deviation of the observations (RSR), and Nash-Sutcliffe efficiency (NSE) (D. N. Moriasi et al., 2007). The statistical techniques like PBIAS, RSR and NSE were estimated using the below mentioned equations 4-6;

$$PBIAS = \frac{\sum_{i=1}^{m}(Y_{i,\exp.} - Y_{i,pred.})}{\sum_{i=1}^{m}(Y_{i,\exp.})} \times 100 \quad (4)$$

$$RSR = \frac{\sqrt{\sum_{i=1}^{m}(Y_{i,\exp.} - Y_{i,pred.})}}{\sqrt{\sum_{i=1}^{m}(Y_{i,\exp.} - Y_{i,mean})}} \quad (5)$$

$$NSE = 1 - \frac{\sum_{i=1}^{m}(Y_{i,\exp.} - Y_{i,pred.})^2}{\sum_{i=1}^{m}(Y_{i,\exp.} - Y_{i,mean})^2} \quad (6)$$

Where m, $Y_{i,exp.,}$ $Y_{i,pred.}$, and $Y_{i,mean}$ denotes the number of experimental data, experimental data, predicted data, and mean of experiment data, respectively.

RESULTS AND DISCUSSIONS

Model Reliability

Adaptive Group method of data handling (GMDH) is an evolutionary algorithm used for computer based mathematical modelling to solve different problems of Complex nature. GMDH is implemented in different fields like prediction, optimization, pattern recognition, and discovery of Knowledge etc (Banik, Bandyopadhyay, Biswal, & Majumder, 2019) (Banik, Biswal, Majumder, & Bandyopadhyay, 2018). The flow phenomena through helical shaped membrane is a complex one, and the GMDH algorithm is used to develop the mathematical model to predict permeate flux. Equation 7 shows the model equation with activation function constructed by using GMDH algorithm. Where Independent parameters like pore size, feed velocity, and operating pressure is selected as input parameters and permeate flux is selected as output of constructed model. The equation 7 illustrates that permeate flux of helical membrane depends on linear interaction of pore size and operation pressure, and 2-way interaction of pore size and feed velocity. Weights in the model equation are used to pronounce the optimum variables.

$$\textit{Permeate Flux }(\%) = -0.016 + [(\textit{Pore Size}) \times 13.6] + [(\textit{operating Pressure})^{-2\times 2678.7}] + [(\textit{Pore Size}) \times (\textit{Feed Velocity}) \times 0.01] \quad (7)$$

The normalized plot of permeate flux (%) of helical shaped membrane is illustrated by using figure 1. Where the blue line in the background represents the actual variables used to construct the GMDH model. Black line shows the training and testing data set which is used as learning dataset for GMDH algorithm. Predicted variables are illustrated by red line in figure 1. Residual plot (actual dataset-predicted dataset) of membrane is illustrated using figure 2. In figure 2 blue and red lines shows the training and predicted residual respectively. The adequacy of the constructed model is assessed by using the value of correlation coefficient. The value of correlation coefficient is calibrated from figure 3 and was found to be 0.908. As the calibrated R^2 value is close to 1, it denotes that predicted variables hold good agreement with the experimental variables. Figure 3 also illustrates that predicted variables follow experimental variables with higher accuracy. From the post processed results of GMDH algorithm, mean absolute error (MAE), root mean square error (RMSE), and standard deviation are calculated and the values are found to be 0.00025, 0.000032, and 4.1×10^{-06}. The calibrated results are close to zero which denotes the acceptability of the constructed model.

Uncertainty Analysis

The uncertainty analysis gives an idea about the developed model and its overall performance. The parameters like PBIAS, NSE, and RSR are used to evaluate the model performance. Table 1 illustrates the uncertainty analysis of developed GMDH model. Table 2 shows the calibrated value of NSE, PBIAS, and RSR which are found to be 0.98, 2.8×10^{-11}, and 3.2×10^{-235} respectively. The calibrated values are found to be in permissible range. Hence, it can be concluded that GMDH model is acceptable and workable for predicting permeate flux.

Figure 1. Normalized plot of permeate flux (%) of helical shaped membrane

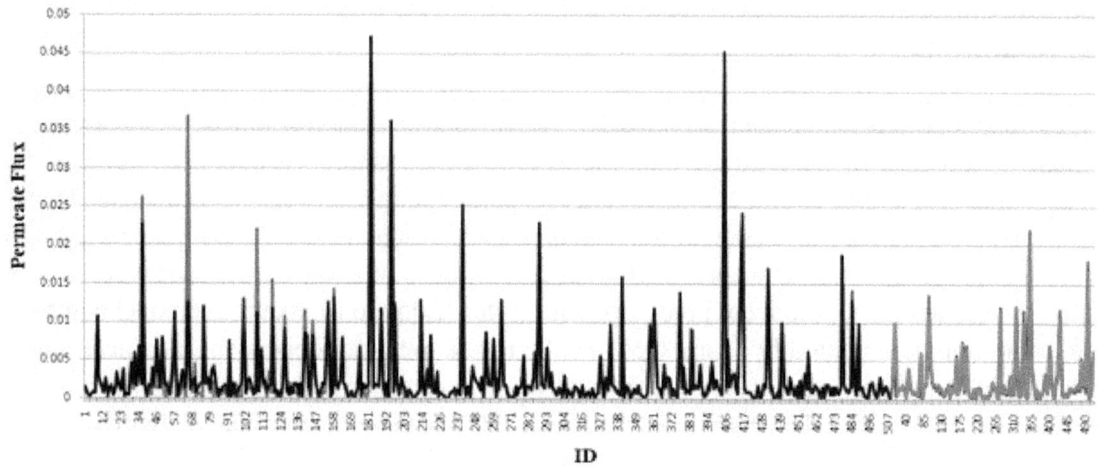

Figure 2. Residual Plot of helical shaped membrane

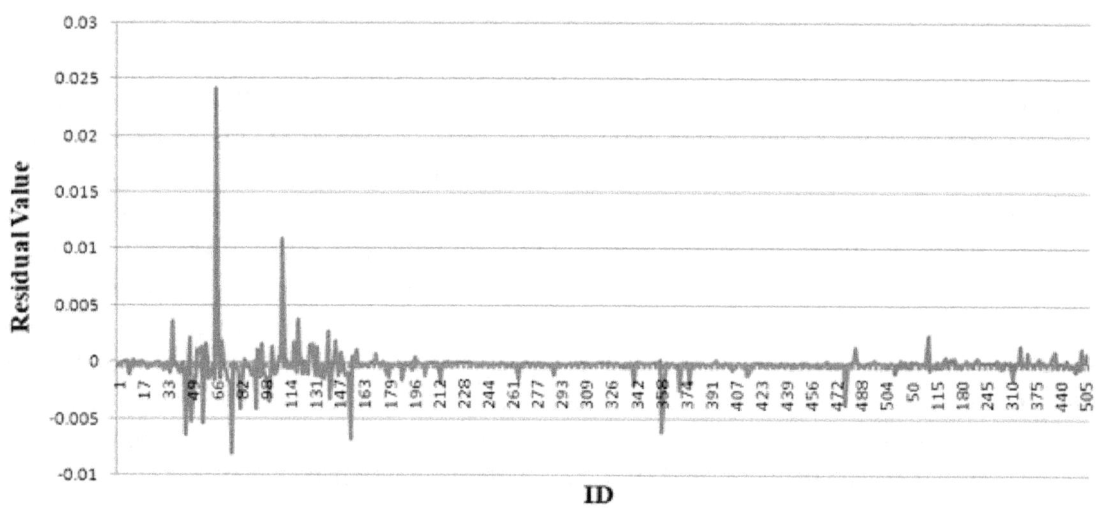

Comparative Study Between GDMH and MLR Model

The uncertainties present in GMDH model are compared with Multi linear regression model to evaluate the acceptability of the proposed model. The parameters like PBAIS, RSR and MAE are considered for the comparative analysis. Table 2 illustrates that GMDH model predicts the results with minimum error compared to the developed MLR model. The average error value of GMDH and MLR model are found to be 0.00025 and 0.00101 respectively. From the study it can be concluded that GMDH has reliable performance and adequate for prediction of permeate flux.

Figure 3. Plot between actual and predicted variables

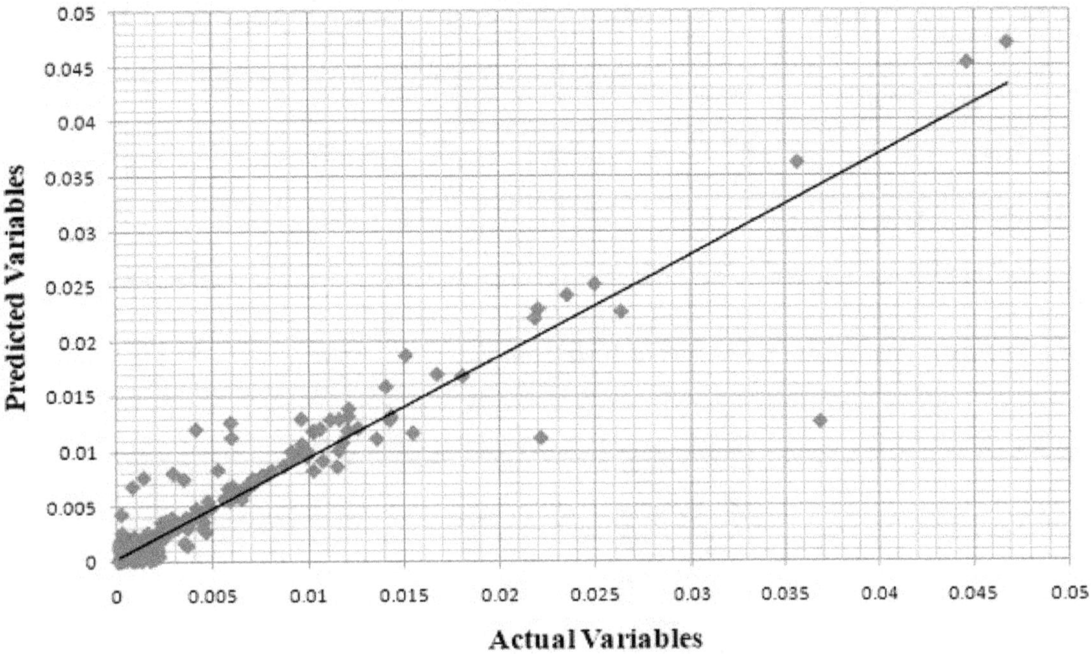

Table 1. Uncertainty analysis of GMDH model

SL. No.	Model evaluation methods	Calculated Value	Best Value
1.	NSE	0.98	1
2.	PBIAS	2.8×10^{-11}	0
3.	RSR	3.2×10^{-235}	0

Sensitivity Analysis

Sensitivity analysis is a method of determining the individual effect of input parameters on output of the developed model. Local one at a time method is used for sensitivity analysis, and the method includes partial derivative of output to input of the model (Saltelli, 2002). The partial derivative values are illustrated in bar graph, where the length of the graph shows the effect of input variable on the model outcome. The bar with maximum length illustrates that corresponding independent variable has maximum effect on output. Figure 4 shows the sensitivity analysis of developed GMDH model and it shows that height of the bar correspond to pore size is maximum which shows that pore size has maximum effect on model output.

Table 2. Comparative Uncertainty analysis between GMDH and MLR model

SL. No.	Model	Methods	Calibrated Value	Best value	Average
		PBIAS	$2.8x10^{-11}$	0	
1	GMDH	RSR	$3.2x10^{-235}$	0	0.00025
		MAE	0.00025	0	
		PBIAS	$1.4x10^{-04}$	0	
2	MLR	RSR	$6.9x10^{-11}$	0	0.00101
		MAE	0.00087	0	

Figure 4. Sensitivity Analysis of Constructed GMDH model

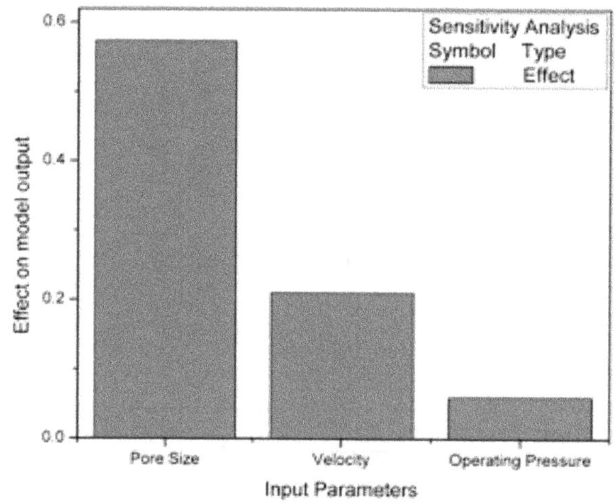

Optimization, Optimum parameters and Experimental Analysis

Optimization is a method of finding optimum operating conditions within the predefined boundary (Tsiptsis, Liimatainen, Kotnik, & Niiranen, 2019)(Mironyuk, Smirnov, Sokolov, & Proshkin, 2019; Panchenko, Chirskiy, & Kharchenko, 2019)(Geleta & Manshahia, 2019). In the present study, operating parameters like feed velocity, pore size, and operating pressure are considered as input parameters for developing GMDH model. The permeate flux of helical membrane is maximized by implementing maximization objective function [f (max)]. The membrane illustrates permeate flux of 76.23% corresponding to the predicted optimum conditions (Pore size= 2.21µm; feed velocity=0.37 m/sec; operating pressure=1.31×10^{-03} KPa). When the model predicted results are replicated in laboratory, membrane shows permeate flux of 74.81%. From experimental analysis, it can be concluded that GMDH predicted results hold good agreement with experimental results.

CONCLUSION

The investigation was conducted to enhance the permeate flux of helical shaped membrane using GMDH algorithm. The research illustrates that parameters such as pore size, feed velocity, and operating pressure affects the permeate flux of membrane. The values of the parameters such as MAE, RMSE, and standard deviation were calculated. The calculated values were close to best value which shows the acceptability of the constructed model. The study shows that predicted results follow the actual results with high accuracy which was concluded from the value of correlation coefficient (R^2=0.908). A GMDH model was developed to study the effect of three input variables on model output. The study uses uncertainty analysis to estimate the error present in the developed model. The study shows that values of NSE, PBIAS, and RSR were close to the best values which represent the acceptability and workability of the model. The study also compared the error present in GMDH and MLR model and shows that GMDH pronounce results with better accuracy than MLR model. Thus, justifying the use of GMDH algorithm to predict and optimize the membrane separation process. Sensitivity analysis illustrates the individual effect of input parameters on model outcome. Pore size was found to be the most sensitive parameter followed by feed velocity during the separation process. The optimum solution was obtained using GMDH algorithm, the optimal values of pore size, feed velocity, and operating pressure were calculated and found to be 2.21µm, 0.37 m/sec, and 1.31×10^{-03} KPa respectively which results in 76.23% (predicted) of permeate flux. Experimental validation illustrates the actual permeate flux and found to be 74.81%. Results obtained can used to increase the permeate flux and minimize the pumping cost of helical shaped membrane.

FUTURE RESEARCH DIRECTIONS

The present study deals with enhancing permeate flux of helical shaped membrane using GMDH algorithm. The study only shows the linear effect of input parameters on model outcome. It does not illustrate the 2-way interactive and square effect on the model outcome which will be covered in future studies. Moreover, future studies will focus on lowering the operating pressure thus lowering the pumping cost. To create an actual marketable product, cost of the product is very important in today's price sensitive markets. So, cost optimizations study will be carried out to minimize the cost of the membrane installation and operation thus achieving the ultimate goal of marketable product.

REFERENCES

Banik, A., Dutta, S., Bandyopadhyay, T. K., & Biswal, S. K. (2018). Prediction of maximum permeate flux (%) of disc membrane using Response Surface Methodology (RSM). *Canadian Journal of Civil Engineering*, *46*(6), 299–307. doi:10.1139/cjce-2018-0007

Banik, A., Bandyopadhyay, T. K., Biswal, S. K., & Majumder, M. (2019). Prediction of Maximum Efficiency of Vertical Helical Coil Membrane Using Group Method of Data Handling (GMDH) Algorithm. In Intelligent Computing and Optimization (pp. 489–500). Springer.

Banik, A., Bandyopadhyay, T. K., & Biswal, S. K. (2017). Computational fluid dynamics simulation of disc membrane used for improving the quality of effluent produced by the rubber industry. *International Journal of Fluid Mechanics Research, 44*(6), 499–512. doi:10.1615/InterJFluidMechRes.2017018630

Banik, A., Biswal, S. K., Majumder, M., & Bandyopadhyay, T. K. (2018). Development of an adaptive non-parametric model for estimating maximum efficiency of disc membrane. *Int. J. Convergence Computing, 3*(1), 3–19. doi:10.1504/IJCONVC.2018.091111

Basu, S., Mukherjee, S., Kaushik, A., Batra, V. S., & Balakrishnan, M. (2015). Integrated treatment of molasses distillery wastewater using microfiltration (MF). *Journal of Environmental Management, 158*, 55–60. doi:10.1016/j.jenvman.2015.04.037 PMID:25956444

Chew, C. M., Aroua, M. K., & Hussain, M. A. (2017). A practical hybrid modelling approach for the prediction of potential fouling parameters in ultrafiltration membrane water treatment plant. *Journal of Industrial and Engineering Chemistry, 45*, 145–155. doi:10.1016/j.jiec.2016.09.017

Choudhury, P., Mondal, P., Majumdar, S., Saha, S., & Sahoo, G. C. (2018). Preparation of ceramic ultrafiltration membrane using green synthesized CuO nanoparticles for chromium (VI) removal and optimization by response surface methodology. *Journal of Cleaner Production, 203*(6), 511–520. doi:10.1016/j.jclepro.2018.08.289

De Souza, D. I., Dottein, E. M., Giacobbo, A., Siqueira Rodrigues, M. A., De Pinho, M. N., & Bernardes, A. M. (2018). Nanofiltration for the removal of norfloxacin from pharmaceutical effluent. *Journal of Environmental Chemical Engineering, 6*(5), 6147–6153. doi:10.1016/j.jece.2018.09.034

Ejraei, A., Aroon, M. A., & Ziarati Saravani, A. (2019). Wastewater treatment using a hybrid system combining adsorption, photocatalytic degradation and membrane filtration processes. *Journal of Water Process Engineering, 28*(August), 45–53. doi:10.1016/j.jwpe.2019.01.003

Geleta, D. K., & Manshahia, M. S. (2019). Optimization of Hybrid Wind and Solar Renewable Energy System by Iteration Method. In Intelligent Computing & Optimization (pp. 98–107). doi:10.1007/978-3-030-00979-3_10

Guo, J., Zhu, X., Dong, D., Wang, K., Guan, Y., & Wang, L. (2019). The Hybrid process of preozonation and CNTs modification on hollow fiber membrane for fouling control. *Journal of Water Process Engineering, 31*(February), 100832. doi:10.1016/j.jwpe.2019.100832

Ivakhnenko, A. G. (1971). Polynomial Theory of Complex Systems. *IEEE Transactions on Systems, Man, and Cybernetics. SMC, 1*(4), 364–378. doi:10.1109/TSMC.1971.4308320

Kuo, K.-P., & Cheryan, M. (1983). Ultrafiltration of Acid Whey in a Spiral-Wound Unit: Effect of Operating Parameters on Membrane Fouling. *Journal of Food Science, 48*(4), 1113–1118. doi:10.1111/j.1365-2621.1983.tb09172.x

Li, S., & Zhang, X. (2011). The study of PAFSSB on RO pre-treatment in pulp and paper wastewater. *Procedia Environmental Sciences, 8*(November), 4–10. doi:10.1016/j.proenv.2011.10.003

Mironyuk, S. S., Smirnov, A., Sokolov, A. V., & Proshkin, Y. (2019). Optimization of Spectral Composition and Energy Economy Effectiveness of Phyto-Irradiators With Use of Digital Technologies. In Handbook of Research on Energy-Saving Technologies for Environmentally-Friendly Agricultural Development (pp. 191–212). doi:10.4018/978-1-5225-9420-8.ch008

Mokhtar, N. M., Lau, W. J., Ismail, A. F., & Veerasamy, D. (2015). Membrane distillation technology for treatment of wastewater from rubber industry in Malaysia. *Procedia CIRP*, *26*, 792–796. doi:10.1016/j.procir.2014.07.161

Moriasi, D. N., Arnold, J. G., Van Liew, M. W., Bingner, R. L., Harmel, R. D., & Veith, T. L. (2007). Model Evaluation Guidelines for Systematic Quantification of Accuracy in Watershed Simulations. *Transactions of the ASABE*, *50*(3), 885–900. doi:10.13031/2013.23153

Nourbakhsh, H., Emam-djomeh, Z., Omid, M., Mirsaeedghazi, H., & Moini, S. (2014). Prediction of red plum juice permeate flux during membrane processing with ANN optimized using RSM. *Computers and Electronics in Agriculture*, *102*, 1–9. doi:10.1016/j.compag.2013.12.017

Panchenko, V., Chirskiy, S., & Kharchenko, V. V. (2019). Application of the Software System of Finite Element Analysis for the Simulation and Design Optimization of Solar Photovoltaic Thermal Modules. In Handbook of Research on Smart Computing for Renewable Energy and Agro-Engineering (pp. 106–131). doi:10.4018/978-1-7998-1216-6.ch005

Piryonesi, S. M., & Tavakolan, M. (2017). A mathematical programming model for solving cost-safety optimization (CSO) problems in the maintenance of structures. *KSCE Journal of Civil Engineering*, *21*(6), 2226–2234. doi:10.100712205-017-0531-z

Radhakrishnan, S., & Vijayarajan, V. (2019). Optimized Deep Learning System for Crop Health Classification Strategically Using Spatial and Temporal Data. In Deep Learning Techniques and Optimization Strategies in Big Data Analytics (pp. 233–250). doi:10.4018/978-1-7998-1192-3.ch014

Saltelli, A. (2002). Sensitivity analysis for importance assessment. *Risk Analysis*, *22*(3), 579–590. doi:10.1111/0272-4332.00040 PMID:12088235

Soleimani, R., Shoushtari, N. A., Mirza, B., & Salahi, A. (2013). Experimental investigation, modeling and optimization of membrane separation using artificial neural network and multi-objective optimization using genetic algorithm. *Chemical Engineering Research & Design*, *91*(5), 883–903. doi:10.1016/j.cherd.2012.08.004

Tsiptsis, I. N., Liimatainen, L., Kotnik, T., & Niiranen, J. (2019). Structural optimization employing isogeometric tools in Particle Swarm Optimizer. *Journal of Building Engineering*, *24*, 100761. doi:10.1016/j.jobe.2019.100761

Vaskov, A. G., Tyagunov, M. G., Shestopalova, T. A., Deryugina, G. V., & Ishchenko, I. (2018). Structure and parameter optimization of renewable- Based hybrid power complexes. In Handbook of Research on Renewable Energy and Electric Resources for Sustainable Rural Development (pp. 352–382). doi:10.4018/978-1-5225-3867-7.ch015

Zhao, Y., Li, P., Li, R., & Li, X. (2019). Direct filtration for the treatment of the coagulated domestic sewage using flat-sheet ceramic membranes. *Chemosphere, 223*, 383–390. doi:10.1016/j.chemosphere.2019.02.055 PMID:30784745

Zhou, Y., Zhao, H., Bai, H., Zhang, L., & Tang, H. (2012). Papermaking Effluent Treatment: A New Cellulose Nanocrystalline/Polysulfone Composite Membrane. *Procedia Environmental Sciences, 16*, 145–151. doi:10.1016/j.proenv.2012.10.020

Chapter 10
Transmission Risk Optimization in Interconnected Systems:
Risk-Adjusted Available Transfer Capability

Nimal Madhu M.
https://orcid.org/0000-0002-5981-2033
Asian Institute of Technology, Thailand

Jai Govind Singh
https://orcid.org/0000-0002-0162-3360
Asian Institute of Technology, Thailand

Vivek Mohan
https://orcid.org/0000-0002-3304-7150
National Institute of Technology Tiruchirappalli, India

Weerakorn Ongsakul
Asian Institute of Technology, Thailand

ABSTRACT

Available transfer capability is a key indicator of transmission reliability and varies with the variation in power flow pattern through the network. ATC determination considering the uncertainties in renewable generation and demand is of key significance for the safe and economic operation of power system, especially in a competitive market environment. A two-stage, risk-adjusted, generation dispatch minimizing the variation in ATC, caused by the changes in renewable energy power output and the change in load, is discussed. The solution strategy is designed for a network operator, considering the ease of use and practicality. A combined transmission-distribution system with solar, wind, and conventional dispatchable energy sources is developed, and ATC for the systems is estimated combining continuation power flow and power transfer sensitivity factor methods. The joint probability distribution function of ATC is derived using individual discrete probabilities renewable power generation and loads. Risk, quantified as the variance of ATC, is minimized using stochastic weight trade-off non-dominated sorting particle swarm optimization, considering various goals of the network operator, for example, maximizing overall system performance and minimizing the renewable energy risk.

DOI: 10.4018/978-1-7998-3970-5.ch010

INTRODUCTION

A world-wide paradigm shift of the electric power production and distribution platform, from monopolistic to decentralized operation, is observed in recent times. An upsurge in the number of independent power producers both on high and low voltage sides, have intensified the competitive nature of the market, escalating the operational complexity of transmission system operator or owner (Unni, Ongsakul, & Madhu M., 2019). Available transfer capability (ATC) is a crucial financial index for the transmission operator, which physically indicates the amount of transfer capability left over after allowing the currently committed bilateral transactions. It also provides the scope of future commercial activities (Li, Li, Ni, & Wu, 2003; Wu, 2007). Mathematically, ATC is computed as given in (1) (NERC, 1996).

$$ATC = \text{`}TTC\text{'} - CBM - TRM - \text{`}ETC\text{'}. \tag{1}$$

Where, TTC is Total transfer capacity and CBM is capacity benefits margin, which are not constant as per the infrastructure and policies of the power system. TRM represents Transmission reliability margin and ETC denotes Existing transfer commitments, which are the base indicators for ATC, since the rest are constants.

MOTIVATION

The most popular of the renewable energy resources, solar PV and wind power generation systems exhibit stochastic nature (Daus, Kharchenko, & Yudaev, 2018; Vasant, Zelinka, & Weber, 2019). The estimation of ATC should be subjected to the variations in generation and load, especially if renewable energy generation is involved. The works including the uncertainties of renewable generations as well as loads at the same time, are scarce. Many of the studies adopt a formulation that is time consuming and hence, less suitable for short-term dispatch. Hence, an attempt to formulate a network operator compliant, ATC estimation algorithm, that can consider the uncertainty associated with renewable energy sources and load, which can be used for short-term operational decision making.

BACKGROUND/RELATED WORKS

Popular deterministic methods available for calculating ATC include, continuation power flow, repeated power flow, power transfer sensitivity factors, optimal power flow (Wu, 2007) and DC-power flow methods (Chiang, Flueck, Shah, & Balu, 1995). Recent literature have introduced the application of machine learning techniques (Vaithilingam & Kumudini Devi, 2013) and hybrid mutation PSO (Farahmand et al., 2012) to determine ATC. Stochasticity associated with generations, including solar and wind systems and load induce both financial and technical risk into the scheduled operation scenario of TSO, which is not considered in the above literature. The effect of these deviations from the forecast is reflected in the ATC values of corresponding lines, affecting the various bilateral transactions. Probabilistic ATC estimation also gained ample research interests owing to these reasons. In (Rodrigues & Da Silva, 2007; Wei, Li, & Zhou, 2015), the stochastic effects of load change and equipment failure on ATC is considered.

Advanced methods for ATC estimation are presented in recent literature. In (Ghawghawe & Thakre, 2006), a combination of sensitivity analysis and power transfer distribution factors is used to determine ATC. The method is suitable for planning domain and real time control and the study is carried out on a 6-bus system. A multi-output feed forward neural network is also tested for online ATC estimation (Prathiba, Moses, Devaraj, & Karuppasamypandiyan, 2015). The data for training ANN is made with the help of a repeated power flow (RPF) method. Pattern search optimization in conjunction with Newton Raphson technique is used for ATC estimation in (Shukla, Lakshmi, & Singh, 2017). IEEE 24-bus RTS is used for testing in both (Prathiba et al., 2015; Shukla et al., 2017). The effect of the entities in the system causing stochasticity in the scheduling and dispatch process like renewable generation, load demand, component outage, etc. are not considered in the above literatures. While in (Pandey, Pandey, Tapaswi, & Srivastava, 2010), load variation and single line outage contingency situation is considered, and a Levenberg-Marquardt algorithm neural network-based ATC estimation approach is discussed. The effect of variation in renewable generation and the effect of their penetration is not considered. A distributed computing environment is required for the fast computing of ATC in this method.

MAIN FOCUS OF THE CHAPTER[1]

Discussion of Issues, Controversies, and Limitations

Effect of the uncertain nature of renewable energy systems, which is frequent and is prone to escalation with their higher penetration levels (Ozkan, Küçük, Buhan, Demirci, & Karagoz, 2020) should be addressed while estimating ATC. Also, in the case of transmission-distribution systems, the presence of uncertain sources on both the sides, increase the unpredictability in the transfer of electric power, mutually and internally. Moreover, the power transfer capability in case of fully fledged microgrid power markets with prosumers and their internal transactions, where distribution system operator is tasked to provide feasible real-time power dispatch information to local controllers (Ghawghawe & Thakre, 2006), demand more attention to these impacts and can have extensive effects on such systems. Existing ATC estimation methods either stress on the speed of calculation ignoring the above-mentioned details or are suitable only to the day-ahead or planning domains. Hence, a feasible solution, addressing the shortcomings and making use of proven ideas, in a practical way is addressed here.

An optimal power dispatch strategy, that helps to hedge the operational risk of the network operator, is discussed considering forecast errors of renewable energy and loads. A risk parameter, quantified as the variance in ATC values with renewable and load forecast error, is formulated that can be used as a financial or technical feasibility assessment measure. A combination of continuation power flow (CPF) and power transfer sensitivity factors (PTSF) methods is formulated as a two-stage process. The first stage of the planning horizon is aimed at the creation of a database of PTSF parameters for every possible uncertainty combination, using CPF based ATC estimation process. In the second stage, the obtained PTSF parameters are used to estimate ATC values and associated risk, in a faster way compared to CPF or RPF, providing a minimal risk power dispatch, which can be used as an analysis tool for very short (15 minutes or even lesser) time horizons. 'Minimal risk' being one operational objective, similar operating strategies targeting operational efficiency or self-sufficiency or operational cum financial benefits are quantified and optimized here using stochastic weight trade-off, non-dominated sorting particle swarm optimization (Biswas, Vasant, Laruccia, Vintaned, & Myint, 2020), the choice being that of the system operator.

TWO-STAGE APPROACH[2]

This approach spans over planning and short-term operating horizons. In the planning horizon, the ATC sensitivity factor database is created, the sensitivity with respect to the dispatch of associated generations. The sensitivity factor database thus created is then used for estimating ATC values pertaining to most prominent forecast values, which constitute the operating horizon. Suitable objective functions can be used in the operating horizon to choose the best possible dispatch scenario, the operational objectives being the choice of the operator.

Stage 1: Planning Domain

Uncertainty Modelling in Renewables

Statistical data is used to predict the probable forecast deviations during any period. Different approaches can be employed so that the forecast errors could be employed. Here, an assumption is made that the enough information regarding the upper and lower limits of possible errors in forecast are available with the network operator. Since it is highly probable for a network operator to have such information in their possession, for the discussion focused in this research, the said information is taken as available input data. The maximum deviations of 40% in wind generation, 20% in solar generation and 3% in load forecast are assumed, following the study in (Mohan, Singh, & Ongsakul, 2015). Better values of limiting deviations can be obtained, provided real data is available to explore on.

Load forecast error results in a different dispatch and hence, a different ATC profile. Uncertainties in solar and wind power production intensify this effect. Dispatchable units can be used as a compensating measure for this deviation in real-time. The range of deviations from the forecasted values in generation and load are sampled into specific intervals and for all the possible combinations of these deviation samples, the PTSF values for critical/important network lines are calculated using equation (2) where, for line 'm', $PTSF_{m,i}^{k}$ .s the sensitivity factor for the kth combination of uncertainty, $d(ATC_m)$ is the change in ATC, and dP_{Gi}. the change in ith dispatchable generation.

$$PTSF_{m,i}^{k} = \frac{d\left(ATC_m\right)}{dP_{Gi}}. \quad (2)$$

Power flow and CPF needs to be run for all the combinations of samples of uncertainty mentioned above. To accommodate a system with both HV and LV/MV levels, a sophisticated power flow method (Nimal Madhu M, Sasidharan, & Singh, 2016) applicable to all voltage levels, irrespective of the transmission line characteristics is chosen. The method and associated constraints are all available in (Nireekshana, Kesava Rao, & Sivanaga Raju, 2016). The baseline scenario corresponds to CPF results (including ATC) for forecasted renewable energy and load. CPF gives the maximum feasible power transfer while, PF results give power transfer corresponding to forecasted condition, the difference of which is the baseline

ATC (Chalermchaiarbha & Ongsakul, 2013). This database is used to create PTSF values of monitored lines. The flowchart for database creation using discrete samples is given in Figure 1.

Joint Probability Distribution Functions

A joint PDF for the variable of interest, ATC is estimated using individual discrete PDFs of solar and wind generations and the loads, which is the random variables inflicting the stochasticity. Three discrete uncertainty sample sets, of size k_S, k_W, and k_L. corresponding to solar, wind and load powers, respectively, are considered as given in (3)-(6). For example, if the sample set for solar power deviation from forecast is [-20%, -10%, 0%, 10%, 20%] then k_s.is 5. y_s^i, y_w^i.and y_L^i .re the probabilities of occurrence of the deviation in the ith sample set corresponding to solar, wind and load, respectively.

$$S = \{(P_s^i, y_s^i\} \text{ .or } i{=}1{:}k_s. \tag{3}$$

$$W = \{(P_w^i, y_w^i\} \text{ .or } i{=}1{:}k_w. \tag{4}$$

$$L = \{(P_L^i, y_L^i\} \text{ .or } i{=}1{:}k_L. \tag{5}$$

$$\sum_i y_s^i = \sum_i y_w^i = \sum_i y_L^i = 1. \tag{6}$$

$$K = k_s * k_w * k_L. \tag{7}$$

A superset of samples, containing all the possible occurrences of forecast deviations, calculated as in (7), can be obtained on combining the three sample sets. Each sample in this superset has a joint probability of occurrence, equaling the product of individual discrete probabilities of random variables in the considered sample. y_L^i .in (8), denotes the discrete probability for the occurrence of ith load sample P_L^i. Hence, the joint probability for an occurrence can be obtained using (8).

$$PTSF_{m,i}^k = \frac{d(ATC_m)}{dP_{Gi}}. \tag{8}$$

Therefore, a consolidation of y_j^i .provides a joint PDF for the ATC in the considered interval. In other words, the joint probability for an uncertainty combination i=5, gives the probability of estimated ATC for the same scenario. Larger number of samples in the interval can achieve a better solution but may increase the computation time. This interval shall be specified based on the statistical data and the probabilities based on the frequency of occurrence. The discrete renewable forecast deviations considered in this research are included in Table 1.

Table 1. Forecast errors and corresponding probabilities for stochastic entities in the system

Load	Error (%)	-3	-2	-1	0	1	2	3		
	Probability	0.02	0.08	0.15	0.5	0.15	0.08	0.02		
Solar	Error (%)	-20	-10	0	10	20				
	Probability	0.1	0.25	0.45	0.15	0.05				
Wind	Error (%)	-40	-30	-20	-10	0	10	20	30	40
	Probability	0.05	0.07	0.1	0.15	0.3	0.15	0.1	0.05	0.03

Stage 2: Operation Domain

Minimal Risk ATC Estimation

The forecast deviations are predictable to a smaller interval in a shorter time span. On finalizing the interval, the PTSF values can be used to estimate ATC corresponding to each uncertainty combination, as a linear increment from the baseline scenario as given in (9) and (10). δatc_m^k .s the change in ATC of line 'm' for uncertainty combination 'k', δP_{Gi} is the change in active power dispatch of conventional generator 'i' from forecasted schedule, ATC_m^k .s the estimated ATC and ATC_{base} .s the baseline ATC. Variance in calculated ATC values corresponding to the 'K' uncertainty samples, are given in (11).

$$\delta atc_m^k = \sum_i PTSF_{m,i}^k * \delta P_{Gi}. \tag{9}$$

$$ATC_m^k = ATC_{base} + \delta atc_m^k. \tag{10}$$

$$var_m = \frac{1}{K} * \sum_i^K \left(ATC_{base} - ATC_m^i \right)^2. \tag{11}$$

$$risk_m^k = \frac{ATC_{base} - ATC_m^k}{ATC_{base}}. \tag{12}$$

Any of the estimated ATC values could be a viable solution but may not correspond to a minimum risk to TSO/DSO. A generalized risk factor, $risk_m^k$ .orresponding to ATC_m^k. is defined as the normalized deviation of estimated ATC from baseline ATC as in (12). A positive value of individual risk indicates that the actual uncertainties have escalated the power flow in the corresponding line, while the reverse indicates a favorable reduction. Hence, for every uncertainty in the sample set, the operational risk to TSO, quantified by the positive risk factor, is minimized altering the dispatch strategy. In addition, three different scenarios are considered to see the effect of various operational decisions or ideology on the results.

Figure 1. Flowchart for sensitivity database creation

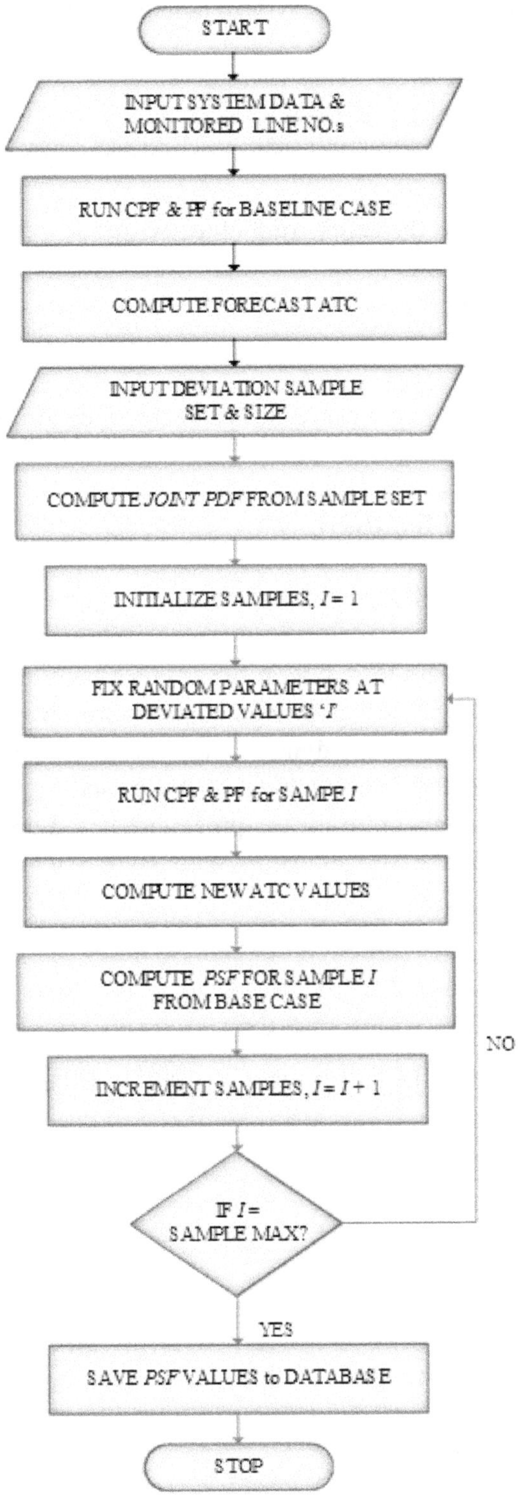

(i) Improving the system operational characteristics:
MO optimization (MOO) with line ATC, system active power loss and bus voltage deviation as objectives.

Objective: {Max(*ATC*), Min(P_{Loss}); Min(*dV*)} (13)

(ii) Risk hedging scenario:
MOO with penalty and ATC as the conflicting objectives.

Objective: {Max(*ATC*), Min(*Risk*)} (14)

(iii) Overall benefits:
MOO with penalty, ATC and system power loss as objectives to improve system efficiency considering benefits.

Objective: {Max(*ATC*), Min(P_{Loss}), Min(*Risk*)} (15)

Stochastic Weight Trade-Off PSO

Stochastic weight trade-off mechanism to maintain the balance between global and local characteristics, improving the search efficiency. The governing equations are given in (16))-(21) [21]. r_1 to r_4 are random variables in (0,1), $c_1(k)$ & $c_2(k)$ are dynamic acceleration coefficients, $e(k)$ is the stochastic momentum trade-off factor, P_{frd} is the freak factor probability and P_{ltg} the lethargy factor probability [11]. Figure 2 shows the functional diagram for the procedure based on the created database, making use of SWT_NSPSO. On replacing the MO algorithm with SWT_PSO in Figure 2, the single objective formulation can be realized.

$$v^k = v^k + P(r_1)\,sign(r_1)\,v^{frk} \tag{16}$$

$$v^{k+1} = e(k)r_2 sign(r_3)v^k + (1-r_2)c_1(k)r_4\left(pbest - x^k\right) + (1-r_2)c_2(k)(1-r_4) + \left(gbest - x^k\right) \tag{17}$$

$$P(r_1) = \begin{cases} 1, & r_1 \le P_{frk} \\ 0, & r_1 > P_{frk} \end{cases} \tag{18}$$

$$e(k) = \left(e_{min} - e_{max}\right)\frac{k}{k_{max}} + e_{max} \tag{19}$$

$$c_1(k) = \left(c_{1,min} - c_{1,max}\right)\frac{k}{k_{max}} + c_{1,max} \tag{20}$$

Figure 2. Functional diagram for MOO approach based on database (Madhu, Mohan, Singh, & Suresh, 2018)

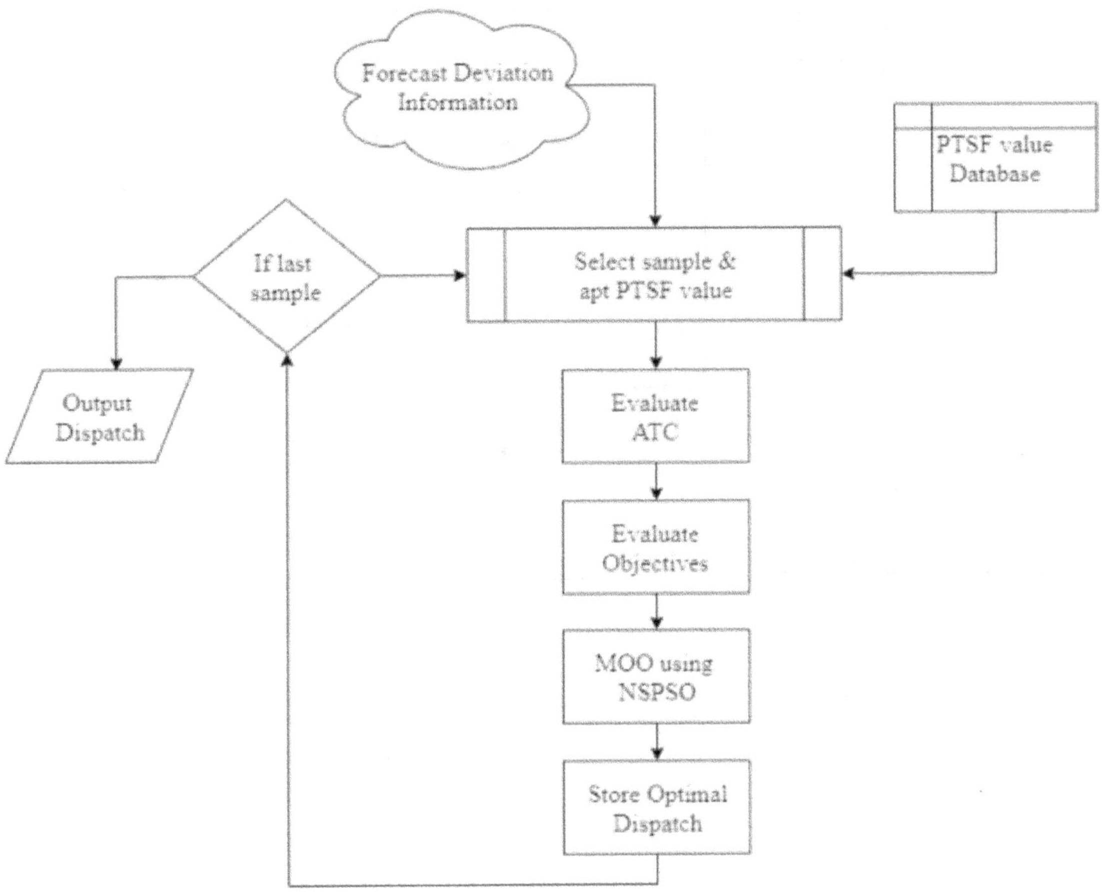

$$c_2(k) = \left(c_{2,max} - c_{2,min}\right)\frac{k}{k_{max}} + c_{2,min} \tag{21}$$

SOLUTIONS AND RECOMMENDATIONS

The test system used for this study is shown in Figure 3. The transmission side of the final network is realized using IEEE 9-bus system [14], while the distribution side of the network comprises of three IEEE 13-bus radial [18] systems, acting as grid connected microgrids. The microgrids are connected at different buses in the network. This system is depicted as TX-1 in Figure 3. TX-2 also follows the same HV-MV interconnected structure and is connected to TX-1 at bus no: 6 as shown in case of TX-1. The entire test system consists of 96 buses, with 96 interconnecting lines, active power load of 683 MW and a reactive load of 330 MVAr. Microgrids have renewable based distributed generation units. The system generation data is given in Table 2. The notations used in Figure 3 can be found in this table. TX denotes transmission or HV system, while MG denotes the distribution feeder with DGs powered by Soar and Wind generations. The multi-objective optimization process is carried out with SWT_NSPSO.

Figure 3. Transmission-Distribution combined test system

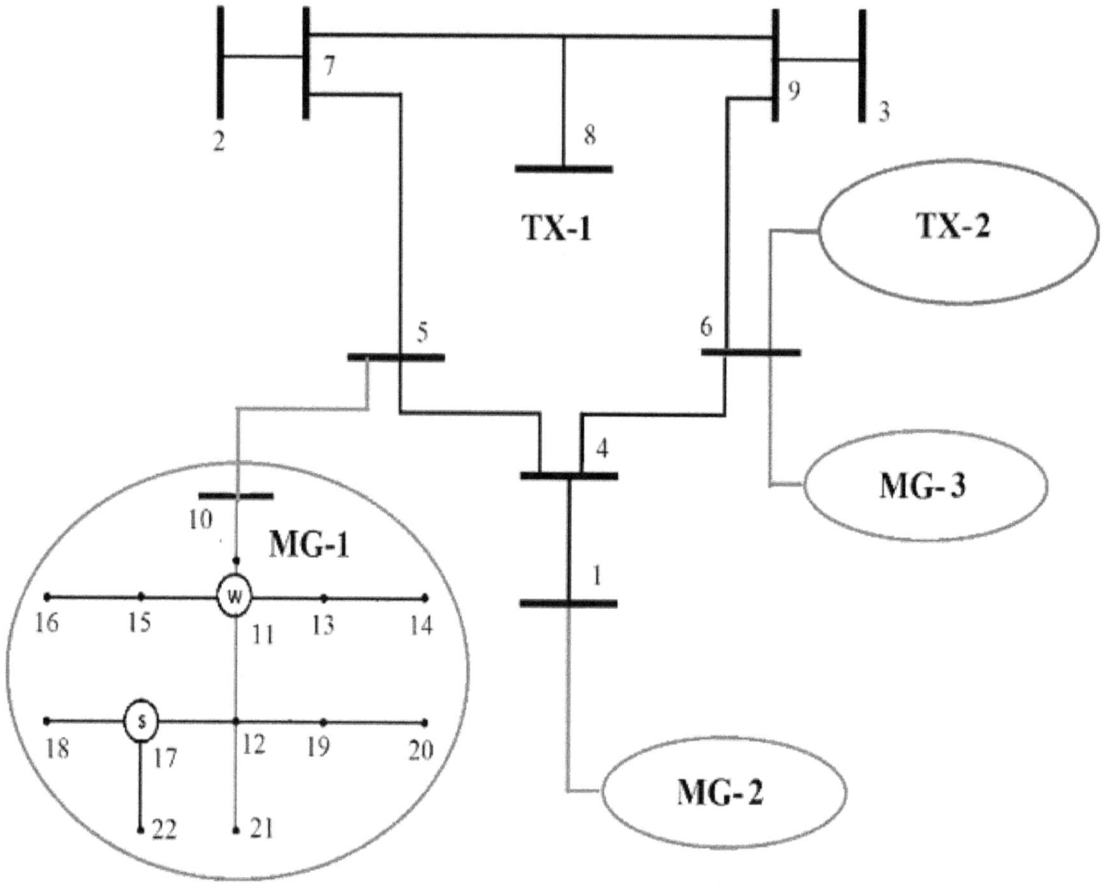

Table 2. Generator installed capacities in the system

Transmission (MW)				Microgrid with TX-1 (MW)				Microgrid with TX-2 (MW)		
Type	PG1	PG2	PG3	Type	MG1	MG2	MG3	MG1	MG2	MG3
TX-1	160	90	70	Solar	6	3	3	3	1.5	2
TX-2	150	95	85	Wind	3	1.2	1.5	2	5	3

Scenario 1: Optimization of ATC, Power Loss and Voltage Deviation

This scenario considers a MOO problem with objectives as (i) maximizing ATC values of lines, (ii) minimizing active power loss (22) and (iii) reducing the voltage deviation (23) of the buses. The relevance of choosing these objectives is because they provide improvement in operational efficiency, providing better operational parameters. The active power loss occurring in the system during the base case or forecasted case is calculated to be 0.2709 p.u. which may vary if the forecast is erroneous. From Figure 4, it is evident that the power loss is minimized considerably such that the maximum value of power loss during all the uncertainty occurrences is 0.244 p.u.

$$dV^k = V^k_{base} - V^k \tag{22}$$

$$Ploss = Txn.gen - total\ load + MG.gen \tag{23}$$

Here, dV^k is the voltage deviation, V^k is the bus voltage the current scenario and V^k_{base} is the bus voltage at bus 'k' in the base case scenario. The base case scenario denotes the system operational details under zero forecast errors.

Figure 4. Active power loss estimated for various forecast error combinations for Scenario-1

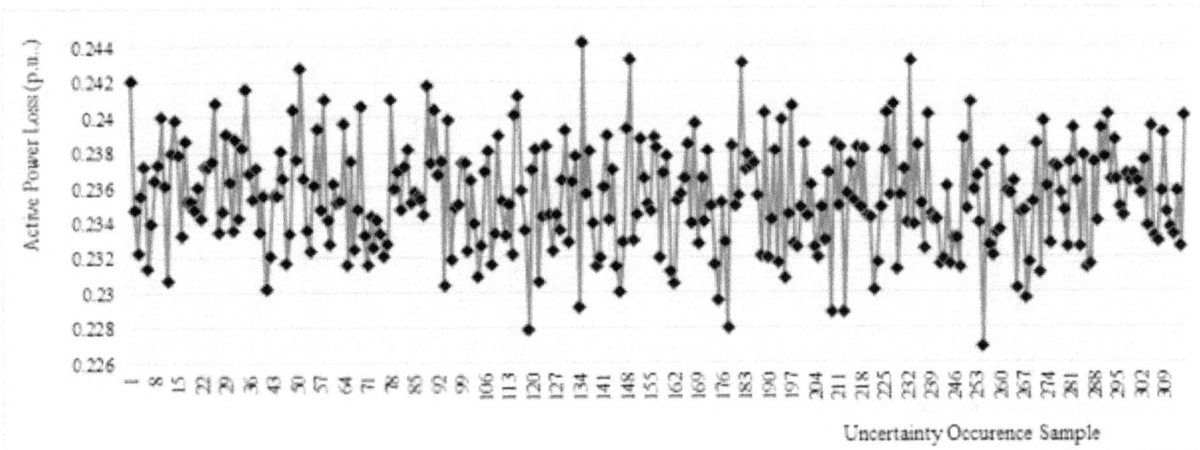

Figure 5 depicts the comparison of expected ATC values from scenario-1 with the base case values for selected lines. The selected lines are the ones that showed maximum variance. It can be observed that the process is efficient for the expected ATC is always higher than forecasted available ATC, providing more options for financial or physical transactions for the operator/owner.

Figure 5. Variance in ATC under different penetration levels

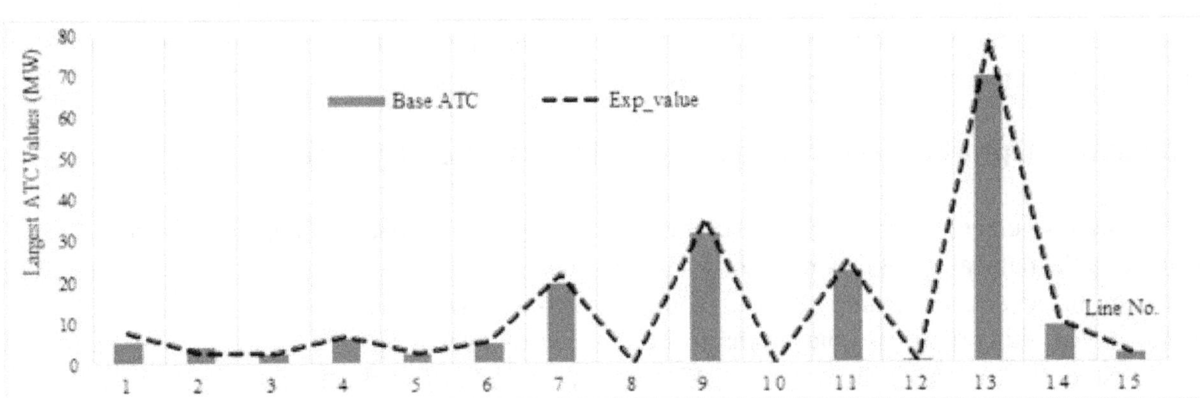

Scenario 2: Optimization of Total Risk, and ATC

The minimization of risk and the maximization of ATC for lines that showed maximum variance in base case, for all uncertainty scenarios is carried out. The generalized risk factor, defined in (11), can also be taken as a financial indicator if incorporated with suitable penalties or incentives. Risk reduction is a conflicting problem within itself as the reduction of line flow in one line may increase the same in one or more lines. That is, the risk associated with one line could be the ATC margin of another. ATC comparison for this scenario is given in Figure 6.

Figure 6. ATC Comparison for Scenario 2

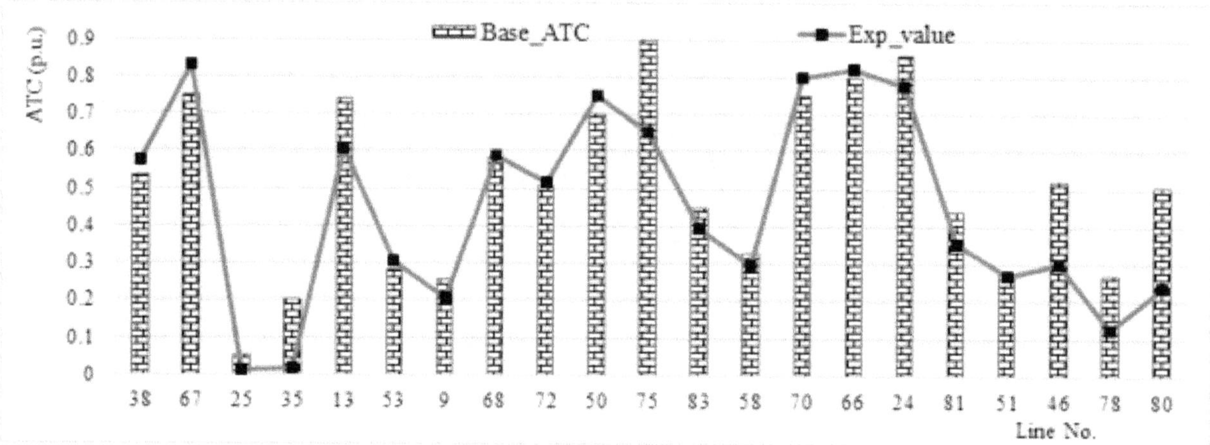

Since, risk is modelled as normalized ATC deviation here, the results obtained from this formulation should provide much better values of ATC variance than the base case. This fact is validated from the result provided in Fig.13 where the line power flows obtained in the current scenario is compared with the base case. It can be observed that most lines have a lower power flow in them in opposition to the base case, which is the result of increased ATC and minimized risk. But in some lines increase in power flow can be observed since the power demand is to be satisfied and the alleviation of flow one line is compensated by the rest. Though trade-off is involved among different network paths, Figure 7 clearly depicts that, with the proposed method, ATC is improved for the whole system by choosing a suitable dispatch and that the transmission margin is improved.

Scenario 3: Optimization of Total Risk, ATC, and Power Loss

This scenario is aimed at providing an efficient, risk hedged and safe operation, in every uncertain state included in the database. In this approach, where the total risk, ATC of all lines and system power loss are co-optimized, a better operating point in comparison to scenario 2 should be obtained. This approach has more conflicting objectives than scenario 2 and hence an overall optimal solution beneficial to the whole system is expected. In Figure 8, the ATC variance, used as a quantification of risk is compared between base case, and scenarios 2 and 3. Though lower than the base case, most of the variance values

Figure 7. Variance comparison for forecasted case & scenario 2

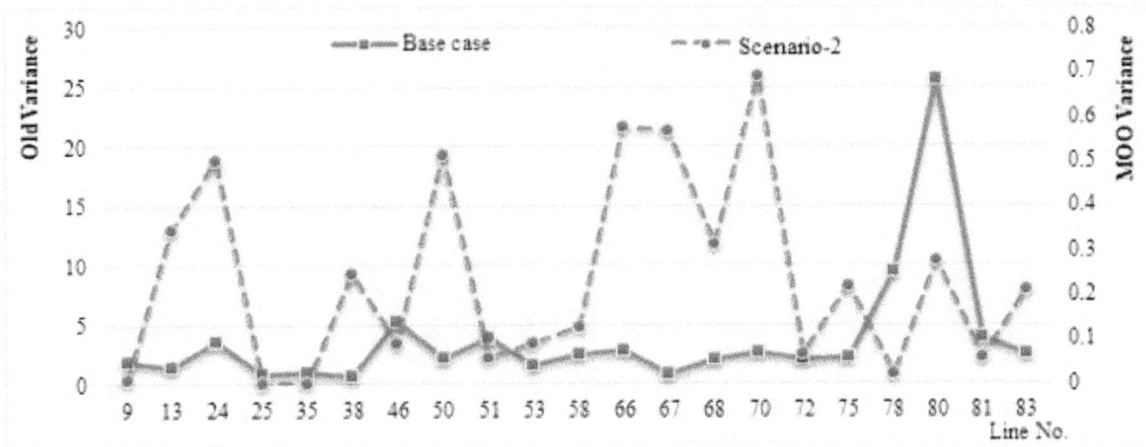

obtained in scenario 3 is visibly higher than that in scenario 2. This is due to the fact the addition of an extra objective, viz. the minimization of power loss has made the algorithm to propagate in the direction satisfiable to all the objectives than just optimizing risk and ATC, which is the case considered in scenario 2.

In Figure 9, the active power loss occurring during the various uncertainty sample combinations of generation and load, obtained in scenarios 2 and 3 are compared. The incorporation of the additional objective to minimize loss in scenario 3 should provide a better value of losses compared to scenario 2, even if the variance values of ATC is compromised. This fact, seen in Figure 9 , can be used to validate the effectiveness of the method. The average of the power loss values for each uncertainty sample situation, obtained in scenario 2 is approximately 0.28 p.u., with the loss values spiking above 0.3 p.u. in several cases. This is higher than the forecast loss of 0.2709 p.u., since the main priority in scenario 2 is to optimize risk and ATC. At the same time, the loss values obtained in scenario 3, in all the cases, lie within the range of 0.23 to 0.25 p.u. Hence, it can be concluded that though there is slight trade-off in terms ATC, the current scenario can be adopted as the overall beneficial strategy giving prime operational and financial benefits.

LIMITATIONS AND ADVANTAGES

The limitation of the discussed two-stage formulation is that, once there is a change in the network, for example, addition of a new line to increase power transferring capacity or a new FACTS device to improve the voltage regulation and reduction of systems losses, stage-1 needs to be repeated. The advantage is in the fact that, once the stage-1 is carried out there is no change in the network infrastructure, stage-2, which is fast and has non-intensive computational requirements, only needs to be run to identify the optimal dispatch. Though it should be considered a limitation that, stage-1 of the formulation is computationally intensive and time consuming, the fact that it only needs to be run once in the planning domain and not afterwards, is to be considered an advantage. This also makes the approach suitable for short-term dispatch as well.

Figure 8. Comparison of ATC variance for forecasted and scenarios 2 and 3

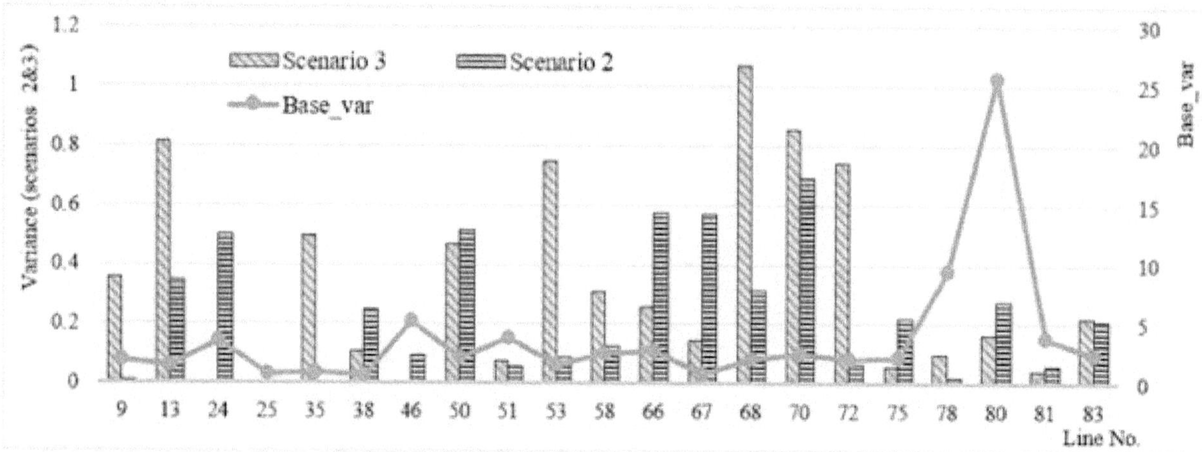

Figure 9. Scenario 3 – Comparison of power loss with scenario 2

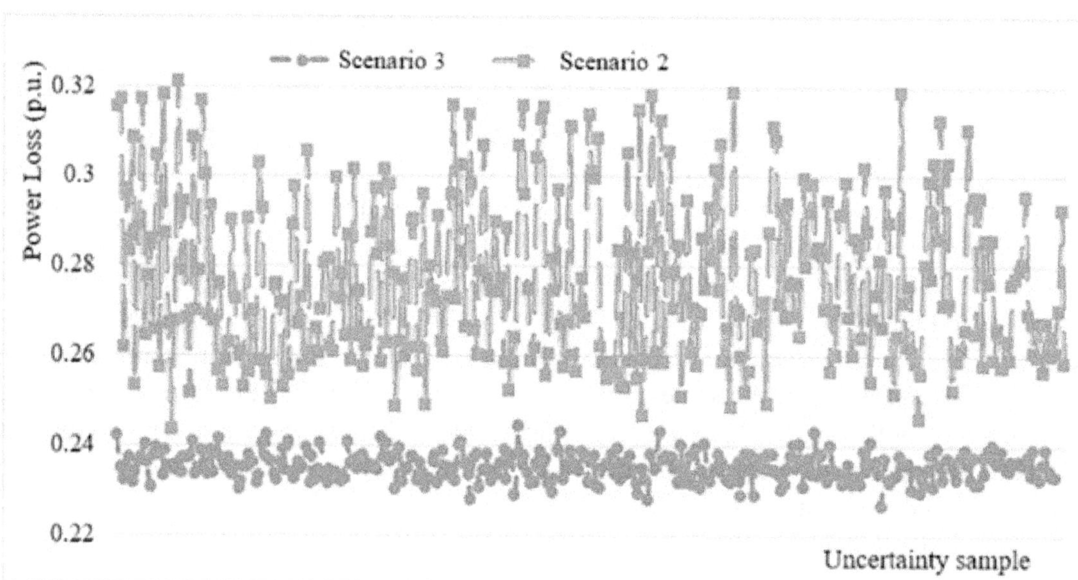

FUTURE RESEARCH DIRECTIONS

The discussed approach is not including real data now. Though the change in data shall alter the results, it shall not affect the methodology. But a more realistic mode of formulation can be obtained. The application of Monte Carlo Simulation for the modelling the uncertainty can be looked at.

CONCLUSION

A two-stage risk hedging strategy for transmission/grid owner/operator is presented in the paper where forecast errors in load demand and renewable energy are considered as sources of uncertainties (leading to financial and operational risk). The problem is formulated with multiple objective formulations and tested on an interconnected transmission-distribution/microgrid systems with renewable generations on both voltage levels. A larger system is made for in order to check the effectiveness of the approach in a harsher and complex environment. Risk is modelled as normalized deviation of current ATC value from the deterministic ATC corresponding to the forecast. In the first stage, using CPF, ATC is estimated for all forecast error combinations, and PTSF parameters are calculated for network lines, thus creating a database of ATC and sensitivity factors for different uncertainty occurrences. The lines exhibiting maximum variance in ATC from the forecasted values, indicating highest vulnerability, in this stage, are selected to be monitored in stage 2. An optimal dispatch strategy minimizing the risk in ATC, is carried out for all the forecast scenarios in stage 2, such that for every scenario, there is a dispatch strategy, expected ATC and its variance with associated probability. In the multi-objective formulation, the effect of various operational strategies of the system operator which prioritize between system efficiency, financial and operational benefit, on the solutions is assessed. The presented approach is successful at obtaining minimal ATC variance or risk, reducing losses and of course, improving the ATC values, while it is also observed that with increased renewable penetration an escalation in the risk for TSO/DSO is also inevitable.

REFERENCES

Biswas, K., Vasant, P. M., Laruccia, M. B., Vintaned, J. A. G., & Myint, M. M. (2020). Review on Particle Swarm Optimization Approach for Optimizing Wellbore Trajectory. In *Deep Learning Techniques and Optimization Strategies in Big Data Analytics* (pp. 290–307). Hershey, PA: IGI Global. doi:10.4018/978-1-7998-1192-3.ch017

Chalermchaiarbha, S., & Ongsakul, W. (2013). Stochastic weight trade-off particle swarm optimization for nonconvex economic dispatch. *Energy Conversion and Management*, *70*, 66–75. doi:10.1016/j.enconman.2013.02.009

Chiang, H. D., Flueck, A. J., Shah, K. S., & Balu, N. (1995). CPFLOW: A practical tool for tracing power system steady-state stationary behavior due to load and generation variations. *IEEE Transactions on Power Systems*, *10*(2), 623–634. doi:10.1109/59.387897

Daus, Y., Kharchenko, V., & Yudaev, I. V. (2018). Solar Radiation Intensity Data as Basis for Predicting Functioning Modes of Solar Power Plants. In *Handbook of Research on Renewable Energy and Electric Resources for Sustainable Rural Development* (pp. 283–309). Hershey, PA: IGI Global. doi:10.4018/978-1-5225-3867-7.ch012

Farahmand, H., Rashidinejad, M., Mousavi, A., Gharaveisi, A. A., Irving, M. R., & Taylor, G. A. (2012). Hybrid Mutation Particle Swarm Optimisation method for Available Transfer Capability enhancement. *International Journal of Electrical Power & Energy Systems*, *42*(1), 240–249. doi:10.1016/j.ijepes.2012.04.020

Ghawghawe, N. D., & Thakre, K. L. (2006). Application of Power Flow Sensitivity Analysis and PTDF for Determination of ATC. In *2006 International Conference on Power Electronic, Drives and Energy Systems* (pp. 1–7). 10.1109/PEDES.2006.344304

(2019). Intelligent Computing & Optimization. InVasant, P., Zelinka, I., & Weber, G.-W. (Eds.), *Advances in Intelligent Systems and Computing* (1st ed., pp. 98–107). Springer International Publishing.

Li, W. X., Li, Z. M., Ni, Y. X., & Wu, F. F. (2003). Available transfer capability calculation with post-contingency generation rescheduling/load curtailment. In *2003 Sixth International Conference on Advances in Power System Control, Operation and Management ASDCOM 2003 (Conf. Publ. No. 497)* (Vol. 2, pp. 563–568). 10.1049/cp:20030648

Madhu, N. M., Mohan, V., Singh, J. G., & Suresh, R. (2018). Risk Adjusted Co-optimization of ATC in High-Low Voltage Interconnected Power Systems. In *2018 IEEE International Conference on Power Electronics, Drives and Energy Systems (PEDES)* (pp. 1–6). 10.1109/PEDES.2018.8707502

Mohan, V., Singh, J. G., & Ongsakul, W. (2015). An efficient two stage stochastic optimal energy and reserve management in a microgrid. *Applied Energy*, *160*, 28–38. doi:10.1016/j.apenergy.2015.09.039

NERC. (1996, June). Available Transfer Capability Definitions and Determination. *North American Electric Realiability Council*, 1–42.

Nimal Madhu, M., Sasidharan, N., & Singh, J. G. (2016). A droop control incorporated dc equivalent power flow method for distribution and low voltage systems. *Electric Power Systems Research*, *134*, 56–65. doi:10.1016/j.epsr.2015.12.033

Nireekshana, T., Kesava Rao, G., & Sivanaga Raju, S. (2016). Available transfer capability enhancement with FACTS using Cat Swarm Optimization. *Ain Shams Engineering Journal*, *7*(1), 159–167. doi:10.1016/j.asej.2015.11.011

Ozkan, M. B., Küçük, D., Buhan, S., Demirci, T., & Karagoz, P. (2020). Large-Scale Renewable Energy Monitoring and Forecast Based on Intelligent Data Analysis. In *Handbook of Research on Smart Computing for Renewable Energy and Agro-Engineering* (pp. 53–77). Hershey, PA: IGI Global. doi:10.4018/978-1-7998-1216-6.ch003

Pandey, S. N., Pandey, N. K., Tapaswi, S., & Srivastava, L. (2010). Neural Network-Based Approach for ATC Estimation Using Distributed Computing. *IEEE Transactions on Power Systems*, *25*(3), 1291–1300. doi:10.1109/TPWRS.2010.2042978

Prathiba, R., Moses, B. B., Devaraj, D., & Karuppasamypandiyan, M. (2015). Multi-output On-Line ATC Estimation in Deregulated Power System Using ANN BT - Advances in Intelligent Informatics. Cham: Springer International Publishing.

Rodrigues, A. B., & Da Silva, M. G. (2007). Probabilistic Assessment of Available Transfer Capability Based on Monte Carlo Method With Sequential Simulation. *IEEE Transactions on Power Systems*, *22*(1), 484–492. doi:10.1109/TPWRS.2006.887958

Shukla, D., Lakshmi, E. S., & Singh, S. P. (2017). Estimation of ATC using PS-NR. In *2017 6th International Conference on Computer Applications In Electrical Engineering-Recent Advances (CERA)* (pp. 111–116). 10.1109/CERA.2017.8343311

Unni, A. C., Ongsakul, W., & Madhu, M. N. (2019). Fuzzy-based novel risk and reward definition applied for optimal generation-mix estimation. *Renewable Energy*.

Vaithilingam, C., & Kumudini Devi, R. P. (2013). Available transfer capability estimation using Support Vector Machine. *International Journal of Electrical Power & Energy Systems*, *47*, 387–393. doi:10.1016/j.ijepes.2012.10.054

Wei, J., Li, G., & Zhou, M. (2015). Numerical bifurcation and its application in computation of available transfer capability. *Applied Mathematics and Computation*, *252*, 568–574. doi:10.1016/j.amc.2014.12.003

Wu, Y.-K. (2007). A novel algorithm for ATC calculations and applications in deregulated electricity markets. *International Journal of Electrical Power & Energy Systems*, *29*(10), 810–821. doi:10.1016/j.ijepes.2007.06.014

Chapter 11
Hybrid Neural Networks for Renewable Energy Forecasting:
Solar and Wind Energy Forecasting Using LSTM and RNN

Firuz Ahamed Nahid
Asian Institute of Technology, Thailand

Weerakorn Ongsakul
Asian Institute of Technology, Thailand

Nimal Madhu M.
https://orcid.org/0000-0002-5981-2033
Asian Institute of Technology, Thailand

Tanawat Laopaiboon
Asian Institute of Technology, Thailand

ABSTRACT

One of the key applications of AI algorithms in power sector involves forecasting of stochastic renewable energy sources. To manage the generation of electricity from solar or wind effectively, accurate forecasting models are imperative. In order to achieve this goal, a sophisticated hybrid neural network formulation is discussed here in this chapter. long-short-term memory and recurrent neural networks combination is formulated for very short-term forecasting of wind speed and solar radiation. In intervals of 15 and 30 minutes, time series forecasts are made that are ahead by multiple steps. For maximum energy harvest, both point wise and probabilistic forecasting approaches are combined. Historic data is collected for solar radiation, wind speed, temperature, and relative humidity, and are used to train the model. The proposed model is compared with convolutional and LSTM neural network models individually in terms of RMSE, MAPE, MAE, and correlation, and is identified to have better forecasting accuracy.

DOI: 10.4018/978-1-7998-3970-5.ch011

INTRODUCTION

Increasing demand of electric power across the world and the issue of global warming are having a complimenting effect, pushing the power generation trend around the globe toward environment friendly energy resources. Besides, the sustainable development goals make it imperative to develop and extract energy form renewable resources, rather than relying on the non-renewable conventional kind, whose availability and affordability are going downhill, daily. Then again, there are issues like CDM (Clean Development Mechanism), climate change, insufficient and unreliable supply of power in developing countries etc., all advising the urgency of identifying alternate and nature-friendly sources of development.

Among all the alternate power production options, nested under renewable energy, wind & solar power are the most promising substitutes. These sources being economic, as well as, being available throughout the clock (wind) (Li, Wu, & Liu, 2018), minimal maintenance requirement and ease of installation (solar), have an upper hand above the other sources. As per (Hu & Chen, 2018), wind power is also one of the most cost-effective sources, that has a huge potential to compete with the traditional fossil fuel-based power plants and is eco-friendly too. These pros have provided a rapid boost to solar & wind-based power generation throughout the world, with a growth rate of 28% per year (Varanasi & Tripathi, 2016).

Background

Though a renewable source of energy, wind as well as solar, in opposition to hydro and biomass, has introduced a certain sense of uncertainty in the power production (Lu, 2019). Their variable nature reduces the suitability of the generator as a dispatchable source, and thus upset the stability of the power system (Boudour, 2016). This also affects the power quality of the sources, security of grid and market economics (Al-falahi, Jayasinghe, & Enshaei, 2017). This stochastic nature, which is absent in the conventional generators, necessitate the requirement of a new paradigm shift in the fields of power system modelling and analysis to control and operation. A review of models applied for solar and wind related forecasting are included here.

Though controlling this stochastic nature is not possible, forecasting these renewable sources in advance, shall enable us to efficiently manage and coordinate the generation of power from the wind farms. In other words, improve the dispatching capability of the renewable power sources, making them controllable (Lee, 2016) and finally, replacing the fossil fuel-based power generation completely. A proper forecast will also help in reducing the cost and increasing the revenue from the electricity market (Ummels, Gibescu, Pelgrum, Kling, & Brand, 2007). There are a lot of forecasting methods available for prediction that can be categorized into statistical (AR, MA, ARMA, ARIMA) (Cadenas & Rivera, 2010; Shukur & Lee, 2015) physical (NWP) (Lynch, OMahony, & Scully, 2014; Wang & Li, 2016) artificial intelligence (AI) (neural networks) (Kaur, Kumar, & Segal, 2016; Marugán, Márquez, Perez, & Ruiz-Hernández, 2018; Yadav, Singh, & Chaturvedi, 2017) and hybrid (combination of NNs or NN with other approaches) (Alencar, Affonso, Oliveira, & Filho, 2018; Doucoure, Agbossou, & Cardenas, 2016; He, Wang, & Lu, 2018; Li, Shi, Han, Duan, & Liu, 2019; Qu, Mao, Zhang, Zhang, & Li, 2019; Sun, Zhou, Liu, & He, 2019).

In (Shukur & Lee, 2015), a time series forecasting of solar irradiance for a day ahead using ARMA model is discussed. A wide variety of time series models for the hour ahead forecasting is discussed using moving average techniques, models based on exponential smoothing techniques, and ARMA models. Global solar radiation performance study is conducted with artificial neural networks (ANN) and Angstroms linear regression approach (Wang & Li, 2016) and ANN is identified as a suitable predictor algorithm. The comparison is also conducted with applied empirical models which are Autoregressive (AR), moving average (MA), autoregressive moving average (ARMA), and autoregressive integrated moving average (ARIMA) in (Kaur et al., 2016)and ANN model is found to be the better option.

Deep neural networks (DNN) are a different class of machine learning model. The main difference between Classical and Deep network scheme is the number of the hidden layer and the training process. Using more hidden layers, DNN can extract higher order of interrelation. Deep Learning is used to forecast solar radiation (Li et al., 2019) with Deep Belief Network (DBN) as training model. This model retained better performance compared with extreme learning machine model (ELM) while forecasting in short-term for 15 minutes. In (Qu et al., 2019)the method is compared with Long-Short Term Memory (LSTM) and MLP, in forecasting solar radiation, and LSTM gave better performance.

The hybrid models are designed to combine the benefits of all the constituent approaches resulting in better performance. A hybrid model combining SARIMA with neural network is applied for multi-step ahead forecasting in (Alencar et al., 2018). The root-mean-square error (RMSE) on using the model, for 6-hours ahead prediction, varies from 0.316 to 1.740 while the same for day-ahead (24 hr.) vary from 0.86 to 4.04. The method is applied for wind data for a farm in Brazil. In (Doucoure et al., 2016), a hybrid method combining NN with wavelet decomposition is proposed. The approach is used for a 5-hour ahead wind speed forecasting. RMSE varied for the obtained results from 7.06 to 13.77. A hybrid method (Li et al., 2019) constituting three algorithms, Wavelet Packet Filter, Wavelet Packet Decomposition and Elman Neural Network, is used for a 9-hour ahead forecasting.

A hybrid forecasting model combining Back Propagation Neural Network with Hybrid Decomposition Technique, is used for resource forecasting in wind farms located in Shandong, China (Qu et al., 2019). The interval length is one hour, and the forecast is made until five steps ahead, with good results. Ensemble Empirical Mode Decomposition, Kernel-Based Fuzzy C-Means Clustering and Wavelet Neural Network are combined (He et al., 2018) to realize data processing, clustering and forecasting, and is used for seasonal wind speed forecast. Proposed in (Sun et al., 2019) is a short-term, multi-step ahead, wind forecasting model, which is a combination of Fast Ensemble Mode Decomposition (FEEMD) and Regularized Extreme Learning Machine (RELM). The results obtained from the forecast show an RMSE range of 0.64 to 1.34.

MAIN FOCUS OF THE CHAPTER[1]

Issues, Controversies, Problems

Though a variety of hybrid models are experimented with, as mentioned in the literature discussed, which brought considerable improvement to the existent wind forecast models, relatively higher values of error can be observed based on the criteria being used, RMSE or MAPE. These errors show the considerable mismatch occurring between the predicted and actual wind speeds and it must be reduced. Also, the uncertainty that may exist in the forecasted wind speeds, existent due to various factors, is also not focused on. Hence, in this chapter, a hybrid, multi-step, short-term solar radiation & wind speed forecast model is proposed. The hybrid model combines Recurrent Neural Network (RNN) and Long Short-Term Memory (LSTM) Neural Network. The forecasting process uses historic data of environmental factors related to the forecasted parameter, like temperature and relative humidity, collected from the AIT meteorological measurement station, Thailand. To determine the accuracy of the discussed model, the forecasted and actual values are compared in terms of various parameters, Root Mean Squared Error (RMSE), Mean Absolute Error (MAE), Mean Absolute Percentage Error (MAPE) and Correlation Coefficient.

Band of Prediction

Prediction models based on conventional methods can generate a single value from the targeted value, providing information about the error occurred in the forecasting process to the system operators. However, the models cannot provide information about the uncertainty related with the intermittent wind speed which could affect the wind power generation. As a result, the power system operators must make decisions based on the single predicted value, while the uncertainty in the forecasting still exists and make the decision-making a complex process.

By contrast, a forecasting model that provides a prediction interval or band of future values is more trustworthy and capable of providing more information regarding the uncertainty of the forecasted values. A band of prediction will provide a range of future values which will cover the predicted data with a specified confidence level. Thus, a band of forecasted value will allow the entities to manage the risk associated in handling the system.

There are many models like Boot strap, Mean-variance or Bayesian which performs well in the field of determining the probabilistic forecasting despite having some disadvantages. Both Bayesian and Boot strap models require massive computational necessities, which is not a requirement for Mean-Variance method but has lesser accuracy in prediction bands. There also exist some statistical models which are also capable of generating the upper and lower bands in the forecasting process. Kernel density function, Regression Analysis etc. are also capable in deciding the upper as well as lower bound although the mostly depends on the point prediction.

The probable upper and lower ranges of the band follow (1) and (2) respectively. The prediction bands with $(1 - \alpha)$ coverage area follows (3).

$$P_u = y' + \bar{\epsilon} + \epsilon' * \omega . \tag{1}$$

$$P_l = y' + \bar{\epsilon} - \epsilon' * \omega . \tag{2}$$

$$P_b\left(f'(x) - \omega(x) \le y' \le f'(x) + \omega(x)\right) = 1 - \alpha . \tag{3}$$

Where, P_u and P_l are respectively the upper and lower levels of the prediction band, y'.indicates the predicted values, $\bar{\epsilon}$ .indicates the mean of error, ϵ' .represents the standard deviation of the error and ω is the prediction interval.

Data and Processing

Wind Speed Data

Wind speed data for twenty-seven months, recorded in span of five minutes interval, is collected from meteorology station of Asian Institute of Technology (AIT), Thailand. The collected dataset, constituted by wind speed (***m/s***)., temperature (degree Celsius) and relative humidity (%)., is then cleansed and converted into half hourly average as per the experimental requirement, as given in Figure 1. Table 1 illustrates the statistical measures of the collected data.

Table 1. Statistical parameters of the collected data

Parameter	Data Type	Value
Wind speed (*m/s*).	Minimum Maximum Standard deviation	0.04 4.45 0.98
Temperature (degree C)	Minimum Maximum Standard deviation	14.7 40.47 3.83
Relative humidity (%).	Minimum Maximum Standard deviation	20.64 123.05 23.64

Solar Radiation Data

The dataset for solar radiation is over a span of 1 year and in 15-minute intervals. The parameters included in this dataset are date, time of the day, solar irradiance (W/m^2), energy from radiation (kWh/m^2), ambient temperature (degree Celsius), and relative humidity (%). Figure 2 depicts the collected data.

Data Cleansing

As the collected data are measures of different entities and the range of their values also vary, the variables with higher magnitude may have greater impact, in comparison with the low valued variables. Hence, all the data must be mapped into a single range and as illustrated in (4), all the variables are rescaled and normalized into the same range. Here, x represents the variable value and y, the normalized values.

$$y = \frac{x - min(x)}{\max(x) - min(x)}. \quad (4)$$

Figure 1. Half-hourly averaged values of collected data

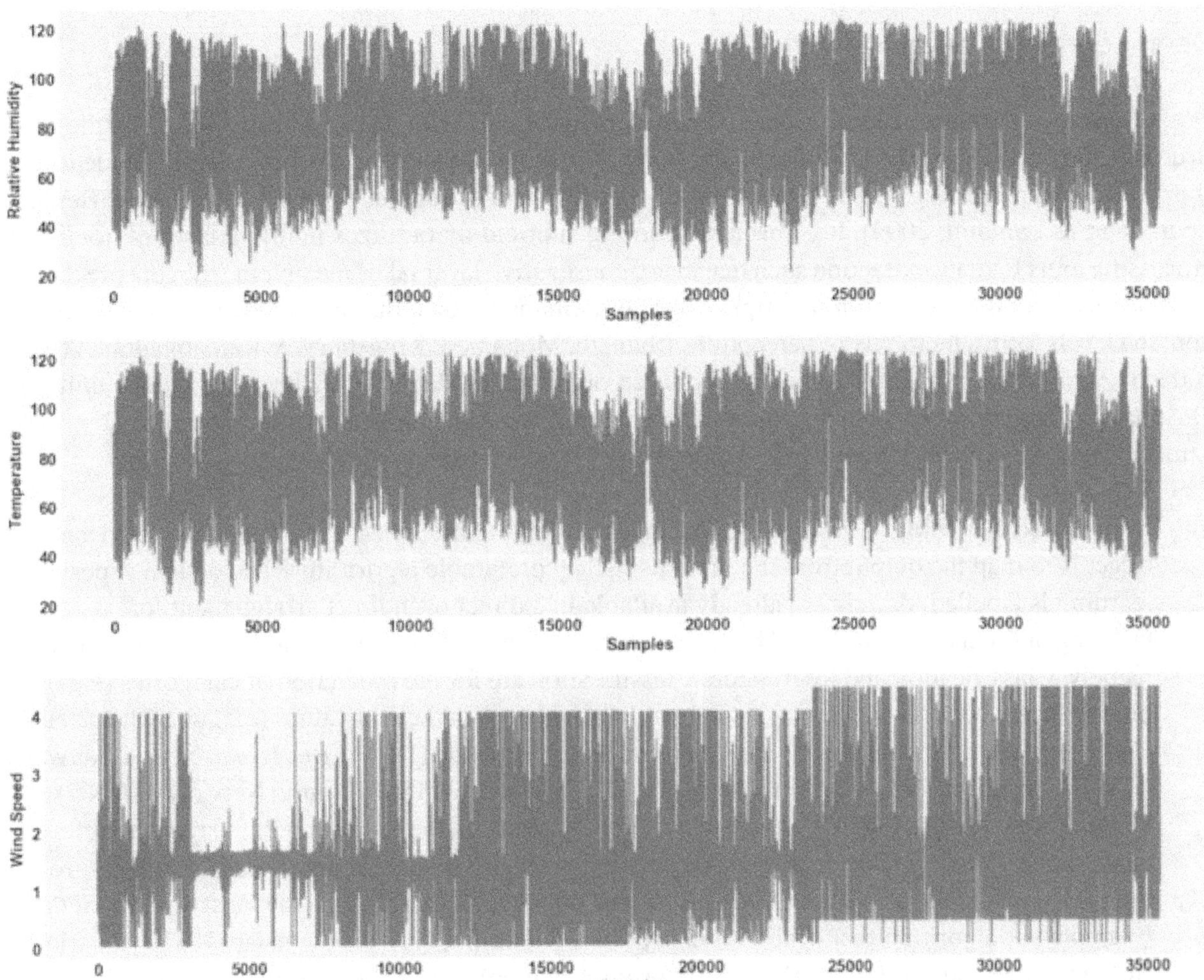

Figure 2. 15 min. interval solar irradiance data

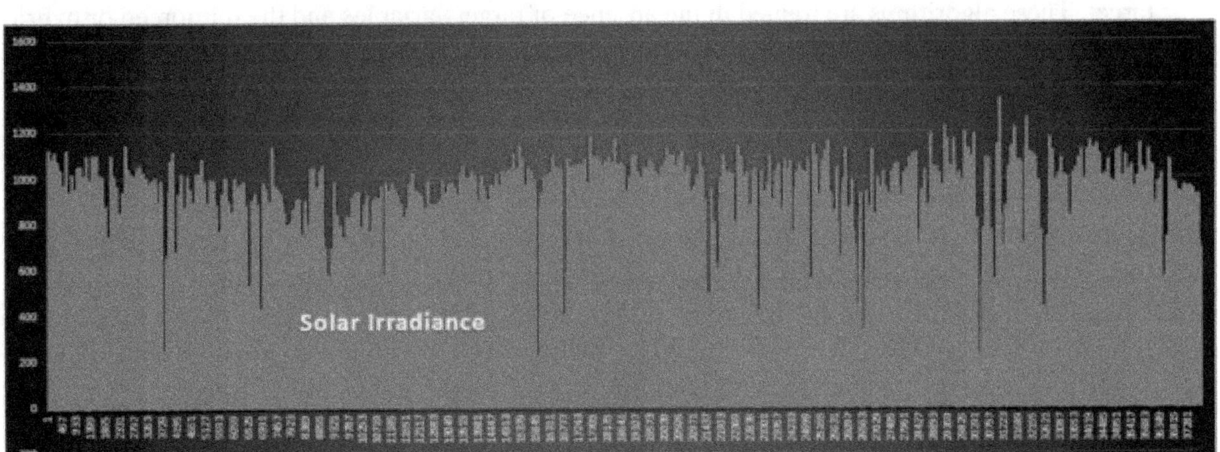

Modelling of R-LSTM Model

Deep Learning Neural Network

Deep Learning (DL) is a class of machine learning which deals with algorithms motivated by the construction and function of the artificial neural networks (Hinton, Osindero, & Teh, 2006). The learning approach in deep learning can be supervised or unsupervised or semi supervised (Larochelle, Bengio, Louradour, & Lamblin, 2009). It is machine learning subfield that utilizes multiple layer of nonlinear processing units for data extraction such that every consecutive layer takes the output from the preceding layer as input. It also learns from multiple stages of demonstrations, that link to different levels of notion; the levels form a hierarchy of perceptions (Sainath, Mohamed, Kingsbury, & Ramabhadran, 2013). A deep neural network holds an input layer plus an output layer, detached by layers of hidden units, as illustrated in Figure 3. Learning approach in deep neural network can be classified in three categories, namely, supervised, unsupervised and reinforced or hybrid learning (Larochelle et al., 2009).

1. **Supervised Learning:** The variables for input and output are found in an organized form and the target is to map the output from the input using any preferable algorithms. The data in supervised
2. learning is labelled, or they are already available in a direct or indirect arrangement for
3. utilizing in supervised learning. The supervised learning models are also classified into two models
4. namely, Classification and Regression. Classification are for the prediction of categories (e.g. red, blue or disease etc.) while the regression problems deal with the real values (e.g. weight, age etc.). The examples of this category are Deep Stacking Network (DSN), Recurrent Neural Network (RNNs), and Convolutional Neural Network (CNN), and Time Delay Neural Network (TDNN) etc.
5. **Unsupervised Learning:** The input variables are known, but the corresponding target variables are unavailable. Used for Clustering and Association problems. Clustering problems deal with grouping of data while the association problems focus on patterns that defines the lager amount of data. Examples of algorithm that deal with unsupervised learning type problems are RBM (Restricted Boltzmann Machine), DBM (Deep Boltzmann Machine), DBN (Deep Belief Network), Apriori algorithm, K-means algorithm etc. (Liu, Mi, & Li, 2018).
6. **Hybrid or Reinforced Learning:** Combination of both the supervised and unsupervised learning. Through trial and error learning method these reinforced algorithm are forced towards optimal target. These algorithms are trained in the absence of target variables and the reinforced or hybrid agent in the algorithm decides what to perform in that environment. They are made to learn using their experience while enough training dataset is absent (Sainath et al., 2013).

Recurrent Neural Network

RNN is a type of ANN – Artificial Neural Network that is usually utilized for the forecasting of time series data. It utilizes the feedback provided by one or more units of its network as input in selecting the succeeding output (Cao, Ewing, & Thompson, 2012). In RNN, the hidden neurons connect the hidden layer from previous time step to current time step, which is why it is called recurrent. As a result, RNN models perform better in comparison with other ANNs in conserving data from preceding events. The input layer provides weights for the first layer. Similarly, each layer gets the weight from their preceding ones, and in this way the network forms a loop enables the persistence of data from past events. The

Figure 3. Illustration of Deep Neural Network (Larochelle et al., 2009)

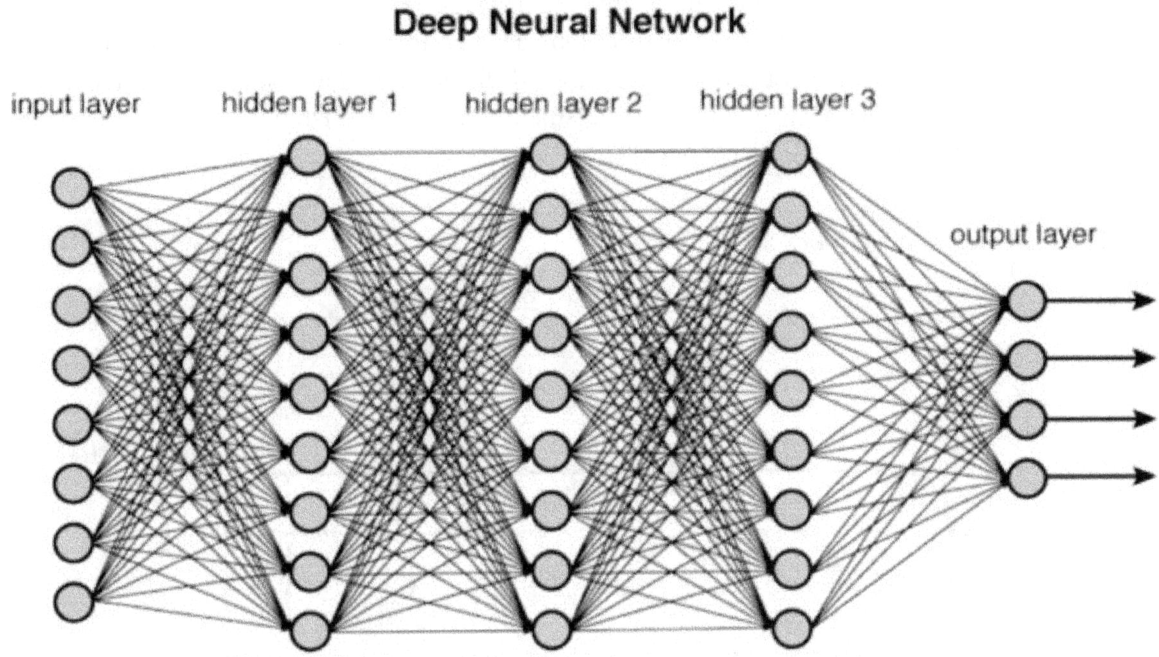

layers of the network perform necessary operations and yield specific output as directed and then pass the information to the next layer as an input (Kumar, Goomer, & Singh, 2018). RNN, at times, requires a large quantity of historic information to perform well. The functioning of RNN can be expressed as in (5)-(6). Here, $x(t)$ and $y(t)$ are input and output, respectively, W_{IH}, W_{HH}, W_{HO} denotes weight metrics, and f_H and f_O are activation functions. Also, $h(t)$ is the context unit of the dynamic system, which indicates a set of information relating to the past behavior of the network, that could help in the portrayal of a future behavior. The pictorial representation of simple RNN is given in Figure 4.

$$h(t+1) = f_H\left(W_{IH}x(t) + W_{HH}h(t)\right). \tag{5}$$

$$y(t+1) = f_O\left(W_{HO}h(t+1)\right). \tag{6}$$

Figure 4. Simple RNN architecture (Salman, Heryadi, Abdurahman, & Suparta, 2018)

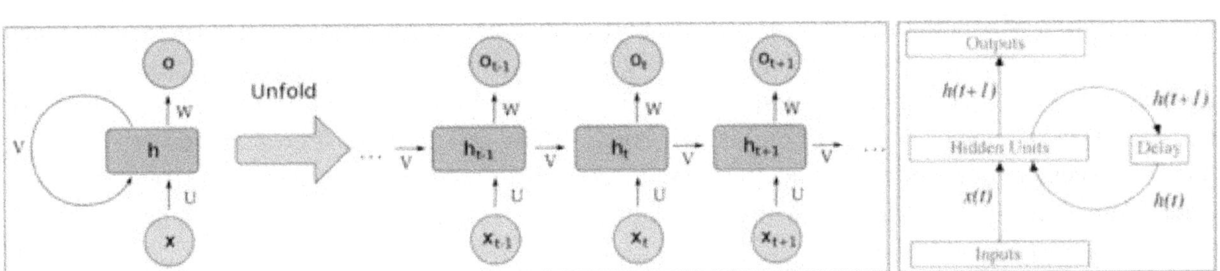

In a regular RNN, memory neurons are more sensitive since the error signals from later time steps are not capable of computing the gradient far enough, to propagate back and influence earlier time steps or layers. This is called the vanishing gradient problem. This issue occurs during the back-propagation phase, when the gradients are calculated and fed backwards from output layer in later time step to inputs layer in earlier time step. The feedback loop having smaller gradient vanish quickly, preventing RNNs from learning longer-term temporal dependencies.

Figure 5. LSTM memory cell block

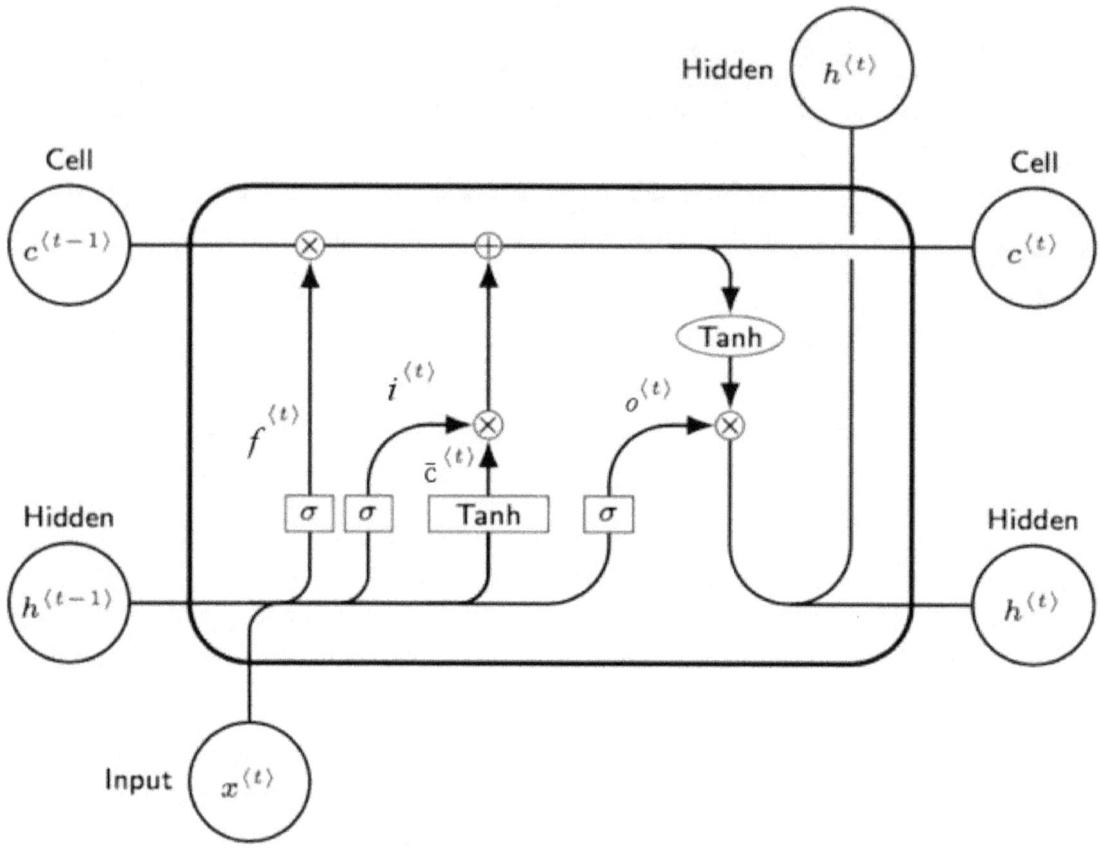

Long Short-Term Memory (LSTM)

LSTM is explicitly a subfield of RNN architecture (Salman et al., 2018), which is more stable and efficient in dealing with both long-term, as well as short-term dependency problems. It is very useful when the gap between the past and the required future values are substantial. Thus, LSTM could be used as a replacement to RNN, which conveniently remedies the vanishing gradient problem. LSTM network houses a special memory block in the hidden layers (Bengio, Boulanger-Lewandowski, & Pascanu, 2013). In these memory blocks, specific memory cells are present, that store the networks temporal states. This is the what makes LSTM better at storing long term information. At the same time, there are multiplicative units in the memory blocks known as gates, that control the flow of information in the hidden layers. The gates inside the memory block are of three categories as below.

a. Input gate - Handles the flow of inputs towards the memory cell.
b. Output gate: Controls the output stream of cell activations toward the remaining network.
c. Forget gate/Internal state: Resets the cells memory and scales the internal state of the cell.

Typical RNN is constituted by neurons units that calculate the neurons value by computing the input's linear combination term and applied activation function to them. In LSTM unit of neurons are replaced by memory-cells that feed into itself, a constant value of weight, across various time steps. Any gradient flowing into this self-recurrent memory cell unit, during back propagation phase, is also retained by means of internal stage. Hence, the small gradient does not vanish quickly, consequently solving the vanishing gradient problem. The gate of the modern LSTM network also contains a special connection from cells to gate known as peephole. It helps in learning the accurate timing of output. Figure 5 depicts the structure of a LSTM cell.

The input gate learns to shield the constant error flowing from irrelevant inputs into the memory cell, and the output gate protects unrelated memory stuffing in the cell. The forget gate learns to control the timing of values memorized by the cells. Altogether, the cells help in controlling the inflow and outflow of information in network.

Figure 6. LSTM unfolded network model (Salman et al., 2018)

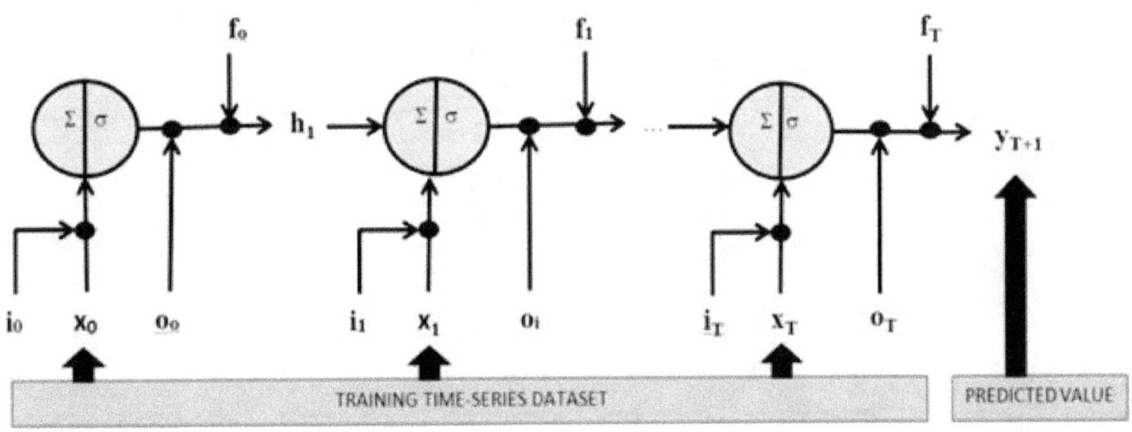

To make the analysis using LSTM neural network easier, very often the network is unfolded through time (t). Figure 6 represents the unfolded architecture of LSTM neural network model. As depicted in the figure, the LSTM network receives the input and passes it through the input, output and forget gates; thus, producing a recurrent signal.

Considering a sequence of input $x=(x_1,\ldots,x_T)$ for an LSTM network and the network computes the sequence of output $y=(y_1,\ldots,y_T)$ by determining the network activation units as in (7) – (12) .

$$i_t = \sigma\left(W_i x_t + U_i h_{t-1} + b_i\right). \tag{7}$$

$$Z_t = \tanh\left(W_z x_t + U_Z h_{t-1} + b_Z\right). \tag{8}$$

$$f_t = \sigma\left(W_f x_t + U_f h_{t-1} + b_f\right). \tag{9}$$

$$C_t = i_t Z_t + f_t C_{t-1}. \tag{10}$$

$$O_t = \sigma\left(W_0 x_t + U_0 h_{t-1} + V_0 C_t + b_0\right). \tag{11}$$

$$h_t = O_t \tanh\left(C_t\right). \tag{12}$$

Figure 7. Deep Network Architecture (Chen, Zeng, Zhou, Du, & Lu, 2018)

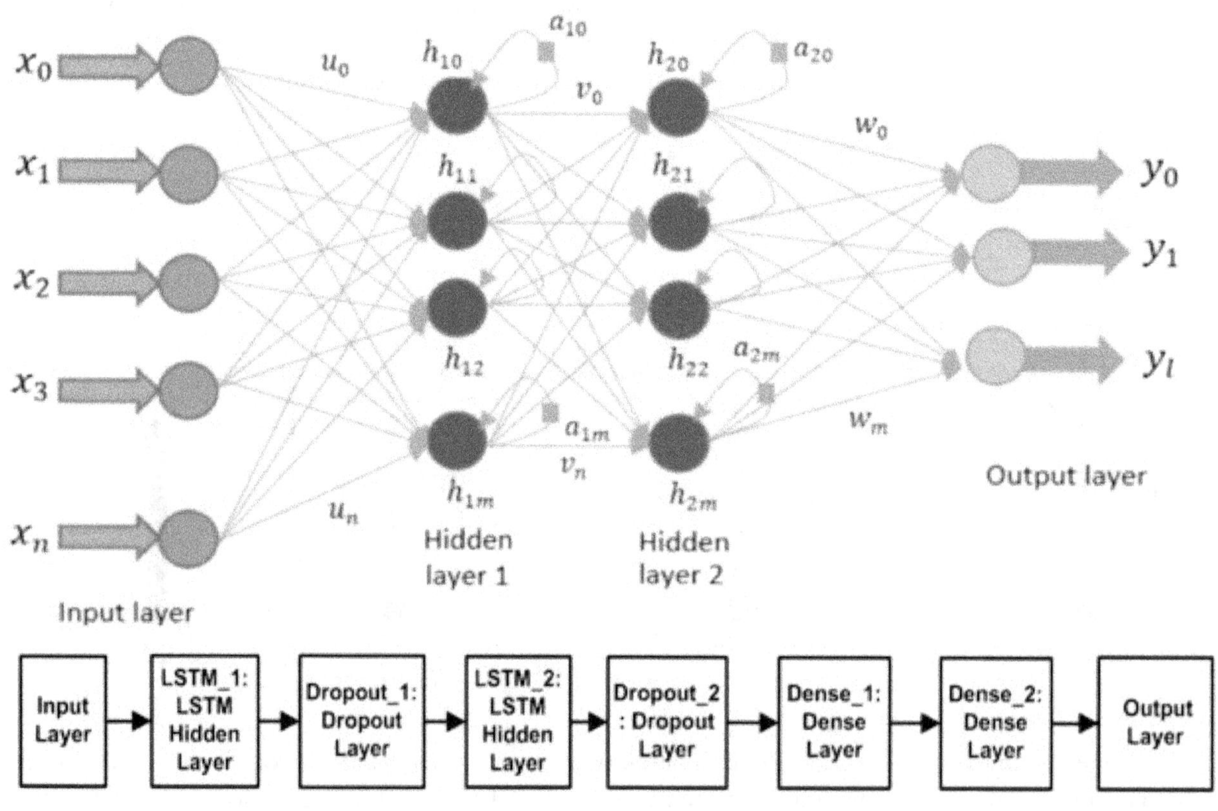

Where, *x* and *y* are the input and output signals; *i*,*o*, and *f* represents the input, output and gate signals, respectively; *h* stands for the recurrent signal; *b* is the bias metric; *U* denotes the connection weight between the input layer to the hidden layer at current time step; *W* denotes the connection weight between hidden layer of previous time step and hidden layer at current time step *O* represents the output gate activation vector; σ (sigmoid) and tan *h* are activation functions.

Figure 8. R-LSTM forecasting model flow chart

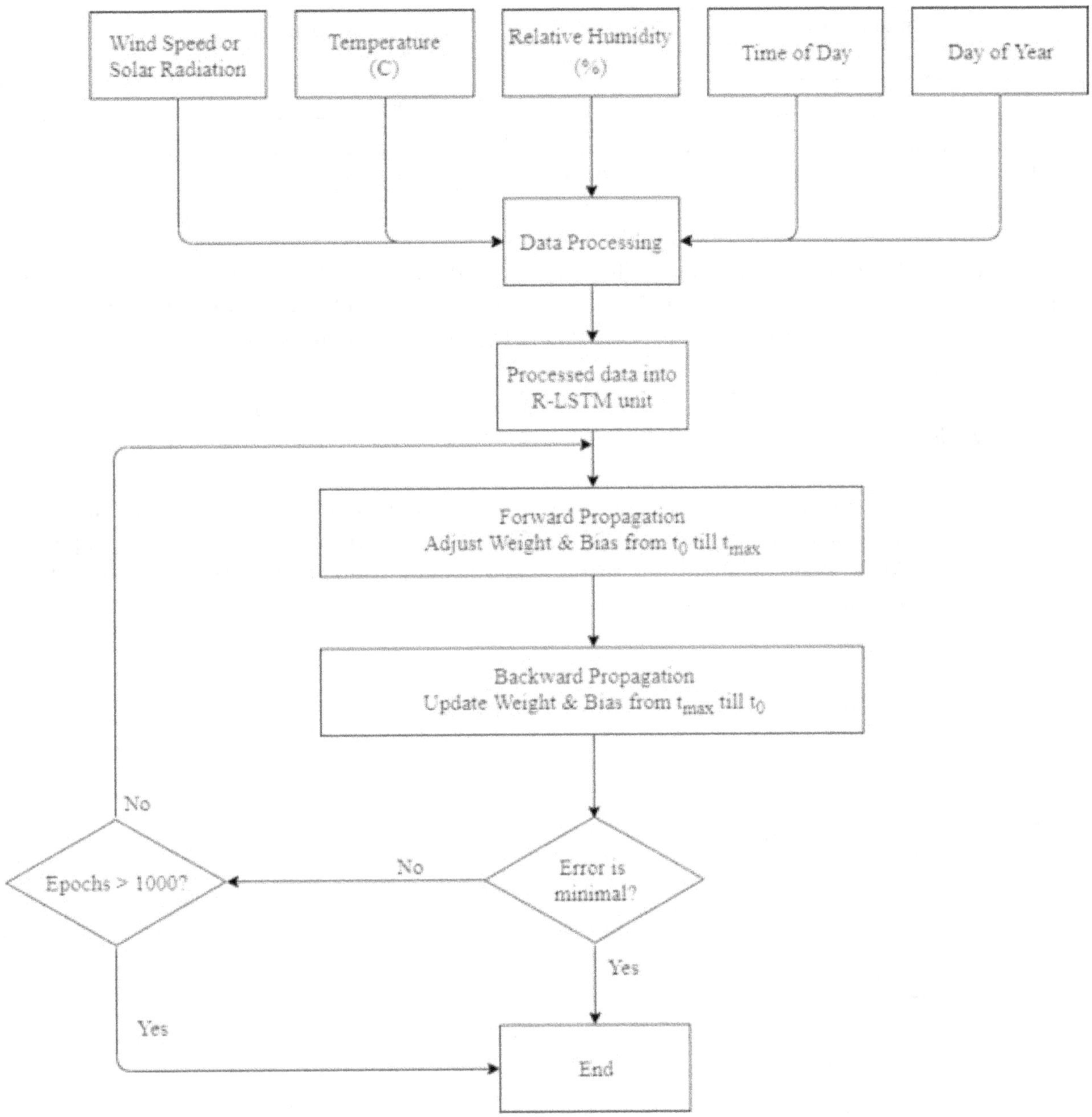

LSTM: RNN Combined Model

Deep learning is an architecture or model that attempt to learn in multiple levels. The major difference identification between the Deep network and simple network is a multi-hidden layer with complex connection. Since, LSTM recurrent network shall be able to remember long term dependencies, applying LSTM memory cell units to recurrent neural network structure can help create a Deep NN structure, or a Deep-LSTM-RNN combined operation could be performed. Replacing the RNN neurons with LSTM

Table 2. List of input variables for the R-LSTM model

Input variable	Description
X_0	Wind speed in *m/s*
X_1	Relative Humidity in percentage
X_2	Temperature in (C)
$X_3 - X_{14}$	Lagged series data for wind speed in *m/s*

memory cells shall realize the said model. The structure of the Deep LSTM network architecture, used in the discussed model, is shown in Figure 7, which consist of one input layer, one output layer and multi hidden layers. The process of R-LSTM is illustrated in Figure 8. The input parameters of the model are depicted in Table 2.

Accuracy Metrics

The forecast accuracy of the R-LSTM hybrid model is evaluated using four popular evaluation criteria namely (Yamashita, Nishio, Do, & Togashi, 2018), Mean Absolute Error (MAE), Mean Absolute Percentage Error (MAPE), Root Mean Squared Error (RMSE), and Correlation Coefficient (R), which are mathematically illustrated in (13)-(16), respectively.

$$MAE = \frac{1}{n}\left(\sum_{i=1}^{n}\left|y_i - y_i^{'}\right|\right). \tag{13}$$

$$MAPE = \frac{1}{n}\left(\sum_{i=1}^{n}\frac{\left|y_i - y_i^{'}\right|}{y_i} * 100\right). \tag{14}$$

$$RMSE = \sqrt{\frac{1}{n}\left(\sum_{i=1}^{n}\left(y_i - y_i^{'}\right)^2\right)}. \tag{15}$$

$$R = \frac{\sum_{i=1}^{n}\left(y_i^{'} - \overline{y_i^{'}}\right)\sum_{i=1}^{n}\left(y_i - \overline{y}\right)}{\sqrt{\sum_{i=1}^{n}\left(y_i^{'} - \overline{y_i^{'}}\right)^2 \sum_{i=1}^{n}\left(y_i - \overline{y}\right)^2}}. \tag{16}$$

SOLUTIONS AND RECOMMENDATIONS

As mentioned before, the R-LSTM model is designed for very-short time series forecasting of wind speed. The discussion of the results shall primarily be based on the wind speed forecast, using the data secured from AIT, Thailand. The forecasting of solar shall only be used as a supplementary result. The wind forecast part of the study is carried out for three cases. The first two cases are for a dataset of 17500 datapoints and the third for 38000 data. This shall aid in testing the performance of the model with large and small sized datasets. The details of the datasets used are given Table 3.

Table 3. Details of wind speed datasets used for R-LSTM performance evaluation

Cases	Dataset size	Training data size	Testing data size
Case – 1	17567	15810	1756
Case – 2	17520	15768	1752
Case – 3	38349	34514	3834

Performance of R-LSTM Model

The forecasting is carried out for one to six steps ahead. For example, if the step size is 15 minutes, the forecast results will include six sets, starting from 15 mins. ahead to 90 mins. ahead of the present data. Figure 9 shows the forecasted results for Case – 3, the bigger dataset. The green shadowed area represents the predicted band. It can be deduced that for smaller prediction horizons (up to 3rd step), the model performance is quite satisfactory. However, it can also be seen that, as the model attempts to predict further, the performance degrades. The distribution of the error in prediction, corresponding to Figure 9 is depicted in Figure 10, and the frequency at which mean or near zero errors occur in the prediction, reduces with the increase in number of steps ahead the prediction is. The lowest of the errors, with mean around zero, is concurrent with the first step of prediction, while the maximum error with the six step-ahead case.

Table 4 elaborates the accuracy metric details while forecasting using the dataset in Case - 3. RMSE varies between 0.58 to 0.86 and the MAPE, between 4.83 to 6.40. The highest value obtained for correlation is in the first step (88%), while it degrades as the number of forecasting steps increases. Depending on the obtained correlation, the acceptable forecasting performance is up to the three steps. The correlation between the forecasted and actual values for case-3 is depicted in Figure 11. The training accuracy of the model is determined using Mean Squared Error (MSE). The training error found during the experiment is around 0.0082. The single-step ahead forecasting result, focusing on 1-day, for the purpose of demonstrating the efficiency of the model, is given in Figure 12.

This model proposed also focuses on considering a probabilistic platform/approach to the forecasting by demonstrating uncertainties and possible risk in the wind speed. The green band in Figure 9, shows the forecasted band for the wind speed value. In Figure 11, two bands having 95% (green dotted line) and 90% (red solid line) confidence interval are shown depicting the risk associated with or the confidence in the predicted value.

Figure 9. Actual and Forecasted Wind Speed (a to f represents 1 to 6-step ahead forecast respectively)

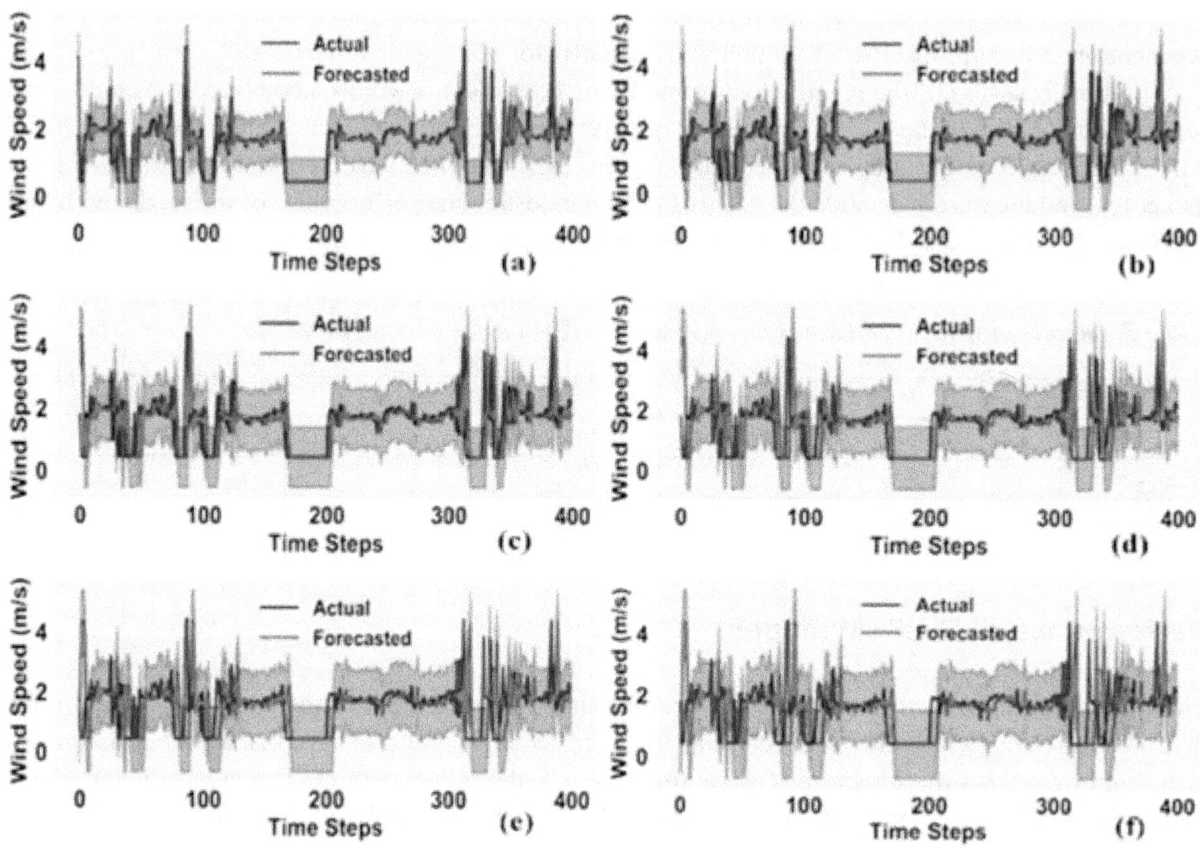

Table 4. Accuracy metrics values in Case - 3

Steps Ahead	MAE		MAPE		RMSE		Correlation	
	LSTM	R-LSTM	LSTM	R-LSTM	LSTM	R-LSTM	LSTM	R-LSTM
1	0.274	0.235	7.252	4.836	0.598	0.581	0.859	0.878
2	0.342	0.315	10.914	5.630	0.789	0.719	0.787	0.814
3	0.447	0.363	10.298	5.940	0.842	0.781	0.687	0.711
4	0.502	0.423	11.010	6.259	0.886	0.823	0.593	0.621
5	0.542	0.452	12.864	6.357	0.9	0.851	0.515	0.546
6	0.567	0.503	12.969	6.4	0.914	0.869	0.454	0.486

Comparison of Forecast Results

The forecast results obtained from R-LSTM model, is compared with other prominent NN models like LSTM and Convolutional Neural Network, for all the three cases. The errors in predictions using the said models, for Cases 1 through 3 are given in Figure 13 - Figure 15, respectively. Despite the differences in the datasets used, the hybrid model performs better than other two models.

Figure 10. Error Distribution of forecasted values for Case – 3

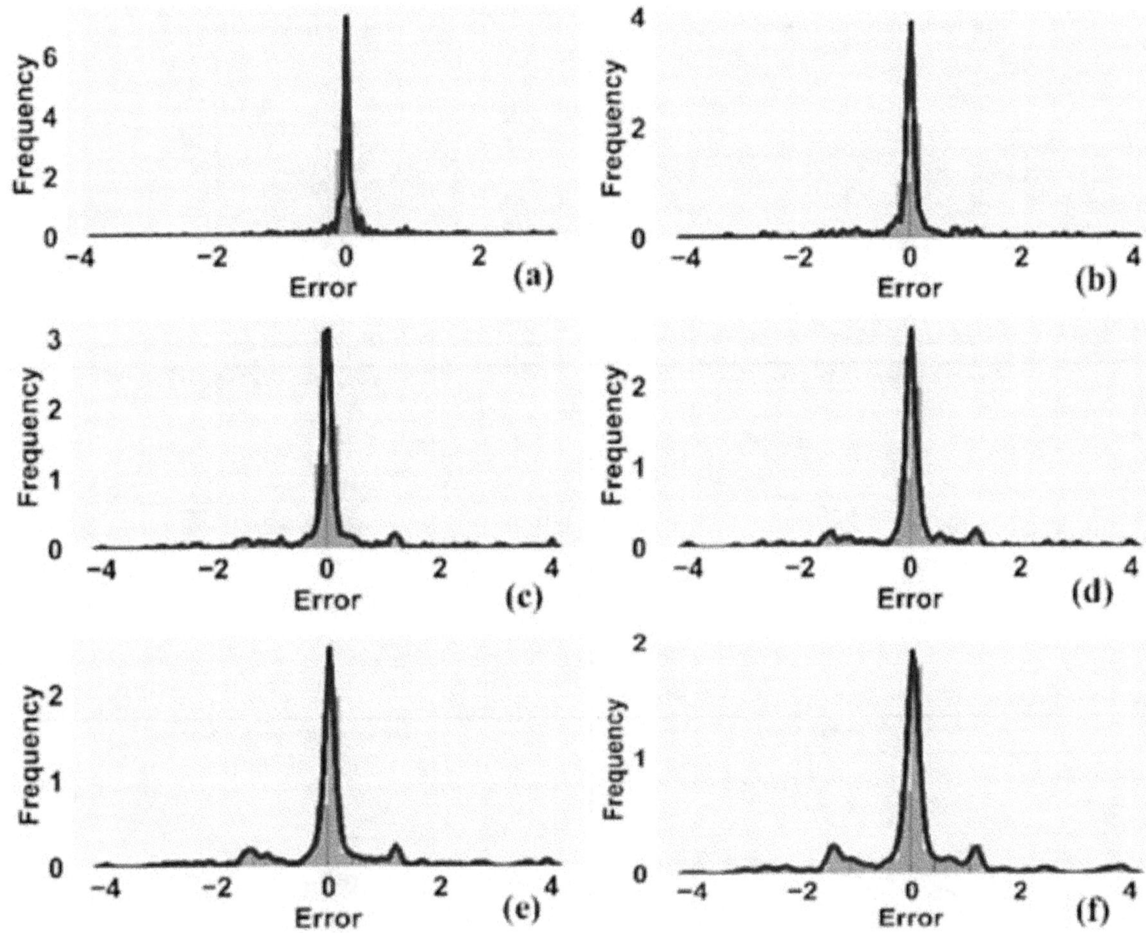

The highest level of accuracy is achieved in Case-2, where the MAPE varied between 2 to 4.65. The RMSE of the proposed model varied between 0.713-1.198, from 0.52 to 0.773, and 0.581 to 0.869, for Case-1 to 3, respectively.

In Case-1, the hybrid model gives us the highest correlations between the actual and predicted wind speed values, around 93% in the first step ahead prediction. Using the hybrid model, better outputs are obtained, especially considering RMSE, for the longer prediction horizons (steps 3rd to 6th), in comparison with either of CNN or LSTM model.

It is interesting to note that for short prediction intervals, 1 or 2 steps ahead, all the three approaches compared produce similar correlation values. But as the prediction step count increases, R-LSTM gives better correlation than the rest of the methods, and in some cases, the margin of improvement in prediction achieved is significant. The MAPE comparison is yet another indication of the superiority of the proposed method. Except for Case-2, in the remaining scenarios, the MAPE values obtained from R-LSTM is much lower than the rest, the difference remaining throughout the six steps of predictions. In Case-3, MAPE is quite constant despite the increase in the steps of prediction, while the same for the other algorithms increase with the prediction step.

Figure 11. Insert the Correlation figure for Case-3

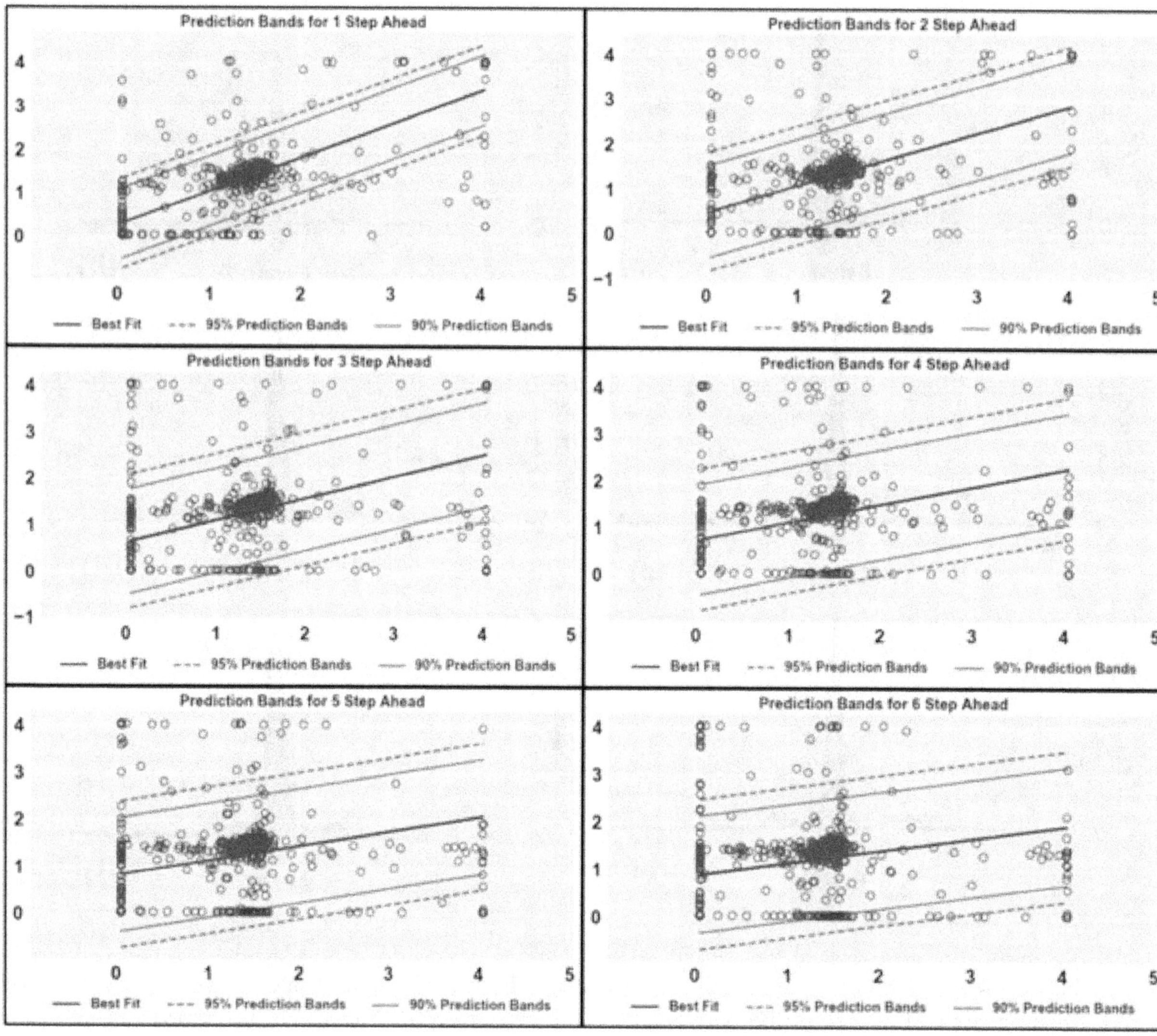

Forecasting Solar Radiation

The discussed R-LSTM hybrid model is used for very short-term wind speed forecasting. The data for solar radiation (37,187 data points) used consists of 15-minute interval measurements, over a period of one year. The list of input data is given in Table 5.

The results obtained for the very-short term solar irradiance forecast is depicted in Figure 16. It is evident that the results are impressive for shorter prediction intervals, while it degrades with the increase in the number of steps. The inference from this experiment is that the hybrid NN model formulated is not only suitable for the forecasting of wind speed on a very short-term scale, but also for solar irradiance forecasting.

Figure 12. Insert One step ahead zoomed in figure here

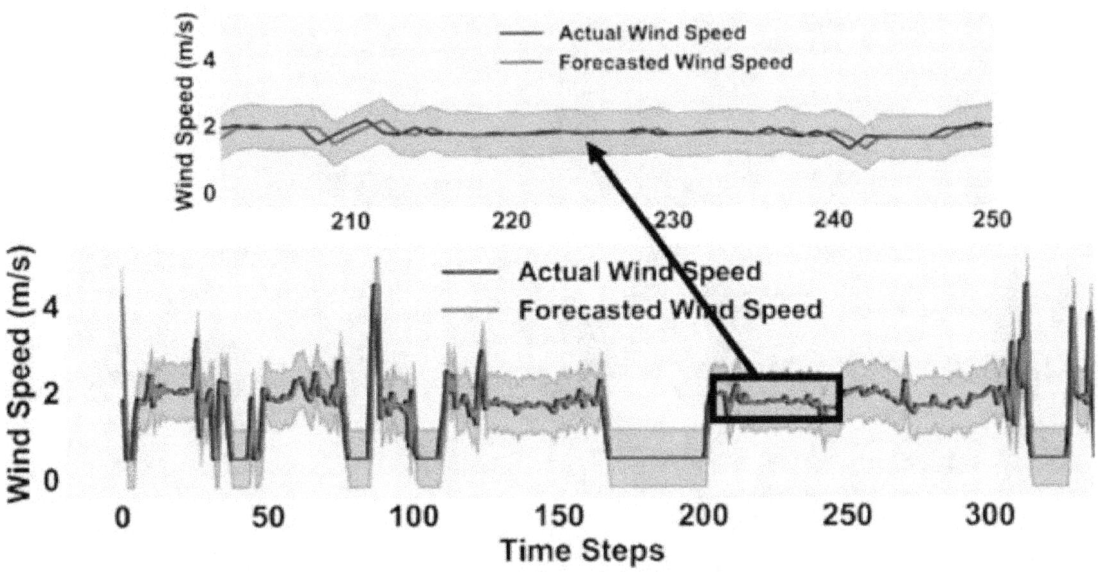

Figure 13. Comparison of Accuracy Metrics for Case-1

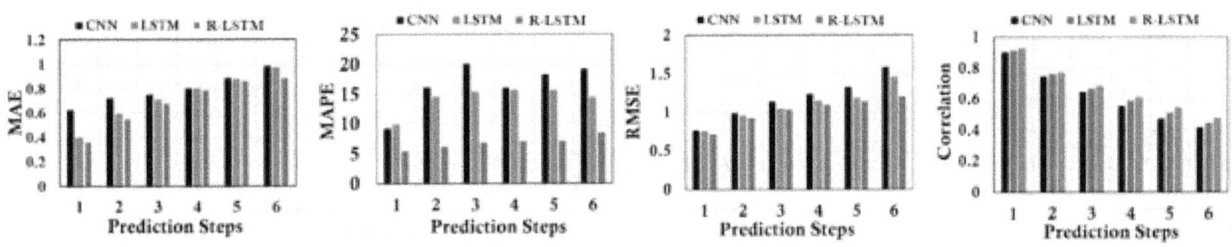

Figure 14. Comparison of Accuracy Metrics for Case-2

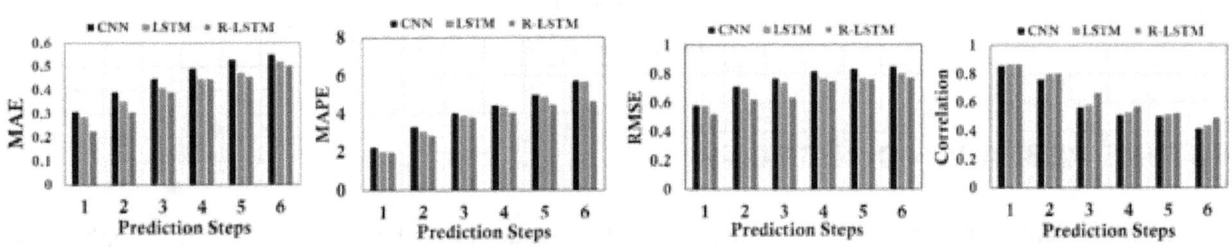

Figure 15. Comparison of Accuracy Metrics for Case-3

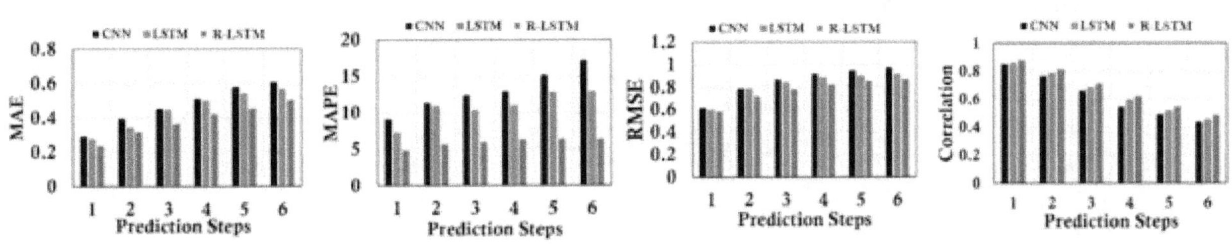

Figure 16. Forecast for three 15-minute steps ahead and the corresponding correlations

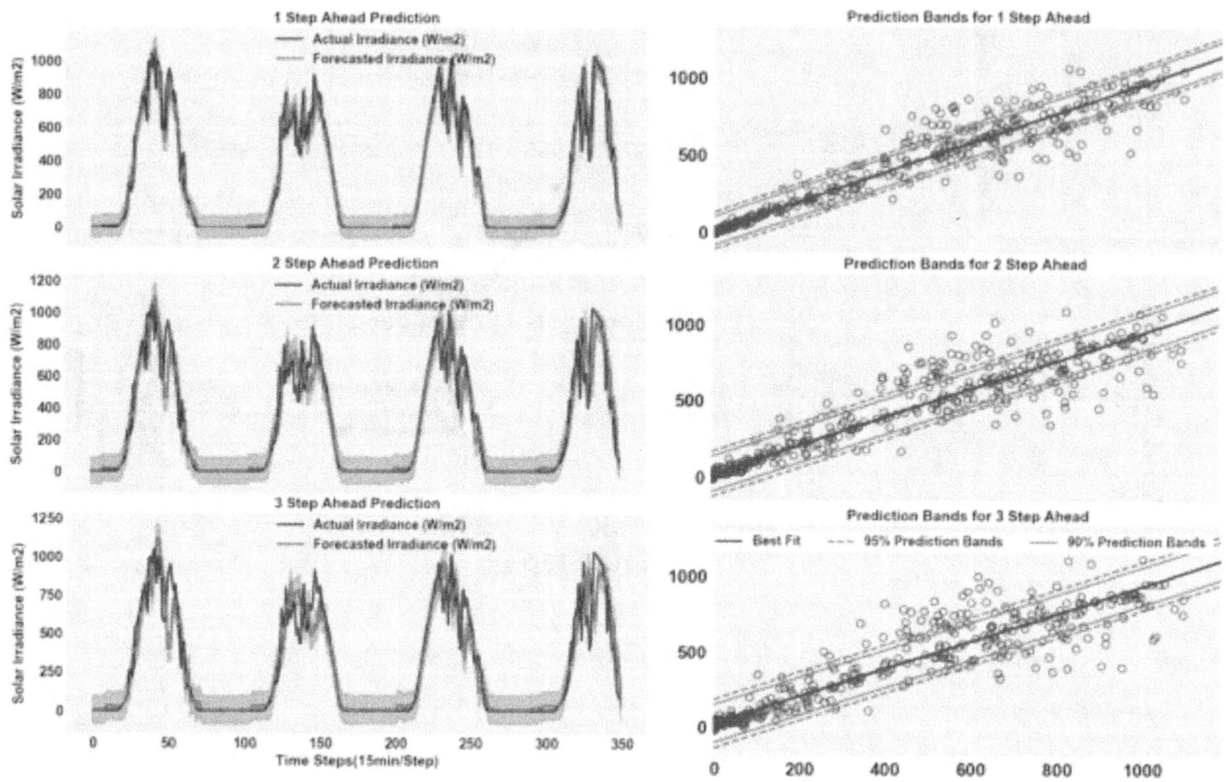

Table 5. List of input variables for solar radiation forecasting

Input variable	Description
X_0	Solar irradiance W/m^2
X_1	Relative Humidity in percentage
X_3	Temperature in (C)
$X_3 - X_{14}$	Lagged series data for solar irradiance W/m^2

FUTURE RESEARCH DIRECTIONS

From the application perspective, the suitability of this research is to be investigated while applying to practical scenarios, as in, microgrid scheduling or power market bidding. Also, the application of such hybrid forecasting algorithms for load forecasting can also be investigated and on a wider time scale viz., short-term and long term.

CONCLUSION

The application of hybrid neural network model for very-short term forecasting is illustrated. The combination to ANN models viz., LSTM and RNN is used for realizing the hybrid NN. With the hybrid model, a multi-step ahead prediction approach, is formulated. The performance of the R-LSTM model is compared with two deep learning based neural network models. The performance evaluation is conducted mainly using wind speed data. The suitability toward very-short term solar radiation forecasting is also evaluated. Results from the tests demonstrate that the proposed hybrid model performs better as per the various accuracy measuring metrics used. The hybrid model also incorporates a probabilistic approach by considering the uncertainties and possible risk in the forecast. The bands of forecast included in the model can useful in power generation balancing and reserve scheduling, trading in energy market, efficient management of power system etc. The correlation values obtained are conclusive that the hybrid model performs satisfactorily until the third step (an hour and half) of prediction, suggesting room for improvement in the model for longer prediction horizons. By combining both the probabilistic and point wise forecasting approaches, the model shall make the operation and control of the system easier and efficient for the system operators or utilities.

REFERENCES

Al-falahi, M. D. A., Jayasinghe, S. D. G., & Enshaei, H. (2017). A review on recent size optimization methodologies for standalone solar and wind hybrid renewable energy system. *Energy Conversion and Management*, *143*, 252–274. doi:10.1016/j.enconman.2017.04.019

Alencar, D. B., Affonso, C. M., Oliveira, R. C. L., & Filho, J. C. R. (2018). Hybrid Approach Combining SARIMA and Neural Networks for Multi-Step Ahead Wind Speed Forecasting in Brazil. *IEEE Access: Practical Innovations, Open Solutions*, *6*, 55986–55994. doi:10.1109/ACCESS.2018.2872720

Bengio, Y., Boulanger-Lewandowski, N., & Pascanu, R. (2013). Advances in optimizing recurrent networks. In *2013 IEEE International Conference on Acoustics, Speech and Signal Processing* (pp. 8624–8628). 10.1109/ICASSP.2013.6639349

Boudour, M. (2016). LFC enhancement concerning large wind power integration using new optimised PID controller and RFBs. *IET Generation, Transmission & Distribution, 10*(16), 4065-4077. Retrieved from https://digital-library.theiet.org/content/journals/10.1049/iet-gtd.2016.0385

Cadenas, E., & Rivera, W. (2010). Wind speed forecasting in three different regions of Mexico, using a hybrid ARIMA–ANN model. *Renewable Energy*, *35*(12), 2732–2738. doi:10.1016/j.renene.2010.04.022

Cao, Q., Ewing, B. T., & Thompson, M. A. (2012). Forecasting wind speed with recurrent neural networks. *European Journal of Operational Research*, *221*(1), 148–154. doi:10.1016/j.ejor.2012.02.042

Chen, J., Zeng, G.-Q., Zhou, W., Du, W., & Lu, K.-D. (2018). Wind speed forecasting using nonlinear-learning ensemble of deep learning time series prediction and extremal optimization. *Energy Conversion and Management*, *165*, 681–695. doi:10.1016/j.enconman.2018.03.098

Doucoure, B., Agbossou, K., & Cardenas, A. (2016). Time series prediction using artificial wavelet neural network and multi-resolution analysis: Application to wind speed data. *Renewable Energy*, *92*, 202–211. doi:10.1016/j.renene.2016.02.003

He, Q., Wang, J., & Lu, H. (2018). A hybrid system for short-term wind speed forecasting. *Applied Energy*, *226*, 756–771. doi:10.1016/j.apenergy.2018.06.053

Hinton, G. E., Osindero, S., & Teh, Y.-W. (2006). A Fast Learning Algorithm for Deep Belief Nets. *Neural Computation*, *18*(7), 1527–1554. doi:10.1162/neco.2006.18.7.1527 PMID:16764513

Hu, Y.-L., & Chen, L. (2018). A nonlinear hybrid wind speed forecasting model using LSTM network, hysteretic ELM and Differential Evolution algorithm. *Energy Conversion and Management*, *173*, 123–142. doi:10.1016/j.enconman.2018.07.070

Kaur, T., Kumar, S., & Segal, R. (2016). Application of artificial neural network for short term wind speed forecasting. In *2016 Biennial International Conference on Power and Energy Systems: Towards Sustainable Energy (PESTSE)* (pp. 1–5). 10.1109/PESTSE.2016.7516458

Kumar, J., Goomer, R., & Singh, A. K. (2018). Long Short Term Memory Recurrent Neural Network (LSTM-RNN) Based Workload Forecasting Model For Cloud Datacenters. *Procedia Computer Science*, *125*, 676–682. doi:10.1016/j.procs.2017.12.087

Larochelle, H., Bengio, Y., Louradour, J., & Lamblin, P. (2009). Exploring Strategies for Training Deep Neural Networks. *Journal of Machine Learning Research*, *10*, 1–40. Retrieved from https://dl.acm.org/citation.cfm?id=1577069.1577070

Lee, H.-H. (2016). Effective power dispatch capability decision method for a wind-battery hybrid power system. *IET Generation, Transmission & Distribution, 10*(3), 661-668. Retrieved from https://digital-library.theiet.org/content/journals/10.1049/iet-gtd.2015.0078

Li, Y., Shi, H., Han, F., Duan, Z., & Liu, H. (2019). Smart wind speed forecasting approach using various boosting algorithms, big multi-step forecasting strategy. *Renewable Energy*, *135*, 540–553. doi:10.1016/j.renene.2018.12.035

Li, Y., Wu, H., & Liu, H. (2018). Multi-step wind speed forecasting using EWT decomposition, LSTM principal computing, RELM subordinate computing and IEWT reconstruction. *Energy Conversion and Management*, *167*, 203–219. doi:10.1016/j.enconman.2018.04.082

Liu, H., Mi, X., & Li, Y. (2018). Smart deep learning based wind speed prediction model using wavelet packet decomposition, convolutional neural network and convolutional long short term memory network. *Energy Conversion and Management*, *166*, 120–131. doi:10.1016/j.enconman.2018.04.021

Lu, X. (2019). Time-periodic model of wind speed and its application in risk evaluation of wind-power-integrated power systems. *IET Generation, Transmission & Distribution, 13*(1), 46-54. Retrieved from https://digital-library.theiet.org/content/journals/10.1049/iet-gtd.2018.5619

Lynch, C., O'Mahony, M. J., & Scully, T. (2014). Simplified Method to Derive the Kalman Filter Covariance Matrices to Predict Wind Speeds from a NWP Model. *Energy Procedia, 62*, 676–685.

Marugán, A. P., Márquez, F. P. G., Perez, J. M. P., & Ruiz-Hernández, D. (2018). A survey of artificial neural network in wind energy systems. *Applied Energy, 228*, 1822–1836. doi:10.1016/j.apenergy.2018.07.084

Qu, Z., Mao, W., Zhang, K., Zhang, W., & Li, Z. (2019). Multi-step wind speed forecasting based on a hybrid decomposition technique and an improved back-propagation neural network. *Renewable Energy, 133*, 919–929. doi:10.1016/j.renene.2018.10.043

Sainath, T. N., Mohamed, A., Kingsbury, B., & Ramabhadran, B. (2013). Deep convolutional neural networks for LVCSR. In *2013 IEEE International Conference on Acoustics, Speech and Signal Processing* (pp. 8614–8618). 10.1109/ICASSP.2013.6639347

Salman, A. G., Heryadi, Y., Abdurahman, E., & Suparta, W. (2018). Single Layer & Multi-layer Long Short-Term Memory (LSTM) Model with Intermediate Variables for Weather Forecasting. *Procedia Computer Science, 135*, 89–98. doi:10.1016/j.procs.2018.08.153

Shukur, O. B., & Lee, M. H. (2015). Daily wind speed forecasting through hybrid KF-ANN model based on ARIMA. *Renewable Energy, 76*, 637–647. doi:10.1016/j.renene.2014.11.084

Sun, N., Zhou, J., Liu, G., & He, Z. (2019). A hybrid wind speed forecasting model based on a decomposition method and an improved regularized extreme learning machine. *Energy Procedia, 158*, 217–222. doi:10.1016/j.egypro.2019.01.079

Ummels, B. C., Gibescu, M., Pelgrum, E., Kling, W. L., & Brand, A. J. (2007). Impacts of Wind Power on Thermal Generation Unit Commitment and Dispatch. *IEEE Transactions on Energy Conversion, 22*(1), 44–51. doi:10.1109/TEC.2006.889616

Varanasi, J., & Tripathi, M. M. (2016). Artificial Neural Network based wind speed & power forecasting in US wind energy farms. In *2016 IEEE 1st International Conference on Power Electronics, Intelligent Control and Energy Systems (ICPEICES)* (pp. 1–6). 10.1109/ICPEICES.2016.7853622

Wang, L., & Li, J. (2016). Estimation of extreme wind speed in SCS and NWP by a non-stationary model. *Theoretical and Applied Mechanics Letters, 6*(3), 131–138.

Yadav, M. R., Singh, K. G., & Chaturvedi, A. (2017). Short-term wind speed forecasting of knock airport based on ANN algorithms. In *2017 International Conference on Information, Communication, Instrumentation and Control (ICICIC)* (pp. 1–8). 10.1109/ICOMICON.2017.8279089

Yamashita, R., Nishio, M., Do, R. K. G., & Togashi, K. (2018). Convolutional neural networks: An overview and application in radiology. *Insights Into Imaging, 9*(4), 611–629. doi:10.100713244-018-0639-9 PMID:29934920

KEY TERMS AND DEFINITIONS

Deep Neural Network (DNN): A class of machine learning model. The main difference between Classical and Deep network scheme is the number of the hidden layer and the training process. Using more hidden layers, DNN can extract higher order of interrelation.

Long Short-Term Memory (LSTM): LSTM is explicitly a subfield of RNN architecture, which is more stable and efficient in dealing with both long-term, as well as short-term dependency problems. It is very useful when the gap between the past and the required future values are substantial.

Recurrent Neural Network (RNN): RNN is a type of ANN, usually used for the forecasting of time series data. It utilizes the feedback provided by one or more units of its network as input in selecting the succeeding output. In RNN, the hidden neurons connect the hidden layer from previous time step to current time step, which is why it is called recurrent.

Chapter 12
Multi-Fuel Power Dispatch Considering Prohibited Operating Zones and Tie-Line Flow Limits Using Ant Lion Optimizer

Ganesan Sivarajan
Government College of Engineering, Salem, India

Jayakumar N.
Government Polytechnic College, Uthangarai, India

Balachandar P.
Government Polytechnic College, Valangaiman, India

Subramanian Srikrishna
https://orcid.org/0000-0002-9888-5382
Annamalai University, India

ABSTRACT

The electrical power generation from fossil fuel releases several contaminants into the air, and these become excrescent if the generating unit is fed by multiple fuel sources (MFS). The ever more stringent environmental regulations have forced the utilities to produce electricity at the cheapest price and the minimum level of pollutant emissions. The restriction in generator operations increases the complexity in plant operations. The cost effective and environmental responsive operations in MFS environment can be recognized as a multi-objective constrained optimization problem. The ant lion optimizer (ALO) has been chosen as an optimization tool for solving the MFS dispatch problems. The fuzzy decision-making mechanism is integrated in the search process of ALO to fetch the best compromise solution (BCS). The intended algorithm is implemented on the standard test systems considering the prevailing operational constraints such as valve-point loadings, CO_2 emission, prohibited operating zones and tie-line flow limits.

DOI: 10.4018/978-1-7998-3970-5.ch012

MULTI-FUEL POWER GENERATION DISPATCH (MFPGD)

In practical conditions of power system operations, different Fuel Sources (FS) like coal, natural gas and oil supply certain generating units. The cost function for each fuel type is derived and is segmented as Piecewise Quadratic Cost Function (PQCF) for a generating unit fed by Multiple Fuel Sources (MFS). These generating units face with the dilemma of finding out the most economical fuel to fire. Further, the operational complexity is increased while considering the valve-point discontinuities and prohibited operating zones. Now-a-days, Emission Control (EC) is likewise an important objective, which must be weighed along with fuel price. The emission function can also be approximated like the Fuel Cost (FC). Therefore, the process becomes trickier when the conflicting objectives (total operating cost and pollutant emission) are taken in concert. The solution process is detailed in Figure 1.

SOLUTION METHODS

The solution approaches addressing this problem can be categorized into mathematical and heuristic methods. The research reports addressed the multi-fuel power dispatch problems are briefed in this section.

Figure 1. Solution procedure to multi-objective multi-fuel generation dispatch

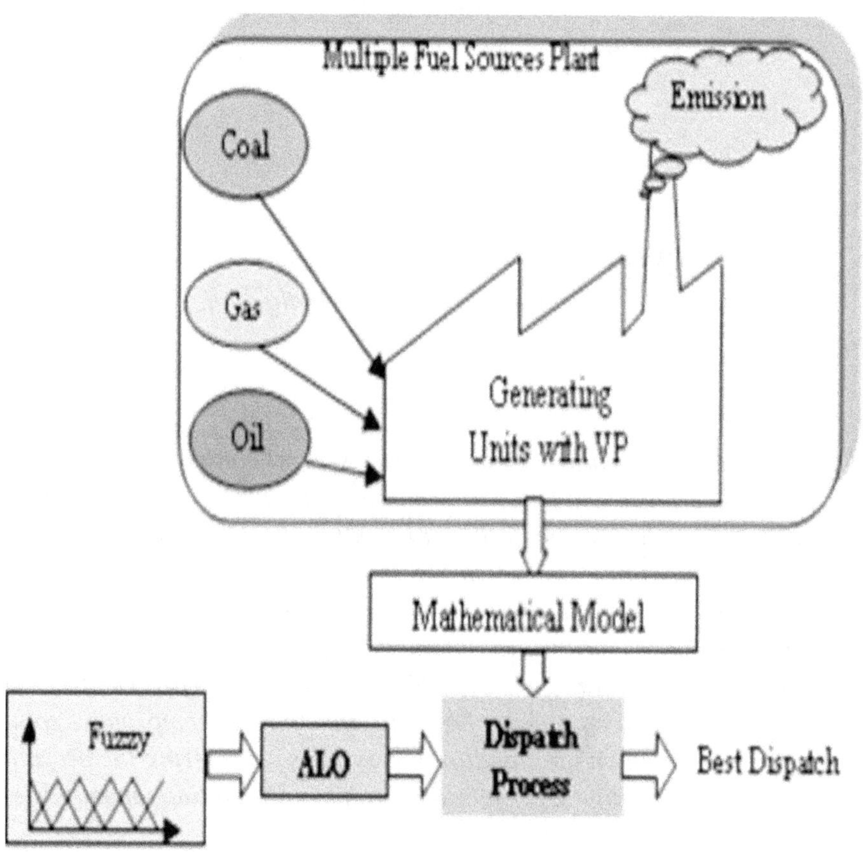

The classical optimization methods, including Hierarchical Method (HM) and artificial neural network models such as Hopfield Neural Network (HNN) and Adaptive HNN (AHNN) models have been reported to address the economic operation of MFS (Shoults & Mead, 1984; Lin & Viviani, 1984; Park et al., 1993; Lee et al., 1998). The main drawback of these methods is the exponentially growing time for large scale systems with non-convex constraints.

The meta-heuristic search techniques such as Genetic Algorithm (GA) (Baskar et al., 2003), Evolutionary Programming (EP) (Jayabarathi et al., 2005), Particle Swarm Optimization (PSO) (Park et al., 2005), Artificial Immune System (AIS) (Panigrahi et al., 2007), Differential Evolution (DE) (Noman & Iba, 2008), Artificial Bee Colony Algorithm (ABC) (Hemamalini & Simon, 2010) and Biogeography Based Optimization (BBO) (Bhattacharya & Chattopadhyay, 2011) have been reported for solving ED with PQCF. The modified versions of heuristic search techniques such as hybrid Real Coded GA (RCGA), fast EP, improved fast EP, Improved GA – Multiplier Updating (IGA-MU), New PSO-Local Random Search (NPSO-LRS), penalty parameter less PSO/DE and New Adaptive PSO (NAPSO) have been reported to solve multi-fuel power dispatch problem (Baskar et al., 2003, Jayabarathi et al., 2005; Park et al., 2005; Chiang, 2005; Selvakumar & Thanushkodi, 2007; Manoharan et al., 2008; Niknam et al., 2011). The improved version of PSO has been reported to solve the ED problem considering the valve-point effects (Polprasert et al., 2013). Further, improved versions of HNN and mathematical methods such as Augmented Lagrange HNN (ALHNN), Enhanced ALHNN (EALHNN), Quadratic Programming – Augmented Lagrange Hopfield Network (QP-ALHN), Hopfield Lagrange Network (HLN), Auction based Algorithm (AA) and Dimensional Steepest Decline (DSD) have also been reported to determine cost effective dispatch schedules (Vo & Ongsakul, 2012; Dieu et al., 2013; Dieu & Schenger, 2013; Thang, 2013; Binetti et al., 2014; Zhan et al., 2015). The Teaching Learning Based Optimization (TLBO) algorithm and Chaotic Global Best ABC (CGBABC) algorithm have been applied for the economic solution considering tie line flows and MFS (Basu, 2014; Secui, 2015).

Recently, the population based soft computing techniques like Kinetic Gas Molecule Optimization (KGMO) (Basu, 2016), Crisscross Optimization (CCO) (Meng & Yin, 2016), Grey Wolf Optimization (GWO) (Pradhan et al., 2016), Backtracking Search Algorithm (BSA) (Modiri-Delshad et al., 2016), Predator-Prey Optimization (PPO) (Singh et al., 2016), Synergic PPO (SPPO) (Singh et al., 2016), Lightning Flash Algorithm (LFA) (Kheshti et al., 2017) have been reported for cost effective multi-fuel power dispatch schedules. Further, the modified versions of meta-heuristic algorithms such as DE-PSO (Parouha & Das, 2016), Colonical Competitive DE (CCDE) (Ghasemi et al., 2016), Opposition based Greedy Heuristic Search (OGHS) (Singh & Dhillon, 2016), Surrogate Worth Trade-off Method (SWTM) (Singh et al., 2017), Pseudo-inspired Choatic Bat Algorithm (PCBA) (Shukla & Singh, 2017), Ant Lion Optimizer (ALO) (Balachandar et al., 2017; Balachandar 2017), Double Weighted PSO (DWPSO) (Kheshti et al., 2018) and Adaptive Predator – Prey Optimization (APPO) (Singh et al., 2018) have been reported for the economical real power scheduling considering multiple energy sources. An improved version of TLBO and group leader optimization technique have also been reported for the ED solutions (Banerjee et al., 2016; Roy et al., 2017). Lin & Wang, 2019 suggested Improved Stochastic Fractal Search (ISFS) algorithm for solving ED problems considering tie-line flow limit.

The mathematical approaches suffer from the drawback of trapping in local solutions and their applications are limited to small-scale linear MFS problems. Withal, the meta-heuristic methods also accept a few drawbacks like algorithmic parameter settings, premature phenomena, trapping into infeasible solution and are computationally expensive. Hence, it is of great significance to improve the existing optimization techniques or exploring new optimization techniques to solve MFS problem. This chapter details the formulation of various types of MFPGD problems and implementation of ALO for solving MFPGD problems.

MULTI-OBJECTIVE MFPGD (MO-MFPGD)

Profuse solution methods have been addressed the multi-fuel power dispatch problems aiming the minimum cost as the only operational objective. Since, the clean air amendment forces the electric power utilities to maintain the pollutant level within the predefined limits, a multi-objective problem formulation has been desirable that optimizes the total fuel cost and emission in a single framework. The economic-environmental compromising operation is less concentrated in the area of multi-fuel power generation dispatch. The environmental issues must be integrated into the operational model to get it desirable for practical power system conditions. Thang (2013) has reported a model considering environmental issues with MFS. Dieu et al, 2013 has incorporated the Prohibited Operating Zone (POZ) as an operational constraint in the MFS environment.

MATHEMATICAL MODEL OF MO-MFPGD

The mathematical model for performing cost – environmentally compromising operation of thermal power plants is given in this section. In this formulation, the decision variables are real power outputs of online generators.

State-of-the Art Models

Objective 1: Minimization of Total Fuel Cost

The total fuel cost of thermal power plant (FC) is the sum of fuel costs of online generating units and is expressed as,

$$FC = Min\sum_{i=1}^{N} F_i(P_i) \ \$/h \tag{1}$$

The fuel cost of a generating unit 'i' considering valve-point loadings and MFS (j) is expressed as,

$$F_i(P_i) = \begin{cases} a_{i1} + b_{i1}P_i + c_{i1}P_i^2 + \left|e_{i1} \times \sin(f_{i1} \times (P_{i1}^{\min} - P_i))\right| & fuel\,1, P_i^{\min} \le P_i \le P_{i1} \\ a_{i2} + b_{i2}P_i + c_{i2}P_i^2 + \left|e_{i2} \times \sin(f_{i2} \times (P_{i2}^{\min} - P_i))\right| & fuel\,2, P_{i1} \le P_i \le P_{i2} \\ \vdots & \vdots \\ a_{ij} + b_{ij}P_i + c_{ij}P_i^2 + \left|e_{ij} \times \sin(f_{ij} \times (P_{ij}^{\min} - P_i))\right| & fuel\,j, P_{ij-1} \le P_i \le P_i^{\max} \end{cases} \tag{2}$$

Objective 2: Minimization of Pollutant Emissions

This objective is realized by summing up the release of pollutants from online generating units into the atmosphere and is expressed as,

$$EC = Min\sum_{i=1}^{N} E_i\left(P_i\right)\, kg/h \tag{3}$$

The emission characteristics of a generating unit with multiple fuel options is also expressed as a piece wise segment function and is stated as,

$$E_i(P_i) = \begin{cases} \alpha_{i1} + \beta_{i1}P_i + \gamma_{i1}P_i^2 & fuel\,1, P_i^{\min} \le P_i \le P_{i1} \\ \alpha_{i2} + \beta_{i2}P_i + \gamma_{i2}P_i^2 & fuel\,2, P_{i1} \le P_i \le P_{i2} \\ \vdots & \vdots \\ \alpha_{ij} + \beta_{ij}P_i + \gamma_{ij}P_i^2 & fuel\,j, P_{ij-1} \le P_i \le P_i^{\max} \end{cases} \tag{4}$$

Development of Bi-Objective Model

The most realistic operational model can be formulated by combining the design objectives FC and EC detailed in Equations (1) and (3) in a single framework. Thus, the model is proposed as in Equation (5) to optimize the conflicting objectives simultaneously subject to variety of constraints.

$$\min F_i(P_i) = (FC, EC) \tag{5}$$

Constraints

The system and operational constraints are as follows:

Power Balance

The total generation by all the generators must be equal to the total power demand (P_d) and transmission line loss (P_L).

$$\sum_{i=1}^{N} P_i = P_d + P_L \tag{6}$$

$$P_L = \sum_{i=1}^{N}\sum_{j=1}^{N} P_i B_{ij} P_j \tag{7}$$

Generation Limits

The real power generation of each generator is to be controlled inside its upper (P_i^{max}) and lower (P_i^{min}) operating limits.

$$P_i^{\min} \le P_i \le P_i^{\max} \quad i = 1, 2, \ldots N \tag{8}$$

Prohibited Operating Zones

The restricted operating regions of generating units decomposes the entire feasible operating regions into a number of feasible sub-regions and the operating point of a generator should lie in any one of the sub-regions as follows:

$$P_i^{min} \le P_i \le P_{i,1}^l$$

$$P_{i,j-1}^u \le P_i \le P_{i,j}^l, \quad j = 2, 3, \ldots, n_i \tag{9}$$

$$P_{i,n_i}^u \le P_i \le P_i^{max}$$

ANT LION OPTIMIZER

The ant lions are a class of net winged insects in nature. The lifecycle of ant lions comprises the stages as: larvae and adult. A larva is the longest period in their lifecycle and ant lions mostly hunt during this period. An ant lion larva digs a cone shaped pit in sand by moving along a circular path, then the larvae hides underneath the bottom of the cone and waits for the prey to be trapped in the pit. Once the ant lion realizes a prey in the trap, it tries to catch intelligently by throw sands towards the edge of the pit to slide the prey into the bottom of the pit. After consuming the prey, the ant lions throw leftovers outside the pit and amend the pit for next hunt.

The ALO mimics the interactions between the ant lions and ants in the trap. The ants are allowed to move over the search space and ant lions hunt those using traps to become fitter. These activities are mathematically modelled and are detailed in the literature (Mirjalili, 2015). The main steps involved in the ALO are random walk of ants, trapping in ant lion's pits, building traps, entrapment of ants in preys, catching in prey and rebuilding of traps.

Random Walks of Ants

To model the interactions between ant lions and ants in the trap, ants are necessitated to move over the search space and ant lions are consented to hunt them and become fitter using traps. A random walk is chosen for modelling ants' movement, since during the search for food the ants move stochastically in nature. Therefore, to facilitate the random walks inside the search space they are normalized using Equation (10).

$$X_i^k = \frac{(X_i^k - r_i)(m_i - q_{i,k})}{(m_{i,k} - r_i)} + q_i \tag{10}$$

Trapping in Ant Lion's Pits

The assumption considered in ALO is that "The random walks of ants are affected by ant lion's traps" (Mirjalili, 2015). The above assumption is mathematically modelled as:

$$q_{i,k} = AL_{j,k} + q_k \ ; \ m_{i,k} = AL_{j,k} + m_k \tag{11}$$

Building Trap

In this phase, a roulette wheel operator is used to select the ant lions based on their fitness during optimization. This mechanism offers high possibilities to the fitter ant lions for grasping ants.

Exploration of Search Space

To prevent the trapped ants from escaping the radius of ants' random walks hyper sphere is reduced adaptively. To mathematically model the above behaviour, the following equations, which shrink the radius of updating ant's positions and mimic the sliding process of ant inside the pits, are used.

$$q_k = {q_k}/{R} ; m_k = {m_k}/{R} \tag{12}$$

Where, $R = 10S\ (k/itermax)$ and S = 2 if $k > 0.1\ itermax$; = 3 if $k > 0.5\ itermax$; = 4 if $k > 0.75\ itermax$; = 5 if $k > 0.9\ itermax$; = 6 if $k > 0.95\ itermax$. The accuracy level of exploitation depends on the constant S.

Catching Prey and Re-Building the Pit

The final stage of hunting behaviour is when an ant reaches the bottom of the pit and is caught in the ant lion's jaw. After this stage, the ant lion pulls the ant inside the sand and consumes its body. This behaviour is modelled using the following equation:

$$AL_{j,k} = A_{i,k} \text{ if } f(A_{i,k}) > f(AL_{j,k}) \tag{13}$$

Elitism

It is assumed that every ant randomly walks around a selected ant lion from the roulette wheel and the elite simultaneously as follows:

$$A_{i,k} = {R_{A,k} + R_{B,k}} \Big/ 2 \tag{14}$$

FUZZY GUIDED ANT LION OPTIMIZER

The MO-MFPGD model considers the multi-fuel options with practical issues; the model aims to optimize the fuel cost and emission levels simultaneously in a single framework and the ALO is applied to solve the multi-fuel generation dispatch problems. The MFS problem is formulated as a multi-objective search, aiming at searching for a set of control variable settings that are comparatively equally good for multiple objectives. In order to select a suitable representative solution for the multiple objectives the fuzzy decision-making mechanism concept is used. Thus, Fuzzy Guided ALO (FGALO) can be used to address the MO-MFPGD problems.

FGALO Implementation for MO-MFPGD

The computational flow of FGALO for solving multi-objective power dispatch considering multiple fuel options is illustrated in Figure 2 and the algorithmic steps are as follows.

Step 1: Read the system data and initialize the algorithmic parameters such as search agents (*Ps*), maximum number of iterations ($iter_{max}$), number of variables (*Nd*) and its limits.

Step 2: The decision variables such as real power outputs of generating units are randomly generated within the lower and upper bounds to initialize the first population of ant and ant lions using Equations (15) and (16).

$$P_i^j = P_j^{\min} + rand * (P_j^{\max} - P_j^{\min})\; i = 1,2,......Ps;\; j = 1,2,...Nd \tag{15}$$

$$P^{AL}{}_i^j = P_j^{\min} + rand * (P_j^{\max} - P_j^{\min})\, i = 1,2,......Ps;\; j = 1,2,...Nd \tag{16}$$

The population matrix of ants and ant lions are formed as matrices as in Equations (17) and (18) respectively.

$$Pop^A = \begin{bmatrix} P_1^1 & P_1^2 & \cdots\cdots & P_1^{Nd-1} & P_1^{Nd} \\ P_2^1 & P_2^2 & \cdots\cdots & P_2^{Nd-1} & P_2^{Nd} \\ \cdots & \cdots & \cdots\cdots & \cdots & \cdots \\ P_{PS-1}^1 & P_{PS-1}^2 & \cdots\cdots & P_{PS-1}^{Nd-1} & P_{PS-1}^{Nd} \\ P_{PS}^1 & P_{PS}^2 & \cdots\cdots & P_{PS}^{Nd-1} & P_{PS}^{Nd} \end{bmatrix} \begin{matrix} \rightarrow & f_1(FC), f_1(EC) \\ \rightarrow & f_2(FC), f_2(EC) \\ \rightarrow & \cdots\cdots \\ \rightarrow & f_{PS-1}(FC), f_{PS-1}(EC) \\ \rightarrow & f_{PS}(FC), f_{PS}(EC) \end{matrix} \tag{17}$$

$$Pop^{AL} = \begin{bmatrix} P^{AL1}_{1} & P^{AL2}_{1} & \dots\dots\dots & P^{ALNd-1}_{1} & P^{ALNd}_{1} \\ P^{AL1}_{2} & P^{AL2}_{2} & \dots\dots\dots & P^{ALNd-1}_{2} & P^{ALNd}_{2} \\ \dots. & \dots. & \dots\dots\dots & \dots. & \dots. \\ P^{AL1}_{PS-1} & P^{AL2}_{PS-1} & \dots\dots\dots & P^{ALNd-1}_{PS-1} & P^{ALNd-1}_{PS-1} \\ P^{AL1}_{PS} & P^{AL2}_{PS} & \dots\dots\dots & P^{ALNd-1}_{PS} & P^{ALNd-1}_{PS} \end{bmatrix} \begin{matrix} \rightarrow & f_1(FC^{AL}, EC^{AL}) \\ \rightarrow & f_2(FC^{AL}, EC^{AL}) \\ \rightarrow & \dots.. \\ \rightarrow & f_{PS-1}(FC^{AL}, EC^{AL}) \\ \rightarrow & f_{PS}(FC^{AL}, EC^{AL}) \end{matrix} \quad (18)$$

Step 3: Perform multi-objective strategy.

3.1: Compute the objective functions (i.e. *FC* and *EC*) using Equation (5) and normalize the objective function values by decision making mechanism using Equation (19).

$$\mu_i = \begin{cases} 1, & if\ F_i \le F_i^{\min} \\ (F_i^{\max} - F_i)/(F_i^{\max} - F_i^{\min}), & if\ F_i^{\min} \langle F_i \langle F_i^{\max} \\ 0, & if\ F_i \ge F_i^{\max} \end{cases} \quad (19)$$

3.2: Evaluate the normalized membership value (μ^k) using Equation (20).

$$\mu^k = \left(\sum_{i=1}^{N_{obj}} \mu_i^k \right) / \left(\sum_{k=1}^{M_{nds}} \sum_{i=1}^{N_{obj}} \mu_i^k \right) \quad (20)$$

3.3: Find the best compromised solution.

Step 4: The ant lion having the best fitness is assumed as elite.

Step 5: Iteration = Iteration +1.

Step 6: Apply Roulette wheel selection to select an ant lion for each ant and perform the following steps for each ant.

Step 7: Update the minimum and maximum bounds of all variables using Equation (11).

Step 8: Create a random walk and normalize it using Equation (10).

Step 9: Update the positions of ants using Equation (13).

Step 10: Repeat the multi-objective strategy.

Step 11: Replace an ant lion with its corresponding ant if becomes fitter.

Step 12: Update elite if an ant lion becomes fitter than elite using Equation (14).

Step 13: Check for maximum iterations reached. Otherwise, go to Step 5.

Step 14: Print the best feasible solution.

TEST CASE STUDIES AND DISCUSSIONS

The optimization procedure is coded in MATLAB 7 and is executed in the personal computer with the hardware configuration of Intel Core i3 2.4 GHz processor and 4 GB RAM. Initially, parameter sensitivity analysis is performed and the optimal algorithmic parameters are identified. As the ALO is a random-

Figure 2. Flowchart for the implementation of FSALO to solve multi-fuel power dispatch

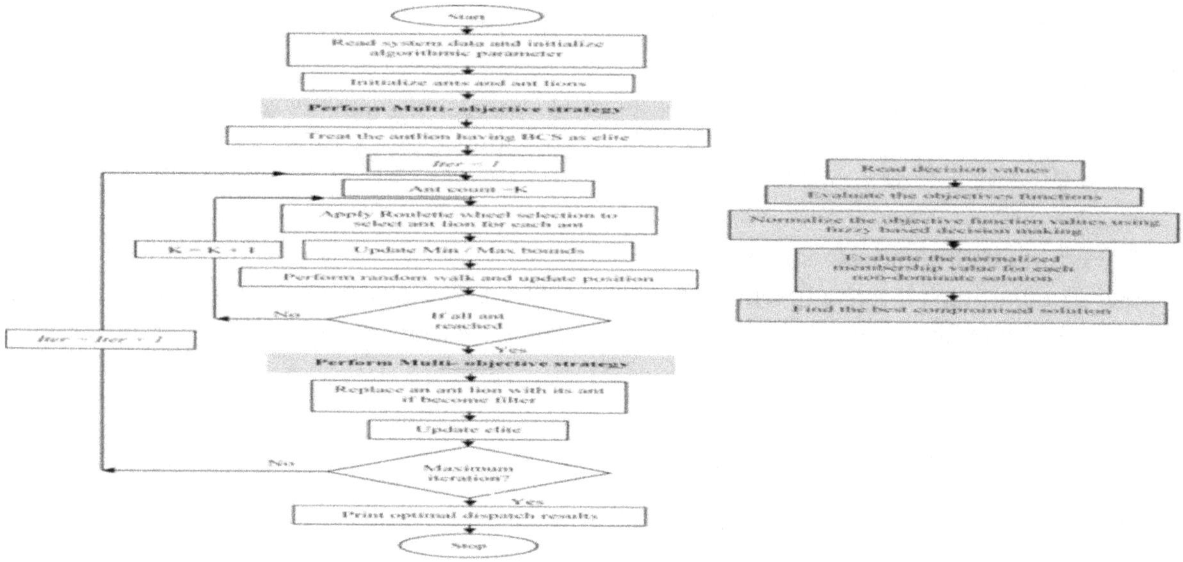

ized procedure, in order to prove its consistency in getting the optimal solutions 50 independent runs have been conducted. The obtained numerical results are compared against earlier reports in terms of solution quality. The input parameters of the algorithm are search agent size = 30 and maximum number of iterations = 200. The simulation studies are detailed in this section and the criterion for selecting the number of search agents is also detailed.

The standard ten-unit system is used for demonstration. Lin & Viviani, 1984 have proposed first this test system that has 3 subsystems and 10 generating units and the system particulars are available in the literature (Lin & Viviani, 1984; Chiang, 2005). The generating units are fueled with two or three fuels and the piecewise quadratic cost functions represents different fuel types. The total system demand is gradually varied in steps of 100 MW from 2400 MW to 2700 MW neglecting transmission loss.

Cost Effective (CE) Schedules

The intended algorithm is applied for the following scenarios:

Scenario 1: CE operation considering PQCF
Scenario 2: CE operation considering PQCF and valve-point loadings and
Scenario 3: CE operation considering POZ.

Scenario 1

The ALO is implemented on the standard 10-unit system neglecting valve-point loadings. The intended algorithm is executed and the obtained best feasible solution including fuel type, the best dispatches of generators and total costs for different load demands are presented in Table 1. The ALO has converged to the total fuel costs of \$481.7223, \$526. 2386, \$574.3808 and \$623.8085 for load demands of 2400

Table 1. Best CE dispatches neglecting for 10-unit system by ALO

Unit No.	P_d = 2400 MW		P_d = 2500 MW		P_d = 2600 MW		P_d = 2700 MW	
	FS	P_i (MW)	FS	P_i (MW)	FS	P_i (MW)	FS	P_i (MW)
P_1	1	189.7403	2	206.5191	2	216.54	2	218.2511
P_2	1	202.3426	1	206.4573	1	210.91	1	211.6626
P_3	1	253.895	1	265.7392	1	278.5	1	280.7217
P_4	3	233.0455	3	235.9532	3	239.1	3	239.6315
P_5	1	241.8293	1	258.017	1	275.5	1	278.4963
P_6	3	233.0457	3	235.953	3	239.1	3	239.6315
P_7	1	253.275	1	268.8636	1	285.7	1	288.5845
P_8	3	233.0456	3	235.9532	3	239.1	3	239.6315
P_9	1	320.383	1	331.4878	1	343.55	3	428.5212
P_{10}	1	239.3973	1	255.0563	1	272	1	274.8669
FC ($/h)	481.7223		526.2386		574.3808		623.8085	

MW, 2500 MW, 2600 MW and 2700 MW respectively. In order to validate the obtained numerical results, the total fuel costs are compared with the earlier reports and the comparison is presented in Table 2. It is worthy to note that the ALO provides an improved CE dispatch schedule for all load demands. As erroneous test data is followed in the reports using ABC (Hemamalini & Simon, 2010) and OGHS (Singh & Dhillon, 2016), they cannot be taken for direct comparison.

Scenario 2

Further, the valve-point effects along with the quadratic fuel cost functions are considered. The obtained best feasible dispatches for different load demands using ALO are presented in Table 3. For the sake of comparison, the total fuel cost for load demand of 2700 MW is compared against the published reports and is presented in Table 4. The reports by using ABC (Hemamalini & Simon, 2010), BBO (Bhattacharya & Chattopadhyay, 2011), NAPSO (Niknam et al., 2011), DPD (Parouha & Das, 2016), KGMO (Basu, 2016), CSO (Meng et al., 2016), OGHS (Singh & Dhillon, 2016) and GWO (Pradhan et al., 2016) cannot be taken for direct comparison due to erroneous test data has been adopted. It is also seen from Table 4 that the ALO affords the exact dispatch schedule that leads to a nominal savings in the fuel cost.

Scenario 3

The POZ has been considered as an additional operational constraint in the optimization frame that increases complexity of the dispatch problem under study. The generating units 3, 5, 7 and 10 having restricted operations and prohibited operating regions are detailed in the literature (Dieu & Schenger, 2013). The ALO is implemented and the best economic dispatch schedules for various load demands are detailed in Table 5. The attained numerical results are compared with the recent reports such as QP-ALHN (Dieu & Schenger, 2013), PSO (Dieu & Schenger, 2013), DE (Dieu & Schenger, 2013), PPO (Singh et al., 2018), SPPO (Singh et al., 2018), APPO (Singh et al., 2018) and the comparison is also presented in Table 6. It is also observed that the intended algorithm attains the competitive results.

Table 2. Fuel cost comparison for Scenario 1

Methods	P_d = 2400 MW	P_d = 2500 MW	P_d = 2600 MW	P_d = 2700 MW
HM	488.50	526.70	574.03	625.18
HNN	487.87	526.13	574.26	626.12
AHNN	481.72	526.23	574.37	626.24
GA	481.723	526.239	574.396	623.809
Hybrid RCGA	481.722	526.238	574.380	623.809
Improved Fast EP	NR	526.246	NR	NR
Fast EP	NR	526.262	NR	NR
Classical EP	NR	526.246	NR	NR
Modified PSO	481.723	526.239	574.381	623.809
IGA-MU	NR	NR	NR	623.8093
AIS	481.723	526.240	574.381	623.809
DE	481.723	526.239	574.381	623.809
ABC	470.9506*	516.2793*	588.5632*	607.7481*
EALHNN	481.723	526.239	574.381	623.809
ALHNN	481.723	526.239	574.381	623.809
HLN	481.7226	526.2388	574.7413	623.8092
EP	NR	NR	NR	626.26
OGHS	NR	NR	NR	623.8082*
QP-ALHN	481.723	526.239	574.381	623.809
SPPO	NR	NR	NR	623.809
APPO	NR	NR	NR	623.809
ALO	481.7223	526.2386	574.3829	623.8085

*-Not feasible NR- Not Reported

Table 3. Best CE dispatches considering valve-point loadings for 10-unit system by ALO

Unit No.	P_d = 2400 MW		P_d = 2500 MW		P_d = 2600 MW		P_d = 2700 MW	
	FS	P_i (MW)	FS	P_i (MW)	FS	P_i (MW)	FS	P_i (MW)
P_1	1	189.283	2	206.283	2	218	2	218.593
P_2	1	200.21	1	206	1	210	1	211.216
P_3	1	254.4623	1	266.2502	1	278.1012	1	280.656
P_4	3	234.0337	3	235.6046	3	237	3	239.3707
P_5	1	241.3677	1	258.3708	1	275	1	279.934
P_6	3	233.0557	3	235.3683	3	239.912	3	239.3707
P_7	1	253.6068	1	268.6968	1	286	1	287.7275
P_8	3	233.4948	3	235.9671	3	239	3	239.5051
P_9	1	320.6885	1	331.6617	1	343	3	427.7583
P_{10}	1	239.7971	1	255.7971	1	274	1	275.865
FC ($/h)	**482.4127**		**526.8142**		**575.0544**		**623.8278**	

Table 4. Fuel cost comparison for Scenario 2

Methods	P_d = 2400 MW	P_d = 2500 MW	P_d = 2600 MW	P_d = 2700 MW
IGA-MU	NR	NR	NR	624.5178
NPSO	NR	NR	NR	624.1624
NPSO-LRS	NR	NR	NR	624.1273
PSO-LRS	NR	NR	NR	624.2297
RGA	482.5114	527.0189	575.1610	624.5081
DE	482.5275	527.0360	575.1753	624.5146
PSO	482.5088	527.0185	575.1606	624.5074
RCGA	NR	NR	NR	623.8281
ABC	NR	NR	NR	609.2250*
BBO	NR	NR	NR	605.6387*
NAPSO	NR	NR	NR	623.6217*
AA	NR	NR	NR	623.9524
DSD	NR	NR	NR	623.8325
DE-PSO-DE	NR	NR	NR	623.8265*
KGMO	NR	NR	NR	608.1096*
CCO	NR	NR	NR	623.8237*
OGHS	NR	NR	NR	623.8240*
GWO	NR	NR	NR	605.6818*
BSA	NR	NR	NR	623.9016
CCDE	NR	NR	NR	623.8288
SPPO	NR	NR	NR	623.8279
APPO	NR	NR	NR	623.827
ALO	482.4127	526.8142	575.0544	623.8278

*-Not feasible NR- Not Reported

Table 5. Best CE dispatches considering POZ for 10-unit system by ALO

Unit No.	P_d = 2400 MW		P_d = 2500 MW		P_d = 2600 MW		P_d = 2700 MW	
	FS	P_i (MW)	FS	P_i (MW)	FS	P_i (MW)	FS	P_i (MW)
P_1	1	189.5489	2	206.4992	2	219.2200	2	221.0370
P_2	1	202.2591	1	206.4769	1	212.0940	1	212.8995
P_3	1	253.5949	1	265.6989	1	281.9526	1	284.2837
P_4	3	232.9593	3	235.9534	3	239.8460	3	240.5154
P_5	1	241.4902	1	258.1081	1	260.0000	1	260.000
P_6	3	232.9991	3	235.9529	3	239.9558	3	240.4955
P_7	1	255.0000	1	268.8637	1	290.2970	1	293.2792
P_8	3	232.9796	3	235.9329	3	239.9362	3	240.4953
P_9	1	320.0990	1	331.5075	1	346.6984	3	436.9954
P_{10}	1	239.0699	1	255.0065	1	270.000	1	269.999
FC ($/h)	**481.7102**		**526.1924**		**574.6999**		**624.310**	

Table 6. Fuel cost comparison for Scenario 3

Methods	Load Demand (MW)			
	2400	2500	2600	2700
DE	482.0683	526.4616	575.1903	624.6675
PSO	482.0510	526.4546	574.9327	624.4452
QP-ALHN	481.7266	526.2388	574.7291	624.3212
PPO	---	---	---	624.403
SPPO	---	---	---	624.321
APPO	---	---	---	624.321
ALO	**481.7102**	**526.1924**	**574.6999**	**624.310**

Cost Effective: Environmentally Responsive Schedules

Further, the FGALO algorithm is applied for close practical operation that aims for the cost effective and environmental responsive operation in the multi-objective framework. The CO_2 emission is considered and emission characteristics are taken from the literature (Thang, 2013). The extreme points in the search space are identified by the ALO algorithm through the single objective optimization procedure. The best feasible dispatch schedules for Environmental Responsive (ER) operation are presented in the Table 7.

The obtained best feasible dispatches using FGALO for various load demands are presented in Table 8. The projected algorithm attains the total fuel costs and emissions corresponding to the best feasible dispatches as \$499.0612 and 4693.51 kg for 2400 MW, \$546.3413 and 5142.899 kg for 2500 MW, \$595.1461 and 5572.487 kg for 2600 MW and \$649.8673 and 6043.63 kg for 2700 MW. Table 9 details the comparisons of Best Compromise Solution (BCS) for all load demands that indicate the FGALO attains the best dispatches for all load demands against the comparative methods like HLN (Thang, 2013), Lambda Iteration Method (LIM) (Thang, 2013), ABC, TLBO and GWO.

Table 7. Best feasible ER schedule using ALO

Unit No.	P_d = 2400 MW		P_d = 2500 MW		P_d = 2600 MW		P_d = 2700 MW	
	FS	P_i (MW)	FS	P_i (MW)	FS	P_i (MW)	FS	P_i (MW)
P_1	1	169.8432	2	196.0725	2	196.0924	2	196.6752
P_2	1	199.9977	1	209.734	1	209.615	1	211.4194
P_3	1	270	1	281.0024	1	298.2023	1	301.2775
P_4	3	248.7711	3	254.0253	3	255.0274	3	257.0464
P_5	1	260.9987	1	297.1178	1	300.0157	1	288.5928
P_6	1	165.9886	1	157.9894	1	158.4795	1	167.92
P_7	1	291.5958	1	271.4834	1	291.9906	2	365.299
P_8	3	245.2193	3	243.5108	3	250.4337	3	258
P_9	1	347.5856	3	388.8667	3	439.9585	3	439.7546
P_{10}	1	200	1	200.1975	1	200.1856	1	214.0187
EC (kg/h)	4692.544		5137.025		5572.284		6042.812	

Table 8. Best feasible CE-ER schedule neglecting valve-point loadings using ALO

Unit No.	P_d = 2400 MW		P_d = 2500 MW		P_d = 2600 MW		P_d = 2700 MW	
	FS	P_i (MW)	FS	P_i (MW)	FS	P_i (MW)	FS	P_i (MW)
P_1	1	166.5633	2	196.0623	2	196.0832	2	196.6752
P_2	1	199.3945	1	205.4360	1	208.1230	1	211.4194
P_3	1	259.5376	1	285.0330	1	296.4113	1	302.2775
P_4	3	245.6166	3	254.0144	3	247.1356	3	256.0464
P_5	1	251.7806	1	297.1287	1	298.1175	1	288.5928
P_6	1	156.1777	1	164.9885	1	167.5686	1	170.42
P_7	1	283.8588	1	271.4525	1	297.8834	2	365.299
P_8	3	244.4792	3	236.6117	3	249.4328	3	255.5
P_9	3	392.6513	3	388.9856	3	438.8587	3	439.7546
P_{10}	1	200.0003	1	200.2887	1	200.3876	1	214.0187
FC ($/h)	**499.0612**		**546.3413**		**595.1461**		**649.8673**	
EC (kg/h)	**4693.517**		**5142.899**		**5572.487**		**6043.63**	

Table 9. BCS comparison for CE-ER operation neglecting valve-point loadings

Methods	P_d = 2400 MW		P_d = 2500 MW		P_d = 2600 MW		P_d = 2700 MW	
	FC ($/h)	EC (kg/h)	FC ($/h)	EC (kg/h)	FC ($/h)	EC (kg/h)	FC ($/h)	EC (kg/h)
HLN	498.202	4694.992	547.0897	5143.397	596.5429	5574.018	649.7493	6050.6
LIM	499.0613	4693.518	547.0907	5143.407	596.7164	5573.464	650.0047	6044.31
ABC	501.4721	4702.458	546.5634	5143.059	595.9169	5587.210	649.9823	6044.83
TLBO	501.3097	4701.675	546.5096	5143.041	595.7641	5586.527	649.9541	6044.65
GWO	500.5019	4701.964	546.3560	5143.029	595.6833	5585.685	649.8996	6043.64
FGALO	**499.0612**	**4693.517**	**546.3413**	**5142.899**	**595.1461**	**5572.487**	**649.8673**	**6043.63**

Further, the POZ is included as an operational constraint in multi-objective framework and the attained BCS using FGALO for various load demands are detailed in Table 10. The total fuel cost and CO_2 emission are compared with other methods such as ABC, TLBO and GWO and the comparison is detailed in Table 11. The comparison indicates that the FGALO provides the best solution in the POZ constrained MO- MFPGD problems.

Explicit Operation

Further, to develop the most realistic operational model, the valve-point effects are incorporated in the operational model, that increases further the complexity to find the best feasible solution in the non-linear solution space. The FGALO algorithm is applied to determine the best operating points of generating units and the obtained best dispatch schedules for load demands varying from 2400 MW to 2700 MW is reported in the Table 12.

Table 10. Best feasible CE-ER schedule considering POZ using ALO

Unit No.	P_d = 2400 MW		P_d = 2500 MW		P_d = 2600 MW		P_d = 2700 MW	
	FS	P_i (MW)	FS	P_i (MW)	FS	P_i (MW)	FS	P_i (MW)
P_1	1	166.5636	2	196.0619	2	196.2401	2	196.714
P_2	1	199.3945	1	205.443	1	205.0951	1	211.3988
P_3	1	259.5376	1	283.0329	1	286.3079	1	282.7181
P_4	3	245.6166	3	244.0142	3	247.0961	3	256.1059
P_5	1	251.7806	1	335.000	1	335.1175	1	335.6131
P_6	1	156.178	1	169.1185	1	167.6198	1	170.39
P_7	1	283.8591	1	261.4525	1	287.9129	2	338.012
P_8	3	244.4789	3	216.6119	3	249.3931	3	255.2451
P_9	3	392.6516	3	388.9871	3	424.7992	3	439.6951
P_{10}	1	200	1	200.278	1	200.4183	1	214.1079
FC ($/h)	499.0612		550.9112		597.7798		650.1644	
EC (kg/h)	4693.517		5214.569		5591.2124		6049.9417	

Table 11. BCS comparison for CE-ER operation considering POZ

Methods	P_d = 2400 MW		P_d = 2500 MW		P_d = 2600 MW		P_d = 2700 MW	
	FC ($/h)	EC (kg/h)	FC ($/h)	EC (kg/h)	FC ($/h)	EC (kg/h)	FC ($/h)	EC (kg/h)
ABC	501.498	4702.09	556.298	5321.88	600.214	5599.41	656.725	6058.341
TLBO	501.459	4701.442	551.170	5217.548	598.001	5593.304	651.107	6065.457
GWO	501.137	4701.239	551.078	5217.375	597.953	5593.289	650.972	6065.1034
FGALO	499.061	4693.517	550.9112	5214.569	597.779	5591.212	650.164	6049.9417

In order to validate the results, the competing algorithms are applied in the multi- objective framework and the attained total fuel costs and emission are presented in Table 13. Comparing with ABC, TLBO and GWO, the intended algorithm provides the best feasible solutions for the chosen power demands.

Multi-Area MFPGD (MA-MFPGD)

Further, the applicability of the ALO for solving real-time power system operation, it is applied to an interconnected power system consisting tie lines between areas. The objective function of the Multi - Area ED (MAED) problem is represented mathematically as,

$$FC = Min \sum_{k=1}^{NA} \sum_{i=1}^{N} F_i(P_i) \quad \$/h \tag{21}$$

Table 12. Best feasible CE-ER schedule considering valve-point loadings using ALO

Unit No.	P_d = 2400 MW		P_d = 2500 MW		P_d = 2600 MW		P_d = 2700 MW	
	FS	P_i (MW)	FS	P_i (MW)	FS	P_i (MW)	FS	P_i (MW)
P_1	1	166.6745	2	196.1725	2	196.0925	2	199.0301
P_2	1	199.2836	1	205.634	1	208.014	1	213.9991
P_3	1	259.6255	1	285.0024	1	296.6024	1	303.4366
P_4	3	245.5248	3	254.0253	3	246.0253	3	257.0655
P_5	1	251.8932	1	297.1178	1	298.2178	1	284.2943
P_6	1	156.0649	1	164.9894	1	167.4794	1	167
P_7	1	283.5863	1	271.4834	1	297.9934	2	368.2974
P_8	3	244.7472	3	236.5108	3	249.5108	3	258
P_9	3	392.5025	3	388.8667	3	439.8667	3	438.8755
P_{10}	1	200.1811	1	200.1975	1	200.1975	1	210.0019
FC ($/h)	**499.0621**		**546.8367**		**595.8044**		**650.0174**	
EC (kg/h)	**4693.517**		**5142.898**		**5571.487**		**6046.5622**	

The minimization of the total fuel cost is taken as the main operational objective and the real power generating units are optimized subject to constraints (6)-(8). Moreover, the power transfer (MW) between the areas is treated as operational constraint and is represented mathematically as,

$$-T_{ip}^{max} \leq T_{ip} \leq T_{ip}^{max} \tag{22}$$

The ALO is implemented on the standard 10-unit system with three areas and the generating units operating on multi-fuel sources are considered. The first area comprises of 4 generating units, the second and third areas consist of 3 generators each. The tie line capacity in three areas is 100 MW and the system particulars are adopted from Basu (2014).

Table 13. BCS Comparison for CE-ER operation considering valve-point loadings

Methods	P_d = 2400 MW		P_d = 2500 MW		P_d = 2600 MW		P_d = 2700 MW	
	FC ($/h)	EC (kg/h)	FC ($/h)	EC (kg/h)	FC ($/h)	EC (kg/h)	FC ($/h)	EC (kg/h)
ABC	501.839	4701.453	546.8756	5143.029	596.5722	5587.210	650.317	6044.810
TLBO	501.713	4701.675	546.8042	5143.041	596.4112	5586.527	650.490	6044.837
GWO	500.943	4701.964	547.0384	5143.059	596.3237	5585.685	650.6610	6044.865
FGALO	499.062	4693.517	546.8367	5142.898	595.8044	5571.487	650.0174	6046.562

The best dispatch of generators, tie line flows and the network losses achieved using ALO are given in Table 14. For validity, the best solutions obtained by the ALO algorithm are compared with other solution methods such as RCGA, EP, DE, TLBO, ABC, DE and CGBABC and the comparison is presented in Table 14. The comparison of results indicates that the ALO has the capability of achieving the minimum fuel cost with less transmission loss. Additionally, Table 14 shows that the tie line flows are within the capacity of 100 MW and hence gratifying (tie line flow constraint) in addition to power balance and generation limit constraints.

Table 14. Optimal economic dispatch obtained by ALO for 10 unit MA-MFPGD

Output (MW)	Fuel [a]	RCGA	EP	TLBO	ABC	DE	CGBABC	ALO
$P_{1,1}$	2	239.0958	223.8491	224.3088	225.9431	225.4448	226.1734	224.3122
P_{P2}	1	216.11642	209.5759	210.6642	211.1594	210.1667	211.162	211.6251
P_{P3}	2	484.1506	496.0680	491.6998	489.9216	491.2844	490.6471	490.6982
$P_{1,4}$	3	240.6228	237.9954	240.6247	240.6232	240.8956	239.2834	240.6345
$P_{2,1}$	1	259.6639	259.4299	249.5648	254.0397	251.0049	252.7939	249.4747
$P_{2,2}$	3	219.9107	228.9422	235.8978	235.4927	238.8603	235.6297	235.7767
$P_{2,3}$	1	254.5140	264.1133	263.7414	263.8837	264.0906	263.9331	263.9515
$P_{3,1}$	3	231.3565	238.2280	237.1327	237.0006	236.9982	236.1947	237.2429
$P_{3,2}$	1	341.9624	331.2982	332.5910	328.7373	326.5394	328.7272	332.4808
$P_{3,3}$	1	248.2782	246.6025	249.4628	248.8607	250.3339	251.1653	249.4617
T_{21}		93.1700	100	100	99.8288	99.4680	99.97125	100
T_{31}		93.8739	100	100	99.7334	100	100	100
T_{32}		43.7824	32.5231	35.4599	31.2615	30.2810	32.413868	35.4617
P_{L1}		17.0297	17.4884	17.30	17.2680	17.2680	17.23723	17.269
P_{L2}		9.7010	10.0085	9.6639	9.7688	9.7688	9.79935	9.6635
P_{L3}		8.9408	8.6056	8.7266	8.5905	8.5905	8.67329	8.7237
Cost reported ($/h)		657.3325	655.1716	653.9977	654.0184	654.0351	654.72320	-
Cost achieved ($/h)		657.3325	655.1716	654.80	654.81	654.82	654.72320	654.66398

[a] Fuel type obtained is same for each of the methods presented in Table 14.

Large Scale System

The ALO is further implemented on 40-unit system which has two areas, each area comprises of twenty generating units. The load demands in area 1 and area 2 are 7500 MW and 3000 MW respectively, thus the total load demand is 10500 MW. The tie-line flow is limited to 1500 MW. This test system particulars are detailed in the literature (Lin & Wang, 2019). The intended optimization algorithm is implemented on this test system and the attained best real power settings are detailed in Table 15. The obtained total fuel cost is compared with the recent reports such as Hybridizing Sum Local Search Optimizer (HSLSO) (Lin & Wang, 2019), Choatic Global Best ABC (CGBABC) (Lin & Wang, 2019), Stochastic Fractal

Search (SFS) (Lin & Wang, 2019) and Improved SFS (ISFS) (Lin & Wang, 2019). The comparison clearly indicates that the ALO provides the best feasible solution which leads to greater saving in the total fuel cost.

Practical Implications

The power companies aim to operate with the profit abide by the norms of the regulatory board. To achieve the economic benefits, the power system analysis must be made by using the accurate operational model. The economic advantages of the operational models can be viewed by means of economic deviation indices. Table 16 details the economic advantages of using the various operational models. The piecewise quadratic model is treated as a reference (Scenario 1), and the percentage of increase in the total fuel cost for the remaining cases such as considering valve-point effects (Scenario 2), multi-objective model neglecting valve-point loadings (Scenario 3) and proposed multi-objective model (Scenario 4) are computed. The power production cost increases drastically if the environmental issues are considered. From Table 16, it is clear that the solutions of the proposed model can bring the profit of about 3-4%, in comparison with the solutions for CE operation neglecting valve-point loadings. Moreover, compared with Thang's model (Thang, 2013), this model also provides nominal profit (about $\leq 1\%$). It confirms that the proposed model provides the most realistic operation; hence it is a highly suitable model for power companies to enhance their operational analysis.

Table 15. Optimal economic dispatch obtained by ALO for 2 Area, 40-unit system

Methods	HSLSO	CGBABC	SFS	ISFS	ALO
$P_{1,1}$(MW)	113.8349	114	113.8755	113.9058	113.8808
$P_{1,2}$(MW)	1119357	114	113.9917	114	114
$P_{1,3}$(MW)	118.4896	120	119.9649	119.9864	119.9964
$P_{1,4}$(MW)	183.7781	190	179.9618	179.9941	179.9939
$P_{1,5}$(MW)	97	97	96.9975	97	97
$P_{1,6}$(MW)	139.9985	140	139.9921	140	140
$P_{1,7}$(MW)	297.7845	300	299.9718	300	300
$P_{1,8}$(MW)	294.0095	300	292.0177	291.3401	293.3499
$P_{1,9}$(MW)	295.4437	300	286.2145	287.0178	285.0168
$P_{1,10}$(MW)	200	270	199.9963	200	200
$P_{1,11}$(MW)	151.6280	168.7998	168.8617	230	230
$P_{1,12}$(MW)	230	168.7998	229.9937	168.8385	168.8379
$P_{1,13}$(MW)	394.2025	304.5196	394.2861	394.2548	394.2944
$P_{1,14}$(MW)	490	394.2794	394.3472	394.2946	394.2944
$P_{1,15}$(MW)	472.1173	484.0392	484.0810	484.0588	484.0188
$P_{1,16}$(MW)	394.0728	484.0392	484.0282	483.9991	483.9989
$P_{1,17}$(MW)	494.7530	489.2794	489.4175	489.3491	489.3489
$P_{1,18}$(MW)	492.1624	499.9643	489.4217	489.3036	489.3036

continues on following page

Table 15. Continued

Methods	HSLSO	CGBABC	SFS	ISFS	ALO
$P_{1,19}$(MW)	510.8584	550	511.3790	511.3242	511.3239
$P_{1,20}$(MW)	517.9578	511.2794	511.2666	511.3414	511.3414
$P_{2,1}$(MW)	524.6083	533.5196	433.5220	433.42	433.42
$P_{2,2}$(MW)	437.6060	523.2794	433.4372	523.0749	523.1749
$P_{2,3}$(MW)	442.0369	523.2794	523.2303	523.2439	523.1439
$P_{2,4}$(MW)	423.4167	523.2794	433.5201	523.2985	523.2985
$P_{2,5}$(MW)	431.5304	523.2794	433.4048	433.3512	433.4212
$P_{2,6}$(MW)	524.2870	433.5196	523.3005	433.4643	433.3943
$P_{2,7}$(MW)	10.8217	10	10.0091	10.0034	10.0034
$P_{2,8}$(MW)	10	10	10.0201	10	10
$P_{2,9}$(MW)	12.1197	10	10.0004	10	10
$P_{2,10}$(MW)	89.3499	87.7999	87.7942	87.8100	87.7900
$P_{2,11}$(MW)	166.0768	159.7331	189.4315	159.6862	159.7059
$P_{2,12}$(MW)	163.0429	190	159.7266	189.3811	189.3791
$P_{2,13}$(MW)	157.8148	190	159.7200	159.7251	159.7249
$P_{2,14}$(MW)	162.3617	90	164.8297	164.6679	164.7679
$P_{2,15}$(MW)	162.1677	90	164.7029	164.8266	164.8362
$P_{2,16}$(MW)	161.9376	164.7998	164.7322	164.6281	164.7291
$P_{2,17}$(MW)	88.0872	110	88.9974	89.1646	89.0646
$P_{2,18}$(MW)	96.9268	89.1141	88.8668	89.1364	89.0371
$P_{2,19}$(MW)	93.6130	96.3964	89.0476	89.1093	89.1090
$P_{2,20}$(MW)	342.1681	242	331.6393	242	242
T_{12}(MW)	-1499.9732	-1500	1499.9337	-1499.9915	-1500
Cost($/h)	**125879.9346**	**125240.4399**	**124750.5796**	**124683.0977**	**124680.65**
Time(s)	-	-	-	-	**9.05**

Table 16. Comparison of operational profits

P_d (MW)	Fuel Cost Deviation (%)		
	Scenario 2	**CE-ER**	**Explicit**
2400	0.14425	3.59935	3.59954
2500	0.10938	3.82007	3.91421
2600	0.11727	3.61524	3.72986
2700	0.00309	4.17737	4.20143

PERFORMANCE EVALUATION AND DISCUSSIONS

Selection of Algorithmic Parameters

The selection for number of search agents (P_s) is a compromise between a wider exploration of the search space and increased computational time. Due to the stochastic nature of the ALO algorithm, many trials with different population size are used to find out the results.

It is inferred from Table 17, the values of operating costs below the average values are high that shows the ALO strategy performs well for all the test cases. Moreover, when $P_s = 30$, the results of operating cost are optimum. Though the operating cost is same, when the P_s is increased, but the computational time gets increased. Hence, for acquiring optimum results and to avoid increase in computational time, $P_s = 30$ is selected. Statistical comparison of results presented in terms of search agents in Table 17 reveals that the performance of the ALO algorithm is sensitive to reach the minimum operating cost. Based on this analysis, it is established that the suitable number of ant lion has found to be 30.

Convergence and Robustness Tests

The convergence behaviors of the ALO for all the test systems are illustrated in Figure 3 and Figure 4. The ALO method can reach to the optimum solution more quickly than the other methods reported in literature. The ALO is thus demonstrated to have a better convergence property. Over 200 iterations with several initial random solutions, the ALO has confirmed it as a trustworthy solution procedure by generating the global best solution.

Like other evolutionary algorithms, ALO uses the stochastic techniques, thus randomness is an intrinsic feature of these techniques. Several runs with different initial ant lions have been conducted to test the performance and consistency of ALO. The spread of best fuel costs for 50 runs are calculated and graphically displayed in Figures 5 and 6 to illustrate the robustness of the ALO.

Figure 3. Convergence characteristics of ALO (a) neglecting valve-point and (b) considering valve-point

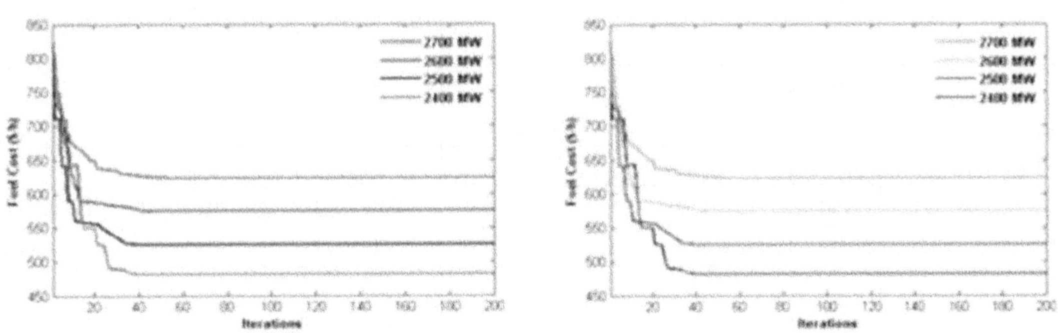

(a) (b)

Table 17. Effect of population size for various demands of 10 unit system

P_d (MW)	*Ps*	Total Fuel Cost ($/h)			SD	Average CPU Time (s)
		Best	Average	Worst		
Scenario 1						
2400	10	481.7345	487.4583	500.14	4.19612	0.039
	30	481.7223	487.4502	500.09	4.19606	0.040
	50	481.7223	487.4505	500.11	4.19609	0.055
	100	481.7223	487.4504	500.10	4.19608	0.214
2500	10	526.4164	530.5077	541.49	2.829199	0.038
	30	526.2386	530.4714	541.30	2.829177	0.040
	50	526.2386	530.5048	541.41	2.829194	0.059
	100	526.2386	530.5045	541.39	2.829189	0.331
2600	10	574.4172	574.4931	575.38	0.10959	0.039
	30	574.3829	574.4798	574.80	0.10922	0.040
	50	574.3829	574.4841	575.12	0.10954	0.103
	100	574.3829	574.4836	575.09	0.10947	0.389
2700	10	623.8612	624.4864	626.04	0.488997	0.038
	30	623.8085	624.3320	625.79	0.488946	0.040
	50	623.8085	624.3752	625.98	0.488988	0.236
	100	623.8085	624.3749	625.99	0.488973	0.412
Scenario 2						
2400	10	482.7543	487.9902	500.7523	4.568490	0.039
	30	482.4127	487.9568	500.3000	4.568484	0.041
	50	482.4127	487.9985	500.7764	4.568488	0.137
	100	482.4127	487.9904	500.7504	4.568486	0.311
2500	10	526.9654	530.9013	541.9845	2.92758	0.040
	30	526.8142	530.8440	541.900	2.92751	0.041
	50	526.8142	530.8734	541.9433	2.92757	0.243
	100	526.8142	530.8645	5419512	2.92755	0.413
2600	10	575.9543	578.3976	575.9934	0.062743	0.039
	30	575.0544	578.2893	575.87	0.062739	0.041
	50	575.0544	578.3276	575.9512	0.062740	0.203
	100	575.0544	578.3104	575.8993	0.062741	0.399
2700	10	623.8512	624.9003	626.0423	0.620985	0.039
	30	623.8278	624.8698	625.8900	0.620981	0.041
	50	623.8278	624.8987	625.9876	0.488984	0.236
	100	623.8278	624.8823	625.9797	0.488973	0.412

Figure 4. Convergence characteristics of ALO for 40-unit system

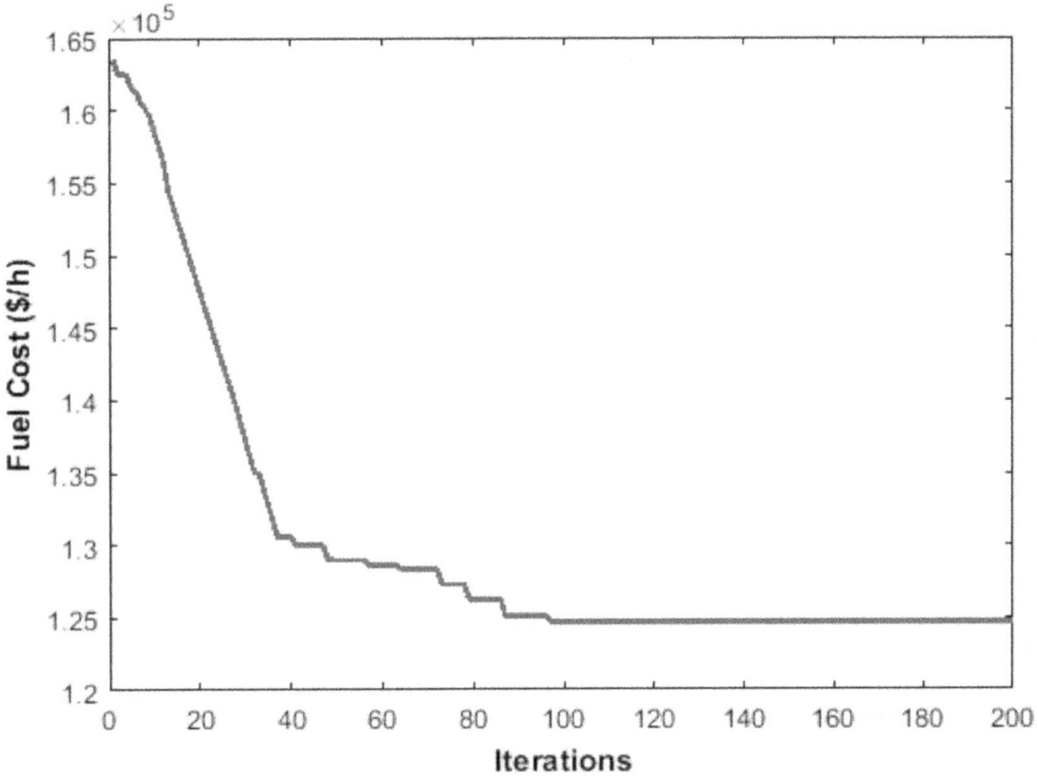

Figure 5. Robustness characteristics of ALO for scenario 2

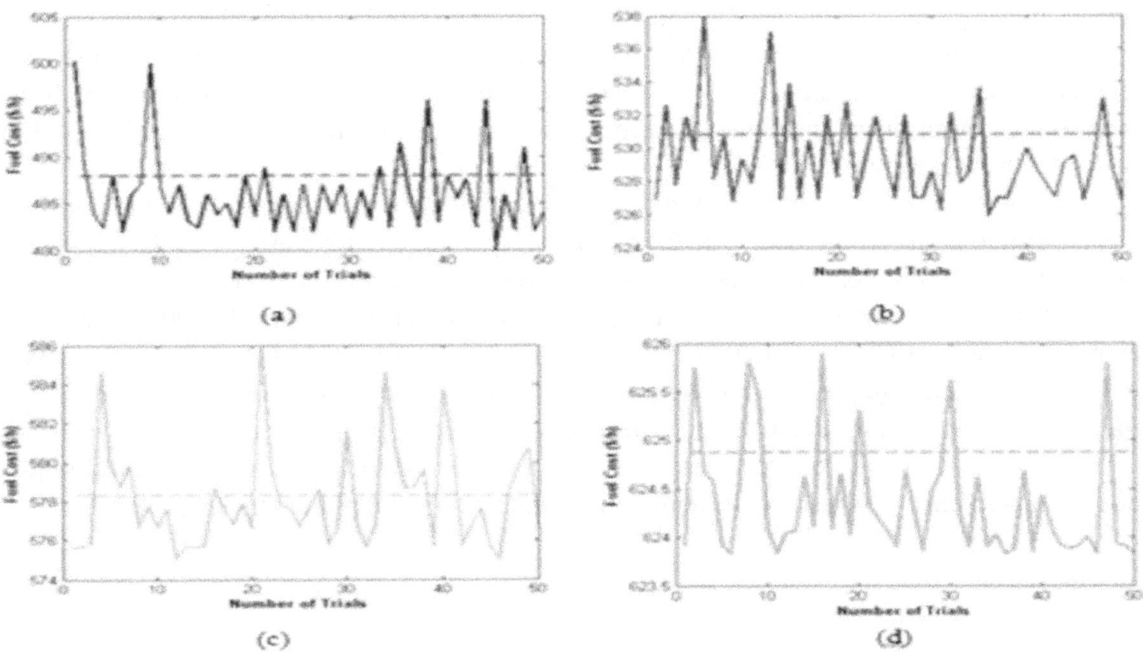

Figure 6. Robustness characteristics of ALO for MA-MFPGD

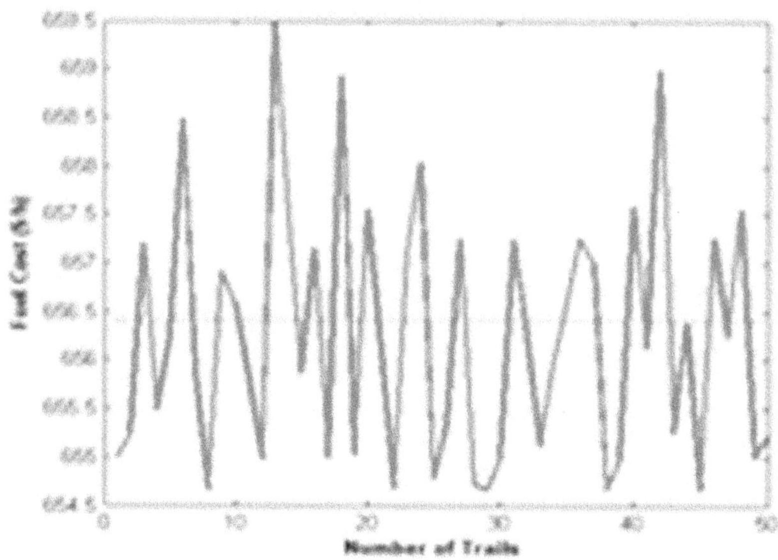

Best Compromise Solution

The ALO has five main steps such as random walk of ants, building traps, entrapment of ants in traps, catching preys and rebuilding traps. These steps are mathematically modelled for equipping the ALO with high exploration and exploitation. The employed random walk and roulette wheel selection mechanism encourage high exploration. This leads to boosting the exploration of the feasible areas of a search space. The main reason for high exploitation is due to the employed adaptive boundary shrinking mechanism and elitism, in which the ants are always guided towards promising feasible regions of the search space. These desirable mechanisms make the ALO to provide very competitive and statistically significant results with other algorithms.

Even though, the solution space is highly non-linear, the projected algorithm is providing well-distributed Pareto fronts for all cases. The fuzzy membership functions are used to evaluate each member of the Pareto optimal set. Then, the BCS that has the maximum value of membership function is extracted. This process is applied for cost effective environmental response and explicit operations in order to determine the BCS. The Pareto optimal fronts attained by the FGALO for 10-unit systems with different load demands are presented in Figures 7 and 8. It is also observed that the method is capable of finding the Pareto front by effectively solving the problem when all the nonlinearities are considered. The overall fuel cost and the compromised solution attained by FGALO is superior for multi-objective optimization problems; which shows its ability to attain the global minimum in a reliable manner.

Figure 7.BCS of ALO for scenarios 1 and 2 (a) P_d=2400 MW (b) P_d=2500 MW (c) P_d=2600 MW and (d) P_d=2700 MW

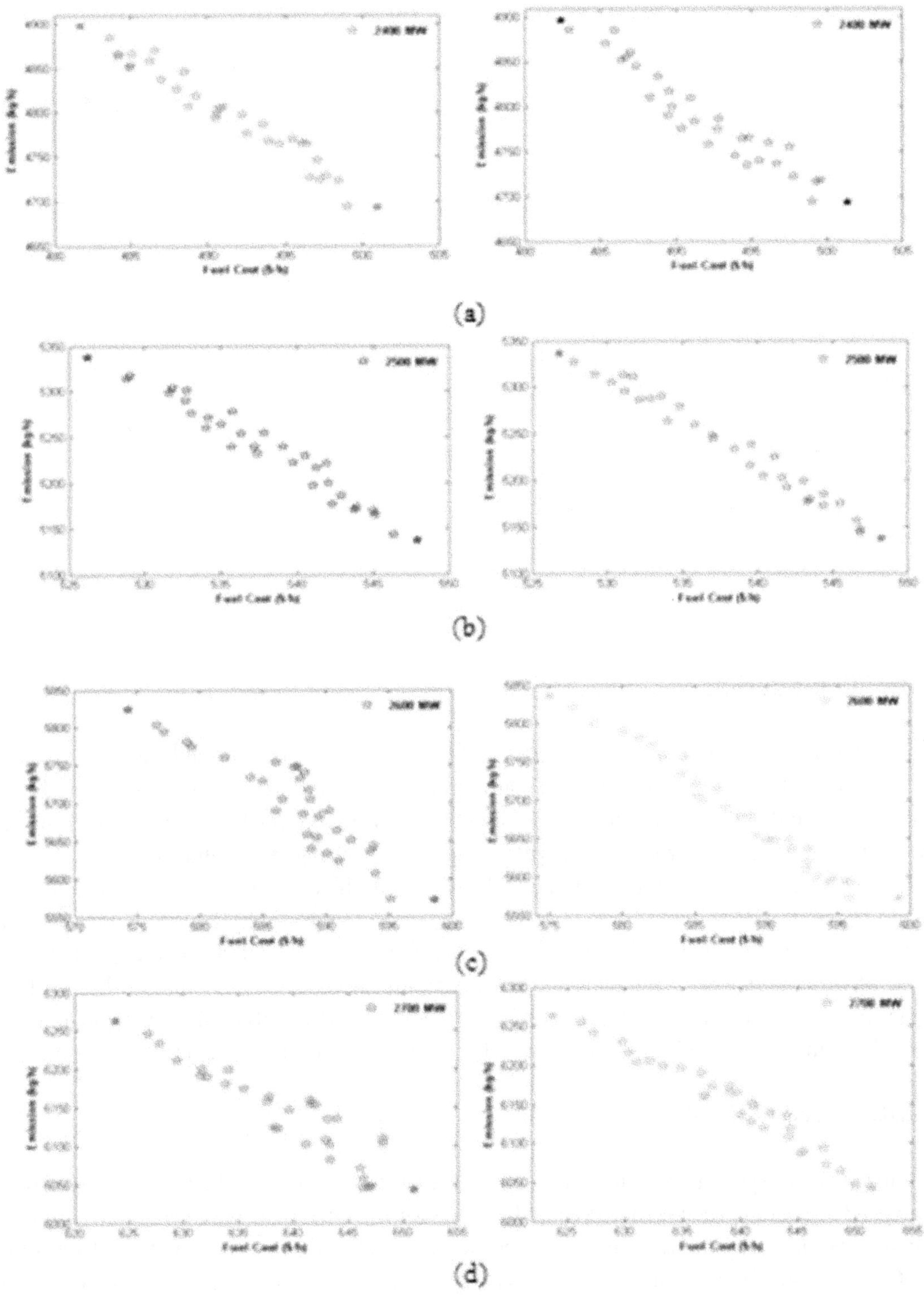

Figure 8.BCS for scenario 3 (a) P_d*=2400 MW (b)* P_d*=2500 MW (c)* P_d*=2600 MW and (d)* P_d*=2700 MW*

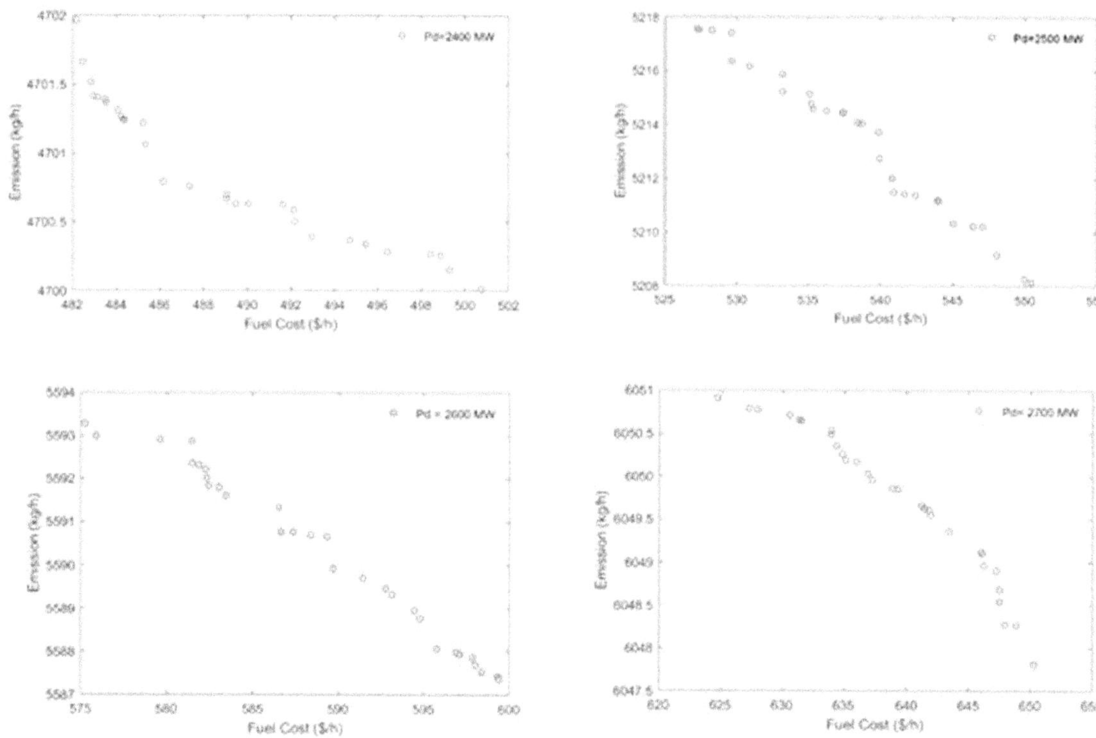

Statistical Measures

To gain further insights into the solution quality of the test cases, the obtained statistical results are reported in Table 18 by means of standard deviation, best, worst and average values. From the statistical results, it is evident that the FC obtained by different trails is close to the best solution, thereby validating that the ALO has the higher probability of attaining reliable and quality solution.

Further, three criteria of goodness are considered and are listed in Table 18. In addition to the well-known indices such as "*Epsilon*", "*Iter*", and "*Sol Iter*" are also considered (Balachandar et al., 2017). For the chosen test systems (Scenario 2), the value of *Epsilon* is 623.8291 $/h for load demand of 2700 MW. The value of variable *Iter* is 52, it means that the algorithm can reach or pass the threshold value after 52 iterations and converged to minimum value in 55 iterations. These variables show the convergence rate of the algorithms and it is a good influential factor for comparing the righteousness of various algorithms.

Multi-Objective Optimization Metrics

As the chosen power system operational objectives are conflict in nature, the fuzzy decision mechanism has been used to coordinate the objectives in the ALO domain. In continuation with the parameter sensitivity study, the multi-objective optimization metrics are used to validate the Pareto distribution attained by the FGALO. The following metrics are used for assessing population dispersion over the decision and objective spaces as well as over the fronts in populations evolved by FGALO (Moncayo–Martínez &

Table 18. Performance indices of ALO

Test System	P_d (MW)	*Epsilon*	*Iter*	*Sol Iter*	Test System	P_d (MW)	*Epsilon*	*Iter*	*Sol Iter*
10-unit Scenario 1	2400	481.7226	32	36	10-unit Scenario 2	2400	482.5114	38	39
	2500	526.2388	33	38		2500	527.0189	37	40
	2600	574.3808	36	43		2600	575.1610	42	45
	2700	623.8092	44	52		2700	624.5081	52	55

Mastrocinque, 2016): Ratio of Non-dominated Individuals (RNI), Spacing (S) and Generational Distance (GD) and RNI is a significant measure that checks the proportion of nondominated individuals in the population. Spacing measures the distribution of solutions throughout the non-dominated solutions (i.e. how well the solutions in the solution set are spread). GD is the way to test the distance (i.e. how far) the true Pareto set is from the solution set. This metrics are calculated from the Pareto fronts obtained for cost effective-environment responsive and explicit operations using FGALO for various load demands. The computed metrics are tabulated in Table 19.

It is clear from Table 19, since the values of RNI are closer to 1 that indicates most of the individuals in the solution set are non-dominated. Moreover, the spacing values are nearly equal to zero which means most of the solutions within the solution set are almost equidistantly spaced. The smaller values of GD indicate the solution set is closer from true Pareto set. It is inferred from these metrics that the FGALO has proven to give the solutions equal or very close to the true Pareto fronts for all load demands of cost effective-environment responsive and explicit operations. It is also shown to be capable of finding many more Pareto solutions covering a large area of the solution space. Therefore, the FGALO method shows its potential when applied for solving multi-fuel power dispatch problem considering conflicting objectives.

Table 19. Multi-objective Optimization Metrics of FGALO

Demand (MW)	CE-ER Operation			Explicit Operation		
	RNI	Spacing	GD	RNI	Spacing	GD
2400	0.7667	2.5369e-4	0.3796	0.6333	1.0840e-4	0.6013
2500	0.7	4.4328e-4	0.5110	0.7333	5.5560e-4	0.4184
2600	0.5667	3.9241e-4	0.7975	0.6	0.0014	0.7407
2700	0.633	5.7049e-4	0.5401	0.6333	8.3111e-4	0.9527

Computational Efficiency

In order to check the computational efficiency of the ALO, it is implemented on the suitably scaled 10-unit system. The obtained total fuel costs are presented in Table 19. From Table 20, it is clear that the ALO approach obtains an optimum solution for multiples of 10 unit, such as 30, 60 and 100 units within 200 iterations. Again, simulation time per iteration in case of ALO is 0.249, 0.352 and 0.401seconds for neglecting valve-point and 0.25, 0.354 and 0.411 seconds for considering valve-point effects. So, the time requirement is quite less and either comparable or better than other previously developed methods. The computational time comparison for various operational scenarios are presented in Table 21, Table 22 and Table 23. The computational time taken by ALO for MA-MFPGD and large-scale system are 7.12 and 9.05 respectively. So, as a whole it can be said that the ALO method is also a computationally efficient method.

Success Rate

The success rate is defined as the ratio of total number of experiments performed to the number of successes that converge to the best solution that is expressed in terms of percentage. The success rate of the intended algorithm for all case studies is above 80% that confirms the algorithm has satisfactory success rate.

Table 20. Total operating cost of different methods for multiples of 10-unit system

Scenario	No. of units	Total Fuel Cost ($/h)				
		CGA	IGA-MU	EALHN	ALHN	ALO
1	30	1873.691	1872.047	1871.4463	1871.427	1871.425
	60	3748.761	3744.722	3742.926	3744.201	3742.71
	100	6251.469	6242.787	6238.210	6238.092	6238.09
2	30	-	-	-	-	1871.46
	60	-	-	-	-	3742.91
	100	-	-	-	-	6238.18

Table 21. Computational time (s) comparison with other methods for CE operation (Scenario 1)

Methods	P_d = 2400 MW	P_d = 2500 MW	P_d = 2600 MW	P_d = 2700 MW
HM	1.08	-	-	-
HNN	~60	~60	~60	~60
AHNN	~4	~4	~4	~4
Hybrid RCGA	6.1	6.1	5.4	6.47
DE	0.083	0.083	0.083	0.083
EALHNN	0.008	0.006	0.005	0.013
HLN	0.124	0.11	0.152	0.225
ALO	8.01	8.02	8.02	8.03

Table 22. Computational time (s) comparison with other methods for CE operation (Scenario 2)

Methods	P_d = 2700 MW
IGA-MU	7.25
NPSO	35
NPSO-LRS	52
PSO	39
NAPSO	2.8
DE-PSO-DE	1.78
CCO	8.321
GWO	2.36
BSA	0.88
ALO	8.2

Table 23. Computational time (s) comparison with other methods for CE-ER operation

Methods	P_d = 2400 MW	P_d = 2500 MW	P_d = 2600 MW	P_d = 2700 MW
HLN	0.8	0.74	1.18	2.04
LIM	77.24	56.72	60.79	53.92
ABC	60.52	60.52	60.52	60.52
TLBO	18.73	18.73	18.73	18.73
GWO	14.52	14.52	14.52	14.52
FGALO	**9.06**	**9.06**	**9.06**	**9.06**

CONCLUSION

The thermal power utilities are facing challenges in cost effective and minimum pollutant emission operations. As these two operational objectives are conflicting in nature, handling of these objectives has become crucial. The utilities require a realistic operational model that comprises of cost effective and environmental concern operational objectives and operational constraints such as generation limits, valve -point loadings, prohibited operation regions and tie line power limits. This chapter outlines the mathematical formulation as follows: economic operation with various operational constraints, combined economic and emission operation and economic operation considering tie-line flow limits.

Among the solution procedures, meta-heuristic algorithms are highly preferable as they are efficient in exploring the search space; handling multiple objectives simultaneously; easy constraint handling mechanisms; and can be implemented for system of any size. ALO is a modern bio-inspired algorithm and is implemented on standard 10-unit system for various kinds of operations. The ALO solves in an exact way, for the different ranges of power demand, the underlying combinatorial problem of determining the fuel that must be used by each power station. The effectiveness of the proposed approach is verified by numerical simulation of different test systems ranging from 10 to 100 units. The results have shown very satisfactory performance when compared to the other algorithms reported in the literature.

The thermal power plants experiencing enormous competitive pressure due to the stringent implementation of energy-saving scheduling approaches. Large-scale thermal power units must adopt effective strategies to reduce energy consumption, reduce emissions and improve the utilization hours in order to create more benefits and survive and develop in the fierce competition. The ALO based ED operation optimizes the units' operation levels and reduce emissions in an effective way to reduce power generation cost and enhance competitiveness. Future work will extend the problem formulation to address the integration of renewable energy sources and energy storage devices that increases the further complexity in the solution procedure.

In a nut-shell, this chapter details the mathematical formulation of multi-fuel power dispatch problems in multi-objective framework and application of ALO for solving various kinds of MFPGD problems.

ACKNOWLEDGMENT

The authors thank Annamalai University, Annamalai Nagar, India and Directorate of Technical Education, Chennai, India for providing the necessary facilities to carry out this research work.

REFERENCES

Balachandar, P. (2017). *Economic – environmental constrained real power scheduling of thermal power plants using bio-inspired algorithm* (Ph.D Dissertation). Annamalai University.

Balachandar, P., Ganesan, S., Jayakumar, N., & Subramanian, S. (2017). Multi-fuel power dispatch in an interconnected power system using ant lion optimizer: Multi-fuel dispatch considering tie-line limits. *International Journal of Energy Optimization and Engineering*, *6*(3), 29–54. doi:10.4018/IJEOE.2017070102

Banerjee, S., Chanda, C., & Maity, D. (2016). A comparative study of improved teaching learning based optimization technique on economic load dispatch problem with generator constraints. *International Journal of Energy Optimization and Engineering*, *5*(4), 1–25. doi:10.4018/IJEOE.2016100101

Baskar, S., Subbaraj, P., & Rao, M. V. C. (2003). Hybrid real coded genetic algorithm solution to economic dispatch problem. *Computers & Electrical Engineering*, *29*(3), 407–419. doi:10.1016/S0045-7906(01)00039-8

Basu, M. (2014). Teaching–learning-based optimization algorithm for multi-area economic dispatch. *Energy*, *68*, 21–28. doi:10.1016/j.energy.2014.02.064

Basu, M. (2016). Kinetic gas molecule optimization for nonconvex economic dispatch problem. *International Journal of Electrical Power & Energy Systems*, *80*, 325–332. doi:10.1016/j.ijepes.2016.02.005

Bhattacharya, A., & Chattopadhyay, P. K. (2011). Solving economic emission load dispatch problems using hybrid differential evolution. *Applied Soft Computing*, *11*(2), 2526–2537. doi:10.1016/j.asoc.2010.09.008

Binetti, G., Davoudi, A., Naso, D., Turchiano, B., & Lewis, F. L. (2014). A distributed auction-based algorithm for the nonconvex economic dispatch problem. *IEEE Transactions on Industrial Informatics*, *10*(2), 1124–1132. doi:10.1109/TII.2013.2287807

Chiang, C. L. (2005). Improved genetic algorithm for power economic dispatch of units with valve-point effects and multiple fuels. *IEEE Transactions on Power Systems, 20*(4), 1690–1699. doi:10.1109/TPWRS.2005.857924

Dieu, V. N., Ongsakul, W., & Polprasert, J. (2013). The augmented Lagrange Hopfield network for economic dispatch with multiple fuel options. *Mathematical and Computer Modelling, 57*(1), 30–39. doi:10.1016/j.mcm.2011.03.041

Dieu, V. N., & Schegner, P. (2013). Augmented Lagrange Hopfield network initialized by quadratic programming for economic dispatch with piecewise quadratic cost functions and prohibited zones. *Applied Soft Computing, 13*(1), 292–301. doi:10.1016/j.asoc.2012.08.026

Ghasemi, M., Taghizadeh, M., Ghavidel, S., & Abbasian, A. (2016). Colonial competitive differential evolution: An experimental study for optimal economic load dispatch. *Applied Soft Computing, 40*, 342–363. doi:10.1016/j.asoc.2015.11.033

Hemamalini, S., & Simon, S. P. (2010). Artificial bee colony algorithm for economic load dispatch problem with non-smooth cost functions. *Electric Power Components and Systems, 38*(7), 786–803. doi:10.1080/15325000903489710

Jayabarathi, T., Jayaprakash, K., Jeyakumar, D. N., & Raghunathan, T. (2005). Evolutionary programming techniques for different kinds of economic dispatch problems. *Electric Power Systems Research, 73*(2), 169–176. doi:10.1016/j.epsr.2004.08.001

Kheshti, M., Ding, L., Ma, S., & Zhao, B. (2018). Double weighted particle swarm optimization to non-convex wind penetrated emission/economic dispatch and multiple fuel option systems. *Renewable Energy, 125*, 1021–1037. doi:10.1016/j.renene.2018.03.024

Kheshti, M., Kang, X., Bie, Z., Jiao, Z., & Wang, X. (2017). An effective lightning flash algorithm solution to large scale non-convex economic dispatch with valve-point and multiple fuel options on generation units. *Energy, 129*, 1–15. doi:10.1016/j.energy.2017.04.081

Lee, K. Y., Sode-Yome, A., & Park, J. H. (1998). Adaptive Hopfield neural networks for economic load dispatch. *IEEE Transactions on Power Systems, 13*(2), 519–526. doi:10.1109/59.667377

Lin, C. E., & Viviani, G. L. (1984). Hierarchical economic dispatch for piecewise quadratic cost functions. *IEEE Transactions on Power Apparatus and Systems, PAS-103*(6), 1170–1175. doi:10.1109/TPAS.1984.318445

Lin, J., & Wang, Z. J. (2019). Multi-area economic dispatch using an improved stochastic fractal search algorithm. *Energy, 166*, 47–58. doi:10.1016/j.energy.2018.10.065

Manoharan, P. S., Kannan, P. S., Baskar, S., & Iruthayarajan, M. W. (2008). Penalty parameter-less constraint handling scheme based evolutionary algorithm solutions to economic dispatch. *IET Generation, Transmission & Distribution, 2*(4), 478–490. doi:10.1049/iet-gtd:20070423

Meng, A., Li, J., & Yin, H. (2016). An efficient crisscross optimization solution to large-scale non-convex economic load dispatch with multiple fuel types and valve-point effects. *Energy, 113*, 1147–1161. doi:10.1016/j.energy.2016.07.138

Mirjalili, S. (2015). The ant lion optimizer. *Advances in Engineering Software*, *83*, 80–98. doi:10.1016/j.advengsoft.2015.01.010

Modiri-Delshad, M., Kaboli, S. H. A., Taslimi-Renani, E., & Rahim, N. A. (2016). Backtracking search algorithm for solving economic dispatch problems with valve-point effects and multiple fuel options. *Energy*, *116*, 637–649. doi:10.1016/j.energy.2016.09.140

Moncayo–Martínez, L. A., & Mastrocinque, E. (2016). A multi-objective intelligent water drop algorithm to minimise cost of goods sold and time to market in logistics networks. *Expert Systems with Applications*, *64*, 455–466. doi:10.1016/j.eswa.2016.08.003

Niknam, T., Mojarrad, H. D., & Meymand, H. Z. (2011). Non-smooth economic dispatch computation by fuzzy and self adaptive particle swarm optimization. *Applied Soft Computing*, *11*(2), 2805–2817. doi:10.1016/j.asoc.2010.11.010

Noman, N., & Iba, H. (2008). Differential evolution for economic load dispatch problems. *Electric Power Systems Research*, *78*(8), 1322–1331. doi:10.1016/j.epsr.2007.11.007

Panigrahi, B. K., Yadav, S. R., Agrawal, S., & Tiwari, M. K. (2007). A clonal algorithm to solve economic load dispatch. *Electric Power Systems Research*, *77*(10), 1381–1389. doi:10.1016/j.epsr.2006.10.007

Park, J. B., Lee, K. S., Shin, J. R., & Lee, K. Y. (2005). A particle swarm optimization for economic dispatch with nonsmooth cost functions. *IEEE Transactions on Power Systems*, *20*(1), 34–42. doi:10.1109/TPWRS.2004.831275

Park, J. H., Kim, Y. S., Eom, I. K., & Lee, K. Y. (1993). Economic load dispatch for piecewise quadratic cost function using Hopfield neural network. *IEEE Transactions on Power Systems*, *8*(3), 1030–1038. doi:10.1109/59.260897

Parouha, R. P., & Das, K. N. (2016). A novel hybrid optimizer for solving Economic Load Dispatch problem. *International Journal of Electrical Power & Energy Systems*, *78*, 108–126. doi:10.1016/j.ijepes.2015.11.058

Polprasert, J., Ongsakul, W., & Dieu, V. N. (2013). A new improved particle swarm optimization for solving nonconvex economic dispatch problems. *International Journal of Energy Optimization and Engineering*, *2*(1), 60–77. doi:10.4018/ijeoe.2013010105

Pradhan, M., Roy, P. K., & Pal, T. (2016). Grey wolf optimization applied to economic load dispatch problems. *International Journal of Electrical Power & Energy Systems*, *83*, 325–334. doi:10.1016/j.ijepes.2016.04.034

Roy, S., Bhattacharjee, K., & Bhattacharya, A. (2017). A modern approach to solve of economic load dispatch using group leader optimization technique. *International Journal of Energy Optimization and Engineering*, *6*(1), 66–85. doi:10.4018/IJEOE.2017010104

Secui, D. C. (2015). The chaotic global best artificial bee colony algorithm for the multi-area economic/emission dispatch. *Energy*, *93*, 2518–2545. doi:10.1016/j.energy.2015.10.012

Selvakumar, A. I., & Thanushkodi, K. (2007). A new particle swarm optimization solution to nonconvex economic dispatch problems. *IEEE Transactions on Power Systems*, *22*(1), 42–51. doi:10.1109/TPWRS.2006.889132

Shoults, R. R., & Mead, M. M. (1984). Optimal estimation of piece-wise linear incremental cost curves for EDC. *IEEE Transactions on Power Apparatus and Systems*, *PAS-103*(6), 1432–1438. doi:10.1109/TPAS.1984.318480

Shukla, A., & Singh, S. N. (2017). Pseudo-inspired CBA for ED of units with valve-point loading effects and multi-fuel options. *IET Generation, Transmission & Distribution*, *11*(4), 1039–1045. doi:10.1049/iet-gtd.2016.1360

Singh, M., & Dhillon, J. S. (2016). Multiobjective thermal power dispatch using opposition-based greedy heuristic search. *Electrical Power and Energy Systems*, *82*, 339–353. doi:10.1016/j.ijepes.2016.03.016

Singh, N. J., Dhillon, J. S., & Kothari, D. P. (2016). Synergic predator-prey optimization for economic thermal power dispatch problem. *Applied Soft Computing*, *43*, 298–311. doi:10.1016/j.asoc.2016.02.042

Singh, N. J., Dhillon, J. S., & Kothari, D. P. (2017). Surrogate worth trade-off method for multi-objective thermal power load dispatch. *Energy*, *138*, 1112–1123. doi:10.1016/j.energy.2017.07.063

Singh, N. J., Dhillon, J. S., & Kothari, D. P. (2018). Multiobjective thermal power load dispatch using adaptive predator–prey optimization. *Applied Soft Computing*, *66*, 370–383. doi:10.1016/j.asoc.2018.02.006

Thang, N. T. (2013). Economic emission load dispatch with multiple fuel options using Hopfiled Lagrange Network. *International Journal of Advanced Science and Technology*, *57*, 9–24.

Vo, D. N., & Ongsakul, W. (2012). Economic dispatch with multiple fuel types by enhanced augmented Lagrange Hopfield network. *Applied Energy*, *91*(1), 281–289. doi:10.1016/j.apenergy.2011.09.025

Zhan, J., Wu, Q. H., Guo, C., & Zhou, X. (2015). Economic dispatch with non-smooth objectives—Part II: Dimensional steepest decline method. *IEEE Transactions on Power Systems*, *30*(2), 722–733. doi:10.1109/TPWRS.2014.2330618

Chapter 13
Oppositional Differential Search Algorithm for the Optimal Tuning of Both Single Input and Dual Input Power System Stabilizer

Sourav Paul
Dr. B. C. Roy Engineering College, India

Provas Kumar Roy
https://orcid.org/0000-0002-3433-5808
Kalyani Government Engineering College, Kalyani, India

ABSTRACT

Low frequency oscillation has been a major threat in large interconnected power system. These low frequency oscillations curtain the power transfer capability of the line. Power system stabilizer (PSS) helps in diminishing these low frequency oscillations by providing auxiliary control signal to the generator excitation input, thereby restoring stability of the system. In this chapter, the authors have incorporated the concept of oppositional based learning (OBL) along with differential search algorithm (DSA) to solve PSS problem. The proposed technique has been implemented on both single input and dual input PSS, and comparative study has been done to show the supremacy of the proposed techniques. The convergence characteristics as well authenticate the sovereignty of the considered algorithms.

INTRODUCTION

Power system stability is authentic as the adeptness of the system to endure in a state of equilibrium under normal operating condition and to achieve an adequate accomplishment to calm afterwards getting subjected to disturbance. Over the accomplished few years, the major concern for many power system

DOI: 10.4018/978-1-7998-3970-5.ch013

researchers is the damping of squat frequency vibrations in the ambit of 0.2-0.3 Hz. Most important types of oscillations observed so far are local-modes, occurring between one machine to the rest of the systems. Apart from these, another type includes inter-area oscillations, occurring between interconnected machines. The effects of fast acting, high gain (to improve transient stability) automatic voltage regulator (AVRs) and the burden of heavy transmission line over weak tie-line are the supreme grounds for small signal disturbances that are realized on the power. The introduction of high gain voltage regulator introduces negative damping in the system. A solution to improve damping is to introduce a stabilizing signal into the excitation system (Abido et.al., 2002). Power system stabilizer (PSS) is acclimated to boost the system stability by providing auxiliary damping. Power system stabilizer (PSS), as emerged as a facile and economical approach in reducing low frequency oscillation in accession to providing of affixed stabilize indicator to the excitation system (Ugalde-Loo et al., 2010, Arrifano et al., 2007). In order to produce extra damping to the excitation control loop of the generating unit, PSS engender an additional stabilizing signal. The frequently used PSS identified as conventional PSS (CPSS), (Talaq, 2012) consists of the lead-lag type factors along with high gain stage and washout block. Frequently used conventional techniques such as root locus and phase compensation are used to refrain the factors of the lead-lag block at a definite operating condition recompense for the system's phase lag. In addition to these, the practical power system of concern is awful non-linear in nature in which the machine parameters changes continuously with elapsed time span and implementation of load. The dynamic response also varies at different points. PSS does not provide adequate damping in the vicinity of changing operating conditions. To tune the factors of PSS, several techniques have been implemented from time to time to reach optimal set of parameters.

Many approaches based on modern control theory have been applied to design different PSS structures such as adaptive controller (Wu & Malik, 2006), Fuzzy logic controller (Hossenizadeh & Kalam, 1999) and extended integral controller (Hoensu & Hyun, 2002). In (Ellithy et al., 2014) proposed the design of the PSSs based on μ-controller to enhance power systems stability and improve power transfer capability using lead-lag PSS structure. Damping torque technique is applied to tune the PSS parameters and the results have been verified by eigen value analysis and time-domain simulations. In (Farahain et al., 2015) online trained fuzzy neural network controller (OTFNNC) derived by the Lyapunov stability has been employed to improve the stability in a power system. The overall dynamic performance of the power system by using a comprehensive analysis of the effects of the different CPSS has been presented in (Kundur, Klein, Rogers & Zywno, 1989). In the past few years, Artificial Neural Network (ANN) techniques have been used for designing PSS (Zhang, Chen, Malik & Hope, 1993; Mahabuba & Khan, 2009). The ANN approach has its own merits and demerits. Even though the performance of the system is improved by the ANN based controller, yet, it suffers from long training time, the selecting number of layers and the number of neurons in each layers. Another techniques like pole shifting is illustrated in (Kothari, Bhattacharya & Nanda, 2002; Vasant, Marmolejo & Litvinchev, 2019; Vasant, Zelinka, Weber & Wilhelm, 2019; El-Sherbiny, Hasan, El-Saady & Yousef, 2003; Ganesan, Vasant & Elamvazuthi, 2016, Vasant, Kose & Watada, 2017,) to design PSS. However the above stated demerits has been overcome by optimization methods(Izquierdo et al., 2017; Kaliannan et al., 2017; Vasant, Zelinka & Weber, 2018). Here, the tuning process is converted to a constrained optimization problem which is solved by using an optimization algorithm.

Recently various evolutionary techniques are explicitly used in various fields of engineering and technology (Marmolejo et al., 2017). These evolutionary techniques are found to be advantageous in solving various complex problems. Due to fast computing ability, these techniques have found its applicability in solving various power system problems. Successful results have been obtained in tuning the PSS parameters by different evolutionary optimization techniques namely particle swarm optimization (PSO) (Peres et al., 2017), honey bee mating optimization (Niknam et al., 2011) etc. to locate the optimal set of factors to successfully plan the PSS as the objective function need not be clear or differentiable. Minimum damping can be achieved as the concerned objective function using the aforesaid evolutionary techniques avoids non-linearity and non-convexity in the problem. (Eke et al., 2015) used orthogonal learning artificial bee colony algorithm for optimal design of PSS parameters considering single machine infinite bus system. The supremacy of the proposed technique has been validated by minimizing the objective function over wide range of different loading condition. Minimization of low frequency fluctuation in power system over different loading condition considering both single machine and multi-machine by the use of hybrid bacteria foraging optimization (BFO) algorithm and particle swarm optimization (PSO) have been proposed by (Panda et al., 2013). (Sambariya et al., 2014) proposed BAT algorithm to optimize the gain and pole-zero parameters of the CPSS. Eigen values based objective function has been considered in the study to generate the stability of single machine infinite bus (SMIB) power system over a wide range of operating conditions. The optimized BA based CPSS (BA-CPSS) results thus obtained has been compared with those obtained using particle swarm optimization based CPSS (PSO-CPSS) controller to show the robustness of the proposed method. (Paul et al. 2015) proposed chemical reaction optimization (CRO) technique to damp the power system low-frequency oscillation and enhance power system stability. The efficacy of the method is tested on a SMIB and compared with other well known optimization techniques.

Hybrid approach using PSO and Takagi-Sugeno (TS) fuzzy for optimal tuning and placement of PSS with the objective of reducing the low-frequency oscillation has been presented by (Kenmarsi et al., 2016). The result of the test has been implemented on a two area four machine grid sample for reducing low frequency oscillation of local and inter-area modes and hence improving the system stability. A multi-objective based objective function concerning the damping factor and damping ratio of the electromechanical modes have been measured by (Chitara et al., 2015). The stated problem has been solved by a meta-heuristic based cuckoo search algorithm in amalgamation with Levy flight behaviour of some birds and fruit flies. The parameters of the PSS are tuned with the objective of shifting the unbalanced or poorly damped electromechnical modes to the left half of the s-plane in order to improve stability. Based on decision making approach on human beings, collective decision optimization (CDO) have been proposed by (Das et al., 2018) to solve eigen value and damping ratio based PSS on interconnected systems by considering WSCC 3-machine 9-bus systems and IEEE 14-bus test system. In addition to the objective function minimization, optimal location of PSS has also been derived in this work. The computed results by the proposed algorithm have been compared with whale optimization algorithm (WOA), crow search algorithm (CSA), differential evolution (DE) and grey wolf optimization (GWO) to validate the superiority of the proposed CDO. (Peres et al., 2017) proposed three metaheuristics namely gravitational search algorithm (GSA), particle swarm optimization (PSO) and bat algorithm (BAT) for the vigorous and co-ordinated architecture of PSS. Here the affability action is modelled as an optimization problem which targeted at maximizing the damping ratio coefficients in closed loop operation.

To analyze the supremacy of the projected method, two criterion systems have been engaged into assessment. (Elazim *et al.*, 2016) proposed cuckoo search (CS) algorithm for the tuning process of PSS parameters. Damping ration and damping factor of oscillatory electromechanical modes have been taken as a suitable objective function for selected PSS design problem and the result thus obtained have been compared with those obtained by other reported algorithms to show the sovereignty of the proposed

CS algorithm. The flexibility of the CS method have been validated by considering different loading conditions and disturbances. A comparative results obtained by backtracking search algorithm (BSA) with those obtained by bacterial foraging algorithm (BFO) and PSO have been proposed by (Islam et al. 2016). Here the authors have used the proposed techniques in multi-machine power system stabilizers. The potential adequancy of Thyristor controlled series capacitor (TCSC) along with PSS in damping inter-area power swings and in extenuating the sub-synchronous resonances has been established by (Naresh et al., 2016). In this research work, two test system have been taken into consideration for co-ordinated design of multiple PSS and TCSC in providing damping in low frequency oscillation. To show the efficacy of the proposed technique in damping inter-area power swings, a comparative study have been done with bacterial swarm optimization (BSO) algorithm, GA, PSO. A genetic algorithm (GA) based method has been proposed by (Andreoiu et al., 2002) to tune the parameters of a PSS. Their projected approach integrates the classical factor optimization approach, linking the solution of a Lyapunov equation, within a genetic hunt process. The solution, appropriately obtained, has been claimed to be globally optimal and robust. In the work of (Kamwa et al., 2005) two modern dual-input PSSs, the IEEE PSS2B and PSS4B, have been systematically studied to pin-point the main dis-similarity in their behaviour from the view point of their intrinsic design characteristics. The comparison of single input CPSS and dual input IEEE PSSs has not been focused in this work. Dual input IEEE-PSS3B has also not been dealt with. Objective function based on eigen-value has been painstaking in this study for the stability of SMIB system over a multiple series of operating scenario. The PSSs parameter tuning problem is formulated as an optimization problem which is solved by cuckoo search (CS) algorithm (Elazim et al., 2016).

An eigen value based objective function involving the damping ratio, and the damping factor of lightly damped electromechanical modes is considered for the PSSs design problem. The superiority of the proposed CS based PSSs (CSPSSs) has been validated by comparing the obtained results with GA, and CPSS under different loading conditions and disturbances. Though, the above-stated optimization techniques have successfully been applied in various field of power system although, they suffer from many demerits like poor convergence rate and low exploitation ability. Additionally, the performance of the aforesaid optimization techniques is highly determined by some of their input control parameters. For example, CA requires a very long run time depending on the size of the system under study. Also, it suffers from optimal settings of algorithm parameters and gives rise to repeat revisiting of the same suboptimal solutions. In case of PSO, the premature convergence nature may trap the algorithm into the local optimum, hence reducing the chance of reaching the global optimal solution. Also the performance of PSO algorithm is highly susceptible to the initial value of weighting strategy of the velocity vector. In BFO algorithm, during the process, the performance of the algorithm is highly determined by the random search direction that may leads to delay to reach the global optimal point.

It is clearly evident from the literature perspectives that the majority of these techniques, described above, are centred on angular speed deviation as single input variable to classical phase compensation PSS. Also complexity of the algorithm, abundant calculation accountability and memory accumulator requirement also effects the performance of some of these techniques. Some suffer from robustness attributable to the limitation of allotment the control parameters of stabilizer, bound amount of optimization functions and call for absolute time on-line fast alteration stabilizer parameters. Further, as per "no-free-lunch" theorem, no single optimization technique is capable enough to solve all the wide range of different optimization problems. This encourages the present author to design a input control parameter free algorithm to solve the desired objective function. Hence, there is ample opportunity for further development to eradicate the stated problems and to reach the global optimal solutions.

The prime emphasis of the work presented in this chapter is to die out rotor oscillations of single machine power systems equipped with PSS. Furthermore, establishment of the fact that the transient performance of dual input IEEE-PSS is better than the conventional phase compensation single input PSS with the help of different optimizing techniques have been in this chapter. Rest of the chapter is organized as follows: In Section 2, SMIB model and various PSS models are given. Problem formulation for PSS study and the system constraints has been given in Section 3. Section 4 is devoted for the proposed optimization techniques used in this chapter. Chapter 5 is devoted for the solution methodology where the optimization technique applied to PSS is explained. The simulation results and discussion by considering different loading condition has been shown in Section 6 and then finally the Section 7 concludes the chapter.

SMIB MODEL AND VARIOUS PSS MODELS

In this present work, The SMIB system (Kundur, 2006). Have been considered as illustrated in Figure 1. The philosophical origin of the PSS description have been illustrated with the help of block diagram representation as outlined in Figure 2 (Kundur, 2006). Speed deviation is used as logical signal to control generator excitation for the prospect of PSS in introducing damping torque component. The block illustrated representation of SMIB system containing synchronous generator, high gain thyristor exciter, AVR and PSS is exhibited in Figure 3 (Kundur, 2006). The synchronous generator with AVR, IEEE type ST1A thyristor excitation system and equivalent transmission line reactance are represented by a two-axis, fourth order model. However, the design of conventional single-input PSS (CPSS) illustrated in Figure 4 has some deficiencies when applied to SMIB system which can be overcome using dual-input PSS. In this chapter, in addition to the optimal designing of CPSS (Kundur, 2006), three varities of dual input PSSs (Kundur, 2006), namely PSS2B, PSS3B and PSS4B shown in Figures 5-7 are performed.

CPSS represent ingle input PSS, with input and output denoted by $\Delta\omega_r$ and ΔV_{pss} respectively. Contrary to single input PSS, dual input PSS includes $\Delta\omega_r$ and ΔT_e as the two input and ΔV_{pss} as output. The three varities of dual input PSS structures considered in the present work are PSS2B, PSS3B and PSS4B. This ΔV_{pss} acts as a appended control signal to the excitation system. The CPSS mainly constituted of a gain stage (K_{pss}) which determines the amount of damping introduced by the stabilizers, a washout stage, which allow only high frequency oscillations to pass and blocking the dc offsets. The final stage consists of lead-lag compensators to compensate for the phase lag introduced by the AVR and the field circuit of the generators. The lead-lag parameters (T_1–T_4) are tuned in such a way that speed oscillations give a damping torque on the rotor. The interaction between the torsional modes of vibration and PSS should be avoided as possible. But it is revealed from the literature perspective that all the modern excitation system are very prone to such interaction as these devices yield relatively high gain at high frequency. Particularly, at lightly loading conditions of the synchronous generators where the inherent mechanical damping is relatively small, stabilizer-torsional instability with a high-response excitation system may result. Shaft damage does not occur due to such instability but it can cause saturation of the stabilizer output yielding the stabilizer to be ineffective and possibly causing saturation of the voltage regulator, which in turn, results in loss of synchronism and tripping of the unit. Thus, it is imperative that stabilizers do not induce torsional instabilities. A modified form of the model PSS2A published in IEEE standard 421.5 is the computer representation model of PSS2B (1996). To model the stabilizers that incorporate a more complex phase lead function, an additional block with lag time constant T_5 and lead time constant T_{10} can be used. All the different form of PSS (i.e. PSS2B, PSS3B and PSS4B0 as depicted in the Figures 5-7 consists of two inputs i.e. $\Delta\omega_r$ and ΔT_e. Time constants (T_{d2}, T_{d4} in PSS3B model and T_2, T_4 in PSS4B model) represent rotor angular speed and washout time constants for electrical torque, respectively. For PSS3B model, the stabilizing signal ΔV_{pss} results from the vector summation of processed signals for electrical torque and angular frequency deviation. The function of the limit values ΔV_{pss}^{min} and ΔV_{pss}^{max} is to adjust the minimum and maximum allowed, respectively. The same structure of the angular frequency deviation channel, the formulation and limitation of the stabilizing signal are followed for PSS4B model. The conditioning network for accelerating power requires the system constant T_0 as well as the inertia time constant M (=$2H$) for the combined turbine-generator shaft system.

The synchronous machine model can be represented by four first order linear differential equations (1-4). These equations represent a fourth order generator model. Magnetic saturation is either neglected or considered by using saturated values of mutual inductances. Higher machine's models are also proposed based on the varying degrees of complexity which provides better results:

$$\frac{d\delta}{dt} = \omega - \omega_0 \tag{1}$$

$$\frac{d\omega}{dt} = \frac{1}{2H}(-D\omega + T_m - T_e) \tag{2}$$

$$\frac{dE_q^{'}}{dt} = \frac{1}{T_{do}^{'}}(-E_q^{'} + (x_d - x_d^{'})i_d + E_{fd}) \tag{3}$$

$$\frac{dE_d^{'}}{dt} = \frac{1}{T_{qo}^{'}}(-E_d^{'} + (x_q - x_q^{'})i_q) \tag{4}$$

where, E_{fd} is the voltage proportional to field voltage, $E_d^{'}, E_q^{'}$ are the voltage proportional to damper winding and field flux, i_d, i_1 are the *d-q* components of armature currents, $T_{do}^{'}, T_{qo}^{'}$ are the *d-q* axis transient time constant. A detailed description of all rest of symbols and quantities can be found in (Sauer, 1998). Fig. 1 illustrates the linearized model of SMIB system from the Heffron-Phillips model. The constant k_1, k_2, k_3, k_4, k_5 and k_6 are modelled in (Padiyar, 1996) and are illustrated in (5-10):

$$k_1 = \frac{E_b E_{q0} \cos\delta_0}{(x_e + x_q)} + \frac{(x_q + x_d^{'})}{(x_e + x_d^{'})} E_b i_{q0} \sin\delta_0 \tag{5}$$

$$k_2 = \frac{(x_e + x_q)}{(x_e + x_d^{'})} i_{q0} \tag{6}$$

$$k_3 = \frac{(x_e + x_d^{'})}{(x_d + x_e)} \tag{7}$$

$$k_4 = \frac{(x_d - x_d^{'})}{(x_d^{'} + x_e)} E_b \sin \delta_0 \tag{8}$$

$$k_5 = \frac{-x_q v_{do} E_b \cos \delta_0}{(x_e + x_q) V_{to}} - \frac{x_d^{'} v_{qo} E_b \sin \delta_0}{(x_e + x_d^{'}) V_{t0}} \tag{9}$$

$$k_6 = \frac{x_e}{(x_e + x_d^{'})} \left(\frac{v_{qo}}{V_{to}} \right) \tag{10}$$

where, E_b is the infinite bus voltage; x_d,x_q are the *d*-*q* axis steady state reactance; i_{qo} is the *q*-components of armature current; $i_{qo}^{'}$ is the *d*-axis component of armature current; E_{qo} is the *q*-axis component of the bus terminal voltage; $x_d^{'}$ is the *d*-axis transient reactance; x_e is the line reactance; v_{do},v_{qo} are the *d*-*q* axis component of the terminal voltage; V_{to} is the voltage component of the connected terminal

PROBLEM FORMULATION

In this present study, two different single objective functions and two different multi-objective functions are considered to determine the effectiveness of the proposed ODSA algorithm.

Single Objective Optimization

Case-I: Eigen Value Minimization

The relative stability of the system is determined by position of the eigen values. For a real eigenvalues, the system is corresponding to a non-oscillatory mode. If the real part of the eigenvalue is positive, then the system is unstable. If the eigenvalues are negative, the system is stable in decaying mode. The further the eigenvalues are in the negative s-plane as depicted in Fig. 8, the faster will be the system response. As for the complex form, like in eq (11), the eigenvalues occur in conjugate pairs. The real component 'σ', depending on whether it is positive or negative, increases the oscillation amplitude to complete instability or damps out the oscillation.The parameters of the PSS may be chosen to minimize the following objective functions (Shayeghi, Shayanfar, Jalilzadeh & Safari, 2010):

$$\text{OF}_1 = min(\text{abs}(\sigma_0 - \sigma_i));\ i=1,2,\ldots,N \tag{11}$$

where, σ_i is the real part of the eigenvalues; n is the number of states for which the optimization is carried out. The value of σ_0 determine the relative stability in terms of damping factor margin provided ford constraining the placement of eigenvalues during the process of optimization. The value of σ_0 will be problem dependent.

Case-II: Damping Ratio Minimization

The damping ratio of the oscillation is given by eq (12) determines how fast the oscillation is damped.

$$\zeta = \frac{\sigma}{\sqrt{\sigma^2 + \omega^2}} \tag{12}$$

Minimum damping ratio considered, ζ_0=0.3. Minimization of this objective function will minimize overshoot. Undershoot and settling time

$$OF_2 = \sum_i (\zeta_0 - \zeta_i)\ \ i=1,2,3,\ldots,n \tag{13}$$

where, ζ_i is the damping ratio of the ith eigenvalues and $\zeta_i<\zeta_0$

Multi-Objective Optimization

Case-I: Simultaneous Minimization of Eigen-Value and Damping Ratio

The factors of the PSS (for CPSS: K_{pss}, T_{d1}, T_{d2}, T_{d3}, T_{d4}, T_{d5}, T_{d6}; for IEEE-PSS2B:2K_{s1}, T_1, T_2, T_3, T_4, T_5; for IEEE-PSS3B: K_{s1}, K_{s2}, T_{d1}, T_{d2}, T_{d3}, T_{d4}; for IEEE-PSS4B: K_{s1}, K_{s1}, T_1, T_2, T_3, T_4) are to be so modelled that some amount of relative stability and damping of electromechanical modes of oscillations, minimized undershoot (U_s), minimized overshoot (O_s) and lesser settling time (t_s) of transient oscillations of $\Delta\omega_r$ are achieved, subject to an incremental state perturbation $\Delta\delta$. In order to satisfy the stated criterion, two multi-objective optimization functions, namely, OF_1 and OF_2 are formulated which are to be minimized in assumption are declared as follows:

So, the first considered objective function is formulated as in:

$$OF_1 = PF_1 \times X_{11} + PF_2 \times X_{12} + PF_2 \times X_{13} + X_{14} \tag{14}$$

where,

$$X_{11} = \sum_i (\sigma_0 - \sigma_i)^2;\ \ X_{12} = \sum_i (\zeta_0 - \zeta_i)^2;\ \ X_{13} = \sum_i (\delta_0 - \delta_i)^2 \tag{15}$$

X_{14} is arbitrary chosen high numerical value; σ_i is the real part of the ith eigen value; ζ_i is the damping ratio of the ith eigen value; δ_0 is the imaginary part of ith eigenvalue; PF_1, PF_2 and PF_3 are weighting factors chosen to impart more weight to PF_1 and PF_2 and to reduce high value of PF_3, and to make them mutually competitive during optimization. The value of PF_1 and PF_2 are chosen to be 10 and that of PF_3 is 0.01. By minimizing the objective function $OF_1()$, all the closed loop system poles are forced to be shifted to the left of the $j\omega$ axis (Fig. 8) in addition to reduction in imaginary part, thereby improving the dynamic stability and enhancing the damping ratio ζ_0, for which $\sigma_i << -\sigma_0$, $\zeta_i >> \zeta_0$. Low negative value of σ is intentionally chosen. The purpose is to shift the closed loop poles to the left of $j\omega$ axis to improve the stability of the system to a great extent.

Case-II: Simultaneous Minimization of Overshoot, Undershoot and Settling Ratio

It has been deliberately observed in some cases, the considered objective function (OF_1 ()) does not provide quick tuning of PSS factors. So it become utmost importance to devise a second multi-objective function for the necessary quick tuning of PSS parameters. Thus the second multi-objective function is developed as given in:

$$OF_2 = (O_s \times 10^6)^2 + (U_s \times 10^6)^2 + (t_s)^2 \quad (16)$$

where, O_s, U_s and t_s are the overshoot, undershoot and settling time. All referred to the transient response of $\Delta\omega_r$.

System Constraints

Minimize $OF_1()$ and $OF_2()$ in sequence with the help of any optimization technique to achieve most favourable set of PSS parameters, subject to the limits (Kundur, 1996) The constrained optimization problem for the tuning of PSSs is, thus, formulated as follows:

a) For

CPSS: $175.0 \le K_{pss} \le 230.0$, $0.001 \le T_{di} \le 1.0$, $i=1$ to 6 (17)

b) For

$$\text{IEEE-PSS2B:} \begin{cases} 10 \le K_{s1} \le 30, \\ 0.01 \le T_i \le 1.0, \quad i = 1 \text{ to } 5, \\ T_{10} = 0.0001 \end{cases} \quad (18)$$

Former factors are unchanged.

c) For

$$\text{IEEE-PSS3B:}\begin{cases} -100.0 \le K_{s1} \le -10.0, 10.0 \le K_{s2} \le 100.0, \\ 0.005 \le T_{di} \le 2.0, \quad i = 1 \; to \; 4 \end{cases} \tag{19}$$

No fixed parameters are required

d) For

$$\text{IEEE-PSS4B:}\begin{cases} -100.0 \le K_{s1} \le -10.0, 10.0 \le K_{s2} \le 100.0, \\ 0.005 \le T_{i} \le 2.0, \quad i = 1 \; to \; 4, \; T_{0} = 0.2 \end{cases} \tag{20}$$

EVOLUTIONARY ALGORITHM

Differential Search Algorithm (DSA)

Differential search algorithm (DSA) was first introduced by (Civicioglu, 2012) is novel meta-heuristic optimization technique. DSA is stimulated by the migration actions of living organism in nature. accessibility and capability of food areas existing in nature alters at regular intervals during the year. As a result, living organism exhibit seasonal migration behaviour throughout the year. This allows active bacilli to move from one abode to addition abode area availability and ability of the accustomed assets are more. This is the main motivation of deriving DSA in the literature (Civicioglu, 2012).

Mathematical Modelling of DSA

Initialization

Initial population, called *artificial organism*, in DSA is made by a random process corresponds to the optimization problem. Initialization of artificial organism in DSA is defined by eq. (21) [68].

$$X(i,j) = rand*(ub(j) - lb(j)) + lb(j) \tag{21}$$

where, i=1,2,3,…,n_p, j=1,2,3,…,dim, $X(i,j)$ is the initial population of the jth variable of ith agent, *rand* is a random number generated between [0, 1], n_p and *dim* are defined as the number of search agents (*superorganism*) and a number of control variables of the optimization problem, respectively, $ub(j)$ and $lb(j)$ are the upper and lower bounds of jth variable, respectively.

Random Shuffling of Superorganism

Randomly select '*superorganism*' from the initial population and move towards the targets '*donor*' by random shuffling in order to discover the '*stopoversite*', which is important for successful migration. '*scale*' value is determined by gamma random generator to control the positions of the member of artificial '*superorganism*' by using eq. (22).

$$scale = rand\ g(2*r_1)*(r_2 - r_3) \tag{22}$$

where, *rand g* is the random value chosen from a gamma distribution and r_1, r_2, and r_3 are the random numbers selected between (0,1). In DSA, the '*stopoversite*' of '*superorganism*' is defined as (Civicioglu, 2012). The Pseudo code for DSA is stated below:

$$stopoversite = \sup erorganism + scale * \left(\sup erorganism\big|_{random\_shuffling} - \sup erorganism \right) \tag{23}$$

```
Input: stopover, lb, ub
Output: stopover
   for  i=1:n_p
      for j=1:dim
         if stopover (i,j),lb(j) or stopover)(i,j).ub(j)
         stopover (i,j)     update     rand*(ub(j)-lb(j))+lb(j)
         end
      end
   end
```

Boundary Condition

There are alone two ascendency ambit in DSA, i.e., p_1 and p_2, those are randomly selected members of '*superorganism*' to participate in '*stopoversite*'. Generally, the values of p_1 and p_2 are generated between [0, 0.3]. If any element of the '*stopoversite*' goes beyond the limits of habitat, then a boundary condition is imposed on the elements of '*stopoversite*' to force them into the habitat (Civicioglu, 2012). The general layout of DSA-algorithm is depicted in Figure 9.

Oppositional Based Learning (OBL)

Generally evolutionary optimization techniques start with some initial estimation and subsequently reach towards the optimal solution by performing their control mechanisms. Hence the computational time as based on the distance between the initial guesses and the optimal place plays a vital role as nearer the initial guess, less time is required solutions.

The concept of Oppositional based learning (OBL) as first proposed by (Tizhoosh, 2005) is acknowledged as an important concept in minimizing the computational time in reaching towards the optimal solutions. Thus to reduce the computational time opposite guesses or assumption are also simultaneously estimated along with random guesses. The researchers in the field of OBL claims that by simultaneously checking the opposite solution, the adventitious of starting with a closer or fitter solution can be improved. The OBL concept relies on the opposite point and opposite number which are defined as follows:

Opposite number: Let $x \in [m,n]$ be a real number. Its opposite number x^o is defined by:

$$x^o = m + n - x \tag{24}$$

Opposite point: Let $P=(x_1,x_2,\ldots,x_n)$ be a point in n-dimensional space, where $x_r \in [m_r,n_r]$, $r \in \{1,2,\ldots,n\}$. The opposite point $P^o = (x_1{}^o, x_2{}^o, \ldots, x_n{}^o)$ is defined by its components:

$$x_r{}^o = m_r + n_r - x_r \tag{25}$$

ODSA APPLIED TO PSS

Step 1: Random initialization of the '*artificial organism*' (PSS variables) in the search space with population size of 'n_p' and dimension of '*dim=6*'.
Step 2: Create opposite population *OP* using eq. (24)
Step 3: Evaluate the fitness value for the individual organism using (11, 13,14,16)
Step 4: From '*superorganism*', randomly select a singular member and heading towards the target '*donor*' by random shuffling process, which is described below:

```
for j=1:dim
superorganism=superorganism(:,j)
   for i=1:np
      w=[rand*i];
      t=superorganism(w);
      superorganism (w)=superorganism (i);
      superorganism (i)=t;
      end
   donor(;,j)=X;
end
```

Step 5: Computation of *stopoversite* and *Scale*value using eqs. (22,23)
Step 6: Set two control parameters $p_1=0.3*rand$ *and* $p_2=0.3*rand$ update *stopover* by random selection process
Step 7: Check whether any member of *stopover* is violating the upper limit or not and hence update *stopoversite* accordingly.
Step 8: *OP* is created and its objective value is evaluated based on j_r
Step 9: Choose requisite number of acceptable solution from ($P \cup OP$).
Step 10: Sort *stopoversite* from best to worst value
Step 11: Terminate if the stopping criterion is satisfied or else go to Step 3.

SIMULATION RESULTS AND DISCUSSION

The proposed algorithm is implemented on the linearized model of SMIB system and shows the place of the PSS in the SMIB system from the Heffron-Phillips model. To assess the effectiveness and superiority of the proposed algorithm over a wide range of loading conditions, three different cases such as light (P=0.35, Q=0.55), nominal (P=1.25, Q=0.50) and heavy (P=1.45, Q=0.70) loading are considered as system uncertainties. For SMIB system, Inertia constant, H=5, M=2H, nominal frequency, f_0=50Hz, $0.995 \leq |\bar{E}_b| \leq 1.0$, the angle of $\bar{E}_b = 0^{\circ}$, $0.2 \leq P \leq 1.2$, $-0.2 \leq Q \leq 1.0$, $0.4752 \leq X_e \leq 1.08$, $0.5 \leq E_t \leq 1.1$. It is worthwhile to mention that the transient performance of dual-input IEEE-PSS is better than conventional single input PSS. The simulation results obtained by the proposed ODSA and DSA algorithms are compared with other well known algorithms to test the superiority of the proposed approach. Programs are developed in Matlab 2011b and executed on i3 processor with 2.20 GHz and 4-GB RAM. After several trails, following parameters are found to be best for successful implementation of DSA in PSS problem. Size of superorganism (n_p)= 50, maximum generation count=100, P_1=0.3 and P_2=.03. For ODSA and DSA the parameters has been chosen by taking all possible combinations of 50 trials and finally taking those values which gives optimal results parameter setting effect its performance.

Single Objective Optimization

To investigate the effectiveness of the proposed approach, two different objectives namely, eigen-value minimization and damping ratio minimization are considered individually. In order to verify the superiority, the simulation results of the proposed ODSA and DSA algorithms are compared to those obtained by other well known optimization techniques..

Eigen Value Minimization

The system consists of linearized SMIB system from the Heffron-Phillips model. PSS are installed at respective location to damp out the low frequency oscillation pertaining in power system. The eigen-values of the state space matrix is represented as:

$$\det|A - sI| = 0 \tag{26}$$

where, A is the state matrix; I is the identity matrix; S is the eigen-values.

The algorithms are tested over a wide range of operating conditions namely, light (P=0.65, Q=0.55), nominal (P=1.25, Q=0.50) and heavy (P=1.45, Q=0.70) loading as system uncertainties. The final optimized value of PSS parameter along with the fitness values are depicted in Tables 1-3. From the respectively tables, it can be easily concluded that the proposed ODSA algorithm gives lowest minimum fitness function values compared to other reported algorithms in all the three loading conditions. Also there is an improvement in the overshoot (O_s), undershoot (U_s) and fitness function values by considering the proposed ODSA and DSA approaches. Critical observation of Table I reveals that minimum value of objective function has been obtained with ODSA (**0.004**) compared to DSA (0.0240). Improvement of fitness function with ODSA is 84.58% (DSA), 88.40% (OGSA), 95.69% (GSA), 90.70% (BPPO) and 95.61% (BBO). The similar analysis can be obtained to Table 2-3. The convergence graph as depicted in Fig. 10 also validate the superiority of the proposed ODSA and DSA algorithms in reaching its final value by taking less iterations compared to other reported algorithms.

Damping Ratio Minimization

To discuss the suitability of the proposed ODSA algorithm to cope up with the power system stability, damping ratio minimization has been considered. The damping ratio determines how fast the oscillation is damped. In this study also, the proposed algorithms is tested on above stated loading conditions. The final optimized value of the fitness function along with overshoot and undershoot is given in Tables 4-6. From the simulation results, it can be easily predicted that the there is a considerable improvement in the fitness functions value and thus demonstrate the shifting of the unstable poles to the left side of the s-plane. Also the proposed ODSA algorithm produces stable results in almost all loading condition in comparison to other reported algorithms, which yield meagre response in all loading conditions.

Multi-Objective Optimization

The above analysis is also carried out on dual input PSS (*PSS2B, PSS3B, PSS4B*) to test the superiority and effectiveness of the proposed optimization techniques. It is worthwhile to mention that the transient performance of dual-input IEEE-PSS is better than conventional single input PSS. The simulation results obtained by the proposed ODSA and DSA algorithms are compared with other well known algorithms to test the superiority of the proposed approach.

Case-I

The first objective function including the eigenvalues are been considered in this case. The system comprises of conventional PSS and three other dual input PSS (IEEE-PSS2B, IEEE-PSS3B and IEEE-PSS4B). For optimal PSS design, a set of typical operating conditions adopted from (Shaw, et. Al. 2011) are taken into consideration. The system data for conventional PSS as well as dual input PSS including the synchronizing and damping torque coefficient are the same as (Shaw, et. Al. 2011) The parameters of the conventional PSS (K_{pss}, T_{d1}, T_{d2}, T_{d3}, T_{d4}, T_{d5}, T_{d6}); IEEE-PSS2B (K_{s1}, T_1, T_2, T_3, T_4, T_5); IEEE-PSS3B(K_{s1}, K_{s2}, T_{d1}, T_{d2}, T_{d3}, T_{d4}); IEEE-PSS4B (K_{s1}, K_{s2}, T_1, T_2, T_3, T_4) are so tuned that the degree of relative stability is achieved. Moreover, to check the superiority of the proposed ODSA and DSA algorithms, the simulation results are compared to those obtained by SOA (Shaw, et. Al. 2011) and BFO (Shaw, et. Al. 2011) and some other algorithms reported in the literature. Also to check the flexibility of the proposed approach in minimizing the fitness functions, three different loading conditions has been considered (Shaw, et. Al. 2011). It can be seen from the Tables 7-10 that the values of the objective function is found to be the least for conventional PSS, IEEE-PSS2B, IEEE-PSS3B and IEEE-PSS4B when implemented with proposed ODSA and DSA algorithms. This implies that, on average the quality of solution obtained by ODSA algorithm is better than other methods. Critical observation of Table 7 reveals that minimum value of objective function has been obtained with ODSA (**678.87**) compared to DSA (679.93). Improvement of fitness function with ODSA is 0.16% (DSA), 0.11% (OGSA), 0.20% (GSA), 0.13% (BPPO) and 0.12% (BBO). The similar analysis can be obtained to Table 8-10. The comparative convergence characteristics for the best solution using different algorithms are depicted in Fig. 6.12. From the graph it can be observed that the value of the objective function with the proposed ODSA and DSA algorithms converges to minimum values by taking less iteration in comparison to other methods. This clearly suggests the superiority of the proposed ODSA algorithm compared to other reported approaches.

Case-II

To test the effectiveness of the proposed algorithm and to get sharp tuning of the PSS parameters, the second objective function which minimizes overshoot, undershoot and settling time is taken in this case for the simulation study. In order to verify the effectiveness of the proposed ODSA and DSA algorithms in solving PSS problem, the results obtained by the proposed method are compared to that of SOA (Shaw, et. Al. 2011), BFO (Shaw, et. Al. 2011) and some well known algorithms. It is worthwhile to mention that to test the feasibility of the proposed approaches; five different loading conditions have been used. The optimal value of the fitness values under different loading conditions obtained by different optimization techniques for different PSSs is listed in Tables 11-14. The simulation results of Table 11-14 clearly reveal that the on an average the fitness value obtained by the proposed method are the least among all the algorithms. Also there is a considerable reduction in the overshoot, undershoot and fitness function values in case of ODSA algorithm as compared to other reported algorithms in almost all loading conditions. Also, the proposed fitness function produces better results by implementing the ODSA algorithm amongst other algorithms, thus validate the superiority of the proposed ODSA algorithm.

CONCLUSION

In this study, the dynamic stability of the SMIB system incorporating PSS has been achieved by using a new optimization algorithm, namely ODSA and DSA. Here the abstraction of OBL has been collaborated along with original DSA technique to improve the convergence profile and to achieve higher accuracy of optimal solution. To validate the flexibility of the proposed technique, a multi-dimensional objective function has been taken into account. The supremacy of the proposed methods was validated by considering different types of loading conditions. From the simulation results obtained by different algorithms and under different loading conditions, it can be concluded that the proposed ODSA and DSA has good convergence property and can avoid the shortcoming the premature convergence and produce global or near global optimum solution. Even though the proposed approach yields better results, still the results can further be improved by implementing some other algorithms. Also with the inclusion of different FACTS devices, the results of PSS can be improved further.

REFERENCES

Abido, M.A., & Abdel-Magid, Y.L. (2002). Optimal design of power system stabilizers using evolutionary programming. *IEEE Transaction on Energy Conversion*, *17*(4).

Andreoiu, A., & Bhattacharya, K. (2002). Robust tuning of power system stabilizers using a Lyapunov method based genetic algorithm. *Proceeding of Inst. Electrical Engineering Generation Transmission and Distribution*, *149*(5), 585–592. doi:10.1049/ip-gtd:20020600

Arrifano, N. S. D., Oliveira, V. A., Ramos, R. A., Bretas, N. G., & Oliveira, R. V. (2007). Fuzzy stabilization of power systems in a co-generation scheme subject to random abrupt variations of operating conditions. *IEEE Transactions on Control Systems Technology*, *15*(2), 384–393. doi:10.1109/TCST.2006.886443

Chitara D., Swarnkar, A., Gupta, N., Niazi, K.R., & Bansal, R.C. (2015). Optimal tuning of multimachine power system stabilizer using cuckoo search algorithm. *SCIENCEDIRECT,* 143-148.

Civicioglu, P. (2012). Transforming geocentric cartesian coordinates to geodentic coordinates by using differential search algorithm. *Computers & Geosciences*, *46*, 229–247. doi:10.1016/j.cageo.2011.12.011

Computer models for representation of digital-based excitation systems. (1996). *IEEE Trans. Energy Convers., 11*(3), 607–615.

Dey, P., Bhattacharya, A., & Das, P. (2018). Tuning of power system stabilizer for small signal improvement of interconnected power system. Applied Computing and Informatics.

Eke, I., & Taplamacioglu, M.C. (2015). Robust tuning of power system stabilizer by using orthogonal learning artificial bee colony. *Sciencedirect*, 149-154.

El-Sherbiny, M. K., Hasan, M. M., El-Saady, G., & Yousef, A. M. (2003). Optimal pole shifting for power system stabilization. *Electric Power Systems Research*, *66*(3), 253–258. doi:10.1016/S0378-7796(03)00092-0

Elazim, S. M. A., & Ali, E. S. (2016). Optimal power system stablizers design via cuckoo search algorithm. *International Journal of Electrical Power & Energy Systems*, *75*, 99–107. doi:10.1016/j.ijepes.2015.08.018

Ellithy, K., Said, S., & Kahlout, O. (2014). Design of Power System Stablizers based on μ-controller for power system stability enhancement. *Electric Power and Energy System*, 63, 933-939.

Farahain, M., & Ganjefar, S. (2015). An online trained neural network controller to improve stability of power systems. *Neurocomputing*, *162*, 245–255. doi:10.1016/j.neucom.2015.03.047

Ganesan, Vasant, & Elamvazuthi. (2016). *Advances in Metaheuristics: Application in Engineering Systems*. CRC Press, Taylor & Francis Group.

Hoen-Su, R., Kyoung, M., Lee, J. G., & Hyun, M. Y. (2002). Extended integral based governor for power system stabilizer. *IEEE Power Engineering Society Summer Meeting*, *3*, 1610-1614.

Hossenizadeh, N., & Kalam, A. (1999). A rule based fuzzy power system stabilizer tuned by a neural network. *IEEE Transactions on Energy Conversion*, *14*(3), 773–776. doi:10.1109/60.790950

Islam, N. N., Hannan, M. A., Shareef, H., & Mohamed, A. (2016). An Application of Backtracking Search Algorithm in Designing Power System Stabilizers for Large Multi-machine System. *Neurocomputing*, *237*, 175–184. doi:10.1016/j.neucom.2016.10.022

Izquierdo, C. E., & Marquez, A. E. (2017). A survey of the cuckoo search and its application in real-world optimization problems. Handbook of research on soft computing and nature-inspired algorithms.

Kaliannan J., Baskaran A. & Dey N. (2017). Automatic generation control of thermal-thermal-hydro power systems with PID controller using ant colony optimization. *Renewable and alternate energy: concepts methodology, tools and application.*

Kamwa, I., Grondin, R., & Trudel, G. (2005). IEEE PSS2B versus PSS4B: The limits of performance of modern power system stabilizers. *IEEE Transactions on Power Systems*, *20*(2), 903–915. doi:10.1109/TPWRS.2005.846197

Keumarsi, V., Simab, M., & Shahgholian, G. (2014). An integrated approach for optimal placement and tuning of power system stabilizer in multimachine systems. *International Journal of Electrical Power & Energy Systems*, *63*, 132–139. doi:10.1016/j.ijepes.2014.05.072

Kothari, M., Bhattacharya, L., & Nanda, J. (2002). Adaptive power system stabilizer based on pole shifting technique. *IEE Proceedings. Generation, Transmission and Distribution*, *143*(1), 96–98. doi:10.1049/ip-gtd:19960300

Kundur, P. (2006). *Power System Stability and Control*. Tata: McGraw-Hill.

Kundur, P., Klein, M., Rogers, G. J., & Zywno, M. S. (1989). Application of power system stabilizers for enhancement of overall stability. *IEEE Transactions on Power Systems*, *4*(2), 614–626. doi:10.1109/59.193836

Mahabuba, A., & Khan, M. A. (2009). Small signal stability enhancement of a multi-machine power system using robust and adaptive fuzzy neural network basd power system stabilizer. *International Transactions on Electrical Energy Systems*, *19*(7), 978–01.

Marmolejo, J. A., Velasco, J., & Selley, H. J. (2017). An adaptive random search for short term generation scheduling with network constraints. *PLoS One*, *12*(2), e0172459. doi:10.1371/journal.pone.0172459 PMID:28234954

Naresh, G., Raju, M. R., & Narasimham, S. V. L. (2016). Coordinated deign of power system stabilizers and TCSC employing improved harmony search algorithm. *Swarm and Evolutionary Computation*, *27*, 169–179. doi:10.1016/j.swevo.2015.11.003

Niknam, T., Mojarrad, H. D., Meymand, H. Z., & Firouzi, B. B. (2011). A new honey bee mating optimization algorithm for non-smooth economic dispatch. *Energy*, *36*(2), 896–908. doi:10.1016/j.energy.2010.12.021

Padiyar, K. R. (1996). *Power system dynamics stability and control*. Bangalore: Interline Publishing Private Ltd.

Panda, S., Yegireddy, N. K., & Mohapatra, S. K. (2013). Hybrid BFOA-PSO approach for coordinated design of PSS and SSSC-based controller considering time delays. *International Journal of Electrical Power & Energy Systems*, *49*, 221–233. doi:10.1016/j.ijepes.2013.01.006

Paul, S., & Roy, P. K. (2015). Optimal design of single machine power system stabilizer using chemical reaction optimization technique. *International Journal of Energy Optimization and Engineering*, *4*(2), 51–69. doi:10.4018/IJEOE.2015040104

Peres, W., Junior, I. C. S., & Filho, J. A. P. (2017). Gradient based hybrid metaheuristics for robust tuning of power system stabilizers. *International Journal of Energy Optimization and Engineering*, *95*, 47–72.

Sambariya, D. K., & Prasad, R. (2014). Robust tuning of power system stabilizer for small signal stability enhancement using metaheuristic bat algorithm. *International Journal of Electrical Power & Energy Systems*, *61*, 229–238. doi:10.1016/j.ijepes.2014.03.050

Shaw, B., Banerjee, A., Ghosal, S. P., & Mukherjee, V. (2011). Comparative seeker and bio-inspired fuzzy logic controllers for power system stabilizers. *Elect. Power and Energy Syst*, *33*(10), 1728–1738. doi:10.1016/j.ijepes.2011.08.015

Shayeghi, H., Shayanfar, H. A., Jalilzadeh, S., & Safari, A. (2010). Multi-machine power system stabilizers design using chaotic optimization algorithm. *Energy Conversion and Management*, *51*(7), 1572–1580. doi:10.1016/j.enconman.2010.02.015

Talaq, J. (2012). Optimal power system stabilizers for multi machine systems. *International Journal of Electrical Power & Energy Systems*, *43*(1), 793–803. doi:10.1016/j.ijepes.2012.06.030

Tizhoosh, H. (2005). Opposition based learning: A new scheme for machine intelligence. *Proc. of Int. Conf. on Computational Intelligence for Modeling Control and Automation*, 695-701. 10.1109/CIMCA.2005.1631345

Ugalde-Loo, C. E., Acha, E., & Liceaga-Castro, E. (2010). Fundamental analysis of the electromechanical oscillation damping control loop of the static VAR compensator using individual channel analysis and design. *Power Delivery on IEEE Transaction*, *25*(4), 3053–3069. doi:10.1109/TPWRD.2010.2046341

Vasant, P., Zelinka, & Weber, G-W. (2018). Intelligent Computing & Optimization. In *International Conference on Intelligent Computing & Optimization*. Springer.

Vasant, P., Marmolejo, J. A., Litvinchev, I., & Aguilar, R. R. (2019). Nature-inspired metaheuristic approaches for charging plug-in hybrid electric vehicle. *Wireless Networks*. doi:10.100711276-019-01993-w

Vasant, P., Zelinka, P., Weber, I., & Wilhelm, G. (2019). Intelligeng Computing and optimization. *Proceedings of the 2nd International Conference on Intelligent Computing and Optimization* 10.1007/978-3-030-00979-3

Vasant, K., & Watada, J. (2017). Metaheuristic techniques in Enhancing the Efficiency and Performance of Thermo-Electric cooling devices. *Energies.*

Wu, B., & Malik, O. (2006). Multivariable adaptive control of synchronous machine. *IEEE Power India Conference*.

Zhang, Y., Chen, G. P., Malik, O. P., & Hope, G. S. (1993). An artificial neural network based adaptive power system stabilizer. *IEEE Transactions on Energy Conversion*, *8*(1), 71–77. doi:10.1109/60.207408

APPENDIX

List of Tables

Table 1. Comparative results of eigenvalue minimization for different algorithms under light loading condition

Algorithm	K_{pss}	T_1	T_2	T_3	T_4	Eigen values	O_s ($\times 10^{-4}$)	U_s ($\times 10^{-4}$)	Fitness Values ($\times 10^{-4}$)
	Light Load (P=0.65, Q=0.55)								
ODSA	**19.344**	**0.632**	**1.504**	**0.635**	**0.475**	**-1.500±6.118i**	**5.605**	**12.061**	**0.004**
DSA	**15.087**	**3.350**	**1.216**	**0.353**	**1.373**	**-1.500±6.100i**	**7.134**	**13.960**	**0.024**
OGSA	17.537	4.692	1.772	0.389	1.683	-1.500±6.386i	4.676	13.035	0.032
GSA	18.712	1.901	1.382	0.219	0.275	-1.500±6.779i	4.844	13.461	0.086
BPPO	17.495	1.887	0.658	0.390	1.814	-1.500±5.998i	7.075	14.033	0.040
BBO	16.407	4.977	1.695	0.391	1.758	-1.500±6.198i	7.130	13.940	0.084

Table 2. Comparative results of eigenvalue minimization for different algorithms under nominal loading condition

Algorithm	K_{pss}	T_1	T_2	T_3	T_4	Eigen values	O_s ($\times 10^{-4}$)	U_s ($\times 10^{-4}$)	Fitness Values ($\times 10^{-4}$)
	Nominal Load (P=1.25, Q=0.50)								
ODSA	**18.186**	**2.843**	**1.907**	**0.861**	**1.444**	**-1.500±6.877i**	**5.708**	**13.001**	**0.003**
DSA	**16.471**	**2.277**	**0.943**	**0.614**	**1.541**	**-1.500±6.621i**	**7.828**	**14.225**	**0.004**
OGSA	15.911	2.228	1.647	0.495	0.676	-1.500±6. 202i	7.130	12.238	0.003
GSA	16.654	3.945	1.413	0.605	1.758	-1.500± 4.279i	4.676	13.139	0.074
BPPO	18.948	4.023	1.078	0.381	1.702	1.500±6.626	8.990	15.776	0.009
BBO	17.753	3.363	1.525	0.632	1.544	-1.500±6.789i	9.016	15.623	0.101

Table 3. Comparative results of eigenvalue minimization for different algorithms under heavy loading condition

Algorithm	K_{pss}	T_1	T_2	T_3	T_4	Eigen values	O_s ($\times 10^{-4}$)	U_s ($\times 10^{-4}$)	Fitness Values ($\times 10^{-4}$)
	Heavy Load (P=1.45, Q=0.70)								
ODSA	**18.145**	**2.470**	**1.935**	**1.608**	**1.914**	**-1.500±3.136i**	**3.317**	**9.480**	**0.001**
DSA	**17.277**	**2.225**	**1.837**	**1.728**	**1.859**	**-1.500±3.137i**	**3.314**	**9.480**	**0.003**
OGSA	16.670	2.282	1.967	1.828	1.821	-1.500±3.160i	3.313	9.487	0.007
GSA	18.049	2.563	1.977	1.594	1.917	-1.500±3.107i	3.313	9.471	0.038
BPPO	18.907	2.356	1.758	1.517	1.970	-1.499±3.103i	3.314	9.470	0.021
BBO	15.422	2.156	1.851	1.993	1.843	-1.500±3.160i	3.312	9.486	0.192

Table 4. Comparative results of damping ratio minimization for different algorithms under light loading condition

Algorithm	K_{pss}	T_1	T_2	T_3	T_4	Eigen values	O_S (×10⁻⁴)	U_S (×10⁻⁴)	Fitness Values (×10⁻⁴)
	Light Load (P=0.65, Q=0.55)								
ODSA	**15.523**	**2.888**	**1.489**	**0.478**	**1.524**	**-1.262±6.184i**	**6.620**	**12.127**	**0.001**
DSA	**17.878**	**1.1222**	**0.625**	**0.455**	**1.556**	**-1.242±6.084i**	**7.945**	**13.575**	**0.015**
OGSA	15.8281	1.717	0.838	0.497	1.721	-1.251±6.127i	7.959	13.785	0.035
GSA	16.529	2.498	1.247	0.248	0.887	-1.232± 6.038i	7.995	14.067	0.432
BPPO	19.9878	2.022	1.192	0.434	1.559	-1.262±6.183i	7.916	14.192	0.150
BBO	16.432	2.081	1.641	0.816	1.770	-1.278±6.262i	7.895	14.102	0.986

Table 5. Comparative results of damping ratio minimization for different algorithms under nominal loading condition

Algorithm	K_{pss}	T_1	T_2	T_3	T_4	Eigen values	O_S (×10⁻⁴)	U_S (×10⁻⁴)	Fitness Values (×10⁻⁴)
	Nominal Load (P=1.25, Q=0.50)								
ODSA	**17.987**	**3.887**	**1.771**	**0.734**	**1.840**	**-1.399±6.855i**	**8.714**	**13.663**	**0.016**
DSA	**18.835**	**4.259**	**1.629**	**0.518**	**1.640**	**-1.382±6.770i**	**9.314**	**14.752**	**0.028**
OGSA	18.388	3.778	1.931	0.643	1.479	-1.389±6.806i	9.343	15.684	0.017
GSA	17.312	2.960	1.565	0.588	1.248	-1.367±6.700i	9.405	16.116	0.526
BPPO	16.091	4.063	0.976	0.376	1.665	-1.335±6.537i	9.746	16.019	0.613
BBO	15.777	4.014	1.525	0.645	1.722	-1.383±6.770i	9.364	15.752	0.713

Table 6. Comparative results of damping ratio minimization for different algorithms under heavy loading condition

Algorithm	K_{pss}	T_1	T_2	T_3	T_4	Eigen values	O_S (×10⁻⁴)	U_S (×10⁻⁴)	Fitness Values (×10⁻⁴)
	Heavy Load (P=1.45, Q=0.70)								
ODSA	19.331	2.531	1.674	0.924	1.817	-0.499±2.442i	1.987	9.121	0.011
DSA	19.281	1.758	1.592	0.635	0.594	-0.711±3.482i	1.995	10.873	0.040
OGSA	18.507	3.561	1.735	0.704	1.792	-0.697±3.413i	2.597	6.873	0.054
GSA	16.925	3.873	1.856	0.644	1.516	-0.684± 3.351i	4.801	9.681	0.081
BPPO	19.040	2.959	1.642	0.704	1.612	-0.691±3.386i	6.855	10.876	0.040
BBO	17.179	2.309	1.739	1.234	1.898	-0.722±3.537i	6.585	10.863	0.116

Table 7. Comparative results using different algorithm for CPSS for Case-I under different loading conditions.

Algorithm	Loading Conditions		
	$P=1.2;Q=0.6;$ $X_e=0.4752;E_t=0.5$	$P=1.0;Q=0.6;$ $X_e=0.93;E_t=0.5$	$P=1.0;Q=0.2;$ $X_e=1.08;E_t=0.5$
ODSA	678.867	400.823	711.3878
DSA	679.926	406.485	711.351
OGSA	679.638	411.078	711.148
GSA	680.252	419.901	712.150
BPPO	679.748	416.528	711.831
BBO	769.804	435.760	712.407

Table 8. Comparative results using different algorithm for PSS2B for Case-I under different loading conditions.

Algorithm	Loading Conditions		
	$P=1.2;Q=0.6;$ $X_e=0.4752;E_t=0.5$	$P=1.0;Q=0.6;$ $X_e=0.93;E_t=0.5$	$P=1.0;Q=0.2;$ $X_e=1.08;E_t=0.5$
ODSA	937.392	808.137	770.938
DSA	938.972	810.412	771.848
OGSA	938.304	815.558	771.268
GSA	953.131	817.707	774.374
BPPO	938.101	803.675	768.463
BBO	950.107	819.938	783.642

Table 9. Comparative results using different algorithm for PSS3B for Case-I under different loading conditions

Algorithm	Loading Conditions		
	$P=1.2;Q=0.6;$ $X_e=0.4752;E_t=0.5$	$P=1.0;Q=0.6;$ $X_e=0.93;E_t=0.5$	$P=1.0;Q=0.2;$ $X_e=1.08;E_t=0.5$
ODSA	317.370	358.882	459.218
DSA	357.629	376.532	584.989
OGSA	352.916	364.490	586.139
GSA	393.929	373.745	588.255
BPPO	405.741	438.815	592.415
BBO	465.501	368.049	659.832

Table 10. Comparative results using different algorithm for PSS4B for Case-I under different loading conditions

Algorithm	Loading Conditions		
	$P=1.2;Q=0.6;$ $X_e=0.4752;E_t=0.5$	$P=1.0;Q=0.6;$ $X_e=0.93;E_t=0.5$	$P=1.0;Q=0.2;$ $X_e=1.08;E_t=0.5$
ODSA	183.884	215.456	231.762
DSA	223.689	244.621	242.346
OGSA	218.999	219.989	236.719
GSA	266.544	245.346	248.419
BPPO	256.523	249.615	247.538
BBO	265.530	267.942	254.788

Table 11. Comparative results using different algorithm for CPSS for Case-II under different loading conditions

Algorithm	Loading Conditions				
	$P=0.2; Q=-0.2;$ $X_e=0.4752; E_t=1.1$	$P=0.5; Q=0.2;$ $X_e=0.4752; E_t=1.0$	$P=0.75; Q=0.5;$ $X_e=0.4752; E_t=0.5$	$P=0.95; Q=0.3;$ $X_e=0.4752; E_t=0.5$	$P=1.2; Q=0.6;$ $X_e=1.08; E_t=0.5$
ODSA	2.878	2.861	2.422	2.531	4.311
DSA	3.017	3.322	2.836	2.797	4.695
OGSA	2.949	2.915	2.777	2.767	4.545
GSA	3.311	3.061	3.461	2.811	4.897
BPPO	2.941	2.894	2.833	2.709	4.651
BBO	3.356	3.290	3.122	2.962	4.835

Table 12. Comparative results using different algorithm for PSS2B for Case-II under different loading conditions

Algorithm	Loading Conditions				
	$P=0.2; Q=-0.2;$ $X_e=0.4752; E_t=1.1$	$P=0.5; Q=0.2;$ $X_e=0.4752; E_t=1.0$	$P=0.75; Q=0.5;$ $X_e=0.4752; E_t=0.5$	$P=0.95; Q=0.3;$ $X_e=0.4752; E_t=0.5$	$P=1.2; Q=0.6;$ $X_e=1.08; E_t=0.5$
ODSA	2.838	2.637	2.517	1.765	2.987
DSA	4.260	2.873	2.570	1.810	3.661
OGSA	1.012	2.378	2.500	1.801	3.658
GSA	1.635	2.379	2.976	1.976	3.865
BPPO	2.814	2.731	2.679	1.820	3.875
BBO	3.178	3.114	2.851	1.981	4.065

Table 13. Comparative results using different algorithm for PSS3B for Case-II under different loading conditions

Algorithm	Loading Conditions				
	P=0.2; Q=-0.2; X_e=0.4752; E_t=1.1	*P=0.5; Q=0.2; X_e=0.4752; E_t=1.0*	*P=0.75; Q=0.5; X_e=0.4752; E_t=0.5*	*P=0.95; Q=0.3; X_e=0.4752; E_t=0.5*	*P=1.2; Q=0.6; X_e=1.08; E_t=0.5*
ODSA	0.172	0.810	0.687	0.080	0.471
DSA	0.200	0.826	0.695	0.099	1.237
OGSA	0.256	0.754	0.527	0.098	1.202
GSA	0.257	0.662	0.772	1.543	1.292
BPPO	0.538	0.644	0.680	0.965	2.167
BBO	0.759	0.790	0.739	1.058	2.375

Table 14. Comparative results using different algorithm for PSS4B for Case-II under different loading conditions

Algorithm	Loading Conditions				
	P=0.2; Q=-0.2; X_e=0.4752; E_t=1.1	*P=0.5; Q=0.2; X_e=0.4752; E_t=1.0*	*P=0.75; Q=0.5; X_e=0.4752; E_t=0.5*	*P=0.95; Q=0.3; X_e=0.4752; E_t=0.5*	*P=1.2; Q=0.6; X_e=1.08; E_t=0.5*
ODSA	1.172	0.982	1.121	0.765	1.310
DSA	1.379	1.029	1.142	0.890	1.435
OGSA	1.330	1.027	1.122	0.8882	1.393
GSA	1.556	1.058	1.142	1.034	1.512
BPPO	1.167	1.151	1.085	0.935	1.663
BBO	1.281	1.260	1.169	1.085	1.881

List of Figures

Figure 1. Single-machine infinite bus test system

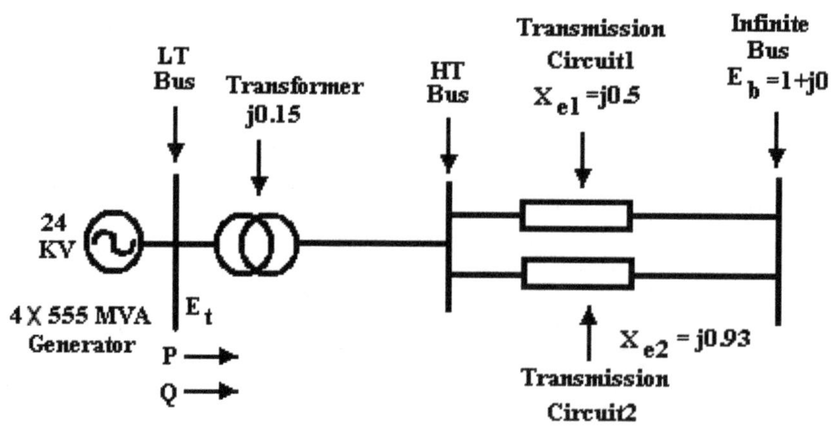

Figure 2. Block diagram representation with AVR and PSS

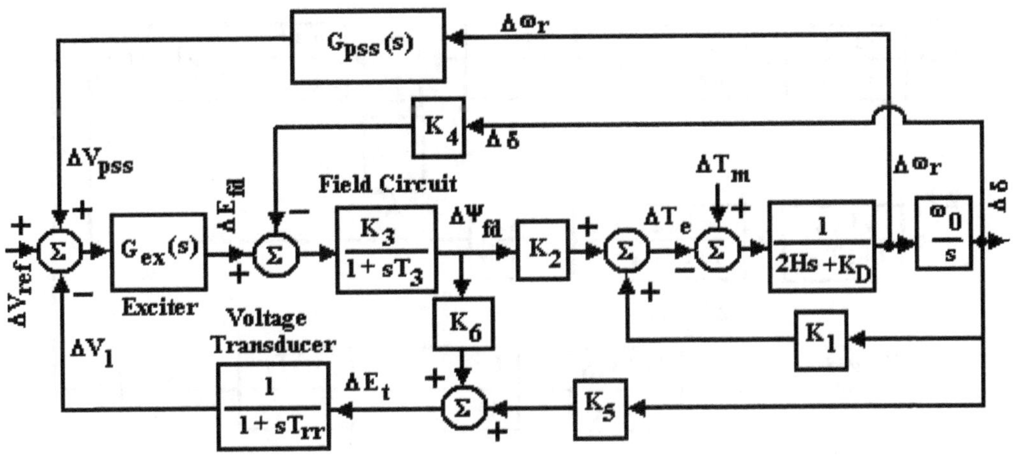

Figure 3 Block diagram representation of SMIB system with AVR, thyristor high gain exciter, synchronous generator and PSS

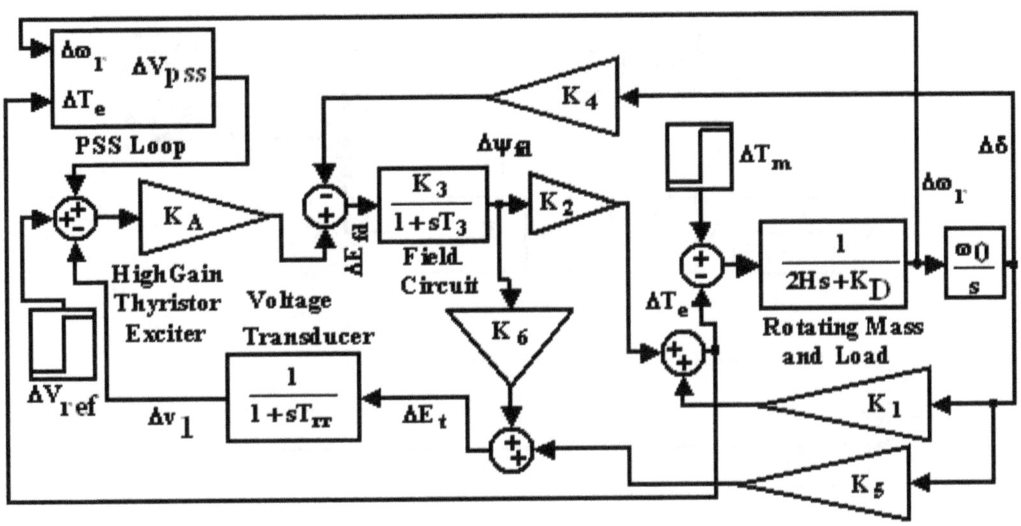

Figure 4. Block diagram representation of CPSS

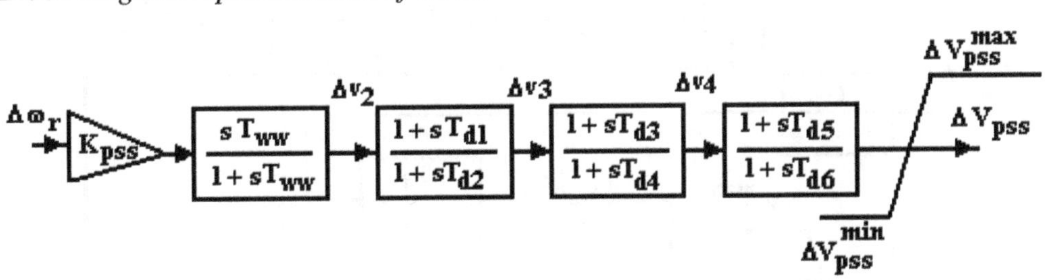

Figure 5. Block diagram representation of dual input IEEE PSS2B

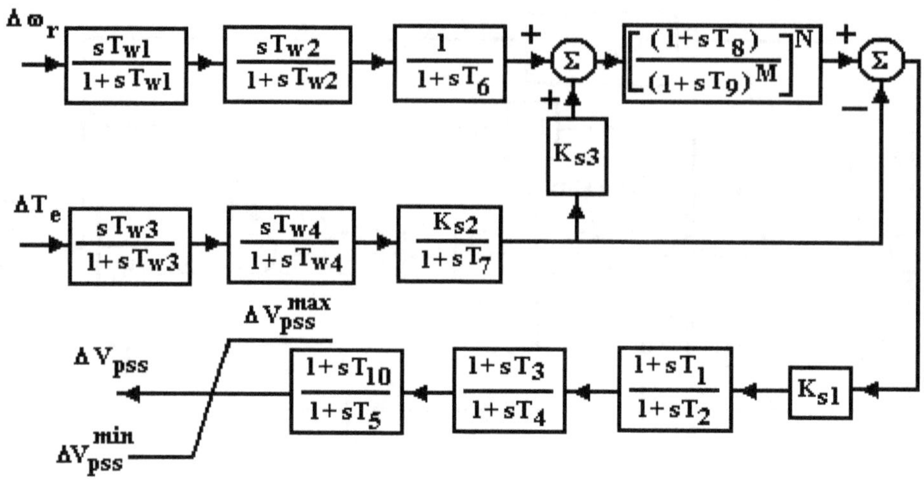

Figure 6. Block diagram representation of dual input IEEE PSS3B

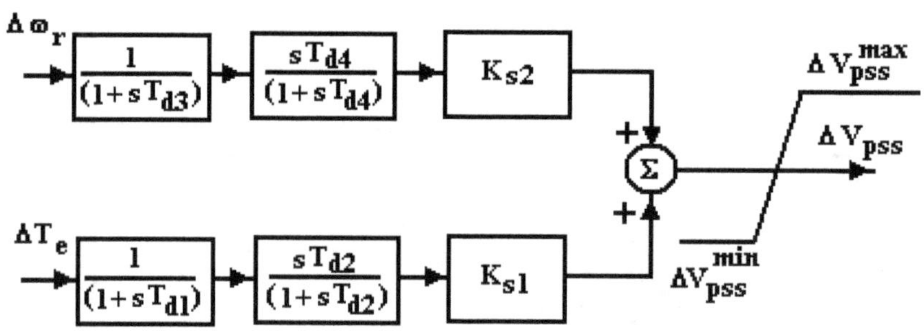

Figure 7. Block diagram representation of dual input IEEE PSS4B

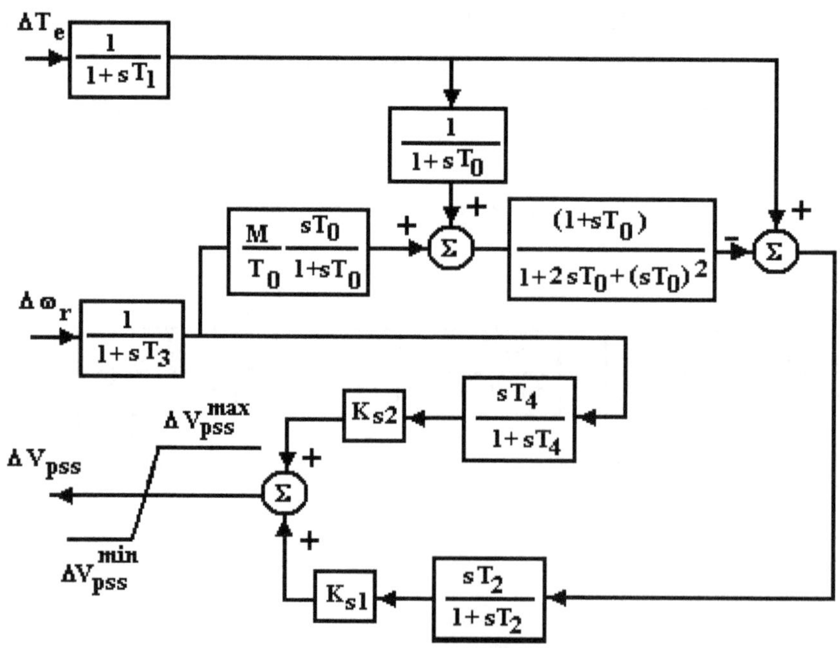

Figure 8. D-shaped sector in the negative half of s-plane

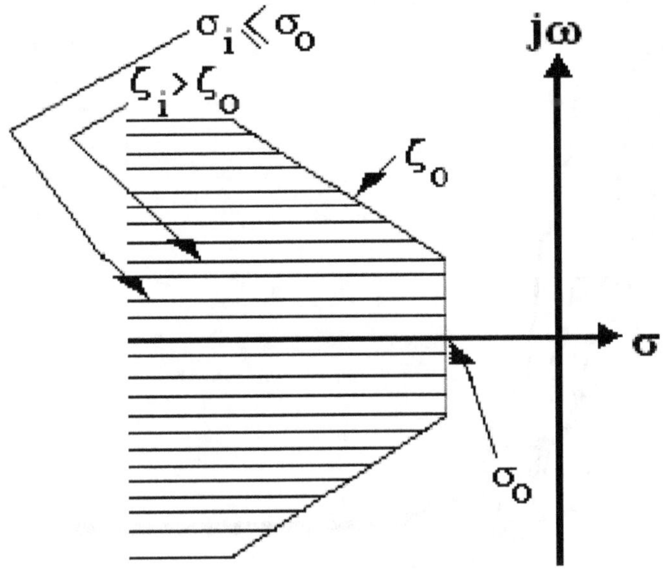

Figure 9. General flowchart of differential search algorithm

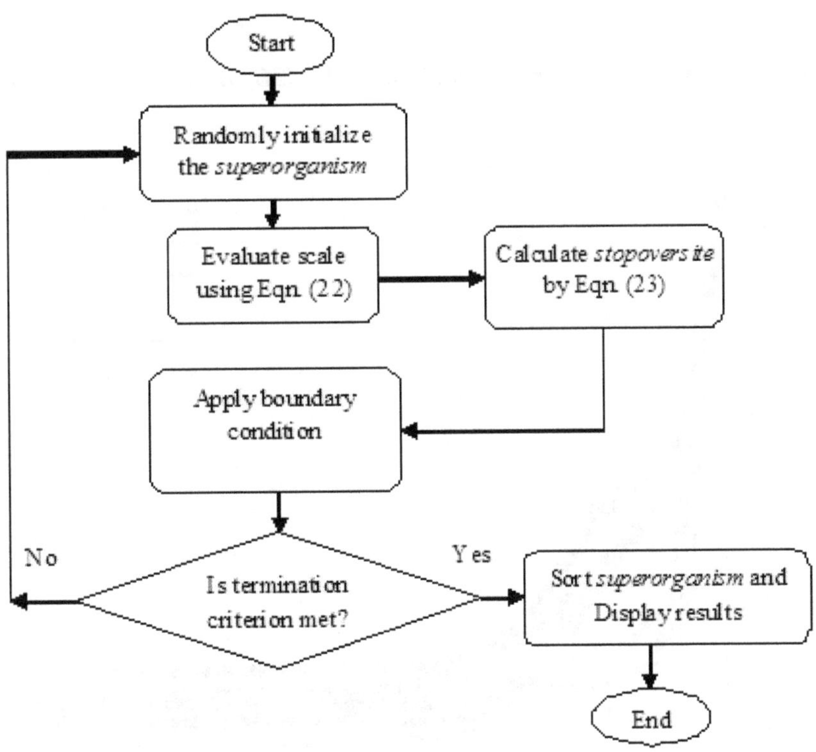

Figure 10. Convergence graph using different algorithms for eigen values under light loading conditions

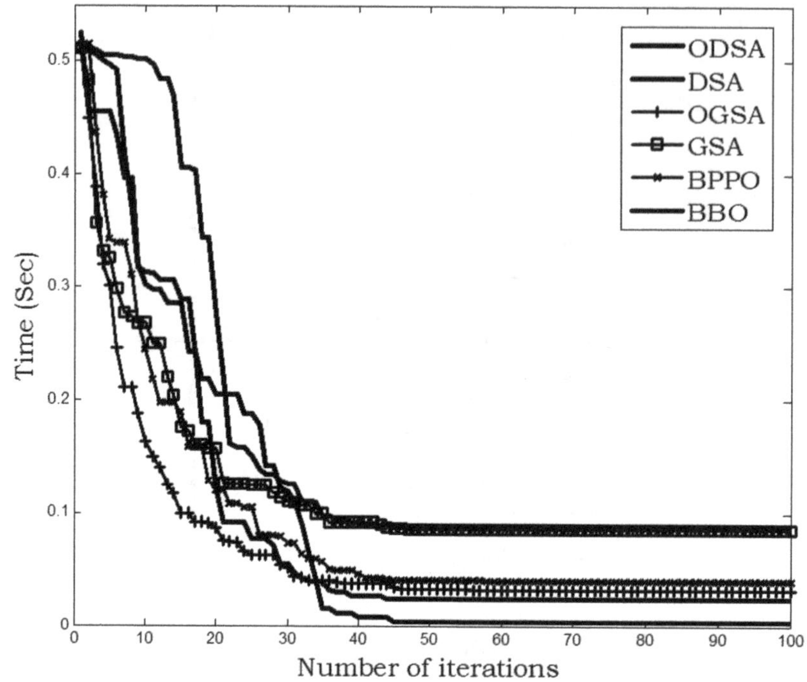

Figure 11. Convergence graph using different algorithms of PSS2B for Case-I ($P=1.2$; $Q=0.6$; $X_e=0.4752$; $E_t=0.5$)

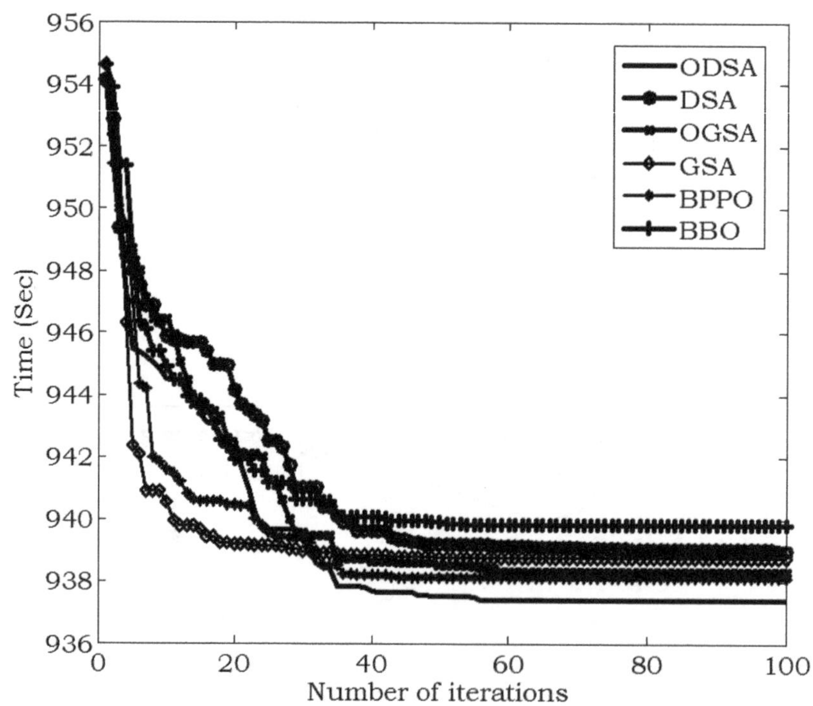

Chapter 14
Extreme Value Metaheuristics and Coupled Mapped Lattice Approaches for Gas Turbine–Absorption Chiller Optimization

Timothy Ganesan
RS Energy, Calgary, Canada

Pandian Vasant
https://orcid.org/0000-0002-0755-1046
University of Technology Petronas, Malaysia

Igor Litvinchev
https://orcid.org/0000-0002-1850-4755
Nuevo Leon State University, Mexico

Mohd Shiraz Aris
TNB Research, Malaysia

ABSTRACT

The increasing complexity of engineering systems has spurred the development of highly efficient optimization techniques. This chapter focuses on two novel optimization methodologies: extreme value stochastic engines (random number generators) and the coupled map lattice (CML). This chapter proposes the incorporation of extreme value distributions into stochastic engines of conventional metaheuristics and the implementation of CMLs to improve the overall optimization. The central idea is to propose approaches for dealing with highly complex, large-scale multi-objective (MO) problems. In this work the differential evolution (DE) approach was employed (incorporated with the extreme value stochastic engine) while the CML was employed independently (as an analogue to evolutionary algorithms). The techniques were then applied to optimize a real-world MO Gas Turbine-Absorption Chiller system. Comparative analyses among the conventional DE approach (Gauss-DE), extreme value DE strategies, and the CML were carried out.

DOI: 10.4018/978-1-7998-3970-5.ch014

INTRODUCTION

As engineering systems become exceedingly complex, there is a growing need for advanced and innovative techniques for optimizing these systems. Metaheuristics have been widely used to tackle such engineering problems – especially in industries related to power generation (Marmolejo *et al.*, 2017). Multiobjective (MO) settings are also becoming commonplace in engineering; where the engineer has to consider multiple target objectives when making critical decisions. MO optimization problems could be broadly divided into two classes. The first is bi-objective problems which have lower complexity as compared to its many-objective counterpart. These problems are endowed with only two objectives and there are well established methods for solving them such as: Non-Dominated Sorting Genetic Algorithm (NSGA-II) (Sadeghi *et al.*, 2014) and Strength Pareto Evolutionary Algorithm (SPEA-2) (Maheta and Dabhi, 2014). As for problems with many objectives, scalarization approaches are among the most effective strategies. Examples of scalarization approaches are the weighted sum (Yang *et al.*, 2013) and the Normal-Boundary Intersection (NBI) (Charwand *et al.*, 2015; Brito *et al.*, 2014) methods. Using scalarization approaches, the multiple objectives are aggregated and the problem is transformed to a single-objective problem. This reduction in complexity then makes the problem easier to solve.

In engineering and other real-world applications, the penalty factor approach is also employed for dealing with situations with multiple objectives. Similar to the weighted sum approach, this method involves the weighted aggregation of the objectives. The distinction with the weighted sum approach is that the penalty factor method converts all the objectives into financial terms forming a single cost function. For instance, in Sheng *et al.*, (2013), an optimization of a distributed generation (DG) power system was carried out using evolutionary algorithms. The penalty factor method was used as a basis to solve the problem; where the DG utilization was maximized while minimizing the system's losses and environmental pollution. Another application using this method could be seen in the works of Daryani and Zare, (2016). In that work, the authors used the Modified Group Search Algorithm (MGSA) in tandem with the penalty factor approach to solve a MO problem in power and emission dispatch.

The Pareto distribution is among the first non-Gaussian distributions encountered in statistics. In economics this distribution shows large amounts of wealth are owned by a smaller percentage of individuals in any society. The idea is often expressed as the 80-20 Rule – 20 percent of the population owns 80 percent of the wealth (Sanders, 1987). Ever since its appearance, the Pareto distribution has found diverse applications – stretching out to other applications besides economics. In quantum statistics the variant of Pareto distributions has been employed to study particle distributions (Biró *et al.*, 2015). These distributions have also seen application in stochastic physical processes; particularly in sub-recoil laser cooling (Bardou, 1995). It has also been used to design and assess software reliability models using failure data (Faqih, 2013; Karagrigoriou and Vonta, 2014). In Fernández *et al.*, (2016), Pareto distributions were used to improve the signal quality of radar systems. In that work the Pareto distribution coupled with a neural network was employed to estimate and eliminate background echoes in radar signals known as 'sea clutter'. Similarly in Lenz (2016), a variant of the Pareto distribution called the Generalized Pareto Distribution in field of microscopy. In that application, the distribution became critical for the construction of an efficient autofocus system.

As with the GPD, the Generalized Extreme Value (GEV) distribution has been utilized in a wide range of applications (Marmolejo and Rodriguez, 2015). For instance in electromagnetic systems, Orjubin (2007) successfully modeled the field in a reverberation chamber. The author used the GEV distribution GEV to determine the maximum field in the chamber. GEC has also been applied to communications technology. In El-Sallabi *et al.*, (2014), the authors employed the GEV to model the radio channel parameter for improving radio channel characterization - showing the significance of extreme values in their application. In power systems, the wind energy potential was assessed using GEV (Ayuketang Arreyndip and Joseph, 2016). In that work, the GEV distribution was employed to analyze the seasonal variation of wind energy potential in Debuncha region, South-West Cameroon. As previously mentioned, extreme value distributions are specifically very handy when dealing with freak probabilistic situations resulting from complex systems. Due to the complex nature of weather systems, GEV has found much application in climate modeling. In the work of Soukissian and Tsalis, (2015), GEV was employed predict extreme wind speeds. Accurate wind speed prediction is critical is critical when it comes to preparing or managing natural disasters. In Cannon *et al.*, (2009), the GEV distribution was coupled with an artificial neural network to model precipitation using non-stationary and complex climatic data. GEV has also been seen to have applications in risk and reliability engineering. In Hirose (1996), the frequency of dielectric voltage breakdown was studied. The authors of that work used the GEV distribution to effectively approximate the voltage breakdown probability.

Coupled Map Lattices (CML) was devised to numerically simulate non-linear dynamics of chaotic systems (Kaneko, 1992; Neufeld and Vicsek, 1995; Neufeld *et al.*, 1996). Since CMLs have been known to exhibit spatiotemporal chaos - they have proven to be very handy to be implemented for simulating systems with such features. For instance in Xingyuan *et al.*, (2018), the authors proposed a spatiotemporal chaotic model which was a logistic-dynamic coupled logistic map lattice (LDCML). The model was shown to be highly effective for modeling spatiotemporal chaotic encryption and secure communications. Deeper exploration into CMLs exhibiting spatiotemporal chaos was carried out in Zhang and Wang, (2014). There the authors proposed a new spatiotemporal dynamics by using a generalized mixed linear-nonlinear coupling. They then showed that the CML is a special case of the proposed dynamics. The authors obtained new chaotic features from the proposed dynamics - which they then applied to cryptography by designing a new encryption scheme. In Zhang, *et al.*, (2018), the spatiotemporal dynamics with mixing degrees of nonlinear chaotic maps for spatial coupling connections based on 2-dimensional CML (2DCML) was investigated. In that work, the authors also investigated the parameter ranges of the proposed system - further simulations showed that the proposed system had a higher percentage of lattices with chaotic behaviors - making highly suitable for applications in cryptography.

Unlike continuous dynamical equations, the computational nature of CMLs make it capable for application to discreet dynamical systems - e.g. predator-prey systems. In Huang *et al.*, (2017), the authors developed a space-time discreet predator-prey model by using a CML. Numerical simulations on the CML model uncovered a variety of spatiotemporal features covering most of the predator-prey types - e.g. regular and irregular patterns of labyrinth, spots, stripes, gaps, spirals, mosaics circles, and many intermediate patterns (in-between). The discreet CML was observed to predict the occurrence of spatiotemporal chaos; showing that the nonlinear mechanisms of the CML was better able to depict and capture the complexity of pattern formation of predator-prey systems. A similar work can be seen in Huang and Zhang (2016) where a novel CML was developed to investigate the spatiotemporal dynamics of a space- and time-discrete predator-prey system. The authors also identified tree Turing instability mechanisms which capture a diversity of complex patterns. The authors found that spatial symmetry breaking on chaotic oscillating states (homogeneous) could result in self-organization of ordered predator-prey systems.

CMLs have also been used to formulate dynamic optimization problems when they represent spatio-temporal fitness landscapes (Richter *et al.*, 2008). In Richter *et al.*, (2008), a CML was employed to analyze such a dynamic fitness landscape (measuring modality, information content, ruggedness and epistasis). These measures were useful in determining the problem hardness and hence its solvability. Despite gauging the difficulty of the optimization problem, such measurements give a very accurate way to gauge the performance of the solution method (evolutionary algorithm).

As for system modeling, in Zhang *et al.*, (2014), a nonlinear coupled reaction-diffusion system in four-dimensional space was modeled using the CML method. In that work, the authors considered chaotic behavior when there is a change in the three variables (known as four dimensional chaos). They characterized the various effects of parameters on chaotic behavior. Two and three dimensional dynamical systems were also simulated in Zhang *et al.*, (2014). A more real-world application model was developed in Huang *et al.*,(2015) - where CMLs were utilized to analyze the cascading failures of two weighted complex networks of urban public transit systems (bus and rail systems). Their analysis uncovered small-world and scale-free network properties of the transit network.

Recently the applications of CMLs have also been extended to encryption technologies (Kumar, 2019; Hao *et al.*, 2017). For instance in Nematzadeh *et al.*, (2018), a hybrid CML and modified genetic algorithm (MGA) approach was developed as a secure encryption method for medical images. The CML was used to generate a number of secure cipher-images; which was then used as the initial population of the MGA. The MGA was then used to increase the entropy of the cipher-images and thus effectively decrease the execution time. In Nematzadeh *et al.*, (2018), the authors showed that the proposed approach is an effective encryption tool which was capable of resisting typical attacks.

This chapter is an extension of the research work conducted in Ganesan *et al.*, (2018) - in this chapter two approaches were utilized to solve optimize the combined Gas Turbine - Absorption Chiller (GT-AC) system. The differential evolution (DE) as well as the CML techniques were employed in tandem with the weighted sum approach. This GT-AC system has four target objectives; GT thermal efficiency, GT heat rate, the absorption chiller's Coefficient of Performance (COP) and the refrigeration load refrigeration load (Q_E). Due to the high complexity of the problem, the DE technique is modified to include extreme value stochastic engines. The Generalized Pareto Distribution and the Generalized Extreme Value Distribution are used to modify the DE approach yielding two techniques; the Generalized Pareto Distribution DE (Pareto-DE) and Generalized Extreme Value DE (Extreme - DE). The results of the application of the extreme value DE and CML techniques to the GT-AC system were obtained. Additionally a comparative analysis on the performances of all the techniques used in this work is presented.

This chapter is organized as follows: Section 2 presents the simulation details of the GT-AC system. The MO problem description of the mentioned system is given in Section 3 while Section 4 provides details on the concept of extreme value DE, CMLs and the measurement metric - hypervolume indicator (HVI). Section 5 consist the findings and the associated detailed analyses. This article ends with some conclusions and perspectives on future works.

COMBINED GT-AC MODEL

The model developed in this work is based on a combined GT - AC system – where the AC is used to cool the compressor inlet air of the GT. The AC system uses lithium bromide solution (LiBr) as a refrigerant. This cooling is aimed to enhance the performance of the GT in terms of: overall thermal efficiency (η_{th})

as well as lowering the heat rate (HR). The GT simulation in this work was modeled based on a V94.2 Siemens GT with a rated speed of 3000 rpm and a rated capacity of 131.5 MW. The GT in the plant as well as in the simulation is operated at a base load 123 MW. The seven day hourly-averaged ambient temperature profile for a day considered in this work is depicted in Table 1:

Table 1. Daily - Average Ambient Temperature

Time (hour)	1	2	3	4	5	6	7	8	9	10
Temperature (°C)	28	28	27	27	26	26	27	27	27	29
Time (hour)	11	12	13	14	15	16	17	18	19	20
Temperature (°C)	30	33	32	33	33	33	32	32	31	30
Time (hour)	21	22	23	24						
Temperature (°C)	31	29	29	29						

The heat exchangers (HEX1 and HEX2) design, pumps and compressor formulations were neglected in this model since they have very minimal effect on the overall optimization. The formulations for the concentration of strong and weak LiBr solution (X_S and X_W) are as follows:

$$X_S = \frac{49.04 + 1.125t_a - t_e}{134.65 + 0.47t_a} \text{ where } X_S = X_1 = X_2 = X_3 \tag{1}$$

$$X_W = \frac{49.04 + 1.125t_g - t_c}{134.65 + 0.47t_g} \text{ where } X_W = X_4 = X_5 = X_6 \tag{2}$$

eak LiBr solution (X_S and X_W) are as follows. For pure water, the concentrations zero off as follows:

$$X_7 = X_8 = X_9 = X_{10} = 0 \tag{3}$$

The pressures for the evaporator and condenser components in the AC are as follows:

$$\log_{10} P_e = 7.8553 - \frac{1555}{t_e + 273.15} - \frac{11.2414\times10^4}{(t_e + 273.15)^2} \text{ where } P_e = P_1 = P_6 = P_9 = P_{10} \tag{4}$$

$$\log_{10} P_c = 7.8553 - \frac{1555}{t_c + 273.15} - \frac{11.2414\times10^4}{(t_c + 273.15)^2} \text{ where } P_c = P_2 = P_3 = P_4 = P_5 = P_7 = P_8 \tag{5}$$

Considering the throttling effect where $H_5 = H_6$ & $H_8 = H_9$, the enthalpies are as follows:

$$H_8 = H^l_{H_2O} = t_c - 25 \quad (6)$$

$$H_{10} = 572.8 + 0.417t_e \quad (7)$$

Transient refrigeration load and its corresponding mass balance formulations are given as the following:

$$\frac{dQ_E}{dt} = \dot{m}_R(H_{10} - H_9) \quad (8)$$

$$\dot{m}_W X_6 + \dot{m}_R X_{10} = \dot{m}_S X_1 = (\dot{m}_W + \dot{m}_R) X_1 \quad (9)$$

$$\therefore m_S = \dot{m}_W + \dot{m}_R \quad (10)$$

$$\dot{m}_W = \left(\frac{Q_E}{H_{10} - H_8}\right)\left(\frac{X_1}{X_4 - X_1}\right) \quad (11)$$

$$\dot{m}_S = \left(\frac{Q_E}{H_{10} - H_8}\right)\left(\frac{X_4}{X_4 - X_1}\right) \quad (12)$$

Since the generator and absorbers have sumps and tubes, some fluid retention occurs in these components. The resulting time delay is modeled according to the formulations built in Kohlenbach and Ziegler, (2008). The tube mass flowrate at the absorber and generator are as follows:

$$\dot{m}_{a,t}(i) = \dot{m}_S + \left[\frac{M_a(i) - M_a(i-1)}{\Delta t}\right] \quad (13)$$

$$\dot{m}_{g,t}(i) = \dot{m}_W + \left[\frac{M_g(i) - M_g(i-1)}{\Delta t}\right] \quad (14)$$

The specific heats and inlet temperatures at the solution HEX are:

$$C_{X_1} = 1.01 - 1.23X_1 + 0.48X_1^2 \quad (15)$$

$$C_{X_4} = 1.01 - 1.23X_4 + 0.48X_4^2 \quad (16)$$

$$t_5 = t_g - E_L(t_g - t_a) \quad (17)$$

$$t_3 = t_a - E_L\left(\frac{X_1}{X_4}\right)\left(\frac{C_{X_4}}{C_{X_1}}\right)(t_g - t_a) \tag{18}$$

where the heat exchanger effectivity, E_L is given by:

$$E_L = \left(\frac{t_g - t_5}{t_g - t_a}\right) = \left(\frac{\dot{m}_S C_{X_1}}{\dot{m}_W C_{X_4}}\right)\left(\frac{t_3 - t_a}{t_g - t_a}\right) \tag{19}$$

The enthalpies at the absorber (inlet and outlet) and enthalpy at the generator outlet are:

$$H_1 = \left(42.81 - 425.92X_1 + 404.67X_1^2\right) + t_a\left(1.01 - 1.23X_1 + 0.48X_1^2\right) \tag{20}$$

$$H_5 = \left(42.81 - 425.92X_4 + 404.67X_4^2\right) + t_5\left(1.01 - 1.23X_4 + 0.48X_4^2\right) \tag{21}$$

$$H_7 = (0.46t_g - 0.043t_c + 572.8) \tag{22}$$

The main component thermal dynamics and balances are given below:

$$\frac{dQ_C}{dt} = \dot{m}_R(H_7 - H_8) \tag{23}$$

$$\frac{dQ_g}{dt} = \dot{m}_W H_5 + \dot{m}_R H_7 - \dot{m}_S H_2 \tag{24}$$

$$\frac{dQ_a}{dt} = \dot{m}_W H_6 + \dot{m}_R H_{10} - \dot{m}_S H_1 \tag{25}$$

$$\Rightarrow \frac{dQ_c}{dt} + \frac{dQ_a}{dt} = \frac{dQ_g}{dt} + \frac{dQ_e}{dt} \tag{26}$$

$$\therefore \frac{dQ_e}{dt} = \frac{dQ_c}{dt} + \frac{dQ_a}{dt} - \frac{dQ_g}{dt} \tag{27}$$

The AC performance is gauged using the real and ideal coefficient of performance (COP) along with the relative performance ratio (RPR):

$$COP = \frac{\text{refrigeration load}}{\text{external heat input}} \tag{28}$$

$$COP = \left(\frac{Q_e}{Q_g}\right) = \left(\frac{dQ_e / dt}{dQ_g / dt}\right) = \frac{(H_{10} - H_8)(X_4 - X_1)}{X_1 H_5 + (X_4 - X_1)H_7 - X_4 H_1} \tag{29}$$

$$COP_i = \frac{T_e(T_g - T_a)}{T_g(T_c - T_e)} \tag{30}$$

$$RPR = \frac{COP}{COP_i} \tag{31}$$

For the GT the subscripts 3, 4, 5 and 6 denote the components; compressor air inlet, combustor inlet, combustor outlet and turbine exhaust outlet. The compression ratio and work done by the compressor is given as follows:

$$P_4 = rP_3 \tag{32}$$

$$\dot{W}_{comp} = \dot{m}_{air} C_{p,air} (T_4 - T_3) \tag{33}$$

The combustion chamber temperatures, pressures, heat generated and efficiency is as follows:

$$T_4 = \frac{T_3}{\eta_{comp}} \left[\left(\frac{P_4}{P_3}\right)^{\frac{k-1}{k}} - 1 \right] + T_3 \tag{34}$$

$$P_5 = P_4 - \Delta P_{comb} \tag{35}$$

$$Q_{in} = \dot{m}_{air} C_{p,air} (T_5 - T_4) \tag{36}$$

$$\eta_{cc} = \frac{\dot{Q}_{in} / LHV}{\dot{m}_{fuel}} \tag{37}$$

The exhaust discharge temperature and the GT power outputs are as follows:

$$T_6 = T_5 - \eta_{turb} T_4 \left[1 - \left(\frac{1}{P_5 / P_6} \right)^{\frac{k-1}{k}} \right] \tag{38}$$

$$\dot{W}_{turb} = \dot{m}_{egy} C_{p,ave} \left(T_5 - T_6 \right) \tag{39}$$

$$\dot{W}_{net} = \dot{W}_{turb} - \dot{W}_{comp} \tag{40}$$

where,

$$\dot{m}_{egy} = \dot{m}_{fuel} + \dot{m}_{air} \tag{41}$$

The GT performance indicators are:

$$SFC = \frac{3600 \dot{m}_{fuel}}{\dot{W}_{net}} \tag{42}$$

$$HR = SFC \cdot LHV \tag{43}$$

$$n_{th} = \frac{3600}{SFC \cdot LHV} \tag{44}$$

PROBLEM DESCRIPTION

A gas turbine (GT) system is the core energy generator in gas power plants. For optimization, waste heat from the GT exhaust stack is fed into an absorption chiller (AC). The AC cools the air before entering the GT's compressor. This cooling reduces the heat rate and increases the GT efficiency. This combined GT-AC system is optimized in a multiobjective setting while considering power limitations enforced by contractual agreements with the customer. The novel optimization problem is formulated where the constraints were based on the chiller design specifications. Due to the dynamic nature of the model, the objective functions are time-averaged. In addition of having four objectives, the complexity of the problem is magnified by the mixed nature of the target objectives (max-min):

$Max \rightarrow$ Overall Thermal Efficiency, n'_{th}
$Min \rightarrow$ Heat Rate, HR'
$Max \rightarrow$ Coefficient of Performance, COP'
$Max \rightarrow$ Refrigeration Capacity, Q'_e

subject to the following constraints:

$$32\leq t_a\leq 38;\ 5\leq t_e\leq 18;\ 480\leq t_g\leq 1000;\ 32\leq t_c\leq 38;\ 0<E_L<1;\ 0\leq M_a\leq 200;\ 0\leq M_g\leq 200;\ Q_e\leq 2040;\ 129\leq \dot{W}_{net}\leq 139 \tag{45}$$

METHODOLOGY

Extreme Value Stochastic Engines

Stochastic engines (or the random number generator segment) of metaheuristic algorithms have been seen to play an important role in the computational process – significantly influencing the quality of the achieved optimal solution. These stochastic engines usually generate random numbers produced from Gaussian or Normal distribution functions. In past works, it has been shown that stochastic engines which deviate from the Gaussian could improve the overall optimization results; especially for highly complex problems (Ganesan *et al.*, 2016a; Ganesan *et al.*, 2016b). Such deviations from the median of standard probability distributions are known as Extreme Value Theory (Bensalah, 2000; Embrechts *et al.*, 1999). This theory explores the outcomes of events from an extreme probabilistic approach.

In most real-world optimization problems, the objective landscape is largely unknown. Similarly, the existence of global optima could not be determined. Conventional metaheuristics assume that a Gaussian random number generator is sufficient to begin the search process. However, for extremely complex objective landscapes this assumption may breakdown. A Gaussian random number generator may not be able to place the start point of the algorithms at best place in the beginning of the search. This bad initialization could then result in a flawed or inefficient search for optimum values by the algorithm.

The Generalized Pareto distribution is a generic version of the conventional Pareto distribution originally conceived in economics as an extreme value (or non-Gaussian) distribution In this work, the Generalized Pareto Distribution was utilized as a stochastic engine for the DE approach (Pikands, 1975; Del Castillo 2009). The cumulative distribution function for the Generalized Pareto Distribution, P for the random variable, $x\in X$ is given as follows:

$$P(x\in X;k,\sigma)=\begin{cases}1-[1-(kx/\sigma)]^{1/k}; & k\neq 0 \quad \sigma>0\\ 1-e^{-x/\sigma} ; & k=0 \quad \sigma>0\end{cases} \tag{46}$$

where σ and k are the scale and shape factors. In this work we consider k = 0; thus we take the lower equation of the piece-wise representation of equation (46).

The Generalized Extreme Value Distribution is combination of three extreme value functions; Gumble, Weibull and Fréchet distributions (Walshaw, 2013). The cumulative distribution function for the Generalized Extreme Value Distribution function used in this work, G for the random variable, $x\in X$ is given as follows:

$$G(x \in X;\mu,\sigma) = \exp\left[-\exp\left(\frac{-x-\mu}{\sigma}\right)\right] \tag{47}$$

where σ and μ are the scale and location parameters respectively. In this work we consider σ=μ=1.

EXTREME VALUE DIFFERENTIAL EVOLUTION

Introduced in the nineties, DE is an evolutionary-type optimization technique (Price *et al.*, 2006). The core of DE is the combination of perturbative methods and the evolutionary approach. DE initiates by producing a minimal population of four individuals denoted, *P*. These individuals are real-coded vectors with some specified size, *N*. The DE algorithm is augmented to enhance its optimization capabilities by the modification of its stochastic engine. The mutation operator functions to diversify the population further - enabling it to spread throughout the search space. First, the population of vectors, x^G_i was generated. Then, one principal parent, x^p_i and three auxiliary parents x^a_i are randomly selected. Differential mutation is performed and the mutated vector, V_i is generated. By recombining V_i with x^p_i (cross-over), the child trial vector, x^{child}_i is created:

$$V_i = x^a_1 + F(x^a_2 - x^a_3) \tag{48}$$

$$CR: V_i \otimes x_i^P \rightarrow x_i^{child} \tag{49}$$

where *F* is the real-valued mutation amplification factor which is usually between 0 and 1. The probability of the cross-over, *CR* is the probability of amount of codes from the mutated vector, V_i with respect to the principal parent, x^p_i being inherited by the child trail vector. The *CR* is an input parameter set by the user. The fitness function for the child trial vector, x^{child}_i is evaluated. Survival selection in the next generation is performed via 'knock-out' competition. The parameter setting employed for the DE technique is given in Table 2:

Table 2. Differential Evolution (DE) Parameter Settings

Parameters	Values
Individual Size, *N*	6
Population Size, *P*	7
Mutation amplification (scaling) factor, *F*	0.3
Stopping Criterion	Convergence check
Cross-over Probability, *CR*	0.6667

In this work the Generalized Pareto and Generalized Extreme Value Distributions are used as stochastic engines in the DE technique. This modification results in the formation of two algorithms; the Generalized Pareto Distribution DE (Pareto-DE) and Generalized Extreme Value DE (Extreme - DE). Figure 1 depicts the components constructing this modified DE framework.

Figure 1. Extreme Value Differential Evolution

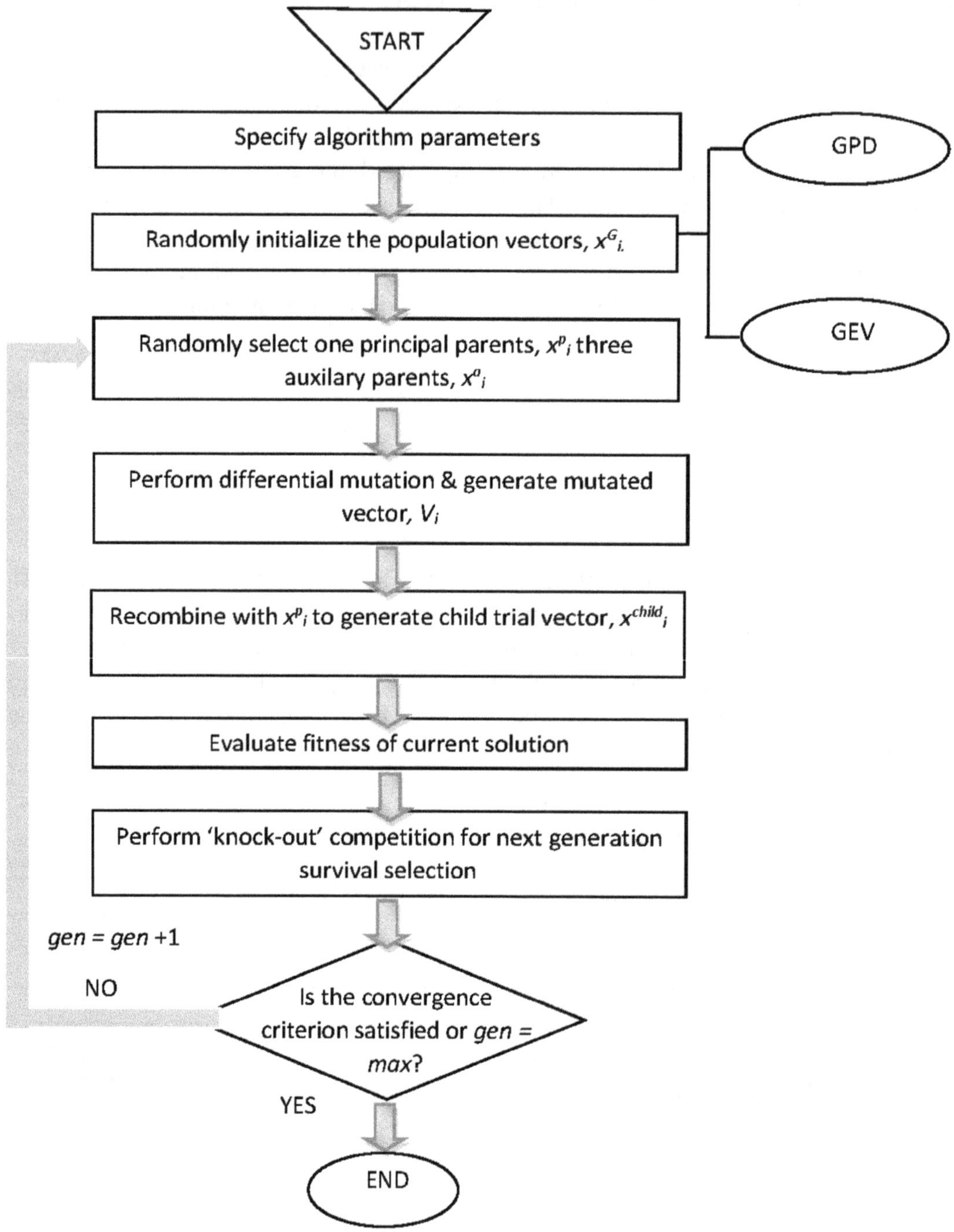

As seen in Figure 1, the two extreme value distributions (GPD and GEV) are supplemented into the existing DE algorithm. The standard random number generator (Gaussian) is replaced using the extreme value distribution. Other components of the DE algorithm (fitness function, selection and mutation) are maintained. This modification varies the DE technique's search capability aiming to improve its propensity in reaching the optima. The same procedure could also be done to any other metaheuristic approach.

COUPLED MAP LATTICE APPROACH

CMLs have been explored extensively simulate non-linear dynamics of chaotic systems. Nevertheless their applications have expanded across various fields: optimization (Zelinka *et al.*, 2007), system identification/modeling (Garcia-Morales *et al.*, 2017) and cryptography (Hao *et al.*, 2017; Wang *et al.*, 2019). Fundamentally a CML consists of interacting elements of a given discreet-time dynamical map - resulting in a multidimensional dynamical system. The CML model is usually constructed in very specific way to simulate a dynamical system. For instance, the dynamical system's field variables (macroscopic) are set on a lattice. The system is then decomposed into independent processes where each process is represented as a procedure in the CML. Then each state variable (lattice point) and coupling term are represented as a nonlinear transformation. Finally each procedure is executed iteratively. CMLs are generally used to demonstrate chaotic behavior - in this work, a logistic map (which is best suited for this purpose) is used where by fixing the constant, $\theta>3.57$, chaotic behavior can be invoked. The iterative logistic map is given as follows:

$$y_{j+1} = \theta \cdot y_j(1 - y_j) \tag{50}$$

where $j \in [1,m]$ is the number of iterations and r is a constant. On the other hand, the recursive map of the CML is given as follows:

$$y_{j+1} = \varepsilon \cdot \lfloor r y_j (1 - y_j) \rfloor_s + (1-\varepsilon) \lfloor r y_j (1 - y_j) \rfloor_{s-1} \tag{51}$$

where $\varepsilon \in [0,1]$ and s denotes the index for the vertices on the lattice. The CML map in equation (49) is a convex map weighted with ε known as the coupling parameter. The second term is from the neighboring lattice points. The convex map approach is the unidirectional coupling. Other coupling methods are mean-field-type global coupling, derivative coupling and inclusion of a conservation law also exist, although in this work unidirectional coupling was employed.CMLs have been employed in the field of optimization and system modeling. In Zelinka *et al.*, (2011), the idea that CMLs could be represented as an analogue to evolutionary optimization techniques was proposed. In that work, the authors presented a novel method for visualizing the dynamics of evolutionary algorithms in the form of complex networks. The analogy between the vertex of a complex network (CML) and the populations in a arbitrary evolutionary algorithm is discussed in detail - where the CML is controlled via chaos control.

Table 3.

Coupled Map Lattice - Optimization Algorithm
START
1. Initialize CML parameters:
2. Logistic map parameter (θ), number of vertices on the lattice (N), coupling parameter (ε)
3. **IF** Iteration=1:
Randomly initialize lattice values, *y* (*j*=1)
4. **IF** Iteration>1 and fit_func=1:
Back-substitute the improved computed lattice values
5. **IF** Iteration>1 and fit_func=0:
Back-substitute the lattice values from the previous iteration (*j*-1)
(so it does not lead the CML algorithm towards non-optimal regions)
6. Compute new lattice values (*y*) using CML (with logistic map)
7. Check fitness of solution candidates (binary decision function)
8. **IF** Candidate solution is **improved**:
8.1. Fitness is **Satisfied**:
fit_func=1
HALT PROGRAM
8.2. Fitness is **NOT Satisfied**:
fit_func=1
Repeat Step 4
9. **IF** Candidate solution is **NOT improved**:
9.1. Fitness is **Satisfied**:
fit_func=0
HALT PROGRAM
9.2. Fitness is **NOT Satisfied**:fit_func=0
Repeat Step 5
END

COUPLED MAP LATTICE APPROACH

The Hypervolume Indicator (HVI) is the only indicator which is strictly Pareto-compliant. It can be utilized for measuring the quality of solution sets generated when performing MO optimization (Zitzler and Thiele, 1998; Beume et al., 2007). Strictly Pareto-compliant is defined such that if there exists two (or more) solution sets to a particular MO problem, then the solution set that dominates the other would return a higher value from the indicator. The HVI measures the hypervolume of the dominated segment of the objective space and thus could be implemented for multi-dimensional scenarios. When using the HVI, a reference point (nadir point) needs to be defined. Relative to this point, the volume of the space of all dominated solutions can be measured. The HVI of a solution set $x_d \in X$ can be defined as follows:

$$HVI(X) = vol\left(\bigcup_{(x_1,\ldots x_d)\in X} [r_1, x_1]\times\ldots\times[r_d, x_d] \right) \tag{24}$$

where $r_1,\ldots,r_d$ is the reference point and vol(.) being the usual Lebesgue measure. In this work the HVI is used to measure the quality of the approximation of the Pareto front generated by the computational techniques strictly Pareto-compliantwhen applied in conjunction with the weighted sum approach.

COMPUTATIONAL OUTCOME

The engineering problem tackled in this work involves four nonlinear objective functions along with the required design and operational constraints. The computational strategies were applied to the problem using the weighted-sum approach as the basis. The weighted sum approach simplifies the problem reducing it to a single aggregate objective function.The aggregate objective function obtained using the mentioned approach using the following inner product:

$$F = \begin{pmatrix} w_1 & w_2 & w_3 & w_4 \end{pmatrix} \begin{pmatrix} n'_{th} \\ HR' \\ COP' \\ Q'_e \end{pmatrix} \quad : \quad \forall w_i \in (0,1) \tag{52}$$

The second and fourth objectives, HR and Refrigeration Capacity, Q'_e have values in the scale of 10^3. To avoid a biased evaluation besides misrepresenting of the aggregated objective function, both objective functions are normalized. The normalization used the high HR value of 8000 kJ/kWhr and Refrigeration Capacity, Q_e of 2000. The following equations depict this normalization procedure:

$$HR' = HR' / 8000 \tag{53}$$

$$Q'_e = Q'_e / 2000 \tag{54}$$

The optimization is carried out such that the best chiller design parameters are selected - which caters accordingly to the GT air inlet cooling requirement. The refrigeration capacity Q'_e of the chiller is maximized such that the design constraints and other requirements are not violated.. This way the chiller would be engineered in a cost-efficient manner. In this work, the DE technique is modified using two extreme value distributions (Generalized Pareto Distribution and Generalized Extreme Value Distribution) yielding two algorithms the Pareto-DE and Extreme–DE respectively. As a benchmark, the conventional DE using a standard Gaussian stochastic engine (termed Gauss-DE) is applied to the problem. This way the three approaches, Pareto-DE, Extreme-DE and Gauss-DE are analyzed and compared. As an extension to this work, the CML (which is analogue to evolutionary algorithms) is also applied to the problem. Due to the complex structure of the GT-AC problem, it is highly possible that the solution space may be better navigated using a chaotic map (CML) as compared to the DE variants. The concept of solution dominance is employed to evaluate the solution quality (Ganesan *et al.*, 2013; Ganesan *et al.*, 2015). For this evaluation, the hypervolume indicator (HVI) is employed where the nadir point (0.1, 1, 0.1, 0.1) was employed (Emmerich *et al.*, 2005).

Individual solution obtained at various scalar values were used to construct the Pareto frontier. It is important to note that although the Pareto–DE technique and Pareto frontier sound similar but in reality these are two distinct concepts. The Pareto-DE is a technique endowed with a stochastic engine which randomly initializes the algorithm based on the Generalized Pareto Distribution. On the other hand the Pareto frontier is way to depict and analyze solutions of MO problems. Executing randomized techniques such as evolutionary algorithms require multiple program executions. In this work, each individual solution was obtained after five independent runs for a single weighting factor and the best solution is selected for analyses. The GT-AC system is modeled as dynamic system running at a 24-hour cycle. Due to this the objective values and other parameters are hourly time-averaged. The individual solutions obtained are ranked for each Pareto frontier (best, median and worst). All techniques employed in this work were designed using the C++ programming language. The software was developed and ran on a personal computer (PC) with an Intel® Core ™ i7 – 3520M processor operating at 2.9 GHz. The ranked solutions obtained using the Gauss-DE approach are given in Table 4. The related Pareto frontier is depicted in three scatter plots in Figure 2:

Figure 2. Pareto frontier constructed using Gauss-DE

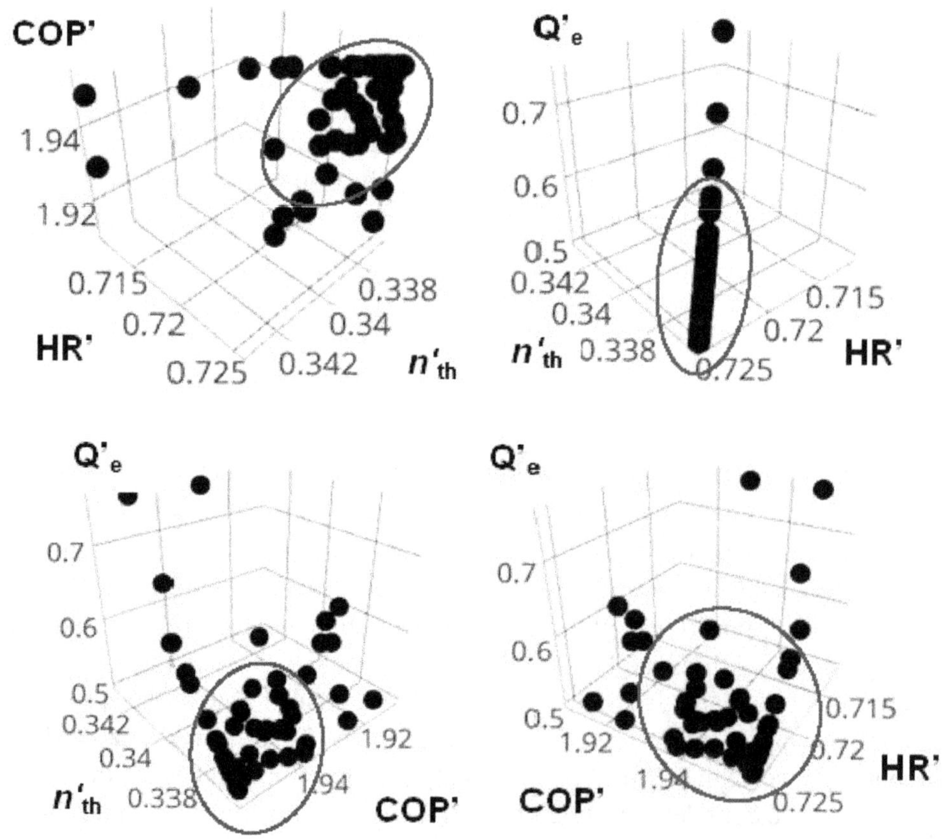

The best, median and worst solution achieved by the Gauss-DE is taken at the weights (0.3, 0.3, 0.3, 0.1), (0.3, 0.4, 0.2, 0.1) and (0.5, 0.1, 0.1, 0.3) respectively. Tables 4 – 6 present the graded individual solutions produced by the CDDE obtained at various degrees of chaos while Figure 3 illustrates the corresponding Pareto frontiers. Table 5 and Figure 3 gives the ranked solutions and the Pareto frontier obtained using the Extreme – DE:

Table 4. The Ranked Individual Solutions Obtained using the Gauss-DE technique

Description		*Best*	*Median*	*Worst*
Objective Functions	n'_{th}	0.34322	0.337385	0.336577
	HR'	0.711423	0.723731	0.725469
	COP'	1.94784	1.94352	1.93796
	Q'_e	0.754	0.53105	0.500247
Decision Variable	t_a	37.2475	36.4172	35.6108
	t_e	7.64611	7.5598	7.61699
	t_g	486.461	485.598	486.17
	t_c	37.3539	37.4402	37.383
	E_L	0.564611	0.55598	0.561699
	M_a	105.709	104.015	103.781
	M_g	105.709	104.015	103.781
	W_{net}	137.288	134.954	134.631
	ΔT	11.3901	8.01133	7.54347
Metric	*HVI*	84.821	52.114	47.778

The associated weights for the ranked individual solutions obtained using the Extreme-DE are best - (0.1 0.4, 0.3, 0.2), median - (0.1, 0.7, 0.1, 0.1) and worst - (0.5, 0.3, 0.1, 0.1). The ranked solutions obtained using the Pareto-DE approach is given in Table 5 while Figure 4 depicts the associated Pareto frontier. The weight values for the ranked individual solutions; best, median and worst obtained using the Pareto-DE are (0.3, 0.2, 0.4, 0.1), median - (0.1, 0.1, 0.2, 0.6) and worst - (0.3, 0.1, 0.5, 0.1).

The solutions obtained using the CML strategy is given in Table 7 while Figure 5 presents the Pareto frontier. The weight values for the ranked individual solutions; best, median and worst obtained using the Pareto-DE are (0.3, 0.2, 0.4, 0.1), median - (0.3, 0.1, 0.4, 0.2) and worst - (0.5, 0.1, 0.1, 0.3).

The most dominant individual solution was achieved by the Pareto-DE followed by the CML, the Gauss-DE and the Extreme-DE techniques respectively. The dominance levels were measured using the HVI. The individual solutions produced by the Pareto-DE outrank that of the CML, Extreme-DE and Gauss-DE by 43.3%, 130.787% and 102.63% respectively. It is observed that in terms of individual solution, CML outweighs both the Gauss-DE and Extreme-DE. The overall frontier dominance attained by all four approaches is given in Figure 6:

Figure 3. Pareto frontier constructed using Extreme-DE

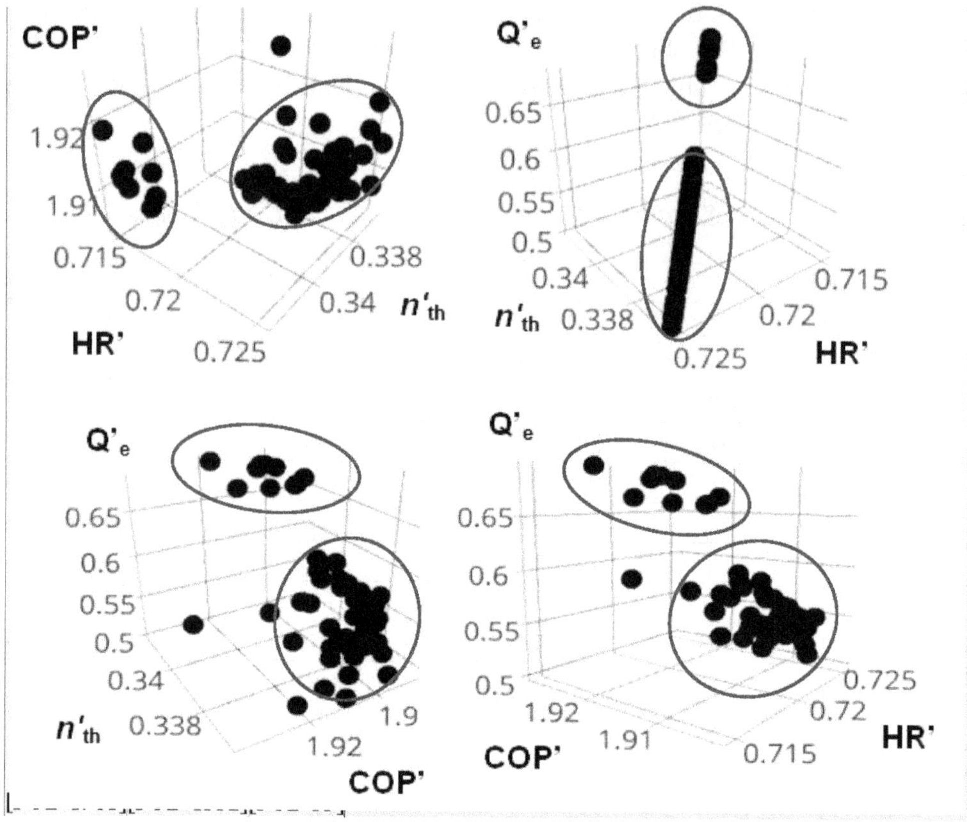

Table 5. The Ranked Individual Solutions Obtained using the Extreme-DE technique

Description		Best	Median	Worst
Objective Functions	n'_{th}	0.341698	0.338127	0.336598
	HR'	0.714591	0.722142	0.725423
	COP'	1.91877	1.90537	1.91441
	Q'_e	0.69358	0.5589	0.50072
Decision Variable	t_a	33.1331	31.1403	32.3352
	t_e	8.01339	8.10903	7.94516
	t_g	490.134	491.09	489.452
	t_c	36.9866	36.891	37.0548
	E_L	0.601339	0.610903	0.594516
	M_a	105.267	104.231	103.787
	M_g	105.267	104.231	103.787
	W_{net}	136.679	135.251	134.639
		10.5092	8.44117	7.55578
Metric	HVI	74.47281	54.81707	47.23357

Table 6. The Ranked Individual Solutions Obtained using the Pareto-DE technique

Description		Best	Median	Worst
Objective Functions	n'_{th}	0.356412	0.342766	0.337216
	HR'	0.685084	0.712365	0.724094
	COP'	1.94309	1.94373	1.94837
	Q'_e	1.25486	0.736584	0.524627
Decision Variable	t_a	37.0995	36.6551	37.0882
	t_e	8.04374	7.69218	7.48779
	t_g	490.437	486.922	484.878
	t_c	36.9563	37.3078	37.5122
	E_L	0.604374	0.569218	0.548779
	M_a	109.537	105.577	103.966
	M_g	109.537	105.577	103.966
	W_{net}	142.565	137.106	134.886
		19.0287	11.1273	7.91349
Metric	HVI	171.8732	81.95635	51.36895

Figure 4. Pareto frontier constructed using Pareto-DE

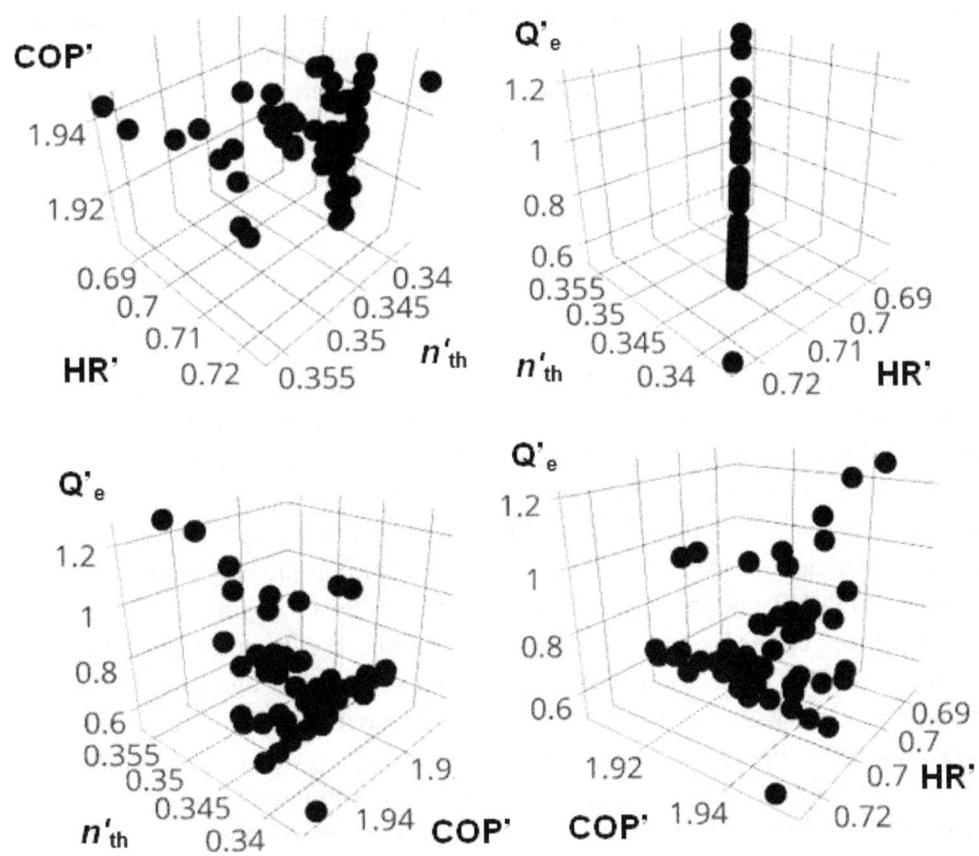

Table 7. The Ranked Individual Solutions Obtained using the CML technique

Description		Best	Median	Worst
Objective Functions	n'_{th}	0.349593	0.343527	0.337296
	HR'	0.698449	0.710786	0.72392
	COP'	1.88361	1.91679	1.94059
	Q'_e	0.993446	0.76337	0.527683
Decision Variable	t_a	28.5522	32.9269	36.0052
	t_e	8.7006	8.08709	7.59845
	t_g	497.006	490.871	485.984
	t_c	36.2994	36.9129	37.4016
	E_L	0.67006	0.608709	0.559845
	M_a	107.558	105.798	103.99
	M_g	107.558	105.798	103.99
	W_{net}	139	137.411	134.919
		15.0807	11.5682	7.96041
Metric	HVI	119.9393	84.88423	51.57086

In Figure 4 that the overall frontier dominance is led by the CML technique followed by the Pareto-DE approach. Although the frontier produced by the Extreme-DE outweighs the Gauss-DE, the difference is very little (approximately 4.516%). The CML on the other hand leaped far ahead from the other three approaches in terms of frontier dominance with the following differences:

$$\text{Pareto-DE} \approx 32.55\% \tag{55}$$

$$\text{Extreme-DE} \approx 57.453\% \tag{56}$$

$$\text{Gauss-DE} \approx 59..292\% \tag{57}$$

The computational time taken for the CML and Pareto –DE to construct the entire Pareto frontiers were 2.585 and 3.765 seconds respectively while the Extreme-DE and Gauss-DE produced the frontiers in 61.147 and 4.875 seconds respectively. Note that the Extreme-DE took the longest computational time; more than a complete minute.

In Figure 2, it can be seen that the solution points on the frontier generated using the Gauss-DE are concentrated on certain regions in the objective space. The solutions are particularly dense at areas with high COP, low n_{th}, high HR and low Q_e (indicated by the circles on Figure 2).The same goes for the Pareto frontier produced by the Extreme-DE (Figure 3). In this case the solutions are conglomerated in two different sections of the objective space – similarly marked by circling the regions on Figure 3. The first section, have solutions with dispersed COP values while having low HR, high Q_e and high n_{th} values. As for the second section, the solution points have well dispersed n_{th} and HR. Nevertheless these points center on regions with high COP and low Q_e. The most well spread Pareto frontier was produced by the Pareto-DE technique (portrayed in Figure 4). As the technique approximates the true frontier, the level of frontier dominance would increase. A diversely spread frontier would cover most of objective space closely estimating the true Pareto frontier. This is why the Pareto-DE has a diversely spread and the second most dominant frontier as compared to the other approaches employed in this work. The Gauss-DE produced the worst frontier since its solutions are narrowly focused on a single region. Although not as diverse as the Pareto-DE, the Extreme-DE could sufficiently diversify its Pareto frontier - not limiting its solution spread to a single region in the objective space. This diversification enables it to outweigh the overall frontier dominance of the solutions produced by the Gauss-DE.

As for the frontier generated by the CML (Figure 5), the solutions seem to be concentrated in specific regions of the objective space as opposed to diverse spread (e.g. frontier generated by the Pareto-DE). Although the solutions produced by the CML are localized to certain regions similar to those produced by the Extreme-DE and Gauss-DE, the solutions are concentrated at the optimal regions of the objective space. Therefore it is highly possible that the structure of the objective space has some inherent chaotic characteristics. The CML's effectiveness could be attributed to this factor - where the CML exploits these chaotic characteristics to reach the optimal regions of the objective space.

Figure 5. Pareto frontier constructed using CML technique

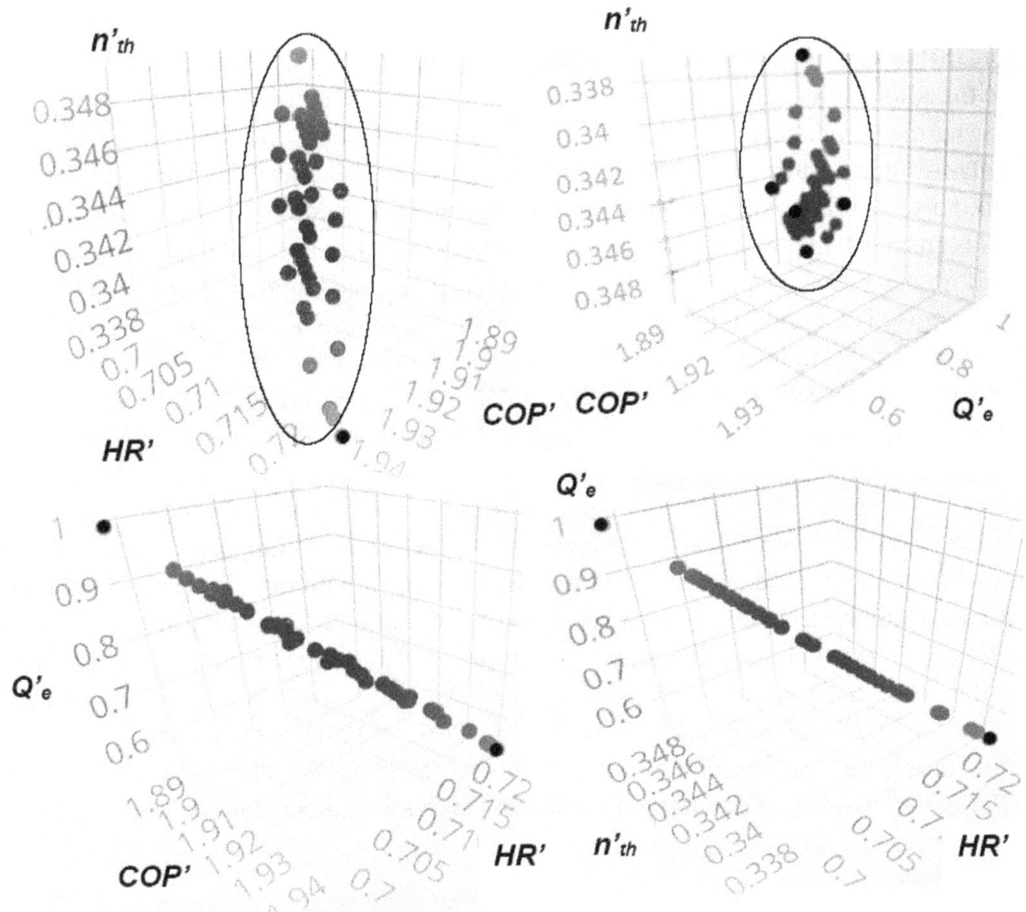

Figure 6. Dominance ranking of the generated Pareto Frontiers

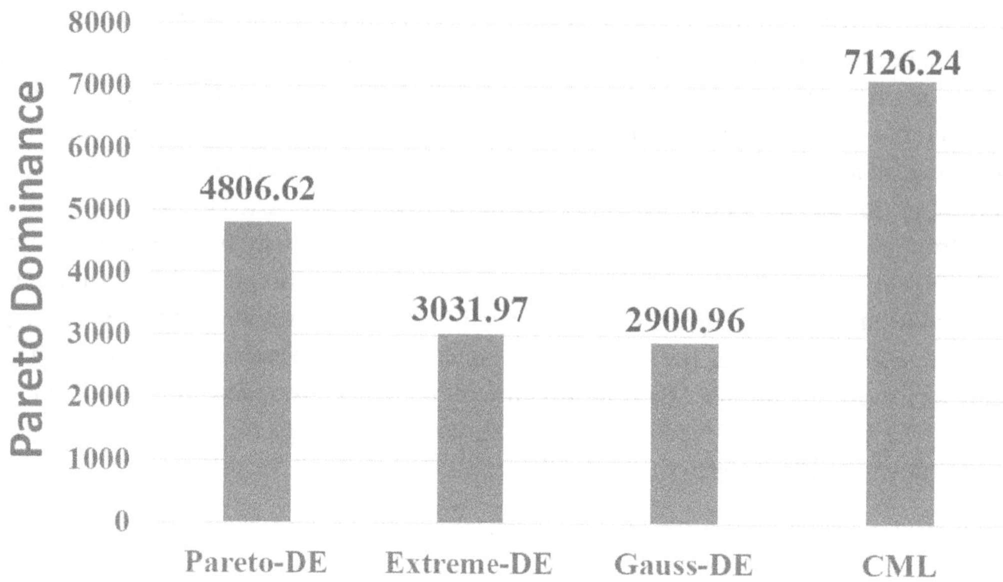

Therefore in this work extreme value stochastic engines and chaos simulators (e.g. CML) are observed to influence the solution quality. However, the selection of the right extreme value distribution is critical for empowering the search capability of the metaheuristic. Execution time is a seminal criterion when evaluating algorithmic performance. The fastest execution time is attained by the CML. This could be attributed to the structural compatibility of the computational method to the objective space - chaotic structure. This ensures that the CML undergoes a seamless exploration of the objective space. The computational effort undergone by the three DE variants also reflect the Pareto dominance achieved by their respective frontiers. The Pareto-DE approach sweeps through the objective space smoothly, diversely spreading its solution throughout. Thus, the Pareto-DE technique does not experience stagnation at any region in the objective space. This may contribute to the speedy convergence taking up minimal execution time as compared to the Extreme-DE and Gauss –DE approaches. The Gauss-DE comes third after the Pareto-DE approach. The Gauss-DE stagnates at only one region of the objective space. Due to its limited spread it completes its exploration swiftly bringing the program to a halt. Albeit the lengthy execution time of the Extreme-DE, its search covers a wider area as compared to the Gauss-DE. The sacrifice in computational efficiency is too costly in comparison with the Pareto-DE algorithm.

The differences in the techniques employed in this work are their respective stochastic engines and chaotic structure. Thus the algorithmic complexity of the extreme value techniques utilized in this work does not differ much from one another. However the CML is the least complex algorithm compared to all the other techniques employed in this work. The CML and all the DE variants performed stable computations during program executions. The solutions generated by all the techniques employed in this work did not violate any of the defined constraints and thus remained feasible.

CONCLUDING REMARKS AND FUTURE RESEARCH DIRECTIONS

Based on the numerical experiments performed, it was observed that the stochastic engines and chaotic structure of the computational techniques do significantly influence the optimization results. The results in this case were gauged using the level of dominance measured using the HVI. Nevertheless the type of stochastic engine used also heavily affects the optimization results. As seen in the previous section, the frontier produced by the CML technique outweighs that of the Pareto-DE, Extreme-DE and Gauss-DE by a large margin. On the other hand, when it comes to individual solutions, the Pareto-DE outrank that of the CML, Extreme-DE and Gauss-DE by 43.3%, 130.787% and 102.63% respectively. This shows that the wrong choice of computational technique could reduce algorithmic performance to such an extent that it could produce results which are worse than conventional approaches (e.g. Gauss-DE). The CML effectively solved the complex real-world MO optimization problem: the combined GT-AC system.

In future works, other extreme value (or non-Gaussian) distributions could be employed to enhance metaheuristic approaches. In this work, the factors of the probability distribution (shape, scale and location) are fixed. The effect of these factors on the algorithmic performance could also be analyzed. This would be particularly useful for tackling real-world problems which are highly complex.

Additionally, future works could also investigate the mathematical aspects of using specific extreme-value distribution to tackle certain classes of optimization problems. This way the mathematical enhancement in the metaheuristic would exactly match the attempted problem. Further studies into the mathematical foundations of extreme value distributions could also provide useful insights in designing extreme value metaheuristics (Nakajima *et al.*, 2017; Ning and Bloomfield, 2017). Other optimization algorithms could be embedded to find for optimal factors for the distributions in the stochastic engines. The strategy of switching stochastic engines could also be implemented using other evolutionary approaches besides DE. This way extreme value distributions may also be tested on metaheuristics such as particle swarm optimization (PSO) (Singh *et al.*, 2015), Bacteria Foraging Optimization (Nasiruddin *et al.*, 2015), Gravitational Search Algotithm (Ganesan and Elamvazuthi, 2017) and other bio-inspired strategies (Plucar *et al.*, 2017).

Hybrid techniques between extreme value stochastic engines and the CML could also be developed and tested. It may be possible that the correct combination of these techniques may amplify the strengths of the individual methods (chaotic structural match, extreme value stochastic enhancement, optimal frontier construction) and cancel out their weaknesses (stagnation at suboptimal regions, slow convergence and inefficient frontier construction). Such advance modifications and real-world implementation of computational techniques could be seen in Vasant et al. (2018), Kharchenko and Vasant (2018) and Islam et al. (2020).

REFERENCES

Ayuketang Arreyndip, N., & Joseph, E. (2016). Generalized Extreme Value Distribution Models for the Assessment of Seasonal Wind Energy Potential of Debuncha, Cameroon. *Journal of Renewable Energy*, *2016*, 2016. doi:10.1155/2016/9357812

Bardou, F. (2005). Fractal time random walk and subrecoil laser cooling considered as renewal processes with infinite mean waiting times. In *Chaotic Dynamics and Transport in Classical and Quantum Systems* (pp. 281–301). Springer Netherlands. doi:10.1007/1-4020-2947-0_12

Bensalah, Y. (2000). *Steps in applying extreme value theory to finance: a review*. Vancouver: Bank of Canada.

Brito, T. G., Paiva, A. P., Ferreira, J. R., Gomes, J. H. F., & Balestrassi, P. P. (2014). A normal boundary intersection approach to multiresponse robust optimization of the surface roughness in end milling process with combined arrays. *Precision Engineering*, *38*(3), 628–638. doi:10.1016/j.precisioneng.2014.02.013

Charwand, M., Ahmadi, A., Heidari, A. R., & Nezhad, A. E. (2015). Benders decomposition and normal boundary intersection method for multiobjective decision making framework for an electricity retailer in energy markets. *IEEE Systems Journal*, *9*(4), 1475–1484. doi:10.1109/JSYST.2014.2331322

Daryani, N., & Zare, K. (2016). Multiobjective power and emission dispatch using modified group search optimization method. *Ain Shams Engineering Journal*.

Del Castillo, J., & Daoudi, J. (2009). Estimation of the generalized Pareto distribution. *Statistics & Probability Letters*, *79*(5), 684–688. doi:10.1016/j.spl.2008.10.021

El-Sallabi, H., Abdallah, M., Chamberland, J. F., & Qaraqe, K. (2014, December). A statistical model for delay domain radio channel parameter affected with extreme values. In *Antennas and Propagation (ISAP), 2014 International Symposium on* (pp. 165-166). IEEE. 10.1109/ISANP.2014.7026582

Embrechts, P., Resnick, S. I., & Samorodnitsky, G. (1999). Extreme value theory as a risk management tool. *North American Actuarial Journal: NAAJ*, *3*(2), 30–41. doi:10.1080/10920277.1999.10595797

Emmerich, M., Beume, N., & Naujoks, B. (2005). An EMO Algorithm Using the Hypervolume Measure as Selection Criterion. *Conference on Evolutionary Multi-Criterion Optimization (EMO 2005)*, 62–76. 10.1007/978-3-540-31880-4_5

Faqih, K. M. (2013). The Performance of Software Reliability Models: A View Point. *International Journal of Performability Engineering*, *9*(4), 375–386.

Ganesan, T., Aris, M. S., & Vasant, P. (2018). Extreme value metaheuristics for optimizing a many-objective gas turbine system. *International Journal of Energy Optimization and Engineering*, *7*(2), 76–96. doi:10.4018/IJEOE.2018040104

Ganesan, T., & Elamvazuthi, I. (2017). A multi-objective approach for resilience-based plant design optimization. *Quality Engineering*, *29*(4), 1–16. doi:10.1080/08982112.2016.1255331

Ganesan, T., Elamvazuthi, I., Shaari, K. Z. K., & Vasant, P. (2013). Multiobjective optimization of green sand mould system using chaotic differential evolution. In *Transactions on Computational Science XXI* (pp. 145–163). Springer Berlin Heidelberg. doi:10.1007/978-3-642-45318-2_6

Ganesan, T., Elamvazuthi, I., & Vasant, P. (2015). Multiobjective design optimization of a nano-CMOS voltage-controlled oscillator using game theoretic-differential evolution. *Applied Soft Computing*, *32*, 293–299. doi:10.1016/j.asoc.2015.03.016

Ganesan, T., Vasant, P., & Elamvazuthi, I. (2016a). Multiobjective optimization using particle swarm optimization with non-Gaussian random generators. *Intelligent Decision Technologies*, *10*(2), 93–103. doi:10.3233/IDT-150241

Ganesan, T., Vasant, P., & Elamvazuthi, I. (2016b). *Non-Gaussian Random Generators in Bacteria Foraging Algorithm for Multiobjective Optimization*. arXiv preprint arXiv:1605.07364

Hao, Z., Xing-yuan, W., Si-wei, W., Kang, G., & Xiao-hui, L. (2017). Application of coupled map lattice with parameter q in image encryption. *Optics and Lasers in Engineering*, *88*, 65–74. doi:10.1016/j.optlaseng.2016.07.004

Hirose, H. (1996). Maximum likelihood estimation in the 3-parameter Weibull distribution. A look through the generalized extreme-value distribution. *IEEE Transactions on Dielectrics and Electrical Insulation*, *3*(1), 43–55. doi:10.1109/94.485513

Huang, A., Zhang, H. M., Guan, W., Yang, Y., & Zong, G. (2015). Cascading failures in weighted complex networks of transit systems based on coupled map lattices. *Mathematical Problems in Engineering*, *2015*, 2015. doi:10.1155/2015/940795

Huang, T., & Zhang, H. (2016). Bifurcation, chaos and pattern formation in a space-and time-discrete predator–prey system. *Chaos, Solitons, and Fractals*, *91*, 92–107. doi:10.1016/j.chaos.2016.05.009

Huang, T., Zhang, H., & Yang, H. (2017). Spatiotemporal complexity of a discrete space-time predator–prey system with self-and cross-diffusion. *Applied Mathematical Modelling*, *47*, 637–655. doi:10.1016/j.apm.2017.03.049

Islam, J., Vasant, P. M., Negash, B. M., Laruccia, M. B., & Myint, M. (2020). A Survey of Nature-Inspired Algorithms With Application to Well Placement Optimization. In *Deep Learning Techniques and Optimization Strategies in Big Data Analytics* (pp. 32–45). IGI Global. doi:10.4018/978-1-7998-1192-3.ch003

Kaneko, K. (1992). Overview of coupled map lattices. *Chaos (Woodbury, N.Y.)*, *2*(3), 279–282. doi:10.1063/1.165869 PMID:12779975

Karagrigoriou, A., & Vonta, I. (2014, September). Statistical Inference for Heavy-Tailed Distributions in Technical Systems. In *Availability, Reliability and Security (ARES), 2014 Ninth International Conference on* (pp. 412-419). IEEE. 10.1109/ARES.2014.62

Kharchenko, V., & Vasant, P. (Eds.). (2018). *Handbook of Research on Renewable Energy and Electric Resources for Sustainable Rural Development*. IGI Global. doi:10.4018/978-1-5225-3867-7

Kohlenbach, P., & Ziegler, F. (2008). A dynamic simulation model for transient absorption chiller performance. Part I: The model. *International Journal of Refrigeration*, *31*(2), 217–225. doi:10.1016/j.ijrefrig.2007.06.009

Kumar, S., Kumar, R., Kumar, S., & Kumar, S. (2019). Cryptographic construction using coupled map lattice as a diffusion model to enhanced security. *Journal of Information Security and Applications*, *46*, 70–83. doi:10.1016/j.jisa.2019.02.011

Lenz, R. (2016). Generalized Pareto Distributions—Application to Autofocus in Automated Microscopy. *IEEE Journal of Selected Topics in Signal Processing*, *10*(1), 92–98. doi:10.1109/JSTSP.2015.2482949

Maheta, H. H., & Dabhi, V. K. (2014, February). An improved SPEA2 Multi objective algorithm with non dominated elitism and Generational Crossover. In *Issues and Challenges in Intelligent Computing Techniques (ICICT), 2014 International Conference on* (pp. 75-82). IEEE. 10.1109/ICICICT.2014.6781256

Marmolejo, J. A., & Rodriguez, R. (2015). Fat tail model for simulating test systems in multiperiod unit commitment. *Mathematical Problems in Engineering*, *2015*, 2015. doi:10.1155/2015/738215

Marmolejo, J. A., Velasco, J., & Selley, H. J. (2017). An adaptive random search for short term generation scheduling with network constraints. *PLoS One*, *12*(2), e0172459. doi:10.1371/journal.pone.0172459 PMID:28234954

Nakajima, J., Kunihama, T., & Omori, Y. (2017). Bayesian modeling of dynamic extreme values: Extension of generalized extreme value distributions with latent stochastic processes. *Journal of Applied Statistics*, *44*(7), 1248–1268. doi:10.1080/02664763.2016.1201796

Nasiruddin, I., Bhatti, T. S., & Hakimuddin, N. (2015). Automatic generation control in an interconnected power system incorporating diverse source power plants using bacteria foraging optimization technique. *Electric Power Components and Systems*, *43*(2), 189–199. doi:10.1080/15325008.2014.975871

Nematzadeh, H., Enayatifar, R., Motameni, H., Guimarães, F. G., & Coelho, V. N. (2018). Medical image encryption using a hybrid model of modified genetic algorithm and coupled map lattices. *Optics and Lasers in Engineering*, *110*, 24–32. doi:10.1016/j.optlaseng.2018.05.009

Neufeld, Z., Vicsek, M., & Vicsek, T. (1996). Complex spatiotemporal patterns in two lattice models with instability. *Physica A*, *233*(3-4), 754–766. doi:10.1016/S0378-4371(96)00188-4

Neufeld, Z., & Vicsek, T. (1995). Spatiotemporal chaos in a coupled map lattice with unstable couplings. *Journal of Physics. A, Mathematical and General*, *28*(18), 5257–5266. doi:10.1088/0305-4470/28/18/015

Ning, B., & Bloomfield, P. (2017). *Bayesian inference for generalized extreme value distribution with Gaussian copula dependence*. arXiv preprint arXiv:1703.00968

Orjubin, G. (2007). Maximum field inside a reverberation chamber modeled by the generalized extreme value distribution. *IEEE Transactions on Electromagnetic Compatibility*, *49*(1), 104–113. doi:10.1109/TEMC.2006.888172

Plucar, J., Grunt, O., & Zelinka, I. (2017, July). Case study: Optimizing fault model input parameters using bio-inspired algorithms. In AIP Conference Proceedings (Vol. 1863, No. 1, p. 070027). AIP Publishing.

Price, K., Storn, R. M., & Lampinen, J. A. (2006). *Differential evolution: a practical approach to global optimization*. Springer Science & Business Media.

Richter, H. (2008). Coupled map lattices as spatio-temporal fitness functions: Landscape measures and evolutionary optimization. *Physica D. Nonlinear Phenomena*, *237*(2), 167–186. doi:10.1016/j.physd.2007.08.016

Sadeghi, J., Sadeghi, S., & Niaki, S. T. A. (2014). A hybrid vendor managed inventory and redundancy allocation optimization problem in supply chain management: An NSGA-II with tuned parameters. *Computers & Operations Research*, *41*, 53–64. doi:10.1016/j.cor.2013.07.024

Sanders, R. (1987). The Pareto principle: Its use and abuse. *Journal of Services Marketing*, *1*(2), 37–40. doi:10.1108/eb024706

Sheng, W., Liu, K. Y., Liu, Y., Meng, X., & Song, X. (2013). A new DG multiobjective optimization method based on an improved evolutionary algorithm. *Journal of Applied Mathematics*, *2013*, 2013. doi:10.1155/2013/643791

Singh, R. P., Mukherjee, V., & Ghoshal, S. P. (2015). Optimal reactive power dispatch by particle swarm optimization with an aging leader and challengers. *Applied Soft Computing*, *29*, 298–309. doi:10.1016/j.asoc.2015.01.006

Vasant, P., Zelinka, I., & Weber, G. W. (2018). Intelligent Computing & Optimization. In *Conference proceedings ICO* (p. 804). Academic Press.

Walshaw, D. (2013). Generalized extreme value distribution. Encyclopedia of Environmetrics.

Xingyuan, W., Le, F., Shibing, W., Zhang, C., & Yingqian, Z. (2018). Spatiotemporal chaos in coupled logistic map lattice with dynamic coupling coefficient and its application in image encryption. *IEEE Access: Practical Innovations, Open Solutions*, *6*, 39705–39724. doi:10.1109/ACCESS.2018.2855726

Zhang, L., Liu, S., & Yu, C. (2014). Chaotic behaviour of nonlinear coupled reaction–diffusion system in four-dimensional space. *Pramana*, *82*(6), 995–1009. doi:10.100712043-014-0753-2

Zhang, Y. Q., He, Y., & Wang, X. Y. (2018). Spatiotemporal chaos in mixed linear–nonlinear two-dimensional coupled logistic map lattice. *Physica A*, *490*, 148–160. doi:10.1016/j.physa.2017.07.019

Zhang, Y. Q., & Wang, X. Y. (2014). Spatiotemporal chaos in mixed linear–nonlinear coupled logistic map lattice. *Physica A*, *402*, 104–118. doi:10.1016/j.physa.2014.01.051

APPENDIX: NOMENCLATURE AND ABBREVIATIONS

Subscripts

1 stream from absorber to pump
2 stream from pump to solution HEX
3 stream from HEX to generator
4 stream from generator to HEX
5 stream from HEX to valve
6 stream from valve to absorber
7 stream from generator to condenser
8 stream from condenser to expansion valve
9 stream from expansion valve to evaporator
10 stream from evaporator to absorber

Absorption Chiller

X_s *strong solution concentration* (kg LiBr/ kg solution)
X_w *weak solution concentration* (kg LiBr/ kg solution)
t_a *absorber temperature* (°C)
t_e *evaporator temperature* (°C)
t_g *generator temperature* (°C)
t_c *condenser temperature* (°C)
P_e *evaporator pressure* (mm Hg)
P_c *condenser pressure* (mm Hg)
H_i *enthalpy of pure water* (kcal/kg)
Q_E *transient refrigeration load* (kcal/hr)
$\dot{m}_R$ *refrigerant mass flowrate* (kg/hr)
$\dot{m}_W$ *weak LiBr solution mass flowrate* (kg/hr)
$\dot{m}_S$ *strong LiBr solution mass flowrate* (kg/hr)
E_L *HEX effectivity*
C_{X_1} *specific heat of the strong solution* (kcal/kg °C)
C_{X_4} *specific heat of the weak solution* (kcal/kg °C)
Q_e *heat input to the evaporator* (kcal)
Q_g *heat input to the generator* (kcal)
Q_c *heat output from the evaporator* (kcal)
Q_a *heat output from the absorber* (kcal)
dQ_e/dt *heat rate - input to the evaporator* (kcal/hr)
dQ_g/dt *heat rate - input to the generator* (kcal/hr)
dQ_c/dt *heat rate - output to the evaporator* (kcal/hr)
dQ_e/dt *heat rate - input to the absorber* (kcal/hr)
COP *coefficient of performance*
COP_i *ideal/max coefficient of performance*
RPR *relative performance ratio*

$\dot{m}_{a,t}(i)$ *outlet mass flowrate at the absorber tube* (kg/hr)
$\dot{m}_{g,t}(i)$ *outlet mass flowrate at the generator tube*(kg/hr)
$M_a(i)$ *mass stored at the absorber sump (shell side) tube* (kg)
$M_g(i)$ *mass stored at the generator sump (shell side) tube*(kg)
Δt *simulation time-interval* (hr)
COP' *time-averaged coefficient of performance*
ΔT *temperature differential for cooling* (hr)

Gas Turbine

p_i *component pressures (kPa)*
r compression pressure ratio
η_{comp} *compressor efficiency*
η_{turb} *turbine efficiency*
η_{cc} *combustion efficiency*
η_{th} *overall thermal efficiency*
$C_{p,air}$ air heat capacity (kJ/kgK)
$C_{p,fuel}$ fuel heat capacity (kJ/kgK)
k_{air} air heat capacity ratio
k_{flue} flue gas heat capacity ratio
T_i *component temperatures* (°C)
$\dot{m}_{air}$ *air mass flowrate (kg/s)*
$\dot{m}_{fuel}$ *fuel mass flowrate (kg/s)*
ΔP_{comb} *combustion pressure loss (kPa)*
LHV fuel lower heat value (kJ/kg)
$\dot{W}_{comp}$ *compressor power input (kW)*
$\dot{W}_{turb}$ *turbine power output (kW)*
$\dot{W}_{net}$ *net power output (kW)*
$\dot{Q}_{in}$ *combustion heat input (kW)*
SFC specific fuel consumption (kg/kWh)
HR heat rate (kJ/kWh)
HR' time-averaged heat rate (kJ/kWh)
n'_{th} *time-averaged overall thermal efficiency*

Differential Evolution and Extreme Value Metaheuristic

N Length of individual string (DE)
P No. of individuals in the population (DE)
x^p_i Principal Parent (DE)
x^a_i Auxiliary Parent (DE)
V_i Mutated vector (DE)
F Mutation amplification factor (DE)
CR The probability of the cross-over (DE)
x^G_{i}. *Population of vectors (DE)*

x^{child}_{i}: *Child trial vector (DE)*
N(t), R(t) Variables for logistic chaotic map
N_{max} Maximum iterations for chaotic logistic map
σ *Scale factor*
k Shape factor
μ *Location factor*
x Random variable

Coupled Map Lattice

ε *Coupling parameter*
j Iteration index
θ *Logistic function constant*
m Maximum number of iteration
s index for the vertices on the lattice
y Mapping variable

Chapter 15
Visual Analytics to Build a Machine Learning Model

Iurii V. Krak
https://orcid.org/0000-0002-8043-0785
Glushkov Cybernetics Institute, Taras Shevchenko National University of Kyiv, Ukraine

Olexander V. Barmak
https://orcid.org/0000-0003-0739-9678
National University of Khmelnytskyi, Ukraine

Eduard Manziuk
https://orcid.org/0000-0002-7310-2126
National University of Khmelnytskyi, Ukraine

ABSTRACT

One of the most interesting and promising areas of development of machine learning is the active involvement of a human in the process of building a model. However, there are problems with the effective integration of humans into a workflow. It is necessary to develop techniques and information technologies that would allow the effective use of human intellectual capabilities, thereby expanding the machine learning tools. This work considers the use of visual analytics with the goal of building a machine learning model by a human and the technique of transferring this model to the machine level. This made it possible to expand the capabilities of machine learning through the active and productive use of human intellectual abilities.

INTRODUCTION

With the development of new technologies and the constant increase in the level of informatization of society, the problem of machine classification is important. The problem of classification can be considered a classic in machine learning. Along with the use of a wide variety of solutions, the search for new approaches remains relevant. Consider an approach based on the use of human intellectual capac-

DOI: 10.4018/978-1-7998-3970-5.ch015

ity to build a classification model. That is, to integrate a human into the process of machine learning. This will extend the ability to build models by utilizing human classification capabilities. This creates a hybrid machine learning system that opens up new avenues for improving results. A large amount of information a human receives visually.

The use of machine learning is widely used in modern areas of information technology (Rahaman & Vasant, 2020), (Gholamy, 2019), (Vasant, 2019), (Ganesan, 2020), (Islam, 2020). Research into the use of artificial intelligence in work technology is being actively developed (Rahim, 2020).

Data visualization is a task that any researcher faces in his work. The problem of data visualization is reduced to the problem of visualizing experimental data or the results of a theoretical study. Traditional tools in this area - graphs and charts cope with the task of visualization when it becomes necessary to display interrelated quantities. Visual display of data is much more informative than other methods of obtaining and perceiving information, more convenient and easier to perceive and understand than, for example, displaying data in tables, diagrams, mathematical matrices, or simply in numbers. Nowadays, tools are being distributed for the visual display of information, data, and for this purpose information technology is widely used, so there is a need to create software for data visualization. New varieties of visual presentation of data have also appeared, some of which come from graphs, charts, histograms and not just visualization on a plane, but visualization in 3D space.

Data visualization - is a visual display of large arrays of numerical and other information, which is made possible through the use of computer graphics. Software systems for data visualization can be easily integrated into information systems. The problem of data visualization is reduced to the problem of visualizing experimental data or the results of a theoretical study.

Information visualization is an interactive study of the visual display of abstract data to enhance human cognition. Abstract data includes both numeric and non-numeric, such as text and geographic information.

INFORMATION VISUALIZATION

The field of information visualization has arisen as a result of studies of human-computer interaction, computer science, graphics, design, psychology and business methods. It is increasingly being used as the most important component in scientific research, digital libraries, data mining, financial data analysis, market research, industrial control. Visualization of information implies that visual representations and methods of interaction take advantage of the ability of the human eye to transmit information to the brain so that users can see, learn and understand a large amount of information at a time. Information visualization is aimed at creating approaches to the transfer of abstract information into intuitive images (Thomas & Cook, 2005).

One of the main elements of machine learning is model building. One of the directions for improving the work of this model is the use of human intellectual abilities. As an example, "human-in-the-loop" approach of Endert A. (Endert, 2014). This trend was developed by VIS4ML (Sacha, 2018) also Interactive Machine Learning (Lee, 2016), (Holzinger, 2019), (Teso & Kersting, 2019), Interactive Intelligence Systems (Sinitsyn, 2019).

In this area of visual analytics, a human is used to iteratively improve the formal model. However, along with the process of building a formal model for a machine, a person forms a mental model in the mind. The concepts of a formal and mental model are described by Andrienko N. (Andrienko, 2018). Today, a human is used to improve the formal model for a machine. The machine cannot use the mental model, but is used exclusively by humans and is transmitted from person to person. Of particular interest and perspective is the development of approaches to using the mental model of the machine.

The purpose of the chapter is the presentation of information technology allowing the machine to use a mental model formed by a person and has a practical direction. The use of the technique of projecting the mental model to the machine level based on the results of this model is demonstrated. This is a new direction in visual analytics and allows you to more closely integrate the result of human intellectual activity, namely the mental model, with the machine. In this direction, the model is trained not by a machine, but by a human, subsequently projected onto the machine level and used by the machine.

Data analysis is an integral part of all applied research and problem-solving in industry. The most fundamental approaches to data analysis are visualization (histograms, surface areas, map tree, parallel coordinated areas, etc.), statistics, data mining) and machine learning methods (clustering, classification, decision trees, etc.). Among all these approaches, the visualization of information, or, in other words, the visual analysis of data, there is one that relies mainly on the cognitive abilities of analysts, and also allows revealing unstructured and effective ideas that are limited only by human imagination and creativity. An analyst does not have to use different sophisticated methods to be able to interpret data visualization.

There are several types of visualization:

1. The usual visual representation of quantitative information in a schematic form. This group includes all known pie and linear diagrams, histograms and spectrograms, tables and various scatter plots.
2. Data during visualization can be converted into a form that enhances the perception and analysis of this information. For example, a map, a polar graph, a timeline and a graph with parallel axes, an Euler diagram.
3. Conceptual visualization allows developing complex concepts, ideas, and plans using conceptual maps, Gantt charts, graphs with a minimal path and other similar types of diagrams.
4. Strategic visualization translates into a visual form of different data on aspects of the work of organizations. These are a performance, life cycle, and organization structure graphs.
5. Metaphorical visualization will help graphically organize structural information with the help of pyramids, trees, and data maps, a striking example of which is the metro map.
6. Combined visualization allows combining several complex graphs into one diagram, as in a map with a weather forecast.
7. Scientific visualization in most cases is focused on the visualization of three-dimensional phenomena (architecture, meteorology, medicine, biology, and so on), with an emphasis on realistically depicted volumes, surfaces, light sources, and therefore similar, possibly with a dynamic component (time).

Can conclude that data visualization is a task that has a lot of input pre-classified or specific data, which for better perception must be displayed in two or three-dimensional space. This task is relevant in many fields of science and information technology because the visual display of data allows to better understand the objects or processes that are being studied.

There are different types of data visualization.

A semantic network is an information model of a subject domain, which has the form of a directed graph, the vertices of which correspond to the objects of the subject domain, and the edges define the relationship between them. Objects can be a concept, events, properties, processes.

A hyperbolic tree (hypertree) is a method of graphical representation of information that is used to visualize large hierarchies. This method is based on the principles of hyperbolic geometry and is often used in web applications. When displaying hierarchical data in the form of trees, complex visualization often arises, since the number of nodes at each next level can increase exponentially.

Multidimensional scaling (MDS) is (Borg & Groenen, 2003), (Gordon, 1983) a class of multidimensional statistical analysis methods in which the studied objects of observation, which are characterized by a large number of signs, are placed in a low-dimensional space based on multidimensional measures of the distance between objects. As usual, this is two or less often three-dimensional space. A frequently used example to illustrate MDS is the restoration of a map of cities based on available information about the distances between cities. The papers of Krak (2019) show the ability to use visual analysis to classify data/

Heat map - is a graphical representation of the data in which individual values hat are contained in the matrix are presented in the form of colors. Fractal maps and tree maps often use similar color-coding systems to represent the values that the interchangeable takes in the hierarchy.

The internet resource paperscape.org was reviewed. This is a classifier of the archive of articles with visualization in the form of a "starry sky", where each element is a circle that designates an article, and its size depends on the number of links to this article. The placement of each element is determined using the Barnes-Hut algorithm, which models the gravitational interaction with n bodies according to Newton's classical gravitational theory. Space modeling is usually performed by dividing the space into cells so that the bodies in neighboring cells are considered individually, and the bodies in distant cells are considered as one large quotient with the center at the point of its center of mass, due to which a significant reduction in calculations is achieved. The algorithm is used in modeling dynamic systems in which the effect on each individual element of the system can be calculated as a superposition of the influence of elements (it is necessary to perform n^* (n-1) calculations), for example, in modeling the behavior of magnetic fluids.

Another example, Tableau is a software product for visualization, data processing, focused on business intelligence. It allows for creating semantic networks, heatmaps, hyper trees, graphs, and charts. The program allows using existing sources of information: Excel spreadsheets, Access databases, text files, json-files.

Data visualization is an effective analysis tool. This is often used to expand knowledge of the subject area of research. Consider the technique of using visual analytics to build a machine model based on grouping data in a multidimensional space and using a human to train a classification model

GROUPING DATA USING MULTIDIMENSIONAL SCALING

The task of MDS in its general form is to identify the structure of the studied set of stimuli. By revealing the structure is understood the allocation of a set of basic factors by which the stimuli differ, and a description of each of the stimuli in terms of these factors. The procedure for constructing a structure is based on the analysis of objective or subjective information about the proximity between the stimuli of this information. In the case of the analysis of subjective data, two tasks are solved simultaneously. On the one hand, the objective structure of subjective data is revealed, on the other hand, factors that influence the decision-making process are determined.

MDS methods can use different types of data: data on the advantages of the subject on a variety of stimuli, data on dominance, on the proximity between stimuli, and data on profiles. As a rule, it is customary to correlate with each type of data a certain group of methods for processing them. However, this ratio should not be too rigid, since it is often not difficult to switch from one data type to another. So, for example, profile data can easily be turned into profile proximity data, for this, only need to use the appropriate metric. Benefit data contains dominance information. On the other hand, calculating the correlations between the columns of the advantage matrix can obtain the proximity matrix between the stimuli, and the correlations between the rows of the same matrix will give us the proximity matrix between the subjects.

The basis of MDS is the idea of a geometric representation of a stimulating set. Let a coordinate space be given, each axis of which corresponds to one of the desired factors. Each stimulus is represented by a point in this space, the projection values of these points on the axis correspond to the values or degrees of factors that characterize the given stimulus. The larger the projection value, the greater the value of the stimulus factor. The measure of similarity between two stimuli is the inverse of the distance between their corresponding points. The closer the stimuli are to each other, the higher the measure of similarity between them (and the lower the measure of difference), the low measure of similarity corresponds to distant points. In order to accurately measure proximity, it is necessary to introduce a metric in the desired coordinate space, the choice of this metric has a great influence on the result of the solution.

The solution of the visualization problem using the MDS method is aimed at developing algorithmically efficient methods for placing points that correspond to the objects understudy in a space of less dimension than the space of the object under study.

For visual representation of the entire structure of a multidimensional data set 1, 2- and 3-dimensional display spaces can be used. For the study, visualization using a 2-dimensional surface was chosen, since it is in this form that a human perceives geometric structures most naturally and the connections between objects look most clearly.

Data visualization image of the multidimensional distribution of data on a two-dimensional plane, in which the basic laws inherent in the initial distribution are qualitatively displayed - its cluster structure, topological features, internal relationships between features, information about the location of data in the source space. As the main applications of visualization methods, can specify the following:

- visual representation of the geometric metaphor of data;
- a concise description of the internal patterns contained in the data set;
- compression of the information contained in the data;
- recovering data gaps;
- solving the problems of forecasting and constructing regression dependencies between features.

Sometimes the initial information is initially presented not in the form of a table of the "object-features" type, but in the form of a square matrix of the distance of the objects from each other. At the intersection of the i-th row of the j-th column, the matrix contains an estimate of the distances from the i-th to the j-th object. Thus, first, each object is assigned coordinates in a multidimensional spacious, and it is difficult to present such information in the form of a geometric metaphor. The task of MDS is to construct a set of data in space or on a plane so that the distances between objects maximally correspond to the distances initially set in the matrix. The resulting coordinate axes can be interpreted as some implicit factors whose values determine the differences between objects. If can provide each object with a pair of coordinates, the result will be a data visualization method.

Based on the visualization system, a nonlinear data classifier can be constructed using the decision tree algorithm. It is possible to classify objects as gatherings from their grouping in visual space. The decision tree is an image of the mapping of rules in a hierarchical, sequential structure, where each object corresponds to a single node that provides a solution. A rule is understood as a logical construction represented as "if ... then ...". As an example, we will use text data. Rules are defined by line segments that cut off groups of texts on the map of the visualization system. As a result, when giving a new text to the test, it can clearly indicate to which semantic category it belongs.

MDS belongs to a group of methods that are widely used to reduce the dimensionality and visualization of multidimensional data (Gordon, 1983).

There are two main algorithms for MDS - metric and nonmetric.

The metric MDS is based on the assumption that the distances in the distance matrix correspond to real distances between objects in the constructed feature space.

In the linear metric scaling method, the principal component method is used, but not to the original distance table, but to the so-called double-centered matrix, in which the average value of the numbers in any row and column is zero.

The double-centered matrix is uniquely calculated from the original. After this, it is possible to determine the dimensionality of the space, which ensures the exact reproduction of the distance matrix, or to determine the effective dimension of the designed feature space, which will ensure the reproduction of the distance matrix with a given accuracy.

In nonlinear methods, the dimension of space is set first and, using gradient methods, the quality functional is optimized, which describes the measure of distortion of the distance matrix. This functionality is called stress.

In non-metric MDS, it is assumed that the distance between objects is given by an ordinary scale, that is, not only the numerical values of pairwise distances are important, but also their rank order. Nonmetric scaling procedures construct such a geometric configuration of points in n-dimensional space that the rank orders of distances are as close as possible to the rank orders of the original distances.

To assess the quality of selected rank coordinates, the stress criterion is applied. Similar to the traditional factor analysis, in MDS, there is an ambiguity in the choice of coordinates, due to the fact that the coordinate system in the resulting space can be arbitrarily rotated - the distances between the objects do not change. As a rule, the rotation is carried out in such a way that the obtained coordinate axes have the most visual interpretation or the value of certain signs is maximally correlated.

The data for MDS is a matrix, which is composed of the distances between each object. Each distance between the *i*-th and *j*-th objects is denoted by δ_{ij}. If objects are defined by multidimensional points $X_i = \{x_{i1}, x_{i2}, \ldots, x_{in}\}$ where $i=1,\ldots,n$, the distance can be determined by the distance between points $\delta_{ij} = d(X_i, X_j)$, where

$$d\left(X_i, X_j\right) = \sqrt{\left(x_{i1} - x_{j1}\right)^2 + \left(x_{i2} - x_{j2}\right)^2 + \ldots + \left(x_{in} - x_{jn}\right)^2} \tag{1}$$

The distance between points in low-dimensional space and corresponding to the *i*-th and *j*-th objects is denoted by $d(X_i, X_j)$.

The goal of MDS is to find low-dimensional points $Y_i = \{y_{i1}, y_{i2}, \ldots, y_{in}\}$ so that the distance between the points in the small space is as close as possible to the distance in the original space. Least square object function (raw Stress), which should be minimized, can be written as

$$\sigma_r(Y)=\sum_{i<j} w_{ij}\left(d\left(Y_i,Y_j\right)-\delta_{ij}\right)^2, \tag{2}$$

where $Y_i = \{y_{i1}, y_{i2}, \ldots, y_{in}\}$ and w_{ij} are nonnegative weights.

Normalized Stress is defined as:

$$\sigma_n(Y)=\frac{\sum_{i<j} w_{ij}\left(d\left(Y_i,Y_j\right)-\delta_{ij}\right)^2}{\sum_{i<j} w_{ij}\delta_{ij}^2} \tag{3}$$

Normalization using the parameter $\sum_{i<j} w_{ij}\delta_{ij}^2$ gives a clear interpretation of the quality of visualization, which is less dependent on the number of objects m and the scale of proximity.

The relative error is determined as follows

$$E(Y)=\sqrt{\sigma_n(Y)}=\sqrt{\frac{\sum_{i<j} w_{ij}\left(d\left(Y_i,Y_j\right)-\delta_{ij}\right)^2}{\sum_{i<j} w_{ij}\delta_{ij}^2}} \tag{4}$$

The reason for using $E(Y)$ rather than the normalized error $\sigma_r(Y)$ is that $\sigma_r(Y)$ in practice is almost always very small, so the value of $E(Y)$ is easier to distinguish (Guttman, 1968). Often $w_{ij}=1$, then equation (4) will look like the following order:

$$E(Y)=\sqrt{\frac{\sum_{i<j}\left(d\left(Y_i,Y_j\right)-\delta_{ij}\right)^2}{\sum_{i<j}\delta_{ij}^2}} \tag{5}$$

The first stress (Stress-1) is behind the equation:

$$\sigma_1(Y)=\sqrt{\frac{\sum_{i<j} w_{ij}\left(d\left(Y_i,Y_j\right)-\delta_{ij}\right)^2}{\sum_{i<j} w_{ij}\delta_{ij}^2}} \tag{6}$$

The classical MDS uses the fact that the coordinate matrix can be obtained by expanding with eigenvalues $B = XX'$, and the matrix B can be calculated from the distance matrix D using double centering (Minsky, 1961).

Steps of the classic MDS algorithm:

1. form a matrix of distances $D^{(2)} = \left[d_{ij}^2 \right]$;
2. apply dual centering: $B = -\frac{1}{2} JD^{(2)}$ using a centering matrix;
3. determine *m*, the largest eigenvalue of (Guttman, 1968) $\lambda_1, \lambda_2 \ldots \lambda_{m,}$ and the corresponding eigenvectors $e1, e2\ldots{}_e m$ of $_t$he matrix B де, where m is *t*he number of dimensions to output;
4. $X = E_m \Lambda_m^{0.5}$, where Em i_s the matrix of m *e*igenvectors, and Λm i$_s$ the diagonal matrix of m *ei*-genvalues B.

There are many MDS variants with different weights and optimization algorithms. One of the MDS options is SMACOF. The principle of majorization is to construct a surrogate function that majorizes a specific objective function. Thus, the principle of majorization proposes to find a simpler and more controlled surrogate function g(*x,y)* that majorizes f(*x*), that is, for all x, g(*x,y)*≥f(*x*), where y *is* some fixed value called the anchor point. The surrogate function should touch the surface in *y*, that is, g(x,y*)*=*f*(x*)*, which with the minimizer *x*∗, g(x,*y*) *ov*er x, *l*eads to the inequality f(x**)*≤*g*(x*,y*)*≤*g(*x,*y*)=*f*(x), *c*alled the sandwich inequality. Majorization is an iterative procedure consisting of:

- select initial value y=y*0;*
- and updates to x(*t*) such as g(*x*(*t*),y)≤g(*y,y);*
- stop if f*(*y*)* – f*(*x(t))≤ε, otherwise y=*x*(t) and go to the previous step.

This procedure can be extended to multidimensional spaces and as long as the unevenness of the sandwich remains, it can be used to minimize the corresponding target function.

The idea of majorization is to iteratively replace the original complex function *f* (*x*) with an auxiliary function g(*x,z*), where *z* is some fixed value (Honarkhah & Caers, 2010). The function g(x,z*) m*ust satisfy some requirements in order to be called the majority function. The auxiliary function should be:

- easier to minimize than f(*x*)*;*
- not less than the initial function, f*(*x*)*≤g*(*x,*z*)*;*
- touch *f*(*x*) at the so-called anchor point, *f*(*z*)=*g*(*z,z*).

The set of $Y=\{Y_1, Y_2, \ldots, Y_m\}$ *d*-dimensional points can be iteratively calculated using the Gutman transform equation:

$$Y(t+1) = V^{+}B(Y(t))Y(t), \tag{7}$$

where *t* is the iteration number and B*(*Y(t)) has the elements:

$$b_{ij} = \begin{cases} -\dfrac{w_{ij}\delta_{ij}}{d\left(Y_i, Y_j\right)}, where\ i \neq j\ and\ d\left(Y_i, Y_j\right) \neq 0, \\ 0, where\ i \neq j\ and\ d\left(Y_i, Y_j\right) = 0. \end{cases} \tag{8}$$

$$b_{ii} = -\sum_{i=1, j \neq i}^{m} b_{ij}. \tag{9}$$

V is a matrix of weights with elements:

$$v_{ij} = -w_{ij}, \text{ where } i \neq j,$$
$$v_{ii} = \sum_{i=1, j \neq i}^{m} w_{ij}. \tag{10}$$

And $V+$ denotes the Moore-Penrose pseudoreverse for the matrix V. If $wij_{=1}$ (7) is simplified to

$$Y(t+1) = \frac{1}{m} B(Y(t))Y(t), \tag{11}$$

As a result, the SMACOF algorithm is as follows:

1. set $t = 0$ and the initial values of low-dimensional points $Y(0)$;
2. calculate the value of the stress function $\sigma_r(Y(t))$ by the equation (1);
3. calculate $Y(t+1)$ by equation (7);
4. calculate $\sigma_r(Y(t))$ by the equation (1);
5. increase the iteration counter by one: t ++.
6. If $\sigma_r(Y(t-1) - \sigma_r(Y(t)) < \varepsilon$ or t is equal to the maximum number of iterations, then the cycle stops (ε is a small positive constant), otherwise go to step 3.

The result is a set of objects that have a pair of coordinates that can be displayed on a plane. The objects obtained are maps of the input data, that is, a training dataset that is pre-classified.

The next step is to separate the categories on the plane and form a decision tree of the linear classifier.

A classifier is called linear, if its discriminant function is linear, that is, it permits representation in the following form:

$$d(\bar{x}) = \bar{W}^T \bar{x} + w_N, \tag{12}$$

where $\bar{x} = (x_0, \dots, x_{n-1})^T$ is the feature vector that determines the image of the object to be classified, $\bar{W} = (w_0, \dots, w_{n-1})^T$ is the vector of the classifier weights, and, is the threshold value.

The process of deciding the class number of the current object is carried out in accordance with the following rule:

$$d(\bar{x}) = \sum_{i=0}^{N-1} w_i x_i + w_N \underset{<}{\overset{>}{}} 0 \rightarrow \bar{X} \begin{cases} \Omega_0 \\ \Omega_1 \end{cases} \tag{13}$$

The task of synthesizing a linear classifier is to find the vector of coefficients $\overline{W}$ and the threshold value w_N.

It is known (Honarkhah & Caers, 2010) that the Bayesian classifier is linear in the case of normal distributions with equal correlation matrices. However, often the equality of correlation matrices is an unacceptable assumption, so there is a need to construct a linear classifier, abandoning normal distributions.

When a linear-divisible function is found that minimizes the total probability of misclassification, it is assumed that in classes Ω_0 and Ω_1 the random variable $d(X)$ has a normal distribution law. The method based on minimizing the root mean square deviation of the solution allows to obtain similar results without this assumption.

Suppose, when constructing the classifier, that a set of K values of the vectors of realizations $\overline{x}^1,\ldots,\overline{x}^K$ is a random vector of signs X, for each of which it is known which class Ω_0 and Ω_1 it belongs to. In this case, they say that a training sample is specified. A new vector $\overline{z}=\left(z_0,\ldots,z_{N-1},1\right)^T$, which is formed as follows. For objects in class Ω_1, the vector $\overline{z}=\left(x_0,\ldots,x_{N-1},1\right)^T$, and for objects of class Ω_0 $\overline{z}=\left(-x_0,\ldots,-x_{N-1},-1\right)^T$.

The vector is supplemented by the weights of the threshold value:

$$\overline{W}=\left(w_0,\ldots,w_N,\right)^T \tag{14}$$

Carrying out such transformations allows to write a linear distribution function in the form:

$$d\left(\overline{z}\right)=\overline{W}^T\overline{z} \tag{15}$$

The construction of the classifier is thus to determine the complemented vector of coefficients $\overline{W}$ such that for any known vectors $\overline{z}^k\left(k=1,\overline{K}\right)$, the following inequality holds:

$$d\left(\overline{z}\right)=\overline{W}^T\overline{z}\geq 0 \tag{16}$$

Let $\gamma(z)$ be the best distribution function, that is, a function that can be regarded as the necessary output of an ideal discriminant function. As a rule, $\gamma(z)$ is not known, but it can be roughly determined based on the training sample from condition (15). For example, can take $\gamma(z)=1$, which corresponds to the case when a correct classification variable is determined on the basis of a training sample:

$$r\left(\overline{x}\right)=\begin{cases}-1,\overline{x}\in\Omega_0,\\ 1,\overline{x}\in\Omega_1\end{cases} \tag{17}$$

Mean square deviation between the necessary and the true value of the separating function is determined by the expression:

$$\varepsilon^2 = M\left(\gamma\left(\bar{Z}\right) - \bar{W}^T \bar{Z}\right)^2 \tag{18}$$

Linear classifiers have a significant drawback, namely the impossibility of distinguishing a curve category, or a broken line. To solve this drawback, it was decided to develop, on the basis of a linear classifier, a nonlinear classifier that divides objects on a plane using broken lines. For classification, it was decided to use the decision tree.

The decision tree is an image of representing rules in a hierarchical, sequential structure, where each object corresponds to a single node that provides a solution.

A rule is understood as a logical construction represented as "if ... then ...".

Let $\{C_1, C_2, \ldots, C_k\}$ some training set be given that contains objects, each of which is characterized by m attributes, and one of them indicates that the object belongs to a certain class.

Let denote the classes, then there are three situations:

1. The set T contains one or more examples belonging to the same class C_k. Then the decision tree for T is the letter that defines the class C_k.
2. The set T does not contain any examples, that is, an empty set. Then it is again a letter, and the class associated with the letter is selected from another set other than T, for example, from the set associated with the parent.
3. The set T contains examples belonging to different classes. In this case, it is necessary to divide the set T into some subsets. For this purpose, one of the signs having two or more distinct values $O_1, O_2, \ldots O_n$ is chosen. T is subdivided into subsets T*1*, T2, ... T*n* where each subset Ti contains all examples that have a value Oi for the selected trait. This procedure will continue recursively until the finite set consists of examples belonging to the same class.

To construct a tree on each inner node, it is necessary to find a condition that would break the set associated with that node into subsets. One of the attributes must be selected as such condition. The general rule for selecting an attribute can be formulated as follows: the selected attribute must split the set so that the resulting subsets consist of objects that belong to one class or are as close as possible to it, that is, the number of objects from other classes in each of these there were as many sets as possible.

CLASS SEPARATION HYPERPLANES

With increased dimensions of space, when it becomes necessary to turn the line into a hyperplane, it is necessary to divide the line into n-1 parts, where n is the dimension of the new space, so that the line has n points on it, and determine the equation of the plane.

The result will be a system of the form:

$$\begin{cases} a_{11}x_1 + a_{12}x_2 + \ldots + a_{1n}x_n = b_1, \\ a_{21}x_1 + a_{22}x_2 + \ldots + a_{2n}x_n = b_2, \\ a_{m1}x_1 + a_{m2}x_2 + \ldots + a_{mn}x_n = b_m, \end{cases} \tag{19}$$

This is a system of m linear equations with n unknowns. Need to seek solutions to such a system based on algebraic transformations of the system. The goal of these transformations is that the given system needs to be replaced with a new, more "simple" type and that the new system should have the same set of solutions that are given.

If two systems satisfy the condition: the set of solutions of one of them coincides with the set of solutions of the second or both are incompatible, then such systems are called equivalent or equivalent (Minsky 1961).

The following elementary transformations are possible:

- permutation of two arbitrary equations of the system;
- multiplication of both parts of any equation of system by a number, different from 0;
- adding to one of the equations of system another equation of this system, multiplied by any number.

In the course of such work the equation, all coefficients may appear in the system of which the free term is 0 (equation of the form $0X=0$). Such is the equation is satisfied with any n-dimensional vector and does not affect the solution of the system. Therefore, it is possible to derive such an equation from the system by reducing the number of system equations. Hence $l \leq m$. Along with such equations, there may be equations in which all coefficients for unknowns are zero, and the free term is not zero (Eq. $0X=b$, $b \neq 0$) Such an equation is not satisfied by any n-dimensional vector (unsolved). So there is no solution to the system in which it exists. That is a system will be incompatible. Therefore, if the system b does not have an equation of the form $0X=b$, that is, it compatible, and can apply the following conversion cycles to it, except unknown x_2, x_3, x_4 (Možina, Demšar, Kattan & Zupan, 2004).

As a result of solving a system of linear homogeneous equations, obtain the equation of the hyperplane. Many of these planes distinguish a category in hyperspace. To determine whether a new article belongs to a category, it is necessary to substitute its coordinates for the equation of planes, as a result, the values can be large, smaller, or level zero. The results are compared with the rules, which are formed in a similar way, substituting the coordinates of the test article.

Since the nonlinear classifier operates in n-dimensional space, there is a need to change the space in the lines and convert them to $n-1$-dimensional hyperplanes. A linear classifier (Yuan, Ho, & Lin, 2012) is one of the simplest ways to solve a classification problem. In this case, a straight line (hyperplane in the N-dimensional space) is sought, which separates all points of one class from points of another class. For this purpose, it is necessary to use the module of MDS but in the opposite direction, that is, it is necessary to increase the dimensionality of space. Figure 1 shows the general scheme of the algorithm.

After changing the dimension and forming the hyperplane, it is necessary to determine the equation of the hyperplane. Since the construction of a hyperplane in n-dimensional space requires n points, so it was decided to add $n-2$ points on each segment of the polyline, where n is the dimension of the space. The result will be a system of linear algebraic equations (18), which is solved by the Gaussian method

After forming an article map, it can be observed that articles belonging to the same category are grouped close together and the group is away from the center of the map.

As a result of category selection, a set of rules is formed by which the following articles will be classified, which will be added, while MDS is no longer fulfilled, as the lines of the bounding category boundaries are transferred to the n-dimensional space (Fig. 2).

Figure 1. The scheme of formation of rules for the nonlinear classifier

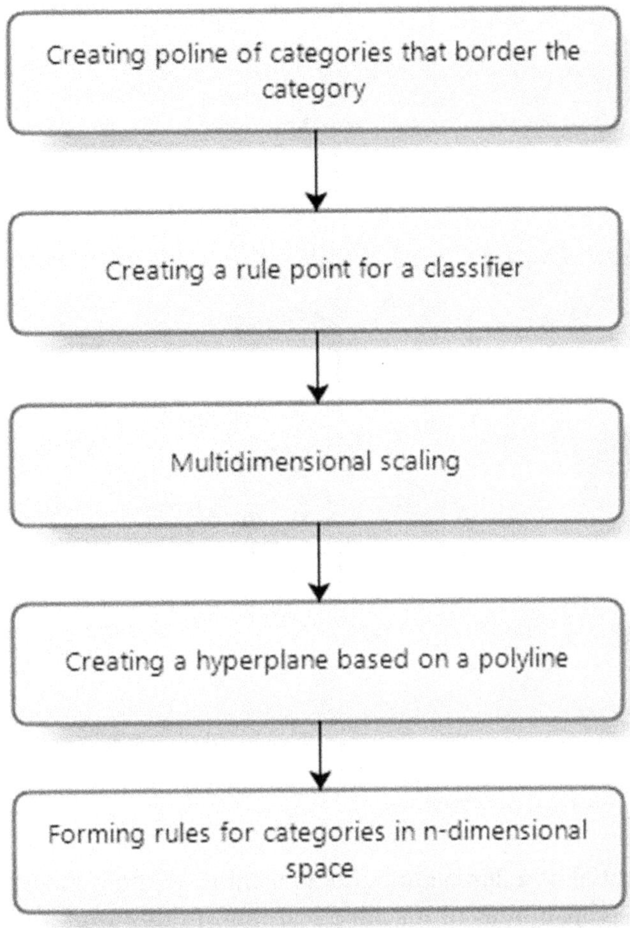

Rules are formed when the coordinates of the article are inserted into the equations of planes, as a result, can get a result greater than 0, which means that the article is to "the left" of the hyperplane, if 0, then to the plane, and if less than 0, then to "the right". When adding an article, it is checked that it belongs to one of the category rules

In the general case, the position of the class objects relative to the contour of the class bounding is necessary. Thus, train the system, denoting the boundaries of classes. A trained system is defined boundaries of the volumes of hyperspace in which objects of classes consist of the necessary generalization tolerances. When displayed on two-dimensional space, it can determine the position of new objects.

The class boundaries are a piecewise broken line. The line is drawn graphically by a human, based on an analysis of the grouping of data and their distribution in two-dimensional space. A machine learning model must have the ability to generalize data. This property is rather difficult to determine since it requires a comprehensive analysis of this and their distribution. In this case, the human plays an important role, since he can analyze and determine the position of the demarcation lines due to the visual analysis. Thus, class boundaries are drawn based on a generalized analytical understanding.

Figure 2. Visual representation of the distribution of documents and the formation of classes

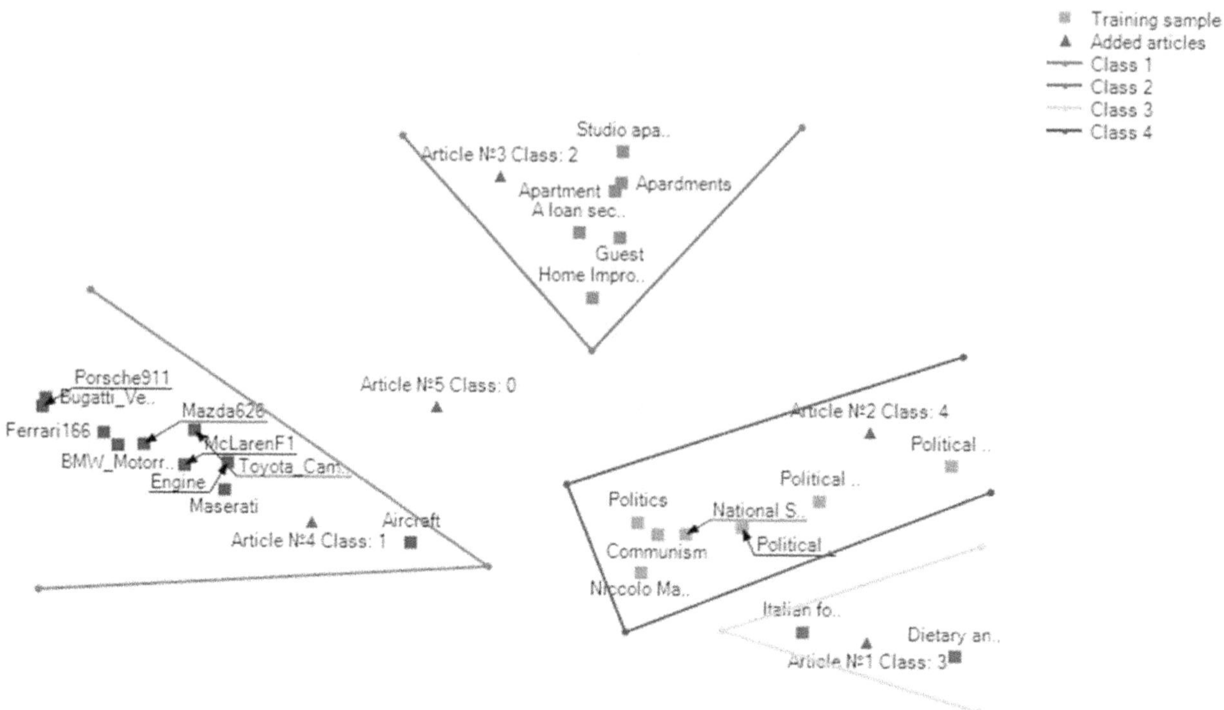

CONCLUSION

Using the intellectual potential of a human to build a machine learning model is a promising area that can expand to improve the capabilities of machine learning. In this work, a technique for integrating a human into machine learning is proposed. The technique is based on the use of an interactive visual analytics tool. As a result of analysis, a human forms an internal understanding of data, its structure, distribution, and existing relationships. This allows sorting the data into classes and defines these classes. Human, based on the formed mental model, graphically defines the boundaries of classes. The location of class boundaries is adjusted by a person with the possibility of forecasting, based on a mental model, the location of new data. One of the main advantages of the proposed approach is that the formation of a human's understanding of data and distribution by classes is formalized and projected onto the machine level. Human using a mental model forms the boundaries of classes. These boundaries are projected into the original hyperspace of attributes and define class hypercubes. The position of new data in hyperspace relative to the boundaries of classes determines the belonging of data to classes

The chapter presents information technology for projecting the mental model to the machine level. The technique of formalizing the mental model is used by projecting the results into the space of formalized representations. The results of the mental model are formalized by human through the graphic boundaries of classes' separation. Further development of the proposed technology is the development of hybrid models in which formal models are corrected by a human based on mental representations.

REFERENCES

Andrienko, N., Lammarsch, T., Andrienko, G., Fuchs, G., Keim, D., Miksch, S., & Rind, A. (2018). Viewing visual analytics as model building. *Computer Graphics Forum, 37*(6), 275–299. doi:10.1111/cgf.13324

Barmak, A. V., Krak, Y. V., Manziuk, E. A., & Kasianiuk, V. S. (2019). Information Technology of Separating Hyperplanes Synthesis for Linear Classifiers. *Journal of Automation and Information Sciences, 51*(5), 54–64. doi:10.1615/JAutomatInfScien.v51.i5.50

Borg, I., & Groenen, P. (2003). Modern multidimensional scaling: Theory and applications. *Journal of Educational Measurement, 40*(3), 277–280. doi:10.1111/j.1745-3984.2003.tb01108.x

Endert, A., Hossain, M. S., Ramakrishnan, N., North, C., Fiaux, P., & Andrews, C. (2014). The human is the loop: New directions for visual analytics. *Journal of Intelligent Information Systems, 43*(3), 411–435. doi:10.100710844-014-0304-9

Ganesan, T., Vasant, P., & Litvinchev, I. (2020). Multiobjective Optimization of a Biofuel Supply Chain Using Random Matrix Generators. In *Deep Learning Techniques and Optimization Strategies in Big Data Analytics* (pp. 206–232). IGI Global. doi:10.4018/978-1-7998-1192-3.ch013

Gholamy, A., Parra, J., Kreinovich, V., Fuentes, O., Anthony, E., Watada, J., ... Jain, L. C. (2019). (to appear). How to best apply deep neural networks in geosciences: Towards optimal 'averaging' in dropout training. *Smart Unconventional Modelling, Simulation and Optimization for Geosciences and Petroleum Engineering, Springer Verlag.*

Gordon, A. D., Coxon, A. P. M., & Davies, P. M. (1983). The User's Guide to Multidimensional Scaling, with Special Reference to the Mds (X) Library of Computer Programs. *The Statistician, 32*(3), 355–356. doi:10.2307/2987947

Guttman, L. (1968). A general nonmetric technique for finding the smallest coordinate space for a configuration of points. *Psychometrika, 33*(4), 469–506. doi:10.1007/BF02290164

Herbrich, R. (2001). *Learning kernel classifiers: theory and algorithms.* MIT Press. doi:10.7551/mitpress/4170.001.0001

Holzinger, A., Plass, M., Kickmeier-Rust, M., Holzinger, K., Crişan, G. C., Pintea, C. M., & Palade, V. (2019). Interactive machine learning: Experimental evidence for the human in the algorithmic loop. *Applied Intelligence, 49*(7), 2401–2414. doi:10.100710489-018-1361-5

Honarkhah, M., & Caers, J. (2010). Stochastic simulation of patterns using distance-based pattern modeling. *Mathematical Geosciences, 42*(5), 487–517. doi:10.100711004-010-9276-7

Islam, J., Vasant, P. M., Negash, B. M., Laruccia, M. B., & Myint, M. (2020). A Survey of Nature-Inspired Algorithms With Application to Well Placement Optimization. In *Deep Learning Techniques and Optimization Strategies in Big Data Analytics* (pp. 32–45). IGI Global. doi:10.4018/978-1-7998-1192-3.ch003

Krak, I., Barmak, O., Manziuk, E., & Kudin, H. (2019). Approach to Piecewise-Linear Classification in a Multi-dimensional Space of Features Based on Plane Visualization. In *International Scientific Conference "Intellectual Systems of Decision Making and Problem of Computational Intelligence"* (pp. 35-47). Springer.

Krak, I., Barmak, O., Manziuk, E., & Kulias, A. (2019). Data Classification Based on the Features Reduction and Piecewise Linear Separation. In *International Conference on Intelligent Computing & Optimization* (pp. 282-289). Springer.

Krak, I. V., Kudin, H. I., Barmak, O. V., Manziuk, E. A., Smolarz, A., & Mamyrbaev, O. (2019). Method and algorithm of the piecewise-hyperplane clusterization using tools of pseudo-inverse matrices. In Photonics Applications in Astronomy, Communications, Industry, and High-Energy Physics Experiments 2019 (Vol. 11176, p. 1117628). International Society for Optics and Photonics. doi:10.1117/12.2537417

Kruskal, J. B., & Wish, M. (1978). *Multidimensional scaling* (Vol. 11). Sage. doi:10.4135/9781412985130

Lee, T., Johnson, J., & Cheng, S. (2016). *An Interactive Machine Learning Framework.* arXiv preprint arXiv:1610.05463

Minsky, M. (1961). Steps toward artificial intelligence. *Proceedings of the IRE*, *49*(1), 8–30. doi:10.1109/JRPROC.1961.287775

Možina, M., Demšar, J., Kattan, M., & Zupan, B. (2004, September). Nomograms for visualization of naive Bayesian classifier. In *European Conference on Principles of Data Mining and Knowledge Discovery* (pp. 337-348). Springer.

Rahaman, M. S. A., & Vasant, P. (2020). Artificial Intelligence Approach for Predicting TOC From Well Logs in Shale Reservoirs: A Review. In Deep Learning Techniques and Optimization Strategies in Big Data Analytics (pp. 46-77). IGI Global. doi:10.4018/978-1-7998-1192-3.ch004

Rahim, K. N. K. A., Elamvazuthi, I., Vasant, P., & Ganesan, T. (2020). Robotic Assistive System: Development of a Model based on Artificial Intelligent Technique. In Robotic Systems: Concepts, Methodologies, Tools, and Applications (pp. 1688-1720). IGI Global.

Sacha, D., Kraus, M., Keim, D. A., & Chen, M. (2018). Vis4ml: An ontology for visual analytics assisted machine learning. *IEEE Transactions on Visualization and Computer Graphics*, *25*(1), 385–395. doi:10.1109/TVCG.2018.2864838 PMID:30130221

Sinitsyn, S., Panchenko, V., Kharchenko, V., Kovalev, A., & Vasant, P. (2019, October). The Concept of Information Modeling in Interactive Intelligence Systems. In *International Conference on Intelligent Computing & Optimization* (pp. 249-259). Springer.

Teso, S., & Kersting, K. (2019). *Explanatory Interactive Machine Learning.* http://www.aies-conference.com/accepted-papers/

Thomas, J., & Cook, K. A. (2005). *Illuminating the path: The R&D agenda for visual analytics national visualization and analytics center*. National Visualization and Analytics Center-US Department of Homeland Security.

Vasant, P., Litvinchev, I., & Marmolejo-Saucedo, J. A. (Eds.). (2019). *Innovative computing trends and applications*. Springer International Publishing. doi:10.1007/978-3-030-03898-4

Yuan, G. X., Ho, C. H., & Lin, C. J. (2012). Recent advances of large-scale linear classification. *Proceedings of the IEEE, 100*(9), 2584–2603. doi:10.1109/JPROC.2012.2188013

Chapter 16
Smart Connected Digital Products and IoT Platform With the Digital Twin

Mohamed Uvaze Ahamed Ayoobkhan
https://orcid.org/0000-0001-9120-4516
Department of Computer Science, Cihan University-Erbil, Kurdistan Region, Iraq

Yuvaraj D.
https://orcid.org/0000-0002-2745-8092
Department of Computer Science, Cihan University-Duhok, Kurdistan Region, Iraq

Jayanthiladevi A.
Computer Science and Information Science, Srinivas University, Mangalore, India

Balamurugan Easwaran
https://orcid.org/0000-0003-2492-9589
Department of Computer and Mathematical Sciences, University of Africa, Toru-Orua, Nigeria

ThamaraiSelvi R.
https://orcid.org/0000-0002-9895-4368
Bishop Heber College, India

ABSTRACT

A digital illustration of a novel prevalence of a physical product helps one to gain larger insight into that product's state performance and behavior digital twin, which is an unequivocal advanced copy of an item, method, or control. This living model creates a thread between the physical and digital worlds. A model of a physical object—a 'twin'—enables you to observe its standing, diagnose problems, and take a look at solutions remotely. It's a dynamic virtual illustration of a tool that is unendingly fed with knowledge from embedded sensors and packages. This provides associate degree correct period of time standing of the physical device. Digital twins drive innovation and performance and offer development technicians prognostic analytics that give firms the flexibility to boost client expertise.

DOI: 10.4018/978-1-7998-3970-5.ch016

INTRODUCTION

Digital Twin conception constitutes the joining of the visible and also the analogues world wherever each mechanical item can get an energetic computerized outline. All though the merchandise improvement life cycle, right from the planning section to the preparation section, organizations will have a whole digital foot print of their product. These 'connected digital things' generate knowledge in real time, and this helps businesses in the higher analyze and predict the issues ahead or provide early warnings, stop time period, develop new opportunities and even arrange higher product for the longer term at lower prices by victimization simulations. These can have a bigger impact on delivering a higher client expertise in business furthermore. Digital Twins which contain net of Things are key in trade and are preponderantly utilized in the commercial net of Things, engineering, and business organization area (Stark et al., 2017). The widespread reach and usage of the web of Things have created the Digital Twins cheaper and accessible for the business world.

A digital twin might be a powerful advanced model of an item, procedure, or person, that breaks down existing business framework information joined with certifiable information. Be that as it may, the value extends past the one advanced model of Associate in Nursing in addition to Associate in Nursing a supporting part of an association's computerized change activities.

Next-Generation Digital Twins

There is a unit innovative and rising technology spanning the digital twin stack making tangible business price for sensible connected product and sensible connected operations use cases. However, every sort differs in some variety of practicality, complexity, integration and technologies. The additional technology systems additional, the 'fuller' the dual becomes, making a digital thread, that adds challenges and opportunities for enterprise adopters and technology suppliers alike. Elementary technologies up-through rising ones span the physical and digital realms and alter progressively valuable digital twin outcomes (Stark et al., 2017).

THE DIGITAL THREAD

The digital twin partly separates the virtual and real parts of the system. However to mirror the work, the virtual and real got to be connected. This can be wherever the digital thread comes in. The digital thread could be an assortment of digital communications that integrates and drives fashionable style, producing, and products support processes, as well as product and method definitions that begin in style engineering and flow through the complete production chain (Boschert, S., & Rosen, R, 2016). Information flows embrace quality needs and compliance, specifications, take a look at results, maintenance schedules, and more, and conjointly embrace feedback loops like lessons learned and planned versus actual process results.

The digital thread serves the digital twin, serving to take care of its model and supply effective information feeds. In turn, the digital twin sits at the intersection of the digital threads and ensures information sets area unit organized and contextualized. Together, this kind of material required for associate degree business four production model (Boschert & Rosen, 2016).

Digital Twins for Everything and Everyone?

Digital twins can evolve from thought to reality for nearly everything, everyone, each service and each network. The digital twin, because the virtual illustration of the physical product, service or person, acts as a mirror of the important world to provide a way to simulate, predict, forecast, service and herbaceous plant. As such, it presents large opportunities for businesses and to boost people's lives.

A digital twin isn't very new. In-turn, for many years, weather forecasts have captured the period of time conditions measured by varied sensors and devices, simulating and visually representing them in digital type to form long-run climate predictions and short-run forecasts.

Now, with the large technology push within the web of Things (IoT), sensors, and cloud solutions, the potential of digital twins isn't any longer simply an inspiration, however a turbulent driver of innovation and it's cheaper than ever.

Capacities of Advanced Digital Twins

In common, virtual models are utilized to screen, analyze and move forward their physical models (Qi et al., 2016). Metaphorically, their capacities can be isolated into three stages:

- To See — at this organize, sensors and gadgets collect information to picture the situation.
- To Think — at this organize, the shrewd program analyzes collected information and, in the event that there are any issues, finds a few conceivable arrangements to each one.
- To Do — at this arrange, brilliantly calculations select the foremost fitting arrangement and actualize them to address the problems

The innovation permits individuals to see the inward issues of physical objects without getting interior due to computer visualization, and illuminate them with no dangers for people's well-being and life (Qi et al., 2016).

Smart Connected Products and IoT platform

Smart Connected Products are comprised of the 'things' in the Internet of Things (IoT) that have connectivity components which allow the product's operating environment to exist in the cloud. The real-time data collected and pushed to the cloud informs the user (for example via smart-phone app) to provide actionable suggestions used for decision-making (Ray, 2016). Shrewd associated items permit an undertaking to move past real-time control to prescient control and eventually lead to independent operation that adjust with client needs and preferences.

Technology disturbance together with developing client desires are changing items at a fast pace. Shrewd associated items raise a modern set of vital choices related to how esteem is made and captured, how the monstrous sum of touchy information created is utilized and overseen, how connections with conventional trade accomplices are re-imagined, and what part companies play as industry boundaries are expanded. From homes and businesses to endeavors, Web of Things have ended up the commonplace in all strolls of life. Interfacing everything and everybody, IoT is making the world more intelligent and way better than ever some time recently. In any case, typically fair the starting. Opening the total potential of Web of Things requires that businesses get it the openings for esteem creation and systematically address the basic challenges.

Happiest Minds Web of Things benefit empowers organizations to convert commerce needs into competitive differentiate by conveying imaginative IoT powered solutions. From coordination the correct sensors and inferring motivated experiences at choosing the best-fit stage, we offer comprehensive IoT administrations to our clients (Zheng et al., 2018). We enable organizations to:

- Connect and scale with efficiency.
- Integrate and change commerce processes.
- Analyze and Act on unused data.
- Design, create, coordinated, convey and oversee conclusion to conclusion IoT process.
- Improve choice making with increased intelligence.
- Seamlessly coordinated IoT arrangements with existing venture design.

IMPORTANCE OF IOT

IoT touches each industry, counting health-care, fund, retail and manufacturing. Smart cities help citizens diminish to squander and vitality utilization and associated sensors are indeed utilized in cultivating to assist screen trim and cattle yields and anticipate development patterns.

As such, IoT is one of the foremost important innovations of standard of living and it'll proceed to choose up steam as more businesses realize the potential of associated gadgets to keep them competitive.

Advantages of IoT

The web of things offers a number of benefits to organizations, empowering them to:

- Monitor their overall commerce processes
- Save time and money
- Improve the client experience
- Enhance worker productivity
- Integrate and adjust trade models
- Generate more revenue and
- Make way better commerce choices.

IoT energizes companies to reconsider the ways they approach their businesses, industries and markets and gives them the instruments to progress their trade methodologies.

WORKING THE MECHANISM OF DIGITAL TWINS

Digital Twins, the virtual partners of the physical resources are made as digitized copies of machines/ gear or physical destinations utilizing sensors (Sierla et al., 2018). These advanced resources can be made indeed some time recently a resource is built physically.

Digital Twin Model

Figure 1 shows a simple twin model. Digital twin demonstration comprises of three primary parts as per Dr. Grieves, which are as stated below:

- Virtual items in virtual space
- Physical items in physical space
- Connected information that couple physical and virtual items all together

Figure 1. Twin Model

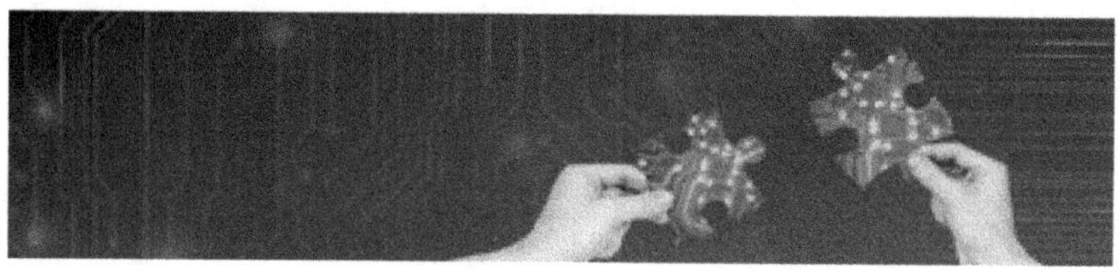

The complete concept behind the computerized twin is to a mass, expansive sum of digital information for virtual items; computerized data that's nearly indistinguishable from the physical partners. When the computerized data assembled, it serves as a twin of the data. This data embedded/immerse inside the physical item by itself and connected to it all through the life-cycle of the system (Stark et al., 2017).

The Unused Item Capabilities

To completely get a handle on how shrewd, associated items are changing how companies work, we must to begin with get it their inborn components, innovation and capabilities—something that our past article inspected. All savvy, associated items, from domestic apparatuses to mechanical gear, share three center elements: physical components (such as mechanical and electrical parts); keen components (sensors, chip, information capacity, controls, computer program, an implanted working framework, and a computerized client interface); and connectivity components (ports, radio wires, conventions, and systems that empower communication between the item and the item cloud, which runs on inaccessible servers and contains the product's outside working system).

Reshaping the Fabricating Company

To make items and get them to clients, producers perform a wide run of exercises, which for the most part take to put in a standard set of utilitarian units: inquire about and advancement (or designing), IT, fabricating, coordination's, showcasing, deals, after-sale benefit, human assets, acquirement, and back. The unused capabilities of savvy, associated items modify each movement in this esteem chain. At the center of what is reshaping the esteem chain is information.

Why Your Items Must be Shrewd and Connected

While the commerce esteem of a keen, associated item will shift by industry, showcase components, and client require quicker, more viable item overhauls and development. Cheaper, more effective upkeep and repair. Because keen, associated items can be checked within the field, that information can be used to altogether streamline the method of upkeep and repair. Because producers can persistently screen items in utilize, they can overhaul their items, fix issues and reconsider functionality—all of which makes strides client fulfillment. The insights from savvy, associated items can moreover advise future item highlights and unused item improvement.

The Impact of Digital Twins on Business Modernization

Exactly when an association passes on an automated twin, there ought to be desire to aggregate information, anyway to in like manner take that data and use it to drive express business results. Automated twins give extensively more than detectable quality into how remotely discovered equipment is functioning. The data assembled from modernized twins, and the value that best in class twins give, can and should be used to propel how the business limits (Buldakova & Suyatinov, 2020). At the most fundamental level, modernized twins give affiliations detectable quality and learning into the movement of their machines with respect to the earth or IoT condition they're a bit of. The data collected from sensors empower associations to have the parts or staff close by before issue become issues. This data is especially stunning — associations can use the data accumulated to be insightful, altering business methodology to be progressively shrewd later on. The decent thought of cutting edge twins is incredibly productive. By understanding a certifiable individual machine, rather than making theories reliant on industry wants, clears a segment of the ambiguity around a machine's errands.

The showings and examination are especially significant aftereffects of using mechanized twins. Subject masters can test circumstances, changes or updates, asking think about how conceivable it is that request to repeat conditions that could develop yet wouldn't be viable to attempt, in fact. All testing is done on the propelled twin first, giving endeavors the flexibility to walk around various circumstances going before choosing a decision that could influence business techniques, exercises and laborers (Serebrenny et al., 2019). The adaptability this can give an association to create and progress without annoying ordinary exercises is extremely valuable.

Decreasing costs in IoT developments and sensors, and the frameworks which support them, have moreover provoked a decrease in cost for cutting edge twins. At the point when cutting-edge twins are made, they can put aside a business money and delegate time. The advantages that can be saved by reducing individual time and cutting down help expenses is exceptionally critical to a relationship after sometimes. Breathing life into advanced twins with the Internet of Things.

The Internet of Things (IoT) has made it feasible for computerized twins to exist. IoT stages conquer any hindrance between the advanced and physical world. How can it work? Shrewd associated items and brilliant associated activities are associated with a cloud-based framework that gets and forms every one of the information's observed by sensors (Brossard et al., 2018). Utilizing the information caught by sensors, the recreation model, or computerized twin, is constantly refreshed and gives originators and designers the understanding they have to improve future item advancement endeavors. By considering the computerized twin under real working conditions, organizations can see the item in real life. Architects can settle on progressively educated decisions during the structure procedure and utilize computerized twins to make their reproductions increasingly exact.

LEVERAGING DIGITAL TWINS TO BREATHE NEW LIFE INTO YOUR PRODUCTS AND SERVICES

A digital twin is a virtual representation of a real-world product or service. That could be anything from a toaster to industrial machinery to complex processes. The virtual representation combines three types of information: business data, contextual data, and sensor data.

Connecting Real and Virtual

Business data covers information such as customer name, location, and service-level agreements. Contextual data include details such as ambient temperature, humidity, and weather events. Sensor data involves things like machine speed, operating temperature, and vibration.

Sensor data is key because while companies have been using digital twins for years, it's only with the Internet of Things (IoT) that they've become cost-effective. Gartner predicts that 6.4 billion things will be connected this year, a 30% jump over 2015. By 2020, at least half of all new business processes will incorporate IoT – transforming live data into new value.

Drilling Down on Digital Twins

How does a digital twin work? Let's say you manufacture industrial drills. A digital twin can help you understand how customers use your drill. The goal is to continuously improve the product to increase customer satisfaction and identify opportunities for new products and services.

For example, you might discover that your drill malfunctions in certain situations. That can enable you to improve product design. Or it can let you help customers modify the way they use the drill to avoid problems or you might discover that customers use your drill not only to make holes but also to cut materials. That might lead you to develop a new product that's purpose-built for cutting or maybe you discover that while customers want holes made, they don't necessarily want to purchase and operate a drill (Datta, 2017). So rather than sell drills, you might offer a hole-drilling service. In other words, instead of charging customers for machinery, they operate, you charge them for holes drilled by machinery you operate for them. Some SAP customers have been quite successful in making this kind of leap from products to services.

Digital Twins Across Industries

Digital twins aren't just for manufacturers. Insurers can apply digital twins in offerings like usage-based car insurance. Retailers can track how customers navigate the store and interact with products on the shelves. Cities can model areas for things like smart lighting. Ports can monitor weather, shipping traffic, containers, and trains and trucks entering and leaving.

Digital twins cover the entire lifecycle of an asset or process. In fact, they can form a foundation for an end-to-end, closed-loop value chain for smart, connected products and services, from design to production, from deployment to continuous improvement. The promise of continuous improvement is why it's increasingly important to integrate digital technologies into all products (Boschert & Rosen, 2016). As you leverage your digital twin to identify opportunities for new or better features, you can implement those improvements quickly and cost-effective through firmware updates. Implementing digital twins involves four steps:

I. Integrate smart components such as sensors, software, computing power, or data storage into new or existing products.
II. Connect the product to a central location where you can capture sensor data and enrich that sensor data with business and contextual data.
III. Analyze that data on an ongoing basis to identify opportunities for product improvements, new products, or even new business models.
IV. Leverage these digital insights to transform your company—for example, by reducing costs through proactive avoidance of business interruptions, or by creating new business opportunities.

THE THREE BASIC FUNCTIONS OF THE DIGITAL TWIN

The value of the digital twin is its ability to understand a product, to predict how the product will perform in production and in the field, and to optimize its performance and processes through its life-cycle.

Using the Digital Twin to Understand Product Design and Production

The data captured and managed in the digital twin offers a wealth of information about the product, its simulations, its composition, its production processes, and its expected behavior out in the field. All of this data constitutes the "understanding" of the product.

Using the Digital Twin for Prediction

The second part of the digital twin's value, according to Raman, is its predictive capabilities. "We talk about predicting product behavior based on conditions and constraints (Lohtander et al., 2018). I want to be able to accurately predict how the product will behave in different conditions and environments. I also want to predict plant operations during its production. If I'm in a brownfield plant versus a greenfield plant, I want to be able to predict what I'll need in equipment and whether there will be critical acquisitions required for production. "The digital twin collects and manages the data required to predict how the product will behave from production through its life in the field". IoT plays a role in prediction over the life of the product.

Optimizing With the Digital Twin

The third area of value for the digital twin is its ability to support optimization. "You can use the digital twin to optimize the product, the plant, and the product's performance in the field. We've all heard of generative design and what it offers in weight reduction, all calculated around strength.

The Digital Twin Can Make Strides Production

A computerized twin could be a virtual reproduction of a physical resource. But it's a lot more than that. The computerized twin could be a combination of IoT, machine learning, and AI put together to form a copy of your physical machine. It'll analyze and anticipate the affect of activities on that machine. An advanced twin draws on the Enormous Information drift. We have entered a period where information is one of the foremost vital assets companies can get, either for their clients and their machines (Tao et al., 2018). The objective presently is to accumulate as much information as conceivable in arrange to distinguish patterns or to respond to given circumstances.

THE DIGITAL GENERATION TWIN

A shrewd manufacturing plant could be a completely digitalized plant show speaking to a generation framework — an advanced twin for generation — that's totally associated to an item life-cycle administration (PLM) information store through sensors, supervisory control and information procurement frameworks, programmable rationale controllers and other robotization gadgets. In fabricating, an advanced twin empowers adaptability to decrease time required for the fabricating prepare and framework arranging, as well as for generation office plan. Breakthrough qualities and key enablers of savvy production lines are added substance fabricating, progressed mechanical technology, adaptable robotization and virtual and expanded reality (Knapp et al., 2017).

Conventionally, creators and producers work independently in numerous frameworks and toss data over the divider. This could make issues as data gets out of match up, making it troublesome for everybody to see the same picture. As a result, groups are able to survey execution and make vital alterations as it were at the late arrange for a physical model. Issues found within the stage can cause delays in generation and altogether increment the fetched to fix.

Additionally, these blunders can be exchanged into get together and establishment informational, which conclusion up on the shop floor or within the field. This not as it were making the item more troublesome to deliver the way in which it was outlined, but moreover can adversely affect the quality of the item itself. Using an advanced twin in fabricating offers a unique opportunity to for all intents and purposes reenact, approve, and optimize the whole generation framework to test how the item will be built in its aggregate utilizing the fabricating forms, generation lines, and computerization placed inside. The method coordination can be consolidated into the computerized generation twin to help groups in planning an compelling sideline coordination arrangement to bolster the generation lines (Tao et al., 2019).

THE VALUE OF A DIGITAL TWIN

Digital twins provide a large business potential moving off from analyzing the past towards predicting the long run. The live illustration of reality via digital twins permits us to evolve from ex-post knowledge gathering and analytic towards the period and ex-ante business practices. Based on the digital twin, formerly impossible practices can now be performed (Söderberg et al., 2017). Examples include real-time simulation, self-healing processes, and increased predictability. Imagine how customer satisfaction and customer loyalty will increase when products will let people know there's a problem before they break down.

Things (Products/Assets)

The digital twin has huge potential spanning across the entire product life-cycle, from R&D Engineering, manufacturing, service and maintenance, to operations. Here are some examples:

The digital twin helps manage the manufacturing of the "segment of one" more efficiently. Each physical product has a live representation in the digital world that can be tailored, designed, and created to individual needs. The individualized demand will be channeled into the digital supply chain. Hence, a drastic change will happen: Production will evolve towards mass customization and individualization.

The maintenance business of products and assets will benefit from reduced downtime and time for inspection. The effort for physical inspection could be significantly reduced. Usage and performance data will provide insight into the physical state of a product or asset. Based on usage patterns combined with machine learning, the optimized, cost-effective maintenance plan will be derived for industrial assets and as well as in your home. The digital twin allows establishing predictive maintenance strategies: Fix it before it breaks!.

In manufacturing operations, the digital twin innovation is a tremendous opportunity to simulate production and plant operations. Leveraging simulation of real scenarios using digital twins will help to optimize plant layout and accelerate production line design.

People

Another area where digital twins offer huge opportunities is the representation of people. They can improve lives in three main areas: Medical health, sports, and employee training and education (Ayoobkhan et al., 2018).

The illustration of a private person, as well as personal information like weight, health information, activity chase information, and medical treatment information will facilitate to determine prognosticate alerts and guide individuals to healthier lifestyles. It'll offer higher insights and transparency on the individual's health state of affairs by providing a lot of information points. Information shared with doctors can help define higher preventive methods and recovery plans and shrink health care prices.

Athletes can benefit from digital twins by designing training programs for their specific needs based on physical fitness level. In addition, the technology can have a dramatic mpact on individual performance in conjunction with IoT such as smart tennis racquet and other sports equipment. In the business world, worker training plays a crucial role in time to plug. Employees may be trained within the digital world 1st, resulting in tremendous resource optimization for on-the-spot coaching.

Services

Using a digital twin offers new opportunities for value-added services, some that are new services that are connected to merchandise and assets. As an example, you can add value by analyzing a product's current state and providing the right service in time to forestall downtime period. Improve service by delivering the proper data at the perfect time on technical data, diagnostics, product state, and spare elements.

In addition, new service models, such as remote service, are now becoming reality. With the digital twin, equipment acts as a service delivery unit and can self-heal, features via software updates or adding new features/services.

Networks

The network of digital twins may be a game-changer. By connecting things, people, and services, each side of business is just about described. Connecting distributed digital twins among firms, across firms, and across the system can permit firms to make virtual connected provide chain networks (Yuvaraj et al., 2019).

Leverage massive knowledge capabilities and this strategy provides unexampled visibility into operative performance conditions and states of health and creates the chance of predicting the future wants during a network of digital twins. Finally, it'll add the tremendous price by:

- Enhancing decision-making capabilities
- Offering new decisions
- Performing corrective actions before a system breaks

Imagine having the chance to manage air quality in massive cities. By connecting knowledge on traffic, climatic conditions, pollution, and health impact, measures is taken to enhance people's lives considerably.

Digital Twin of a Product

Equipping a product with its digital companion might create a variety of advantages:

- Behavioral outlook at any given point of the life-cycle, from development to disposal
- Models of deviation from the as-designed/as-built
- Prompt process adjustments

Nonetheless, judgment calls are always recommended before embarking on a digital twin initiative; while it's easy to imagine that fairly complex products can easily pay off in cost and effort burden, simple consumer goods are unlikely to, at least in the short term.

Network of Digital Twins = Connectivity + Communication + Collaboration

The digital twin as "a virtual representation of a real-life physical product. It could be a car, a chair, a desk, a lamp (Schleich et al., 2017). It could even be a person. Anything that exists in the physical world can be replicated as a digital twin". Figure 2 shows a broad classification of Digital twins.

APPLICATIONS OF DIGITAL TWINS

Digital Twin idea is that next enormous issue in most of the commerce divisions that makes a difference in precisely foreseeing the current state and future of physical resources by analyzing their computerized counter parts. By executing advanced Twins, organizations can pick up superior experiences on item execution, move forward client benefit and make superior operational and vital choices based on these experiences (Wang et al., 2019). We have begun seeing the most important application of Digital Twins within the following sectors. Some of the Digital Twin Applications are as follows:

1. Healthcare
2. Automobile
3. Manufacturing
4. Retail
5. Industrial IoT
6. Smart Cities

Health-Care

Along with the IoT information, Digital twin has a crucial half within the field of health-care from expenses and savings to patient observation, precautionary or preventative maintenance and providing personalized health care. Figure 3 shows a customized health-care system.

Figure 2. Digital Twins broad picture: products, assets, stakeholders

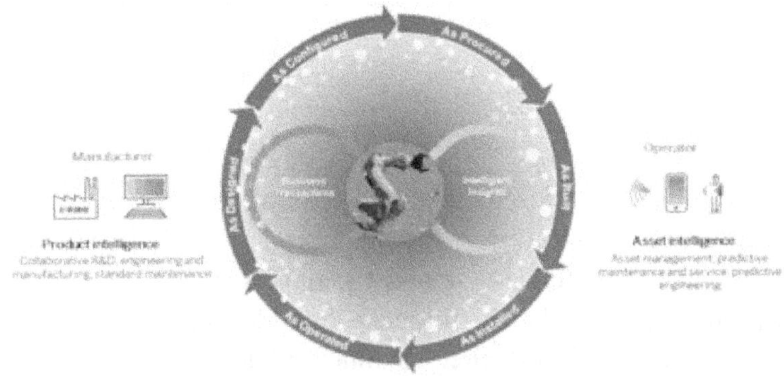

Figure 3. Personalized health-care

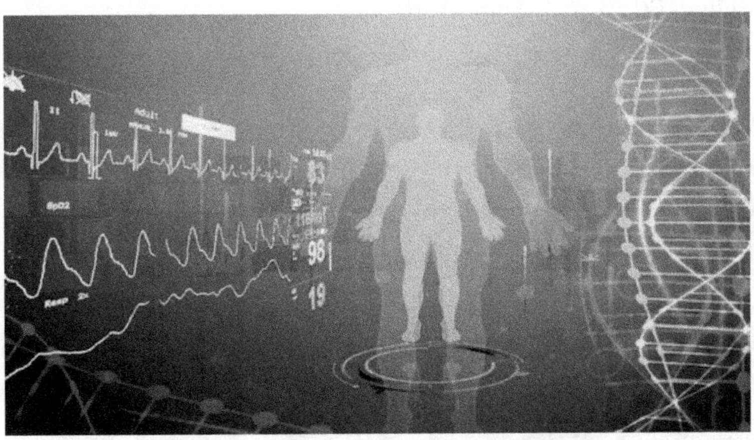

Automobile

The digital twin is employed in the sphere of Automobile to investigate overall vehicle performance and options by/through capturing the behavior and operational knowledge of the vehicle.

It provides custom service to the purchasers. Thus, the appliance of those technologies within the gift day vehicles can offer the generation what they require. Figure 4 shows a virtual automation.

Figure 4. Automation in Digital Twins

Manufacturing

Formerly, once there was a necessity to urge the elaborate info regarding the report of any in operation product then it had been done in person observation that product. However these days, Digital twin is impacting the producing sector. It virtualize the task by making and maintaining a digital illustration. This conjointly makes producing of product additional economical and optimized whereas cutting down prototyping or construction prices. It allows to predict failures additional simply. During this means, Digital twin helps to open new strategies for up the productivity of product. Figure 5 shows a manufacturing unit.

Figure 5. Product Manufacturing in Digital Twins

Retail

Digital twin implementation will play a significant role in increasing the retail client expertise by making virtual twins for patrons. It conjointly helps in higher security implementation, in-store coming up with, associated energy management in an optimized manner. Figure 6 shows an account maintenance robot.

Figure 6. Product Billing and Account Maintenance

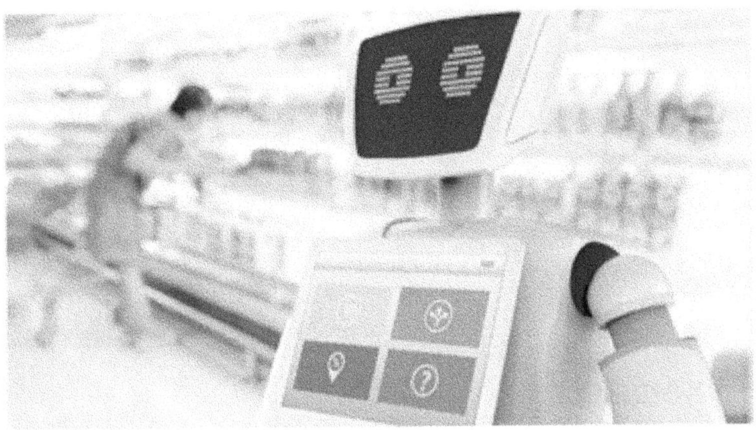

Industrial IoT

Industrial companies will digitally monitor, track and management industrial systems. Besides operational knowledge, digital twin model stores environmental knowledge like location, monetary models, configuration, etc. It helps in foretelling the operations and anomalies.

Smart and Sensible Cities

Figure 7 shows a smart, sensible city. Digital twin beside the IoT information helps in coming up and implementation of smart cities with totally different aspects:

- Reduction of the ecological footprint
- Efficient Management of resources
- To enhance economic development.
- Increase the overall quality of a citizen's life

Digital Twins and Sensible Connected Product: Business Goals and Edges

The digital twin technology also do expecting to expand its reach on the far side the presently predominant.

Figure 7. Automated Smart and Sensible Cites

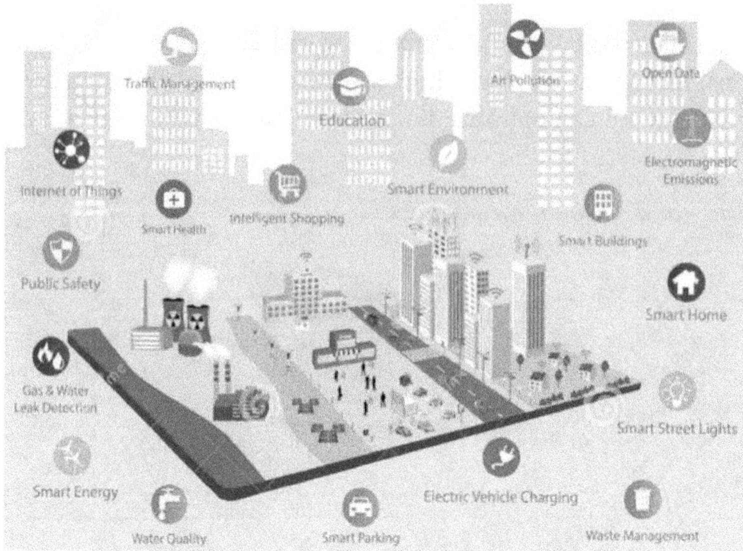

Smart Connected Products

We are within the era of the "smart" product – connected devices that collect and transmit a massive vary in valuable information that usually goes harvested and unrefined. Strip-mined and used effectively, that information will offer antecedent not possible insight into virtually each facet of the method your organization, customers and marketplace behave, permitting you to form vital future enhancements and efficiency to the method you are doing business.

Embracing the Benefits of the Digital Twin

The manufacturing digital twin offers many benefits to production. Primarily it provides complete visibility onto the whole plant to give greater confidence in performance. The view it provides onto processes presents clarity on where to focus and where continuous process improvements can be made. It enables the allocation of smart resources in real time and provides a clean foundation, with the complete, real-time data required for technologies such as virtual and augmented reality. With the continuous process improvement and greater production efficiency, they offer for businesses, the digital twin and Industry 4.0 can be used to defend a company's position in the market (Söderberg et al., 2017). They are, however, also being used to create new lines of business with the premium.

- products and quicker launching
- First Time Right & Dependability
- Increase in the unit of time between failure and success
- Reduction in special maintenance sessions

- Reduced Repair Service visits along with Delivery Cost and Time
- Reduced necessary standby equipment/additional floor area to hide period of time

- Client Fulfillment
- Reduces production period of time
- Real time hep selection on Plant Operations for the client
- Integrated visibility of monetary information victimization worth tree analysis advantages in hep decision
- Improved utilization and accessibility of wealth

- Cost effective Saving
- Minimize maintenance value
- Reduced spare-part inventory
- Reduction in field crew examination miles
- Asset improved support bench-marking & Plant wise utilization
- Integrate with Existing Systems for a Holistic read into Operations reducing Manual processes.

FUTURE OF THE DIGITAL TWIN?

Digital twins supply new and memorable prospects within the designed surroundings. Digital twins will be wont to improve efficiency, optimize processes, discover issues before they occur and pioneer for the longer term. They can be used to create long-term value by solving big problems using a people-centric approach.

Digital twins process a lot of data and generate a lot of useful information. Through APIs, this information can be exposed to building operators. But it can also be exposed to other critical people and services — such as first responders. In this way, digital twins hold the promise of connecting smart building with other smart buildings and also with smart cities.

Digital twins are truly becoming the lifeblood of smart buildings. They are unlocking the promise of IoT and enabling the new future of work.

Virtually Prototyping With Digital Twins

For businesses already using engineering simulation to design products, connecting the simulation to the physical product in the field is necessary to deploy the digital twin solution. For businesses that are new to simulation, the engineering team would first need to build the 3D product model, optimize its performance and replicate the real world in which the product system would operate. Figure 8 shows a virtual prototype. Re-using this data in similar product simulation scenarios for future product development saves time and money.

1. Build a product using 3D CAD.
2. Build a physical prototype of the product.
3. Add sensors to the prototype.
4. Build the sensors in the 3D CAD model.
5. Collect the data.
6. Associate the sensors' data streams to the sensors in your 3D CAD model.

Voila! Now your model can talk.

Figure 8. Virtual Prototyping

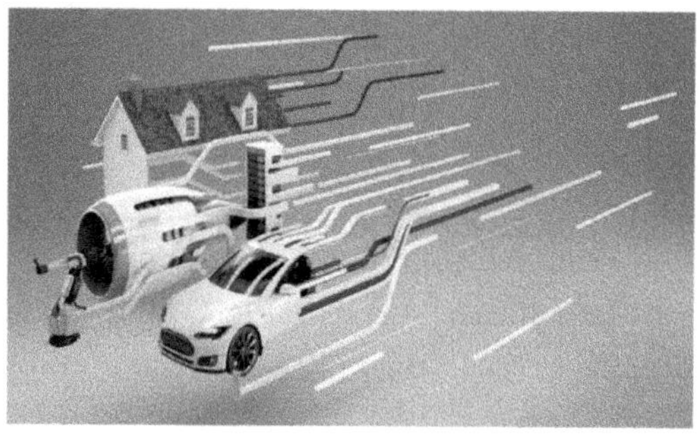

Not solely do digital twins improve future innovation and products development efforts for designers and engineers – they build a stronger relationship between engineering and operations groups. The info collected from sensors is analyzed by the operations team to optimize performance, service, and maintenance over the period of time of a product. The digital twin will facilitate organizations avoid the pricey period of time, repairs, replacements, or keep prior alternative performance problems.

Get Ready for Digital Twins to Revolutionize Manufactured Products

Digital Twins, with the use of IoT technologies to make the concepts of smart products and smart services a reality. The purpose of people deploying them is to be able to configure digitally the asset qualities.

Digital Twins: The Background

As a part of digital deployment an simulates an asset's physical behavior by digital twin. These are used in the industrial design world as a way of optimally adjusting the qualities of a product. Digital twins are also used in optimize the behavior of an asset in the field, through simulations that are made with the use of software tools.

If you want to design a car, but you don't know how well it will perform in real life situations like very tough weather conditions or the industrial designer can design the component with the use of special 3-D design software. This software gives you to create a digital drawing of the component you want to manufacture and simulate real-life situations, like those extreme weather conditions. Applying tension load to your digital component design, or digital twin, in specific directions you can mimic real world stresses on the product, like the force of the wind. We can test the attributes, such as reaction, resistance, and failure thresholds.

After testing, you could go back to your digital drawing, the digital twin, and adjust its shape and qualities before committing to manufacturing it. By this process you are ability to mitigate the chances of product failures before going into production.

Digital Twins Until Today

Digital twins are primarily focused in the following areas:

The raw material and the processes to manufacture the leads to good design and finished good, BOM (Bill of Materials) and BOP (Bill of Processes).

Digital twins concerned solely with the R&D/design/manufacturing processes. The digital twin technologies are the various Computer Aided Design, Engineering and Manufacturing packages (CAD/CAE/CAM), along with their connection with the manufacturing equipment, machines and devices in the field, and dashboards and control panels. Till now, the digital twin has not enabled the tracking of the asset throughout its entire lifecycle, as there was limited relation of the production planning, the warehouse management and the shipping processes of the products (Ahuett-Garza & Kurfess, 2018). The visibility of the top-level operations within the manufacturing premises was limited.

Various new tech developments can really push the benefits of the concept beyond traditional manufacturing and product management, digital twins are gaining traction and interest today.Starting from industrial and product design, quality management and, and extending to the areas of ergonomics, manufacturing and marketing, new tech is enabling novel fields lifecycle management and capabilities for digital twins, such as decentralized manufacturing, mass-customization of products and Interactive industrial design.

The IoT (Internet of Things), because the enabler of the communication for assets or products, will empower the prevailing practicality of digital twins in a every method that:

Bridges the gap between style and merchandise usage, as a result, It will provide services and insight throughout the after-sale process, once the merchandise is within the hands of consumers.

Benefits Includes

- Offering or Enabling the assets as-a-service for the products through continuous communication between the product and its manufacturer.
- Transferring the physical world to the digital realms
- As industrial manufacturers move to new business models that help them achieve higher profitability and discover new revenue streams, the concept of the selling "as-a-service" and "pay-per-use" or "pay-per-outcome" are emerging
- Increased product customization through the participation of consumers or external designers/engineers in the product creation process.

- Real-time optimized warehouse management and logistics.
- Continuous visibility after-shipment and after-sale of the product.
- As industrial manufacturers move to new business models that help them achieve higher profitability and discover new revenue streams, the concept of the selling "as-a-service" and "pay-per-use" or "pay-per-outcome" are emerging.
- The reason for the digital twin enthusiasm is the underlying notion that the world is being radically digitized in all possible ways. They even though currently applied within the environment of the industrial work space, represent in a great way the capabilities of changing the physical world to the digital realm. The main concept of the digital twin, as envisaged in the context of the 4th Industrial Revolution (Industry 4.0), promises to enable full connection between consumers and the products that they use, and that of manufacturers with the products they create and merge these worlds even more.
- As important as it is to have the internet of things platform strategy in place, a digital twin's strategy for IoT- enhanced products and services is proving to be equally important. The concept of the digital twin is already proving to be valuable for product management and is expected to complement or even replace the every concept of the product offering itself.

Bringing Digital Twins Together

The interconnection of physical objects via IoT is going to be replicated with the interlocking of digital twins. The general insights and analytics gleaned from objects and the way they move with each other become exponentially higher, resulting in even a lot of prospects to impact outcomes in complicated operations. These composite digital twins can offer a holistic approach to viewing and analyzing processes throughout a complete organization.

CONCLUSION

The few examples of industries where the digital transformation is done by the digital twin, but the potential is wide-reaching. In the manufacturing and engineering spaces, you can put your years behind you, competitors, by advising your company about digital twins. Digital twins will be a disrupted for many industries and will simultaneously change business models.

REFERENCES

Ahuett-Garza, H., & Kurfess, T. (2018). A brief discussion on the trends of habilitating technologies for Industry 4.0 and Smart manufacturing. *Manufacturing Letters*, *15*, 60–63. doi:10.1016/j.mfglet.2018.02.011

Ayoobkhan, M. U. A., Chikkannan, E., & Ramakrishnan, K. (2018). Feed-forward neural network-based predictive image coding for medical image compression. *Arabian Journal for Science and Engineering*, *43*(8), 4239–4247. doi:10.100713369-017-2837-z

Boschert, S., & Rosen, R. (2016). Digital twin—the simulation aspect. In *Mechatronic Futures* (pp. 59–74). Cham: Springer.

Brossard, M., Erntell, H., & Hepp, D. (2018). Accelerating product development: The tools you need now. *The McKinsey Quarterly*.

Buldakova, T. I., & Suyatinov, S. I. (2020). Assessment of the State of Production System Components for Digital Twins Technology. In *Cyber-Physical Systems: Advances in Design & Modelling* (pp. 253–262). Cham: Springer. doi:10.1007/978-3-030-32579-4_20

Datta, S. P. A. (2017). Emergence of Digital Twins-Is this the march of reason? *Journal of Innovation Management*, *5*(3), 14–33. doi:10.24840/2183-0606_005.003_0003

Knapp, G. L., Mukherjee, T., Zuback, J. S., Wei, H. L., Palmer, T. A., De, A., & DebRoy, T. (2017). Building blocks for a digital twin of additive manufacturing. *Acta Materialia*, *135*, 390–399. doi:10.1016/j.actamat.2017.06.039

Lohtander, M., Ahonen, N., Lanz, M., Ratava, J., & Kaakkunen, J. (2018). Micro Manufacturing Unit and the Corresponding 3D-Model for the Digital Twin. *Procedia Manufacturing*, *25*, 55–61. doi:10.1016/j.promfg.2018.06.057

Qi, Q., Tao, F., Zuo, Y., & Zhao, D. (2018). Digital twin service towards smart manufacturing. *Procedia CIRP*, *72*, 237–242. doi:10.1016/j.procir.2018.03.103

Ray, P. P. (2016). A survey of IoT cloud platforms. *Future Computing and Informatics Journal*, *1*(1-2), 35–46. doi:10.1016/j.fcij.2017.02.001

Schleich, B., Anwer, N., Mathieu, L., & Wartzack, S. (2017). Shaping the digital twin for design and production engineering. *CIRP Annals*, *66*(1), 141–144. doi:10.1016/j.cirp.2017.04.040

Serebrenny, V., Lapin, D., & Mokaeva, A. (2019, July). The Concept of Flexible Manufacturing System for a Newly Forming Robotic Enterprises. In *Lecture Notes in Engineering and Computer Science: Proceedings of The World Congress on Engineering 2019* (pp. 267-271). Springer.

Sierla, S., Kyrki, V., Aarnio, P., & Vyatkin, V. (2018). Automatic assembly planning based on digital product descriptions. *Computers in Industry*, *97*, 34–46. doi:10.1016/j.compind.2018.01.013

Söderberg, R., Wärmefjord, K., Carlson, J. S., & Lindkvist, L. (2017). Toward a Digital Twin for real-time geometry assurance in individualized production. *CIRP Annals*, *66*(1), 137–140. doi:10.1016/j.cirp.2017.04.038

Stark, R., Kind, S., & Neumeyer, S. (2017). Innovations in digital modeling for next generation manufacturing system design. *CIRP Annals*, *66*(1), 169–172. doi:10.1016/j.cirp.2017.04.045

Tao, F., Cheng, J., Qi, Q., Zhang, M., Zhang, H., & Sui, F. (2018). Digital twin-driven product design, manufacturing and service with big data. *International Journal of Advanced Manufacturing Technology*, *94*(9-12), 3563–3576. doi:10.100700170-017-0233-1

Tao, F., Sui, F., Liu, A., Qi, Q., Zhang, M., Song, B., ... Nee, A. Y. C. (2019). Digital twin-driven product design framework. *International Journal of Production Research*, *57*(12), 3935–3953. doi:10.1080/00207543.2018.1443229

Wang, J., Ye, L., Gao, R. X., Li, C., & Zhang, L. (2019). Digital Twin for rotating machinery fault diagnosis in smart manufacturing. *International Journal of Production Research*, *57*(12), 3920–3934. doi:10.1080/00207543.2018.1552032

Yuvaraj, D., Sivaram, M., Mohamed Uvaze Ahamed, A., & Nageswari, S. (2019). Some Investigation on DDOS Attack Models in Mobile Networks. *International Journal of Interactive Mobile Technologies*, *13*(10), 71–88. doi:10.3991/ijim.v13i10.11304

Zheng, P., Lin, T. J., Chen, C. H., & Xu, X. (2018). A systematic design approach for service innovation of smart product-service systems. *Journal of Cleaner Production*, *201*, 657–667. doi:10.1016/j.jclepro.2018.08.101

Chapter 17
Taxonomy of Influence Maximization Techniques in Unknown Social Networks

B. Bazeer Ahamed
https://orcid.org/0000-0003-1559-8386
Al Musanna College of Technology, Sultanate of Oman

Sudhakaran Periakaruppan
SRM TRP Engineering College, India

ABSTRACT

Influence maximization in online social networks (OSNs) is the problem of discovering few nodes or users in the social network termed as 'seed nodes', which can help the spread of influence in the network. With the tremendous growth in social networking, the influence exerted by users of a social network on other online users has caught the attention of researchers to develop effective influence maximization algorithms to be applied in the field of business strategies. The main application of influence maximization is promoting the product to a set of users. However, a real challenge in influence maximization algorithms to deal with enormous amount of users or nodes obtainable in any OSN is posed. The authors focused on graph mining of OSNs for generating 'seed sets' using standard influence maximization techniques. Many standard influence maximization models are used for calculation of spread of influence; a novel influence maximization technique, namely the DegGreedy technique, has been illustrated along with experimental results to make a comparative analysis of the existing techniques.

INTRODUCTION

The conceptualization of Online Social Networks (OSNs) has been one of the most stimulating events in this century. A social network can be considered to be any website or web application which allows for social experience in the form of user interactions. Nowadays, such online social interactions have led to several amusing online activities including posting photos, chatting, tweeting, online shopping, etc. Many

DOI: 10.4018/978-1-7998-3970-5.ch017

popular OSNs such as Facebook, Orkut, Twitter, LinkedIn and YouTube have become more popular day-by-day. Statistics reveal that currently over 75% of all Internet users are also profoundly engaged in using social media sites [1]. Growth in usage of social media emerged mainly due to the heavy increase in the usage of smart phones, tablets, laptops, and other similar gadgets, which has become one of the defining factors in advancement of human life.

Social Network Analysis and Mining

In this new era, where social networking has gained huge popularity, social media marketing has also become a major business strategy. Social media marketing is a form of marketing that makes use of the social networking sites as a marketing tool to increase brand exposure and to expand customer reach. As a result, the vast amount of content-rich data retrieved from such OSNs is considered to be a wealth of information that is analyzed using data mining techniques for various analytical researches. This social data is generated as a result of various implicit and/or explicit user activities such as purchasing of products, rating an item, liking an image, chatting with others, posting tweets, and so on. However, the real challenge lies in how to accurately and efficiently analyze and process such data to extract meaningful and valuable patterns or knowledge. It is also important to note that OSN data used in various researches is noisy, enormous, scattered and dynamic. To analyze such vast, complex, and regularly changing raw data requires proper data mining techniques, commonly known as social network mining. Social Network Analysis and Mining (SNAM) is one of the hot topics in the area of research in data mining and it is still in its infancy.

Influence Maximization

Currently, as OSNs have become a daily activity for majority of the Internet users, the latter largely rely on making decisions based on the influence of such sites. For instance, influence propagation helps a social network user to consciously or unconsciously decide in making a choice on which product to buy, which movie to watch, which community to join, which programmed to participate in, and so on. Patterns of influence can also be studied as an indication of user trust and utilized for computing trust propagation. For instance, a user „X" writes in her timeline about a movie that she recently watched and liked. This update gets communicated to many of her online friends who read her update. When few such friends give a comment on that update, the information is passed on to their friends, and this becomes a chain process. Thus, an update has the potential to be propagated positively through a social network.

A major goal of influence maximization is „viral marketing" or „viral advertising", which is nowadays considered a major marketing technique being adopted by every company and organization to promote their products or brands, or to create awareness about their organization (Ahamed, B. B., & Ramkumar,2016). The basic principle of viral marketing is to initially find a set of few influential users of a particular social network, and convince those users about the goodness of a product (one way of doing so is to supply free samples of the product), so that it can generate a cascade of influence among the users' friends who will, in turn, recommend the same product to other friends, and this will result in a grand promotion of the product. But the major issue lies in initially finding the best influential users in the social network, who are considered as the „seed set" of the network. A similar strategy of viral marketing via social networks is also adopted in political campaigns or personalized recommendations.

Figure 1. The generic influence maximization framework

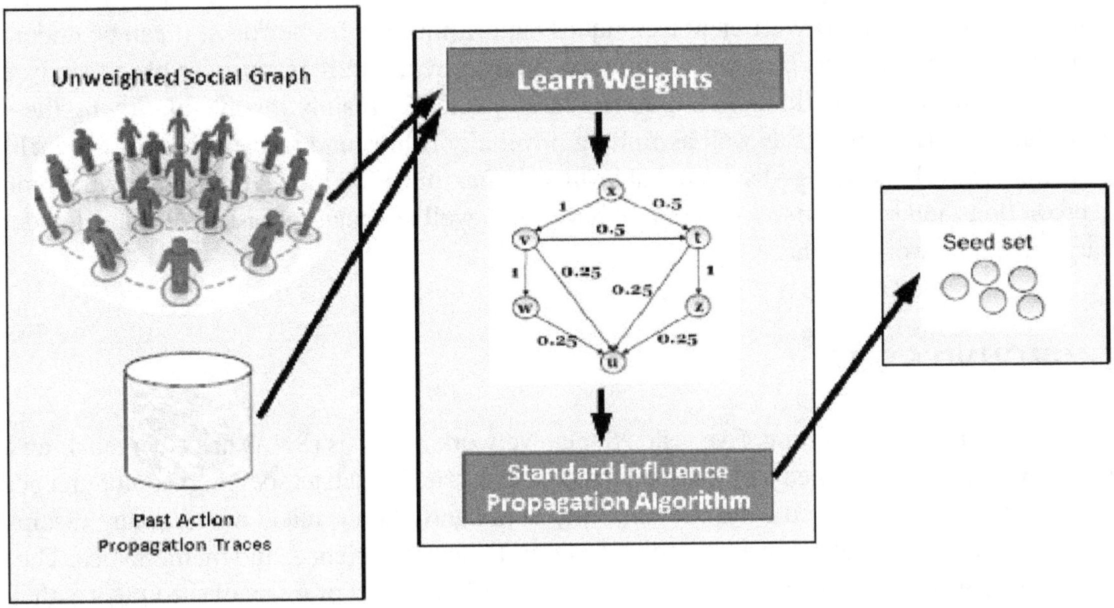

Link Prediction

Another common research area in SNAM is the 'link prediction' problem which considers a static snapshot of the network at a given time 't1' and predicts future links or edges of the network at a future time 't2'. The link prediction problem is a means to predict the user as to 'who can be a possible friend' of that particular user and accordingly provide suggestions for the same. The common example of this is the „Friends Request" feature found in Facebook or „Received Invitations" feature in LinkedIn networking sites. These features allow enhancing the social circle of friends or professionals for a user who might be just a few steps away from the desirable link and this, in turn, enhances the other activities related to social networking sites.

Figure 2. The generic link prediction framework

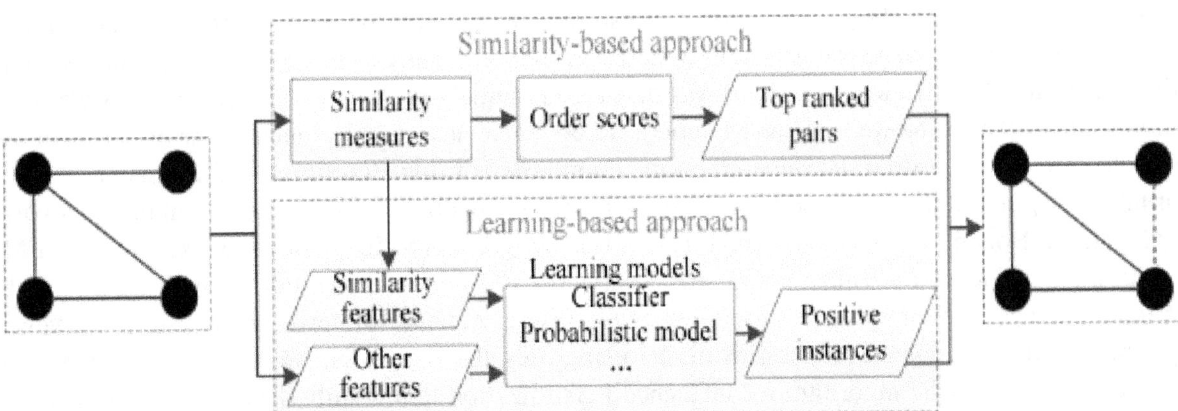

Social media data are largely user-generated content on social media sites which are vast, noisy, distributed, unstructured, and dynamic. These characteristics pose challenges for researchers to carry out data mining tasks to invent new efficient techniques, algorithms and frameworks. It can be understood from the previous discussions that OSN mining is a huge area, and it is not possible to study every aspect of it in limited time period. Hence, our research work will mainly involve improving the existing algorithms and frameworks as well as finding novel algorithms and frameworks for the Influence maximization a novel knowledge-based social influence maximization model that can provide more accurate predictions and solve the low computational cost as well as scalability issues involved in dealing with large social network graphs.

BACKGROUND STUDY

With a history of more than seventy five years, Social Network Analysis (SNA) has emerged as an interdisciplinary technique developed and widened under several fields such as sociology, computer science and mathematics, and studies in SNA are currently abundantly being made across many disciplines, including sociology, psychology, medical science, communication science, and mathematics. The field of SNAM has gained a rapid stimulation in recent years. A key aspect of many of the OSNs is that they are rich in data and provide exceptional challenges and opportunities from the perspective of data mining.The basic computational problem in influence maximization in social networks is that of picking up a "seed set", i.e., a set of initial users which will result in as an optimum set for influencing the rest of the users in the OSN. The problem of influence maximization in social networks was first discussed using a benchmark algorithm by Domingos and Richardson in (Ahamed, B. B., & Ramkumar,T,2015). and (Ahamed, B. B., & Murugan, K,2019) where emphasis is laid on the growing importance of viral marketing in OSNs to promote products. Domingos and Richardson also emphasized on the study of two significant aspects of customers: the intrinsic value of customers (which is based on the products he/she has purchased or likely to purchase) as well as the network value of customers (which is based on how much of a positive influence he/she can create to the maximum members in a network).

The next algorithmic treatment to influence maximization was given by (Chen, W., Wang, C., & Wang, Y.2010) who proposed two related fundamental models for the same: the Independent Cascade (IC) model and the Linear Threshold (LT) model. In these models, a node is considered be in either of two modes – active or passive – at a given timestamp. For instance, if we consider a particular product, those customers who have bought the product can be considered to be active. Also, those customers who have the tendency to buy the product can also be considered to be active. Then each customer's tendency to become active in the near future increases monotonically as more of its neighbors become active. Later, another efficient algorithm was developed, namely the Cost-Effective Lazy Forward selection (CELF) (Liben-Nowell, D., & Kleinberg, J,2007) which generates much faster output than the traditional Greedy algorithm. However, the time complexity of this algorithm is still a major concern as it takes a significantly long time to suggest even fifty influential users in a large complex network. In (Goyal, A., Bonchi, F., & Lakshmanan, L. V2011) suggested the Expectation Maximization (EM) algorithm for influence maximization in social networks using the simple shortest path technique. In EM model, the concept of episode is used, which refers to a sequence of newly active nodes. However, experiments were conducted for the EM model using only the IC model. (Guler, B et al.,2015) also proposed a new heuristic algorithm for influence maximization that mainly deals with the scalability issue and produces considerably good results for even millions of nodes in a network. More recently, Chen et al. proposed another model called Local Directed Acyclic Graph (LDAG) [7] for the LT model but their model cannot adopt to working for all the social media sites to generate seed sets for influence propagation. (Leskovec, J., et al., 2007)also studied the problem of learning influence probabilities using the database approach. In particular, a new model was built, namely the credit distribution model that studies the flow of influence in a network and uses this result to estimate the probable influence spread.

However, to the best of our knowledge, only very few papers such as (Saito et.al,2008) and (Adamic, L., & Adar, E.,2005). have studied influence maximization OSNs from the topic modeling perspective and hence it requires a more in-depth study in this domain from the topic modeling perspective. In (Kundu, A., & Ratha,2019) a novel influence propagation model is introduced that deals with social sensor networks that consider both positive and negative relationship types. In this model, sensor observations are used for the relationship types in the social network for minimizing the end-to-end propagation cost of influencing a target user.The main target for link prediction in OSNs is to surmise new interactions among users of the social network that are likely to take place in the near future. The first discussion on link prediction in social networks was discussed in (Liu, L., et.al,2010) in which two approaches are mentioned to handle the link prediction problem: the first approach is based on the local features of a network. Examples of few local-based approaches of link prediction include Common Neighbors Index, also known as Friend Of A Friend (FOAF) algorithm (Richardson, M., & Domingos, P., 2002) Jaccard Coefficient (Liu, L., et.al,2010) Adamic/Adar Index (Barbieri, N,2013) Preferential Attachment (Chen, J., et.al,2009) and FriendLink algorithm (Domingos, P., & Richardson, M,2001) The other approach is based on global features, which explores the entire path structure of the network. Examples of few global-based approaches of link prediction include Katz Status Index (Papadimitriou, A., et al,2012), Random Walk with Restart (RWR) algorithm (Bullas, J., 2014) and SimRank algorithm (Chen, W.,et. al, 2010) There are also some high-level approaches that can be used along with any above mentioned methods of link prediction, namely low-rank approximation, unseen bigrams and clustering.

Influence Maximization Models

Most of the standard influence maximization techniques consider that the following information is available to produce the seed set:

a). A social network G=(V,E) where V denotes the set of vertices and E denotes the set of edges or links among vertices

b). A standard propagation model to determine how influence diffuses over the social network. The Independent Cascade, the Linear Threshold and the Weighted Cascade models are three of the most fundamental and widely-studied diffusion models, but many extensions of these models can also be considered.

c). A function pu,v: E → [0,1] that indicates the parameter p u,v for assigning a probability to each pair of nodes u and v. The probability value decides the amount of influence node/user u exerts over node/user v.

In case of IC model, each node can be either active or inactive. The tendency of nodes to become active increases monotonically as more and more neighbors become active. When a node v becomes active, it considers its neighbors and tries to activate them too. However, node v has only a single chance of activating each currently inactive neighbor u. The attempt of activation succeeds with probability pvu. As given in Figure 3, if we consider the newly active node V, it has two immediate inactive neighbors U and W. The probability of activating U and W at time t1 by node V is 0.2 and 0.5 respectively. The probability values PV,U and PV,W which are 0.2 and 0.5 respectively are the parameters of the system. Hence, node V is successful in activating W, but not U, considering the minimum threshold probability to be active is 0.3. Now, node W is the newly active node and it, in turn, tries to activate its immediate inactive neighbors U and X. The probability of activating U and W at time t2 by node V is 0.3 and 0.5 respectively. Hence, node W is successful in activating U and X. The process is repeated until no more activation is possible. Considering this example, the spread of influence for node V is three as it is successful in influencing a total of three nodes.

Figure 3. An example of the IC model being executed for a newly active node v

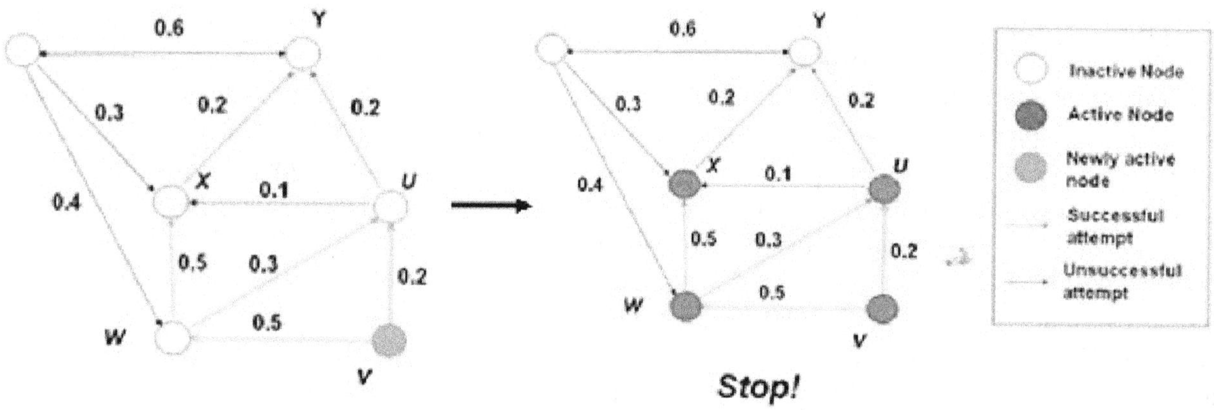

The Linear Threshold (LT) Model

In case of LT model, each node v can be influenced by its immediate neighbor u based on a weight bu,v such that:

$$\sum_{u \text{ neighbour of } v} b_{u,v} \leq 1 \tag{1}$$

The entire process of this diffusion model then proceeds as follows. Each node v at first chooses a threshold, θv, at random from interval [0, 1]. This value signifies the weighted sum of v's neighbors that must become active in order for node v to become active. A node v will be considered active when at least the weighted sum of its neighbors is active. This means, only that particular node v will be activated for which the total weight of its active neighbors is at least θv, as shown in equation 2 for the weight bu,v

$$\sum_{u \text{ active neighbour of } v} b_{u,v} \geq \theta_v \tag{2}$$

In each discrete step, all nodes that became active in the previous steps will remain active for the future steps. For instance, if we consider Figure 4, each node has been assigned a random threshold value marked by blue shade of color. Initially, the active node V is chosen and it further tries to activate nodes U and W. It succeeds in activating node W but not node V. Node W in turn activates node U, and the latter, in turn, activates node X. The process stops once no further activation of nodes is possible

Figure 4. An example of the LT model being executed for a newly active node v

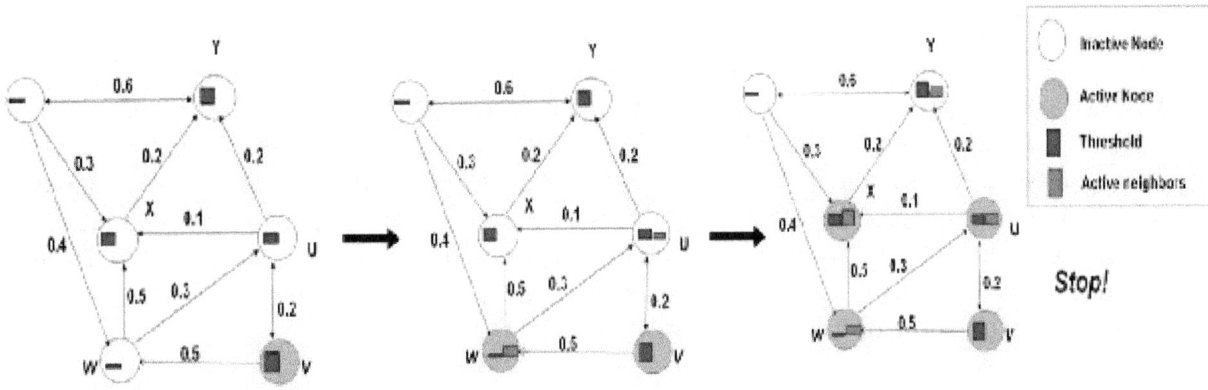

STANDARD INFLUENCE MAXIMIZATION TECHNIQUES

For all the three influence maximization algorithms explained below in this chapter, the OSN considered is represented as an undirected or directed graph G=(V, E), where V denotes a set of vertices or nodes in the network and E represents a set of edges between two vertices to signify a connection between the two vertices/nodes.

One important property used in influence maximization is that given a function f that is submodular, if only non-negative values are taken, then we have

$$f(S \cup \{v\}) \geq f(S) \tag{3}$$

for all nodes v and set S. Thus, the problem of influence maximization can be converted into finding a k-node for set S for which f(S) is maximized

Table 1. Measurement of degree centrality

Node v	1	2	3	4	5	6	7	8	9	10
DC(v)	3	3	3	1	3	1	4	2	3	1
Degree(v)	2	2	2	8	2	8	1	7	2	8

If we consider that there are N nodes in a network, for a node v, its degree centrality (DC) can be defined as:

$$DC(v) = \sum_{i-1}^{N} \sigma(x_i v) \tag{4}$$

where, σ (xi,υ) is defined as follows

σ $(xi,\upsilon) = 1$, iff xi and v are connected; = 0, otherwise

Initially S contains an empty set (step 1) containing no nodes. Then, for each node v, the degree of the node is calculated and stored (steps 2-4). Here, vdi (in step 6) indicates the vertex having ith highest degree centrality value. Once the set S, having top k influential seeds, is determined (step 8), a measure for estimating the total spread of influence in the OSN is carried out using some standard diffusion model. Any of the three standard diffusion models that can be used for calculation of spread of influence are explained in the previous section 4.1. However, which diffusion model will generate more accurate results depend on the nature of the dataset being used.

Algorithm 1: Degree Centrality (G, *k)*

1: Initialize S = ϕ
2: **for** each vertex *v* do
3: compute its degree d_v
4: **end for**
5: **for** i = 1 to k do
6: S = S U $\{v_{di}\}$
7: **end for**
8: output S

The general Greedy approach for influence maximization is also based on using some standard influence propagation models, such as the Linear Threshold (LT) model, the Independent Cascade (IC) model or the Weighted Cascade (WC) model and it finds the spread of influence by carrying out the simulation process for these models [4]. The two parameters required as input to the general greedy algorithm is the OSN graph G and the value of k, which is the number of expected influential nodes in the seed set.As explained in Algorithm 4.2, initially S contains an empty set (step 1) containing no nodes. The entire simulation process is carried out Sim times to consider the average of all results generated in each iteration. The process is repeated k number of times to store top-k nodes in the seed set (step 2). Random(S) (step 6) represents the random simulation process carried out using any of the influence diffusion models to generate the final seed set S containing top-k influential seeds. In each iteration (step 3), exactly one node is added into the set S in such a way that this node v together with the current set S maximizes the spread of influence in G (step 10). Thus, the simulation process is carried out Sim times repeatedly for each vertex v to estimate the spread of influence by considering S U {v} (steps 3–9). Here, sv stores the average spread of influence propagated by vertex v (Step 8) by randomly repeating the process Sim number of times.

Algorithm 2: GeneralGreedy(G, k) [4]

```
Initialize S = ϕ, S_im = 20000
for i = 1 to k do
    for each vertex v ∈ V \ S do
        s_v = 0
        for i = 1 to S_im do
            s_v += |Random(S U {v})|
        end for
        s_v = s_v / S_im
    end for

        S = S U {arg max_v∈V\S {s_v}}
end for
output S
```

Algorithm 3 explains the novel DegGreedy influence maximization technique that aims to provide improved results in terms of accuracy than the High-Degree centrality and the general Greedy algorithms. In fact, the central idea of this DegGreedy algorithm is to apply a hybrid approach of these two existing influence maximization techniques to yield the top-k influential nodes in a network G. The three parameters required as input to the DegGreedy algorithm is the OSN graph G, the total number of expected influential nodes represented as k, and a user-defined value m (value of m should be significantly greater than value of k and is rather dependent on k). As explained in Algorithm 3, initially T contains an empty set (step 1) containing no nodes. Then, the degree of each node in the graph is computed and stored (steps 2-4).

In this novel algorithm, the choice of the values of k and m are very crucial for generating optimum result. The user-defined value of m is an important parameter whose value has been experimentally found to be optimum if it chosen as twice or thrice the value of k for a decently-sized value of k. To be more precise, experiments prove that, if k is a one-digit number, the corresponding value of m can be considered to be thrice the value of k (i.e., 3k). Again, if k is a two-digit number, the corresponding value of m can be considered to be twice the value of k (i.e., 2k). Results have been conducted for higher values of m, by considering m to be 4k and 5k for all the considered datasets. However, the improvement in the accuracy of results is almost negligible. Hence, the optimum value of m can be within the range of 2k to 3k.In the subsequent steps, the set T consists of a set of top-m nodes having the highest degree values (steps 5-7). Now, set T is considered to be the set of nodes forming a social graph G' formed from the original social graph G. Now, considering T as the limited set of nodes to be considered (which is significantly much smaller than the original graph G), we finally find the optimum seed set S containing top-k nodes (steps 8-19). The entire simulation process is carried out 20000 times to consider the average of all results generated from each iteration. Random(S) (step 13) represents the random simulation process carried out using any of the influence diffusion models to generate the final seed set S containing top-k influential seeds.

Algorithm 3: *DegGreedy*(G, k, m)

```
Initialize T = ϕ
for each vertex v do
      compute its degree d_v
end for
for i = 1 to m do
      T = T U {v_di}
end for
Initialize S = ϕ, S_im = 20000
 for i = 1 to k do
   for each vertex v ∈  T \ S do
     s_v = 0
     for i = 1 to S_im do
          s_v += |Random(S U {v})|
     end for
     s_v = s_v / S_im
   end for

     S = S U {arg max_v∈T\S {s_v}}
end for
output S
```

EXPERIMENTAL EVALUATIONS

For carrying out the experiments, three publicly available real-world datasets were used that consists of ties among nodes of social networking sites, namely the dolphin dataset [99], the Santa Fe institute collaboration dataset [100] and the Facebook dataset [108]. A brief description of each of the datasets is given below.

The dolphin dataset consists of an undirected dolphin social network graph having links or connections among various dolphins. The dataset contains 62 distinct nodes and 159 connections or edges to indicate the ties among all dolphins. Figure 5 displays the entire dolphin dataset in the form of a graph. Each node in the graph has been labeled with numbers from 1 to 62. The size and color of nodes differ based on the values of degree of each node. The Santa Fe Institute Collaboration dataset consists of an undirected network having 118 unique nodes and 200 links or edges and this dataset has been often used by researchers in the field of OSN analysis. Figure 5 displays the entire Santa Fe Institute Collaboration dataset in the form of a graph. Each node of the graph has been labeled with numbers from 1 to 118. The size and color of nodes differ based on the values of degree of each node.

The Facebook dataset consists of an undirected network having links or connections among various Facebook users. The dataset contains 63,731 distinct nodes and 8,17,035 edges that indicate the friendship ties among Facebook users. Table 2. Exemplifies the graphical structure of the Facebook dataset. Due to the large number of nodes present in the network, the insignificant nodes having degree less

Table 2. Statistics of the three tested real-time datasets

Dataset	Dolphin	Santa Fe Institute Collaboration	Facebook
Graph Type	Undirected	Undirected	Undirected
#Nodes	62	118	63731
#Edges	159	120	817035
Average Degree	5.129	3.39	25.640
Maximal Degree	12	29	1098
Average Path Length	3.453	5.021	2.832

than six have not been considered in the figure. The size and color of nodes differ based on the values of degree of each node.

Experiments have been conducted using all the three real-time datasets mentioned above, for four various influence maximization algorithms namely, the High-Degree centrality algorithm, the general Greedy algorithm, the DegGreedy centrality algorithm and the random generation algorithm. Out of these four algorithms, the random generation algorithm is taken as a baseline method for comparing with the other three influence maximization algorithms. In case of random generation method, nodes for seed set generation are selected uniformly at random. The diffusion model used for all the techniques in our experiments is the Weighted Cascade (WC) model. The various experiments conducted mainly aim at finding the total number of activated nodes for a given seed set size k. The comparative results illustrate the performance of the four algorithms with respect to the amount of spread of influence using the WC model. The user defined value of m considered for the DegGreedy technique ranged from 2k to 3k depending on whether the value of k taken as input is a two-digit or single-digit value respectively. To

Figure 5. Comparing spread of influence using Random, High-Degree, general Greedy and DegGreedy algorithms

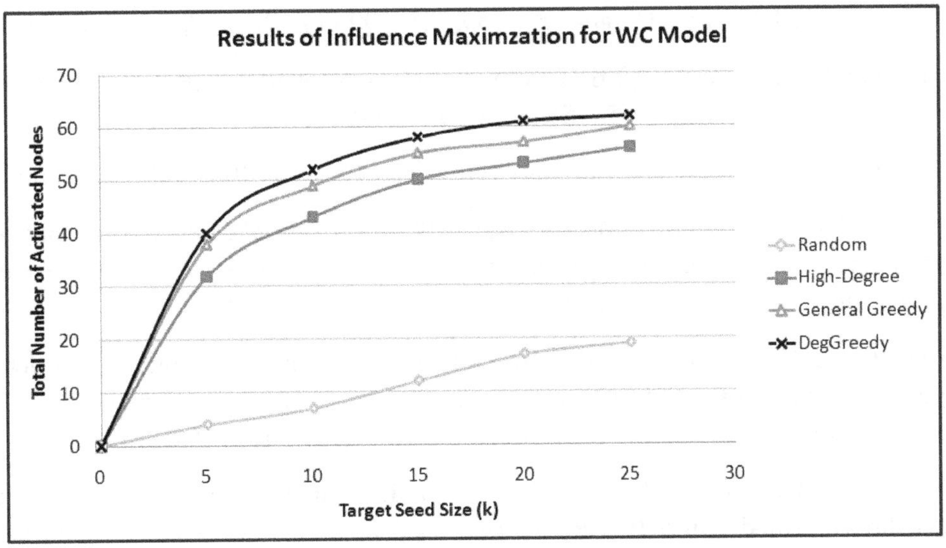

understand the effectiveness of the constraint m, experiments have been conducted for higher values of m (4k and 5k) during the experimental analysis; however it did not give any significantly better results for both the datasets.

CONCLUSION

It can be concluded that the influence maximization problem cannot solely rely on the structural properties of a social graph. Rather, for a better marketing strategy, it is needed to also explicitly reflect on the dynamics of information in the network. We believe that the algorithms discussed in this chapter is considered as basic, standard state-of-the art techniques for studying the concept of influence maximization in social networks and it will help a researcher studying in this field to get a preliminary idea about the same. By adding a constraint m in our proposed DegGreedy algorithm, we have proved that the influence maximization problem can be solved effectively when m is small, and it can be approximated well if the target seed size value k is known. For future work, some comparative study of more recent standard algorithms can be done that will result in the highest gain of influence spread in OSNs.

REFERENCES

Adamic, L., & Adar, E. (2005). How to search a social network. *Social Networks*, *27*(3), 187–203. doi:10.1016/j.socnet.2005.01.007

Ahamed, B. B., & Murugan, K. (2019, October). Study of Socio-Linguistics Online Review System Using Sentiment Scoring Method. In *International Conference on Intelligent Computing & Optimization* (pp. 569-580). Springer.

Ahamed, B. B., & Ramkumar, T. (2015). Deduce User Search Progression with Feedback Session. *Advances in Systems Science and Applications*, *15*(4), 366–383.

Ahamed, B. B., & Ramkumar, T. (2016). An intelligent web search framework for performing efficient retrieval of data. *Computers & Electrical Engineering*, *56*, 289–299. doi:10.1016/j.compeleceng.2016.09.033

Barbieri, N., Bonchi, F., & Manco, G. (2013). Topic-aware social influence propagation models. *Knowledge and Information Systems*, *37*(3), 555–584. doi:10.100710115-013-0646-6

Bullas, J. (n.d.). *Social media facts and statistics you should know in 2014*. jeffbullas.com.

Chen, J., Geyer, W., Dugan, C., Muller, M., & Guy, I. (2009, April). Make new friends, but keep the old: recommending people on social networking sites. In *Proceedings of the SIGCHI Conference on Human Factors in Computing Systems* (pp. 201-210). 10.1145/1518701.1518735

Chen, W., Wang, C., & Wang, Y. (2010, July). Scalable influence maximization for prevalent viral marketing in large-scale social networks. In *Proceedings of the 16th ACM SIGKDD international conference on Knowledge discovery and data mining* (pp. 1029-1038). 10.1145/1835804.1835934

Chen, W., Yuan, Y., & Zhang, L. (2010, December). Scalable influence maximization in social networks under the linear threshold model. In *2010 IEEE international conference on data mining* (pp. 88-97). IEEE.

Domingos, P., & Richardson, M. (2001, August). Mining the network value of customers. In *Proceedings of the seventh ACM SIGKDD international conference on Knowledge discovery and data mining* (pp. 57-66). 10.1145/502512.502525

Goyal, A., Bonchi, F., & Lakshmanan, L. V. (2011). *A data-based approach to social influence maximization.* arXiv preprint arXiv:1109.6886

Guler, B., Varan, B., Tutuncuoglu, K., Nafea, M., Zewail, A. A., Yener, A., & Octeau, D. (2015). Using social sensors for influence propagation in networks with positive and negative relationships. *IEEE Journal of Selected Topics in Signal Processing*, *9*(2), 360–373. doi:10.1109/JSTSP.2014.2386792

Kundu, A., & Ratha, N. K. (2019). *U.S. Patent Application No. 16/399,276.* Washington, DC: US Patent Office.

Leskovec, J., Krause, A., Guestrin, C., Faloutsos, C., VanBriesen, J., & Glance, N. (2007, August). Cost-effective outbreak detection in networks. In *Proceedings of the 13th ACM SIGKDD international conference on Knowledge discovery and data mining* (pp. 420-429). ACM.

Liben-Nowell, D., & Kleinberg, J. (2007). The link-prediction problem for social networks. *Journal of the American Society for Information Science and Technology*, *58*(7), 1019–1031. doi:10.1002/asi.20591

Liu, L., Tang, J., Han, J., Jiang, M., & Yang, S. (2010, October). Mining topic-level influence in heterogeneous networks. In *Proceedings of the 19th ACM international conference on Information and knowledge management* (pp. 199-208). ACM.

Papadimitriou, A., Symeonidis, P., & Manolopoulos, Y. (2012). Fast and accurate link prediction in social networking systems. *Journal of Systems and Software*, *85*(9), 2119–2132. doi:10.1016/j.jss.2012.04.019

Richardson, M., & Domingos, P. (2002, July). Mining knowledge-sharing sites for viral marketing. In *Proceedings of the eighth ACM SIGKDD international conference on Knowledge discovery and data mining* (pp. 61-70). 10.1145/775047.775057

Saito, K., Nakano, R., & Kimura, M. (2008, September). Prediction of information diffusion probabilities for independent cascade model. In *International conference on knowledge-based and intelligent information and engineering systems* (pp. 67-75). Springer. 10.1007/978-3-540-85567-5_9

Compilation of References

(2019). Intelligent Computing & Optimization. InVasant, P., Zelinka, I., & Weber, G.-W. (Eds.), *Advances in Intelligent Systems and Computing* (1st ed., pp. 98–107). Springer International Publishing.

Abido, M.A., & Abdel-Magid, Y.L. (2002). Optimal design of power system stabilizers using evolutionary programming. *IEEE Transaction on Energy Conversion*, *17*(4).

Adamic, L., & Adar, E. (2005). How to search a social network. *Social Networks*, *27*(3), 187–203. doi:10.1016/j.socnet.2005.01.007

Agroecological atlas of Russia and neighboring countries. (n.d.). http://www.agroatlas.ru/ru/content/Climatic_maps/

Ahamed, B. B., & Murugan, K. (2019, October). Study of Socio-Linguistics Online Review System Using Sentiment Scoring Method. In *International Conference on Intelligent Computing & Optimization* (pp. 569-580). Springer.

Ahamed, B. B., & Ramkumar, T. (2015). Deduce User Search Progression with Feedback Session. *Advances in Systems Science and Applications*, *15*(4), 366–383.

Ahamed, B. B., & Ramkumar, T. (2016). An intelligent web search framework for performing efficient retrieval of data. *Computers & Electrical Engineering*, *56*, 289–299. doi:10.1016/j.compeleceng.2016.09.033

Ahuett-Garza, H., & Kurfess, T. (2018). A brief discussion on the trends of habilitating technologies for Industry 4.0 and Smart manufacturing. *Manufacturing Letters*, *15*, 60–63. doi:10.1016/j.mfglet.2018.02.011

Akiyama, T., Miyazaki, T., Ito, H., Nogami, H., Higurashi, E., Ando, S., & Sawada, R. (2015). Comparable accuracy of micro-electromechanical blood flowmetry-based analysis vs. electrocardiography-based analysis in evaluating heart rate variability. *Circulation Journal*, *79*(4), 794–801. doi:10.1253/circj.CJ-14-1199 PMID:25740056

Akl, S. (1992). *Design and Analysis of Parallel Algorithms*. Prentice Hall Inc.

Alavi, H. S. (2017). The Evolution of Human-Building Interaction: An HCI Perspective. *Interaction Design and Architecture(s) Journal - IxD&A, 32*, 3-6.

Alencar, D. B., Affonso, C. M., Oliveira, R. C. L., & Filho, J. C. R. (2018). Hybrid Approach Combining SARIMA and Neural Networks for Multi-Step Ahead Wind Speed Forecasting in Brazil. *IEEE Access: Practical Innovations, Open Solutions*, *6*, 55986–55994. doi:10.1109/ACCESS.2018.2872720

Al-falahi, M. D. A., Jayasinghe, S. D. G., & Enshaei, H. (2017). A review on recent size optimization methodologies for standalone solar and wind hybrid renewable energy system. *Energy Conversion and Management*, *143*, 252–274. doi:10.1016/j.enconman.2017.04.019

Alireza, A. (2013). Developing a discrete harmony search algorithm for size opti- mization of wind–photovoltaic hybrid energy system. *Solar Energy*, *98*, 190–195. doi:10.1016/j.solener.2013.10.008

Al-Shamma, A. A., & Khaled, E. (2012). Optimum sizing of hybrid PV/wind/battery/diesel system considering wind turbine parameters using Genetic Algorithm. *Power and Energy (PECon), IEEE International Conference*, 121-126.

Andreoiu, A., & Bhattacharya, K. (2002). Robust tuning of power system stabilizers using a Lyapunov method based genetic algorithm. *Proceeding of Inst. Electrical Engineering Generation Transmission and Distribution*, *149*(5), 585–592. doi:10.1049/ip-gtd:20020600

Andrienko, N., Lammarsch, T., Andrienko, G., Fuchs, G., Keim, D., Miksch, S., & Rind, A. (2018). Viewing visual analytics as model building. *Computer Graphics Forum*, *37*(6), 275–299. doi:10.1111/cgf.13324

ANSYS. (n.d.). https://www.ansys.com/

Applegate, D. L., Bixby, R. E., Chvatal, V., & Cook, W. L. (2006). *The traveling salesman problem: a computational study*. Princeton University Press.

Arrifano, N. S. D., Oliveira, V. A., Ramos, R. A., Bretas, N. G., & Oliveira, R. V. (2007). Fuzzy stabilization of power systems in a co-generation scheme subject to random abrupt variations of operating conditions. *IEEE Transactions on Control Systems Technology*, *15*(2), 384–393. doi:10.1109/TCST.2006.886443

Ashok, S. (2007). Optimised model for community-based hybrid energy system. *Renewable Energy*, *32*(7), 1155–1164. doi:10.1016/j.renene.2006.04.008

Ayoobkhan, M. U. A., Chikkannan, E., & Ramakrishnan, K. (2018). Feed-forward neural network-based predictive image coding for medical image compression. *Arabian Journal for Science and Engineering*, *43*(8), 4239–4247. doi:10.100713369-017-2837-z

Ayuketang Arreyndip, N., & Joseph, E. (2016). Generalized Extreme Value Distribution Models for the Assessment of Seasonal Wind Energy Potential of Debuncha, Cameroon. *Journal of Renewable Energy*, *2016*, 2016. doi:10.1155/2016/9357812

Babaev, B. D., Kharchenko, V. V., & Panchenko, V. (2019). Development and Research of Phase-Transition and Thermochemical Materials for Heat Accumulation. Handbook of Research on Smart Computing for Renewable Energy and Agro-Engineering, 1-26. Doi:10.4018/978-1-7998-1216-6.ch001

Babaev, B.D., Kharchenko, V., Panchenko, V. & Vasant, P. (2019). Materials and Methods of Thermal Energy Storage in Power Supply Systems. *Renewable Energy and Power Supply Challenges for Rural Regions*, 115-135. DOI: . doi:10.4018/978-1-5225-9179-5.ch005

Bai, X., Li, J., Zhou, L., & Li, X. (2009). Influence of the menstrual cycle on nonlinear properties of heart rate variability in young women. *American Journal of Physiology. Heart and Circulatory Physiology*, *297*(2), H765–H774. doi:10.1152/ajpheart.01283.2008 PMID:19465541

Balachandar, P. (2017). *Economic – environmental constrained real power scheduling of thermal power plants using bio-inspired algorithm* (Ph.D Dissertation). Annamalai University.

Balachandar, P., Ganesan, S., Jayakumar, N., & Subramanian, S. (2017). Multi-fuel power dispatch in an interconnected power system using ant lion optimizer: Multi-fuel dispatch considering tie-line limits. *International Journal of Energy Optimization and Engineering*, *6*(3), 29–54. doi:10.4018/IJEOE.2017070102

Banerjee, S., Chanda, C., & Maity, D. (2016). A comparative study of improved teaching learning based optimization technique on economic load dispatch problem with generator constraints. *International Journal of Energy Optimization and Engineering*, *5*(4), 1–25. doi:10.4018/IJEOE.2016100101

Banik, A., Bandyopadhyay, T. K., Biswal, S. K., & Majumder, M. (2019). Prediction of Maximum Efficiency of Vertical Helical Coil Membrane Using Group Method of Data Handling (GMDH) Algorithm. In Intelligent Computing and Optimization (pp. 489–500). Springer.

Banik, A., Dutta, S., Bandyopadhyay, T. K., & Biswal, S. K. (2018). Prediction of maximum permeate flux (%) of disc membrane using Response Surface Methodology (RSM). *Canadian Journal of Civil Engineering, 46*(6), 299–307. doi:10.1139/cjce-2018-0007

Banik, A., Bandyopadhyay, T. K., & Biswal, S. K. (2017). Computational fluid dynamics simulation of disc membrane used for improving the quality of effluent produced by the rubber industry. *International Journal of Fluid Mechanics Research, 44*(6), 499–512. doi:10.1615/InterJFluidMechRes.2017018630

Banik, A., Biswal, S. K., Majumder, M., & Bandyopadhyay, T. K. (2018). Development of an adaptive non-parametric model for estimating maximum efficiency of disc membrane. *Int. J. Convergence Computing, 3*(1), 3–19. doi:10.1504/IJCONVC.2018.091111

Barbieri, N., Bonchi, F., & Manco, G. (2013). Topic-aware social influence propagation models. *Knowledge and Information Systems, 37*(3), 555–584. doi:10.100710115-013-0646-6

Bardou, F. (2005). Fractal time random walk and subrecoil laser cooling considered as renewal processes with infinite mean waiting times. In *Chaotic Dynamics and Transport in Classical and Quantum Systems* (pp. 281–301). Springer Netherlands. doi:10.1007/1-4020-2947-0_12

Barmak, A. V., Krak, Y. V., Manziuk, E. A., & Kasianiuk, V. S. (2019). Information Technology of Separating Hyperplanes Synthesis for Linear Classifiers. *Journal of Automation and Information Sciences, 51*(5), 54–64. doi:10.1615/JAutomatInfScien.v51.i5.50

Barnhart, C., Hane, C. A., & Vance, P. H. (2000). Using branch-and-price-and-cut to solve origin-destination integer multicommodity flow problems. *Operations Research, 48*(2), 318–326. doi:10.1287/opre.48.2.318.12378

Barnhart, C., Johnson, E. L., Nemhauser, G. L., Savelsbergh, M. W. P., & Vance, P. H. (1998). Branch and price column generation for solving huge integer programs. *Operations Research, 46*(3), 316–329. doi:10.1287/opre.46.3.316

Barthel, P., Schneider, R., Bauer, A., Ulm, K., Schmitt, C., Schomig, A., & Schmidt, G. (2003). Risk stratification after acute myocardial infarction by heart rate turbulence. *Circulation, 108*(10), 1221–1226. doi:10.1161/01.CIR.0000088783.34082.89 PMID:12939209

Baskar, S., Subbaraj, P., & Rao, M. V. C. (2003). Hybrid real coded genetic algorithm solution to economic dispatch problem. *Computers & Electrical Engineering, 29*(3), 407–419. doi:10.1016/S0045-7906(01)00039-8

Basu, M. (2014). Teaching–learning-based optimization algorithm for multi-area economic dispatch. *Energy, 68*, 21–28. doi:10.1016/j.energy.2014.02.064

Basu, M. (2016). Kinetic gas molecule optimization for nonconvex economic dispatch problem. *International Journal of Electrical Power & Energy Systems, 80*, 325–332. doi:10.1016/j.ijepes.2016.02.005

Basu, S., Mukherjee, S., Kaushik, A., Batra, V. S., & Balakrishnan, M. (2015). Integrated treatment of molasses distillery wastewater using microfiltration (MF). *Journal of Environmental Management, 158*, 55–60. doi:10.1016/j.jenvman.2015.04.037 PMID:25956444

Bauer, A., Malik, M., Schmidt, G., Barthel, P., Bonnemeier, H., Cygankiewicz, I., ... Zareba, W. (2008). Heart rate turbulence: standards of measurement, physiological interpretation, and clinical use: International Society for Holter and Noninvasive Electrophysiology Consensus. *Journal of the American College of Cardiology, 52*(17), 1353–1365. doi:10.1016/j.jacc.2008.07.041 PMID:18940523

Bealie, E. M. L. (1979). Branch and Bound Methods for Mathematical Programming systems. *Annals of Discrete Mathematics, 5*, 201–219. doi:10.1016/S0167-5060(08)70351-0

Behre, H. M., Kuhlage, J., Gassner, C., Sonntag, B., Schem, C., Schneider, H. P., & Nieschlag, E. (2000). Prediction of ovulation by urinary hormone measurements with the home use ClearPlan Fertility Monitor: Comparison with transvaginal ultrasound scans and serum hormone measurements. *Human Reproduction (Oxford, England), 15*(12), 2478–2482. doi:10.1093/humrep/15.12.2478 PMID:11098014

Bekele, G., & Palm, B. (2010). Feasibility study for a standalone solar–wind-based hybrid energy system for application in Ethiopia. *Applied Energy, 87*(2), 487–495. doi:10.1016/j.apenergy.2009.06.006

Bektaş, T., & Gouveia, L. (2014). Requiem for the Miller–Tucker–Zemlin subtour elimination constraints? *European Journal of Operational Research, 236*(3), 820–832. doi:10.1016/j.ejor.2013.07.038

Belfkira, R., Zhang, L., & Barakat, G. (2011). Optimal sizing study of hybrid wind/PV/diesel power generation unit. *Solar Energy, 85*(1), 100–110. doi:10.1016/j.solener.2010.10.018

Benatiallah, A., Kadia, L., & Dakyob, B. (2010). Modelling and optimization of wind energy systems. *JJMIE, 4*(1).

Bengio, Y., Boulanger-Lewandowski, N., & Pascanu, R. (2013). Advances in optimizing recurrent networks. In *2013 IEEE International Conference on Acoustics, Speech and Signal Processing* (pp. 8624–8628). 10.1109/ICASSP.2013.6639349

Bensalah, Y. (2000). *Steps in applying extreme value theory to finance: a review.* Vancouver: Bank of Canada.

Berihun, G. (2013). *Modeling and Simulating of a micro hydro wind hybrid power generation system for rural area of Ethiopia* (Mater thesis). Addis Ababa University.

Berman, P., & Karpinski, M. (2006). 8/7 - approximation algorithm for *(1,2)* – TSP. *Proc. 17th ACM-SIAM SODA conference*, 641-648. 10.1145/1109557.1109627

Beyene, G. (2018). *The Challenges and Prospects of Electricity Access in Ethiopia* (Doctoral dissertation). Addis Ababa University.

Beyene, G. E., Kumie, A., Edwards, R., & Troncoso, K. (2018). *Opportunities for transition to clean household energy in Ethiopia: application of the household energy assessment rapid tool (HEART).* Academic Press.

Bhandari, B., Ahn, S. H., & Ahn, T. B. (2016). Optimization of hybrid renewable energy power system for remote installations: Case studies for mountain and island. *International Journal of Precision Engineering and Manufacturing, 17*(6), 815–822. doi:10.100712541-016-0100-2

Bhattacharya, A., & Chattopadhyay, P. K. (2011). Solving economic emission load dispatch problems using hybrid differential evolution. *Applied Soft Computing, 11*(2), 2526–2537. doi:10.1016/j.asoc.2010.09.008

Binetti, G., Davoudi, A., Naso, D., Turchiano, B., & Lewis, F. L. (2014). A distributed auction-based algorithm for the nonconvex economic dispatch problem. *IEEE Transactions on Industrial Informatics, 10*(2), 1124–1132. doi:10.1109/TII.2013.2287807

Binitha, S., & Sathya, S. S. (2012). A survey of bio inspired optimization algo-rithms. *International Journal of Soft Comput-ing and Engineering, 2*(2), 137–151.

Biswas, K., Vasant, P. M., Laruccia, M. B., Vintaned, J. A. G., & Myint, M. M. (2020). Review on Particle Swarm Optimization Approach for Optimizing Wellbore Trajectory. In *Deep Learning Techniques and Optimization Strategies in Big Data Analytics* (pp. 290–307). Hershey, PA: IGI Global. doi:10.4018/978-1-7998-1192-3.ch017

Blimpo, M. P., & Cosgrove-Davies, M. (2019). *Electricity Access in Sub-Saharan Africa: Uptake, Reliability, and Complementary Factors for Economic Impact*. World Bank Publications. doi:10.1596/978-1-4648-1361-0

Borg, I., & Groenen, P. (2003). Modern multidimensional scaling: Theory and applications. *Journal of Educational Measurement*, *40*(3), 277–280. doi:10.1111/j.1745-3984.2003.tb01108.x

Borowy, B. S., & Salameh, Z. M. (1996). Methodology for optimally sizing the combi-nation of a battery bank and PV array in a wind/PV hybrid system. *IEEE Transactions on Energy Conversion*, *11*(2), 367–375. doi:10.1109/60.507648

Boschert, S., & Rosen, R. (2016). Digital twin—the simulation aspect. In *Mechatronic Futures* (pp. 59–74). Cham: Springer.

Boudour, M. (2016). LFC enhancement concerning large wind power integration using new optimised PID controller and RFBs. *IET Generation, Transmission & Distribution*, *10*(16), 4065-4077. Retrieved from https://digital-library.theiet.org/content/journals/10.1049/iet-gtd.2016.0385

Brar, T. K., Singh, K. D., & Kumar, A. (2015). Effect of Different Phases of Menstrual Cycle on Heart Rate Variability (HRV). *Journal of Clinical and Diagnostic Research: JCDR*, *9*(10), CC01–CC04. doi:10.7860/JCDR/2015/13795.6592 PMID:26557512

Brena, R. F., García-Vázquez, J. P., Galván-Tejada, C. E., Muñoz-Rodriguez, D., Vargas-Rosales, C., & Fangmeyer, J. (2017). Evolution of indoor positioning technologies: A survey. *Journal of Sensors*. doi:10.1155/2017/2630413

Brezina, P. R., Haberl, E., & Wallach, E. (2011). At home testing: Optimizing management for the infertility physician. *Fertility and Sterility*, *95*(6), 1867–1878. doi:10.1016/j.fertnstert.2011.01.001 PMID:21324446

Brito, T. G., Paiva, A. P., Ferreira, J. R., Gomes, J. H. F., & Balestrassi, P. P. (2014). A normal boundary intersection approach to multiresponse robust optimization of the surface roughness in end milling process with combined arrays. *Precision Engineering*, *38*(3), 628–638. doi:10.1016/j.precisioneng.2014.02.013

Brossard, M., Erntell, H., & Hepp, D. (2018). Accelerating product development: The tools you need now. *The McKinsey Quarterly*.

Brunetta, L., Conforti, M., & Rinaldi, G. (1997). A branch and cut algorithm for the equicut problem. *Mathematical Programming*, *78*(2), 243–263. doi:10.1007/BF02614373

Bueno, C., & Carta, J. A. (2005). Technical–economic analysis of wind-powered pumped hydrostorage systems. Part II: Model application to the island of El Hierro. *Solar Energy*, *78*(3), 396–405. doi:10.1016/j.solener.2004.08.007

Buldakova, T. I., & Suyatinov, S. I. (2020). Assessment of the State of Production System Components for Digital Twins Technology. In *Cyber-Physical Systems: Advances in Design & Modelling* (pp. 253–262). Cham: Springer. doi:10.1007/978-3-030-32579-4_20

Bullas, J. (n.d.). *Social media facts and statistics you should know in 2014*. jeffbullas.com.

Buonomano, A., Calise, F., & Vicidimini, M. (2016). Design, Simulation and Experimental Investigation of a Solar System Based on PV Panels and PVT Collectors. *Energies*, *9*(7), 497. doi:10.3390/en9070497

Cadenas, E., & Rivera, W. (2010). Wind speed forecasting in three different regions of Mexico, using a hybrid ARIMA–ANN model. *Renewable Energy*, *35*(12), 2732–2738. doi:10.1016/j.renene.2010.04.022

Cao, Q., Ewing, B. T., & Thompson, M. A. (2012). Forecasting wind speed with recurrent neural networks. *European Journal of Operational Research, 221*(1), 148–154. doi:10.1016/j.ejor.2012.02.042

Chalasani, R. M. (1986). Ride performance potential of active sprung weigts - part II: Comprehensive analysis based on a full-car model. American Society of Mechanical Engineers. *Applied Mechanics Division, AMD, 80*, 205–234.

Chalermchaiarbha, S., & Ongsakul, W. (2013). Stochastic weight trade-off particle swarm optimization for nonconvex economic dispatch. *Energy Conversion and Management, 70*, 66–75. doi:10.1016/j.enconman.2013.02.009

Charwand, M., Ahmadi, A., Heidari, A. R., & Nezhad, A. E. (2015). Benders decomposition and normal boundary intersection method for multiobjective decision making framework for an electricity retailer in energy markets. *IEEE Systems Journal, 9*(4), 1475–1484. doi:10.1109/JSYST.2014.2331322

Chedid, R., Akiki, H., & Rahman, S. (1998). A decision support technique for the design of hybrid solar-wind power systems. *IEEE Transactions on Energy Conversion, 13*(1), 76–83. doi:10.1109/60.658207

Chen, W., Yuan, Y., & Zhang, L. (2010, December). Scalable influence maximization in social networks under the linear threshold model. In *2010 IEEE international conference on data mining* (pp. 88-97). IEEE.

Chen, J., Geyer, W., Dugan, C., Muller, M., & Guy, I. (2009, April). Make new friends, but keep the old: recommending people on social networking sites. In *Proceedings of the SIGCHI Conference on Human Factors in Computing Systems* (pp. 201-210). 10.1145/1518701.1518735

Chen, J., Zeng, G.-Q., Zhou, W., Du, W., & Lu, K.-D. (2018). Wind speed forecasting using nonlinear-learning ensemble of deep learning time series prediction and extremal optimization. *Energy Conversion and Management, 165*, 681–695. doi:10.1016/j.enconman.2018.03.098

Chen, W.-J. (2007). An efficient algorithm for scheduling jobs on a machine with periodic maintenance. *International Journal of Advanced Manufacturing Technology, 34*(11-12), 1173–1182. doi:10.100700170-006-0689-x

Chen, W., Wang, C., & Wang, Y. (2010, July). Scalable influence maximization for prevalent viral marketing in large-scale social networks. In *Proceedings of the 16th ACM SIGKDD international conference on Knowledge discovery and data mining* (pp. 1029-1038). 10.1145/1835804.1835934

Cherry, A. S., & Jones, R. P. (1995). Fuzzy logic control of an automotive sprung weights. *IEE Proc. Control Theory Applications, 142*(2), 149–160.

Chew, C. M., Aroua, M. K., & Hussain, M. A. (2017). A practical hybrid modelling approach for the prediction of potential fouling parameters in ultrafiltration membrane water treatment plant. *Journal of Industrial and Engineering Chemistry, 45*, 145–155. doi:10.1016/j.jiec.2016.09.017

Chiang, C. L. (2005). Improved genetic algorithm for power economic dispatch of units with valve-point effects and multiple fuels. *IEEE Transactions on Power Systems, 20*(4), 1690–1699. doi:10.1109/TPWRS.2005.857924

Chiang, H. D., Flueck, A. J., Shah, K. S., & Balu, N. (1995). CPFLOW: A practical tool for tracing power system steady-state stationary behavior due to load and generation variations. *IEEE Transactions on Power Systems, 10*(2), 623–634. doi:10.1109/59.387897

Chitara D., Swarnkar, A., Gupta, N., Niazi, K.R., & Bansal, R.C. (2015). Optimal tuning of multimachine power system stabilizer using cuckoo search algorithm. *SCIENCEDIRECT,* 143-148.

Choudhury, P., Mondal, P., Majumdar, S., Saha, S., & Sahoo, G. C. (2018). Preparation of ceramic ultrafiltration membrane using green synthesized CuO nanoparticles for chromium (VI) removal and optimization by response surface methodology. *Journal of Cleaner Production, 203*(6), 511–520. doi:10.1016/j.jclepro.2018.08.289

Civicioglu, P. (2012). Transforming geocentric cartesian coordinates to geodentic coordinates by using differential search algorithm. *Computers & Geosciences*, *46*, 229–247. doi:10.1016/j.cageo.2011.12.011

Computer models for representation of digital-based excitation systems. (1996). *IEEE Trans. Energy Convers.*, *11*(3), 607–615.

Constant, I., Laude, D., Murat, I., & Elghozi, J. L. (1999). Pulse rate variability is not a surrogate for heart rate variability. *Clinical Science (London, England)*, *97*(4), 391–397. doi:10.1042/cs0970391 PMID:10491338

Dakin, R. J. (1965). A tree search algorithm for mixed integer programming problems. *The Computer Journal*, *8*(3), 250–255. doi:10.1093/comjnl/8.3.250

Daliento, S., & Lancellotti, L. (2010). 3D Analysis of the performances degradation caused by series resistance in concentrator solar cells. *Solar Energy*, *84*(1), 44–50. doi:10.1016/j.solener.2009.08.014

Dandessa, C., Negisho, T., & Bekele, G. (2018). *Assessment of Traditional Medicinal Plants Used for Treatment of Human Diseases in Oromia Region, West Shewa Zone*. Jeldu Woreda, Ethiopia: Asian Plant Research Journal.

Daryani, N., & Zare, K. (2016). Multiobjective power and emission dispatch using modified group search optimization method. *Ain Shams Engineering Journal*.

Datta, S. P. A. (2017). Emergence of Digital Twins-Is this the march of reason? *Journal of Innovation Management*, *5*(3), 14–33. doi:10.24840/2183-0606_005.003_0003

Daus, Y., Kharchenko, V., & Yudaev, I. V. (2018). Solar Radiation Intensity Data as Basis for Predicting Functioning Modes of Solar Power Plants. In *Handbook of Research on Renewable Energy and Electric Resources for Sustainable Rural Development* (pp. 283–309). Hershey, PA: IGI Global. doi:10.4018/978-1-5225-3867-7.ch012

De Souza, D. I., Dottein, E. M., Giacobbo, A., Siqueira Rodrigues, M. A., De Pinho, M. N., & Bernardes, A. M. (2018). Nanofiltration for the removal of norfloxacin from pharmaceutical effluent. *Journal of Environmental Chemical Engineering*, *6*(5), 6147–6153. doi:10.1016/j.jece.2018.09.034

de Zambotti, M., Nicholas, C. L., Colrain, I. M., Trinder, J. A., & Baker, F. C. (2013). Autonomic regulation across phases of the menstrual cycle and sleep stages in women with premenstrual syndrome and healthy controls. *Psychoneuroendocrinology*, *38*(11), 2618–2627. doi:10.1016/j.psyneuen.2013.06.005 PMID:23850226

Deakin, A., Crolla, D., Shovlin, A., & Charlton, N. (1997). *Power Consumption in Ride of a Combat Support Vehicle Slow-Active Suspension* (SAE Technical Paper No. 973205). Warrendale, PA: SAE International. doi:10.4271/973205

Deboli, R., Calvo, A., & Preti, C. (2017). Whole-body vibration: Measurement of horizontal and vertical transmissibility of an agricultural tractor seat. *International Journal of Industrial Ergonomics*, *58*, 69–78. doi:10.1016/j.ergon.2017.02.002

Degtyarev, K.S., Panchenko, V.A. & Mayorov, S.V. (2018). Perspektivy energosnabzheniya infrastrukturnyh ob"ektov na osnove vozobnovlyaemyh istochnikov energii v rossijskoj Arktike [Prospects for energy supply of infrastructure facilities based on renewable energy sources in the Russian Arctic]. *Sovremennye problemy sovershenstvovaniya raboty zheleznodorozhnogo transporta*, 79-91.

Del Castillo, J., & Daoudi, J. (2009). Estimation of the generalized Pareto distribution. *Statistics & Probability Letters*, *79*(5), 684–688. doi:10.1016/j.spl.2008.10.021

Derbaremdiker, A. D. (1985). *Amortizatory transportnykh mashin* [Shock absorbers for vehicles]. Moscow, Russia: Mashinostroyeniye.

Dey, P., Bhattacharya, A., & Das, P. (2018). Tuning of power system stabilizer for small signal improvement of interconnected power system. Applied Computing and Informatics.

Diaf, S., Belhamel, M., Haddadi, M., & Louche, A. (2007). A methodology for op-timal sizing of autonomous hybrid PV/wind system. *Energy Policy*, *35*(11), 5708–5718. doi:10.1016/j.enpol.2007.06.020

Dieu, V. N., Ongsakul, W., & Polprasert, J. (2013). The augmented Lagrange Hopfield network for economic dispatch with multiple fuel options. *Mathematical and Computer Modelling*, *57*(1), 30–39. doi:10.1016/j.mcm.2011.03.041

Dieu, V. N., & Schegner, P. (2013). Augmented Lagrange Hopfield network initialized by quadratic programming for economic dispatch with piecewise quadratic cost functions and prohibited zones. *Applied Soft Computing*, *13*(1), 292–301. doi:10.1016/j.asoc.2012.08.026

Domingos, P., & Richardson, M. (2001, August). Mining the network value of customers. In *Proceedings of the seventh ACM SIGKDD international conference on Knowledge discovery and data mining* (pp. 57-66). 10.1145/502512.502525

Doucoure, B., Agbossou, K., & Cardenas, A. (2016). Time series prediction using artificial wavelet neural network and multi-resolution analysis: Application to wind speed data. *Renewable Energy*, *92*, 202–211. doi:10.1016/j.renene.2016.02.003

Dubey, S., & Tay, A. A. O. (2012). Experimental Study of the Performance of Two Different Types of Photovoltaic Thermal (PVT) Modules under Singapore Climatic Conditions. *Journal of Fundamentals of Renewable Energy and Applications*, *2*, 1–6. doi:10.4303/jfrea/R120313

Duke, M., & Goss, G. (2007). Investigation of tractor driver seat performance with non-linear stiffness and on-off damper. *Biosystems Engineering, 94*(4), 477 – 486. WOS:000245890600005

Duran, J., Esnaola, S., Rubio, R., & Iztueta, A. (2001). Obstructive sleep apnea-hypopnea and related clinical features in a population-based sample of subjects aged 30 to 70 yr. *American Journal of Respiratory and Critical Care Medicine*, *163*(3 Pt 1), 685–689. doi:10.1164/ajrccm.163.3.2005065 PMID:11254524

Ecochard, R., Boehringer, H., Rabilloud, M., & Marret, H. (2001). Chronological aspects of ultrasonic, hormonal, and other indirect indices of ovulation. *BJOG*, *108*(8), 822–829. doi:10.1111/j.1471-0528.2001.00194.x PMID:11510707

Eftichios, K., Dionissia, K., Antonis, P., & Kostas, K. (2006). Methodology for optimal sizing of stand alone photovoltaic/ wind generator systems using genetic algorithms. *Solar Energy*, *80*(9), 1072–1088. doi:10.1016/j.solener.2005.11.002

Ejraei, A., Aroon, M. A., & Ziarati Saravani, A. (2019). Wastewater treatment using a hybrid system combining adsorption, photocatalytic degradation and membrane filtration processes. *Journal of Water Process Engineering, 28*(August), 45–53. doi:10.1016/j.jwpe.2019.01.003

Eke, I., & Taplamacioglu, M.C. (2015). Robust tuning of power system stabilizer by using orthogonal learning artificial bee colony. *Sciencedirect*, 149-154.

Elazim, S. M. A., & Ali, E. S. (2016). Optimal power system stablizers design via cuckoo search algorithm. *International Journal of Electrical Power & Energy Systems*, *75*, 99–107. doi:10.1016/j.ijepes.2015.08.018

Eldos, T., & Al Qasim, R. (2013). *On the performance of the Gravitational Search Algorithm*. West Yorkshire, UK: International Journal of Advanced Computer Science Applications. doi:10.14569/IJACSA.2013.040811

Elhadidy, M. A., & Shaahid, S. M. (1999). Optimal sizing of battery storage for hybrid (wind+ diesel) power systems. *Renewable Energy*, *18*(1), 77–86. doi:10.1016/S0960-1481(98)00796-4

Ellithy, K., Said, S., & Kahlout, O. (2014). Design of Power System Stablizers based on µ-controller for power system stability enhancement. *Electric Power and Energy System*, 63, 933-939.

El-Sallabi, H., Abdallah, M., Chamberland, J. F., & Qaraqe, K. (2014, December). A statistical model for delay domain radio channel parameter affected with extreme values. In *Antennas and Propagation (ISAP), 2014 International Symposium on* (pp. 165-166). IEEE. 10.1109/ISANP.2014.7026582

El-Sherbiny, M. K., Hasan, M. M., El-Saady, G., & Yousef, A. M. (2003). Optimal pole shifting for power system stabilization. *Electric Power Systems Research*, *66*(3), 253–258. doi:10.1016/S0378-7796(03)00092-0

Eltamaly, A. M., & Mohamed, M. A. (2014). A novel design and optimization software for autonomous PV/wind/battery hybrid power systems. *Mathematical Problems in Engineering*, *2014*, 2014. doi:10.1155/2014/637174

Embrechts, P., Resnick, S. I., & Samorodnitsky, G. (1999). Extreme value theory as a risk management tool. *North American Actuarial Journal: NAAJ*, *3*(2), 30–41. doi:10.1080/10920277.1999.10595797

Emmerich, M., Beume, N., & Naujoks, B. (2005). An EMO Algorithm Using the Hypervolume Measure as Selection Criterion. *Conference on Evolutionary Multi-Criterion Optimization (EMO 2005)*, 62–76. 10.1007/978-3-540-31880-4_5

Endert, A., Hossain, M. S., Ramakrishnan, N., North, C., Fiaux, P., & Andrews, C. (2014). The human is the loop: New directions for visual analytics. *Journal of Intelligent Information Systems*, *43*(3), 411–435. doi:10.100710844-014-0304-9

Etamaly, A. M., Mohamed, M. A., & Alo-lah, A. I. (2015). A smart technique for op-timization and simulation of hybrid photo-voltaic/wind/diesel/battery energy systems. *Smart Energy Grid Engineering (SEGE), 2015 IEEE International Conference on*, 1-6.

Faqih, K. M. (2013). The Performance of Software Reliability Models: A View Point. *International Journal of Performability Engineering*, *9*(4), 375–386.

Farahain, M., & Ganjefar, S. (2015). An online trained neural network controller to improve stability of power systems. *Neurocomputing*, *162*, 245–255. doi:10.1016/j.neucom.2015.03.047

Farahmand, H., Rashidinejad, M., Mousavi, A., Gharaveisi, A. A., Irving, M. R., & Taylor, G. A. (2012). Hybrid Mutation Particle Swarm Optimisation method for Available Transfer Capability enhancement. *International Journal of Electrical Power & Energy Systems*, *42*(1), 240–249. doi:10.1016/j.ijepes.2012.04.020

Fehring, R. J. (2005). New low- and high-tech calendar methods of family planning. *Journal of Midwifery & Women's Health*, *50*(1), 31–38. doi:10.1016/j.jmwh.2004.07.001 PMID:15637512

Ferrer-Mileo, V., Guede-Fernandez, F., Fernandez-Chimeno, M., Ramos-Castro, J., & Garcia-Gonzalez, M. A. (2015). Accuracy of heart rate variability estimation by photoplethysmography using an smartphone: Processing optimization and fiducial point selection. *Conference proceedings: ... Annual International Conference of the IEEE Engineering in Medicine and Biology Society. IEEE Engineering in Medicine and Biology Society. Conference*, *2015*, 5700-5703.

Feuerhuber, K., Lindert, S. O., & Schlacher, K. (2013). *Vibration attenuation by semi-active dampers.* Paper presented at 14th International Conference on Computer Aided Systems Theory (EUROCAST), Las Palmas, Spain. 10.1007/978-3-642-53862-9_35

Frolov, K. V. (1995). *Vibratsii v tekhnike: Spravochnik* [Vibrations in the technique: Reference Mechanical Engineering]. Vol. 6). Moscow, Russia: Mashinostroyeniye.

Fukasawa, R., Longo, H., Lysgaard, J., Poggi de Aragao, M., Uchoa, E., & Werneck, R. F. (2006). Robust branch-and-cut-price for the Capacitated vehicle routing problem. *Mathematical Programming Series A*, *106*(3), 491–511. doi:10.100710107-005-0644-x

Fulzele, J. B., Daigawane, M. B., & Daigawane, P. M. (2018). Design of hybrid PV-wind stand-alone renewable energy system: Case study. *Journal of Information and Optimization Sciences*, *39*(1), 345–355. doi:10.1080/02522667.2017.1374744

Ganesan, T., Vasant, P., & Elamvazuthi, I. (2016). *Advances in Metaheuristics: Applications in Engineering Systems.* CRC Press.

Ganesan, T., Vasant, P., & Elamvazuthi, I. (2016b). *Non-Gaussian Random Generators in Bacteria Foraging Algorithm for Multiobjective Optimization.* arXiv preprint arXiv:1605.07364

Ganesan, Vasant, & Elamvazuthi. (2016). *Advances in Metaheuristics: Application in Engineering Systems.* CRC Press, Taylor & Francis Group.

Ganesan, T., Aris, M. S., & Vasant, P. (2018). Extreme value metaheuristics for optimizing a many-objective gas turbine system. *International Journal of Energy Optimization and Engineering*, *7*(2), 76–96. doi:10.4018/IJEOE.2018040104

Ganesan, T., & Elamvazuthi, I. (2017). A multi-objective approach for resilience-based plant design optimization. *Quality Engineering*, *29*(4), 1–16. doi:10.1080/08982112.2016.1255331

Ganesan, T., Elamvazuthi, I., Shaari, K. Z. K., & Vasant, P. (2013). Multiobjective optimization of green sand mould system using chaotic differential evolution. In *Transactions on Computational Science XXI* (pp. 145–163). Springer Berlin Heidelberg. doi:10.1007/978-3-642-45318-2_6

Ganesan, T., Elamvazuthi, I., & Vasant, P. (2015). Multiobjective design optimization of a nano-CMOS voltage-controlled oscillator using game theoretic-differential evolution. *Applied Soft Computing*, *32*, 293–299. doi:10.1016/j.asoc.2015.03.016

Ganesan, T., Vasant, P., & Elamvazuthi, I. (2016a). Multiobjective optimization using particle swarm optimization with non-Gaussian random generators. *Intelligent Decision Technologies*, *10*(2), 93–103. doi:10.3233/IDT-150241

Ganesan, T., Vasant, P., & Litvinchev, I. (2020). Multiobjective Optimization of a Biofuel Supply Chain Using Random Matrix Generators. In *Deep Learning Techniques and Optimization Strategies in Big Data Analytics* (pp. 206–232). IGI Global. doi:10.4018/978-1-7998-1192-3.ch013

Gavanidou, E. S., Bakirtzis, A. G., & Dokopoulos, P. S. (1993). A probabilistic method for the evaluation of the performance and the reliability of wind-diesel energy systems. *IEEE Transactions on Energy Conversion*, *8*(2), 197–206. doi:10.1109/60.222705

Geleta, D. K., & Manshahia, M. S. (2019). Optimization of Hybrid Wind and Solar Renewable Energy System by Iteration Method. In Intelligent Computing & Optimization (pp. 98–107). doi:10.1007/978-3-030-00979-3_10

Geleta, D., & Manshahia, M. S. (2017). Optimization of Renewable Energy Systems: A Review. *International Journal of Scientific Re-search in Science and Technology, 8*(3), 769-795.

Geleta, D. K., & Manshahia, M. S. (2017). Nature Inspired Computational Intelligence: A Survey. *International Journal of Engineering. Science and Mathematics*, *6*(7), 769–795.

Geleta, D. K., & Manshahia, M. S. (2020). Artificial Bee Colony-Based Optimization of Hybrid Wind and Solar Renewable Energy System. In *Handbook of Research on Energy-Saving Technologies for Environmentally-Friendly Agricultural Development* (pp. 429–453). IGI Global. doi:10.4018/978-1-5225-9420-8.ch017

Geleta, D., & Manshahia, M. S. (2018). Opti-mization of Hybrid Wind and Solar Renewable Energy System by Iteration Method. *International Conference on Intelligent Computing and Optimization Springer*, 98-107.

Geropanta, V., & Cornelio Mari, E. (2014). Defining smart space: Scales of interaction between architecture and social practices. Journal: Comunicazionepuntodoc, 10, 165-18.

Ghasemi, M., Taghizadeh, M., Ghavidel, S., & Abbasian, A. (2016). Colonial competitive differential evolution: An experimental study for optimal economic load dispatch. *Applied Soft Computing*, *40*, 342–363. doi:10.1016/j.asoc.2015.11.033

Ghawghawe, N. D., & Thakre, K. L. (2006). Application of Power Flow Sensitivity Analysis and PTDF for Determination of ATC. In *2006 International Conference on Power Electronic, Drives and Energy Systems* (pp. 1–7). 10.1109/PEDES.2006.344304

Gholamy, A., Parra, J., Kreinovich, V., Fuentes, O., Anthony, E., Watada, J., ... Jain, L. C. (2019). (to appear). How to best apply deep neural networks in geosciences: Towards optimal 'averaging' in dropout training. *Smart Unconventional Modelling, Simulation and Optimization for Geosciences and Petroleum Engineering, Springer Verlag.*

Gil, E., Bailon, R., Vergara, J. M., & Laguna, P. (2010). PTT variability for discrimination of sleep apnea related decreases in the amplitude fluctuations of PPG signal in children. *IEEE Transactions on Biomedical Engineering*, *57*(5), 1079–1088. doi:10.1109/TBME.2009.2037734 PMID:20142152

Gil, E., Orini, M., Bailon, R., Vergara, J. M., Mainardi, L., & Laguna, P. (2010). Photoplethysmography pulse rate variability as a surrogate measurement of heart rate variability during non-stationary conditions. *Physiological Measurement*, *31*(9), 1271–1290. doi:10.1088/0967-3334/31/9/015 PMID:20702919

Giovanella, C. (2014) Where's the smartness of learning in smart territories? *Interaction Design and Architecture(s) Journal - IxD&A, 22*, 60-68.

Gohrle, C., Schindler, A., Wagner, A., & Sawodny, O. (2014). Design and Vehicle Implementation of Preview Active Suspension Controllers. *IEEE Transactions on Control Systems Technology*, *22*(3), 1135–1142. doi:10.1109/TCST.2013.2272342

Gomory, R. E. (1958). Outline of an algorithm for integer solutions to linear programs. *Bulletin of the American Mathematical Society*, *64*(5), 275–278. doi:10.1090/S0002-9904-1958-10224-4

Gondzio, J. (2012). Interior Point Methods: 25 Years Later. *European Journal of Operational Research*, *218*(3), 587–601. doi:10.1016/j.ejor.2011.09.017

Gonzalez-Perez, C., & Parcero-Oubiña, C. (2011). A conceptual model for cultural definition and motivation. *Proceedings of the 39th Conference on Computer Applications and Quantitative Methods in Archaeology.*

Goodale, B. M., Shilaih, M., Falco, L., Dammeier, F., Hamvas, G., & Leeners, B. (2019). Wearable Sensors Reveal Menses-Driven Changes in Physiology and Enable Prediction of the Fertile Window: Observational Study. *Journal of Medical Internet Research*, *21*(4), e13404. doi:10.2196/13404 PMID:30998226

Gordon, A. D., Coxon, A. P. M., & Davies, P. M. (1983). The User's Guide to Multidimensional Scaling, with Special Reference to the Mds (X) Library of Computer Programs. *The Statistician*, *32*(3), 355–356. doi:10.2307/2987947

Goyal, A., Bonchi, F., & Lakshmanan, L. V. (2011). *A data-based approach to social influence maximization.* arXiv preprint arXiv:1109.6886

Gozick, B., Subbu, K. P., Dantu, R., & Maeshiro, T. (2011). Magnetic maps for indoor navigation. *IEEE Transactions on Instrumentation and Measurement*, *60*(12), 3883–3891. doi:10.1109/TIM.2011.2147690

Guasti, L., Grimoldi, P., Mainardi, L. T., Petrozzino, M. R., Piantanida, E., Garganico, D., ... Venco, A. (1999). Autonomic function and baroreflex sensitivity during a normal ovulatory cycle in humans. *Acta Cardiologica*, *54*(4), 209–213. PMID:10511897

Guilleminault, C., Connolly, S., Winkle, R., Melvin, K., & Tilkian, A. (1984). Cyclical variation of the heart rate in sleep apnoea syndrome. Mechanisms, and usefulness of 24 h electrocardiography as a screening technique. *Lancet, 1*(8369), 126–131. doi:10.1016/S0140-6736(84)90062-X PMID:6140442

Guler, B., Varan, B., Tutuncuoglu, K., Nafea, M., Zewail, A. A., Yener, A., & Octeau, D. (2015). Using social sensors for influence propagation in networks with positive and negative relationships. *IEEE Journal of Selected Topics in Signal Processing, 9*(2), 360–373. doi:10.1109/JSTSP.2014.2386792

Guo, J., Zhu, X., Dong, D., Wang, K., Guan, Y., & Wang, L. (2019). The Hybrid process of preozonation and CNTs modification on hollow fiber membrane for fouling control. *Journal of Water Process Engineering, 31*(February), 100832. doi:10.1016/j.jwpe.2019.100832

Gurhan, R., &Cay, CI. (2008). Performances of tractor's driver seats with passive damping system under simulated test conditions. *TarimBilimleriDergisi-Journal of Agricultural Sciences, 14*(4), 401 – 408. WOS:000207576500013

Gutin, G., & Punnen, A. P. (2006). The Traveling Salesman Problem and Its Variants. Springer.

Guttman, L. (1968). A general nonmetric technique for finding the smallest coordinate space for a configuration of points. *Psychometrika, 33*(4), 469–506. doi:10.1007/BF02290164

Hadidian, M. J., Arabi, N. S., & Bigdeli. (2016). Optimal sizing of a stand-alone hybrid photovoltaic/wind system using new grey wolf optimizer considering re-liability. *Journal of Renewable and Sustainable Energy, 8*(3). https://earthobservatory.nasa.gov/features/EnergyBalance

Hansson, P. A. (2002). Working space requirement for an agricultural tractor axle suspension. *Biosystems Engineering, 81*(1), 57–71. doi:10.1006/bioe.2001.0007

Hao, Z., Xing-yuan, W., Si-wei, W., Kang, G., & Xiao-hui, L. (2017). Application of coupled map lattice with parameter q in image encryption. *Optics and Lasers in Engineering, 88*, 65–74. doi:10.1016/j.optlaseng.2016.07.004

Hardy, R. L. (1971). Multiquadric Equations of Topography and Other Irregular Surface. *Journal of Geophysical Research, 76*(8), 1905–1915. doi:10.1029/JB076i008p01905

Hayano, J., Tsukahara, T., Watanabe, E., Sasaki, F., Kawai, K., Sakakibara, H., . . . Fujimoto, K. (2013). Accuracy of ECG-based screening for sleep-disordered breathing: a survey of all male workers in a transport company. *Sleep & breathing = Schlaf & Atmung, 17*(1), 243-251.

Hayano, J., Barros, A. K., Kamiya, A., Ohte, N., & Yasuma, F. (2005). Assessment of pulse rate variability by the method of pulse frequency demodulation. *Biomedical Engineering Online, 4*(1), 62. doi:10.1186/1475-925X-4-62 PMID:16259639

Hayano, J., Kiyono, K., Struzik, Z. R., Yamamoto, Y., Watanabe, E., Stein, P. K., ... Carney, R. M. (2011). Increased non-gaussianity of heart rate variability predicts cardiac mortality after an acute myocardial infarction. *Frontiers in Physiology, 2*, 65. doi:10.3389/fphys.2011.00065 PMID:21994495

Hayano, J., Mukai, S., Fukuta, H., Sakata, S., Ohte, N., & Kimura, G. (2001). Postural response of low-frequency component of heart rate variability is an increased risk for mortality in patients with coronary artery disease. *Chest, 120*(6), 1942–1952. doi:10.1378/chest.120.6.1942 PMID:11742926

Hayano, J., Mukai, S., Sakakibara, M., Okada, A., Takata, K., & Fujinami, T. (1994). Effects of respiratory interval on vagal modulation of heart rate. *The American Journal of Physiology, 267*(1 Pt 2), H33–H40. PMID:7914066

Hayano, J., Ohashi, K., Yoshida, Y., Yuda, E., Nakamura, T., Kiyono, K., & Yamamoto, Y. (2018). Increase in random component of heart rate variability coinciding with developmental and degenerative stages of life. *Physiological Measurement, 39*(5), 054004. doi:10.1088/1361-6579/aac007 PMID:29693554

Hayano, J., Sakakibara, Y., Yamada, M., Kamiya, T., Fujinami, T., Yokoyama, K., ... Takata, K. (1990). Diurnal variations in vagal and sympathetic cardiac control. *The American Journal of Physiology*, *258*(3 Pt 2), H642–H646. PMID:2316678

Hayano, J., Taylor, J. A., Mukai, S., Okada, A., Watanabe, Y., Takata, K., & Fujinami, T. (1994). Assessment of frequency shifts in R-R interval variability and respiration with complex demodulation. *Journal of Applied Physiology*, *77*(6), 2879–2888. doi:10.1152/jappl.1994.77.6.2879 PMID:7896636

Hayano, J., Taylor, J. A., Yamada, A., Mukai, S., Hori, R., Asakawa, T., ... Fujinami, T. (1993). Continuous assessment of hemodynamic control by complex demodulation of cardiovascular variability. *The American Journal of Physiology*, *264*, H1229–H1238. PMID:8476100

Hayano, J., Watanabe, E., Saito, Y., Sasaki, F., Fujimoto, K., Nomiyama, T., ... Sakakibara, H. (2011). Screening for obstructive sleep apnea by cyclic variation of heart rate. *Circulation: Arrhythmia and Electrophysiology*, *4*(1), 64–72. doi:10.1161/CIRCEP.110.958009 PMID:21075771

Hayano, J., & Yasuma, F. (2003). Hypothesis: Respiratory sinus arrhythmia is an intrinsic resting function of cardiopulmonary system. *Cardiovascular Research*, *58*(1), 1–9. doi:10.1016/S0008-6363(02)00851-9 PMID:12667941

Hayano, J., Yasuma, F., Watanabe, E., Carney, R. M., Stein, P. K., Blumenthal, J. A., ... Kodama, I. (2017). Blunted cyclic variation of heart rate predicts mortality risk in post-myocardial infarction, end-stage renal disease, and chronic heart failure patients. *Europace*, *19*(8), 1392–1400. PMID:27789562

Hayano, J., & Yuda, E. (2019). Pitfalls of assessment of autonomic function by heart rate variability. *Journal of Physiological Anthropology*, *38*(1), 3. doi:10.118640101-019-0193-2 PMID:30867063

Hemamalini, S., & Simon, S. P. (2010). Artificial bee colony algorithm for economic load dispatch problem with non-smooth cost functions. *Electric Power Components and Systems*, *38*(7), 786–803. doi:10.1080/15325000903489710

Henderson-Sellers, B., & Gonzalez-Perez, C. (2010). Granularity in conceptual modelling: application to metamodels. *Proceedings of the 29th international conference on Conceptual modeling*, 219-232. 10.1007/978-3-642-16373-9_16

He, Q., Wang, J., & Lu, H. (2018). A hybrid system for short-term wind speed forecasting. *Applied Energy*, *226*, 756–771. doi:10.1016/j.apenergy.2018.06.053

Herbrich, R. (2001). *Learning kernel classifiers: theory and algorithms*. MIT Press. doi:10.7551/mitpress/4170.001.0001

Hinton, G. E., Osindero, S., & Teh, Y.-W. (2006). A Fast Learning Algorithm for Deep Belief Nets. *Neural Computation*, *18*(7), 1527–1554. doi:10.1162/neco.2006.18.7.1527 PMID:16764513

Hirose, H. (1996). Maximum likelihood estimation in the 3-parameter Weibull distribution. A look through the generalized extreme-value distribution. *IEEE Transactions on Dielectrics and Electrical Insulation*, *3*(1), 43–55. doi:10.1109/94.485513

Hirsch, J. A., & Bishop, B. (1981). Respiratory sinus arrhythmia in humans: How breathing pattern modulates heart rate. *The American Journal of Physiology*, *241*, H620–H629. PMID:7315987

Hoen-Su, R., Kyoung, M., Lee, J. G., & Hyun, M. Y. (2002). Extended integral based governor for power system stabilizer. *IEEE Power Engineering Society Summer Meeting*, *3*, 1610-1614.

Holzinger, A., Plass, M., Kickmeier-Rust, M., Holzinger, K., Crişan, G. C., Pintea, C. M., & Palade, V. (2019). Interactive machine learning: Experimental evidence for the human in the algorithmic loop. *Applied Intelligence*, *49*(7), 2401–2414. doi:10.100710489-018-1361-5

Honarkhah, M., & Caers, J. (2010). Stochastic simulation of patterns using distance-based pattern modeling. *Mathematical Geosciences*, *42*(5), 487–517. doi:10.100711004-010-9276-7

Hosseini, R., Hosseini, N., & Khorasanizadeh, H. (2011). An Experimental study of combining a Photovoltaic System with a heating System. In *Proceedings of the World Renewable Energy Congress* (pp. 2993-3000). 10.3384/ecp110572993

Hossenizadeh, N., & Kalam, A. (1999). A rule based fuzzy power system stabilizer tuned by a neural network. *IEEE Transactions on Energy Conversion, 14*(3), 773–776. doi:10.1109/60.790950

Huang, A., Zhang, H. M., Guan, W., Yang, Y., & Zong, G. (2015). Cascading failures in weighted complex networks of transit systems based on coupled map lattices. *Mathematical Problems in Engineering, 2015*, 2015. doi:10.1155/2015/940795

Huang, T., & Zhang, H. (2016). Bifurcation, chaos and pattern formation in a space-and time-discrete predator–prey system. *Chaos, Solitons, and Fractals, 91*, 92–107. doi:10.1016/j.chaos.2016.05.009

Huang, T., Zhang, H., & Yang, H. (2017). Spatiotemporal complexity of a discrete space-time predator– prey system with self-and cross-diffusion. *Applied Mathematical Modelling, 47*, 637–655. doi:10.1016/j.apm.2017.03.049

Huikuri, H. V., Perkiomaki, J. S., Maestri, R., & Pinna, G. D. (2009). Clinical impact of evaluation of cardiovascular control by novel methods of heart rate dynamics. *Philosophical Transactions. Series A, Mathematical, Physical, and Engineering Sciences, 367*(1892), 1223-1238.

Hu, Y.-L., & Chen, L. (2018). A nonlinear hybrid wind speed forecasting model using LSTM network, hysteretic ELM and Differential Evolution algorithm. *Energy Conversion and Management, 173*, 123–142. doi:10.1016/j.enconman.2018.07.070

Ibrahim, A., Othman, M. Y., Ruslan, M. H., Mat, S., & Sopian, K. (2011). Recent advances in flat plate photovoltaic/thermal (PV/T) solar collectors. *Renewable & Sustainable Energy Reviews, 15*(1), 352–365. doi:10.1016/j.rser.2010.09.024

Ilinca, A., McCarthy, E., Chaumel, J. L., & Retiveau, J. L. (2003). Wind potential assessment of Quebec Province. *Renewable Energy, 28*(12), 1881–1897. doi:10.1016/S0960-1481(03)00072-7

Islam, J., Vasant, P. M., Negash, B. M., Laruccia, M. B., & Myint, M. (2020). A Survey of Nature-Inspired Algorithms With Application to Well Placement Optimization. In *Deep Learning Techniques and Optimization Strategies in Big Data Analytics* (pp. 32–45). IGI Global. doi:10.4018/978-1-7998-1192-3.ch003

Islam, N. N., Hannan, M. A., Shareef, H., & Mohamed, A. (2016). An Application of Backtracking Search Algorithm in Designing Power System Stabilizers for Large Multi-machine System. *Neurocomputing, 237*, 175–184. doi:10.1016/j.neucom.2016.10.022

Ivakhnenko, A. G. (1971). Polynomial Theory of Complex Systems. *IEEE Transactions on Systems, Man, and Cybernetics. SMC, 1*(4), 364–378. doi:10.1109/TSMC.1971.4308320

Iyengar, N., Peng, C. K., Morin, R., Goldberger, A. L., & Lipsitz, L. A. (1996). Age-related alterations in the fractal scaling of cardiac interbeat interval dynamics. *The American Journal of Physiology, 271*, R1078–R1084. PMID:8898003

Izquierdo, C. E., & Marquez, A. E. (2017). A survey of the cuckoo search and its application in real-world optimization problems. Handbook of research on soft computing and nature-inspired algorithms.

Janda, K. (2009). Buildings Don't Use Energy: People Do. *PLEA2009 - Proceedings of the 26th Conference on Passive and Low Energy Architecture.*

Javadi, M. R., Mazlumi, K., & Jalilvand, A. (2011). Application of GA, PSO and ABC in optimal design of a stand-alone hybrid system for north-west of Iran. In *Electrical and Elec-tronics Engineering* (pp. 1–203). ELECO.

Jayabarathi, T., Jayaprakash, K., Jeyakumar, D. N., & Raghunathan, T. (2005). Evolutionary programming techniques for different kinds of economic dispatch problems. *Electric Power Systems Research, 73*(2), 169–176. doi:10.1016/j.epsr.2004.08.001

Jensen, P. A., & Bard, J. F. (2003). *Operations Research Models and Methods*. John Wiley & Sons, Inc.

Jeong Kim, M., & Won Oh, M. (2013). A method for evaluating the performance of green buildings with a focus on user experience. *Energy and Buildings*, (66), 203-210.

Jia, M., & Srinivasa, R. (2015). Occupant Behavior Modeling for Smart Buildings: A Critical Review of Data Acquisition technologies and Modeling Methodologies. *Proceedings of the 2015 Winter Simulation Conference*, 3345-3355. 10.1109/WSC.2015.7408496

Ji, J., Lu, J., Chow, T., He, W., & Pei, G. (2007). A sensitivity study of a hybrid photovoltaic/thermal water-heating system with natural circulation. *Applied Energy*, *84*(2), 222–237. doi:10.1016/j.apenergy.2006.04.009

Kaabeche, A., Belhamel, M., & Ibtiouen, R. (2010). Optimal sizing method for stand-alone hybrid PV/wind power genera-tion system. *Revue des Energies Renouvelables (SMEE'10) Bou Ismail Tipaza*, 205- 213.

Kaliannan J., Baskaran A. & Dey N. (2017). Automatic generation control of thermal-thermal-hydro power systems with PID controller using ant colony optimization. *Renewable and alternate energy: concepts methodology, tools and application.*

Kalogirou, S. A. (2001). Use of TRYNSYS for modeling and simulation of a hybrid PV-thermal solar system for Cyprus. *Renewable Energy*, *23*(2), 247–260. doi:10.1016/S0960-1481(00)00176-2

Kamwa, I., Grondin, R., & Trudel, G. (2005). IEEE PSS2B versus PSS4B: The limits of performance of modern power system stabilizers. *IEEE Transactions on Power Systems*, *20*(2), 903–915. doi:10.1109/TPWRS.2005.846197

Kaneko, K. (1992). Overview of coupled map lattices. *Chaos (Woodbury, N.Y.)*, *2*(3), 279–282. doi:10.1063/1.165869 PMID:12779975

Karaboga, D., & Basturk, B. (2007). A powerful and e cient algorithm for numeri-cal function optimization: Arti cial bee colony (ABC) algorithm. *Journal of Global Optimization*, *39*(3), 459–471. doi:10.100710898-007-9149-x

Karaboga, D., Gorkemli, B., Ozturk, C., & Karaboga, N. (2014). A comprehensive survey: Artificial bee colony (ABC) algorithm and ap-plications. *Artificial Intelligence Review*, *42*(1), 21–57. doi:10.100710462-012-9328-0

Karagianni, A., Geropanta, V., & Parthenios, P. (2019). *Exploring the ICT Potential to Maximize User - Built Space Interaction in Monumental Spaces The case of the municipal agora in Chania, Crete*. https://ecaade.org/conference/downloads/

Karagrigoriou, A., & Vonta, I. (2014, September). Statistical Inference for Heavy-Tailed Distributions in Technical Systems. In *Availability, Reliability and Security (ARES), 2014 Ninth International Conference on* (pp. 412-419). IEEE. 10.1109/ARES.2014.62

Karlof, J. K. (2005). *Integer Programming: Theory and Practice*. CRC Press Inc. doi:10.1201/9781420039597

Kaur, T., Kumar, S., & Segal, R. (2016). Application of artificial neural network for short term wind speed forecasting. In *2016 Biennial International Conference on Power and Energy Systems: Towards Sustainable Energy (PESTSE)* (pp. 1–5). 10.1109/PESTSE.2016.7516458

Kefayat, M. A., Lashkar, A., & Nabavi, N. (2015). A hybrid of ant colony optimiza-tion and arti cial bee colony algorithm for probabilistic optimal placement and sizing of distributed energy resources. *Energy Conversion and Management*, *92*, 149–161. doi:10.1016/j.enconman.2014.12.037

Kellog, W., Nehrir, M., Venkataramanan, G., & Gerez, V. (1998). Generation Unit Sizing and Cost Analysis for Stand -alone Wind, Phovoltaic, and Hybrid Wind/PV Systems. *IEEE Transactions on Energy Conversion*, *13*(1), 70–75. doi:10.1109/60.658206

Keluža, M., & Vukelic, B. (2017). Analysis of an indoor positioning systems. *Zbornik Veleučilišta u Rijeci., 5*(1), 13–32. doi:10.31784/zvr.5.1.2

Ketola, B., McIntosh, K. R., Norris, A., & Tomalia, M. K. (2008). Silicones for photovoltaic encapsulation. *23rd European Photovoltaic Solar Energy Conference*, 2969-2973.

Keumarsi, V., Simab, M., & Shahgholian, G. (2014). An integrated approach for optimal placement and tuning of power system stabilizer in multimachine systems. *International Journal of Electrical Power & Energy Systems, 63*, 132–139. doi:10.1016/j.ijepes.2014.05.072

Kharchenko, V., Panchenko, V., Tikhonov, P., & Vasant, P. (2018). Cogenerative PV Thermal Modules of Different Design for Autonomous Heat and Electricity Supply. Handbook of Research on Renewable Energy and Electric Resources for Sustainable Rural Development, 86-119.

Kharchenko, V., Nikitin, B., Tikhonov, P., Panchenko, V., & Vasant, P. (2019). Evaluation of the Silicon Solar Cell Modules. *Intelligent Computing & Optimization. Advances in Intelligent Systems and Computing, 866*, 328–336. doi:10.1007/978-3-030-00979-3_34

Kharchenko, V., & Vasant, P. (2018). *Handbook of Research on Renewable Energy and Electric Resources for Sustainable Rural Development*. Hershey, PA: IGI Global. doi:10.4018/978-1-5225-3867-7

Kharchenko, V., & Vasant, P. (2020). *Handbook of Research on Energy-Saving Technologies for Environmentally-Friendly Agricultural Development*. Hershey, PA: IGI Global. doi:10.4018/978-1-5225-9420-8

Kharchenko, V., & Vasant, P. (2020). *Handbook of Research on Smart Computing for Renewable Energy and Agro-Engineering*. Hershey, PA: IGI Global. doi:10.4018/978-1-7998-1216-6

Kheshti, M., Ding, L., Ma, S., & Zhao, B. (2018). Double weighted particle swarm optimization to non-convex wind penetrated emission/economic dispatch and multiple fuel option systems. *Renewable Energy, 125*, 1021–1037. doi:10.1016/j.renene.2018.03.024

Kheshti, M., Kang, X., Bie, Z., Jiao, Z., & Wang, X. (2017). An effective lightning flash algorithm solution to large scale non-convex economic dispatch with valve-point and multiple fuel options on generation units. *Energy, 129*, 1–15. doi:10.1016/j.energy.2017.04.081

Kim, S.-E. (2012). Indoor positioning system using geomagnetic anomalies for smartphones. In *International conference on indoor positioning and indoor navigation (IPIN)*. IEEE.

Kim, Woojin, & Kwang-Ryul. (2017). Indoor positioning system using magnetic field map navigation and an encoder system. *Sensors, 17*(3), 651.

Kim, J., Jeon, W., Ko, Y.-W., Uhmn, S., & Kim, D.-H. (2018). Genetic Local Search for Nurse Scheduling Problem. *Advanced Science Letters, 24*(1), 608–612. doi:10.1166/asl.2018.11770

Kiyono, K., Hayano, J., Watanabe, E., Struzik, Z. R., & Yamamoto, Y. (2008). Non-Gaussian heart rate as an independent predictor of mortality in patients with chronic heart failure. *Heart Rhythm, 5*(2), 261–268. doi:10.1016/j.hrthm.2007.10.030 PMID:18242551

Kiyono, K., Struzik, Z. R., Aoyagi, N., Sakata, S., Hayano, J., & Yamamoto, Y. (2004). Critical scale invariance in a healthy human heart rate. *Physical Review Letters, 93*(17), 178103. doi:10.1103/PhysRevLett.93.178103 PMID:15525130

Knapp, G. L., Mukherjee, T., Zuback, J. S., Wei, H. L., Palmer, T. A., De, A., & DebRoy, T. (2017). Building blocks for a digital twin of additive manufacturing. *Acta Materialia, 135*, 390–399. doi:10.1016/j.actamat.2017.06.039

Kohlenbach, P., & Ziegler, F. (2008). A dynamic simulation model for transient absorption chiller performance. Part I: The model. *International Journal of Refrigeration*, *31*(2), 217–225. doi:10.1016/j.ijrefrig.2007.06.009

Komninos, N. (2008). Intelligent Cities and Globalization of Innovation Networks. *Routledge London*, *2008*, 247.

Komninos, N. (2014). *The Age of Intelligent Cities: Smart Environments and Innovation-for-All Strategies*. London: Routledge. doi:10.4324/9781315769349

Kosmadakis, G., Sotirios, K., & Emmanuel, K. (2013). Renewable and conventional elec-tricity generation systems. *Technologies and diversity of energy systems*. In *Renewable En-ergy Governance* (pp. 9–3). Springer London. doi:10.1007/978-1-4471-5595-9_2

Kothari, M., Bhattacharya, L., & Nanda, J. (2002). Adaptive power system stabilizer based on pole shifting technique. *IEE Proceedings. Generation, Transmission and Distribution*, *143*(1), 96–98. doi:10.1049/ip-gtd:19960300

Krak, I. V., Kudin, H. I., Barmak, O. V., Manziuk, E. A., Smolarz, A., & Mamyrbaev, O. (2019). Method and algorithm of the piecewise-hyperplane clusterization using tools of pseudo-inverse matrices. In Photonics Applications in Astronomy, Communications, Industry, and High-Energy Physics Experiments 2019 (Vol. 11176, p. 1117628). International Society for Optics and Photonics. doi:10.1117/12.2537417

Krak, I., Barmak, O., Manziuk, E., & Kudin, H. (2019). Approach to Piecewise-Linear Classification in a Multi-dimensional Space of Features Based on Plane Visualization. In *International Scientific Conference "Intellectual Systems of Decision Making and Problem of Computational Intelligence"* (pp. 35-47). Springer.

Krak, I., Barmak, O., Manziuk, E., & Kulias, A. (2019). Data Classification Based on the Features Reduction and Piecewise Linear Separation. In *International Conference on Intelligent Computing & Optimization* (pp. 282-289). Springer.

Kruskal, J. B., & Wish, M. (1978). *Multidimensional scaling* (Vol. 11). Sage. doi:10.4135/9781412985130

Kumar, J., Goomer, R., & Singh, A. K. (2018). Long Short Term Memory Recurrent Neural Network (LSTM-RNN) Based Workload Forecasting Model For Cloud Datacenters. *Procedia Computer Science*, *125*, 676–682. doi:10.1016/j.procs.2017.12.087

Kumar, S., Kumar, R., Kumar, S., & Kumar, S. (2019). Cryptographic construction using coupled map lattice as a diffusion model to enhanced security. *Journal of Information Security and Applications*, *46*, 70–83. doi:10.1016/j.jisa.2019.02.011

Kumar, S., Munapo, E., & Jones, B. C. (2007). An Integer Equation Controlled Descending Path to a Protean Pure Integer Program. *Indian Journal of Mathematics*, *49*, 211–237.

Kumar, S., Munapo, E., Lesaoana, M., & Nyamugure, P. (2018). A minimum spanning tree based heuristic for the travelling salesman tourm. *OPSEARCH*, *55*(1), 150–164. doi:10.100712597-017-0318-5

Kundu, A., & Ratha, N. K. (2019). *U.S. Patent Application No. 16/399,276*. Washington, DC: US Patent Office.

Kundur, P. (2006). *Power System Stability and Control*. Tata: McGraw-Hill.

Kundur, P., Klein, M., Rogers, G. J., & Zywno, M. S. (1989). Application of power system stabilizers for enhancement of overall stability. *IEEE Transactions on Power Systems*, *4*(2), 614–626. doi:10.1109/59.193836

Kuo, K.-P., & Cheryan, M. (1983). Ultrafiltration of Acid Whey in a Spiral-Wound Unit: Effect of Operating Parameters on Membrane Fouling. *Journal of Food Science*, *48*(4), 1113–1118. doi:10.1111/j.1365-2621.1983.tb09172.x

Kuzmin, V. A., Fedotkin, R. S., & Kryuchkov, V. A. (2017). Iskusstvennaya neyronnaya set' dlya obosnovaniya parametrov khodovykh sistem traktorov [Artificial neural network to justify the parameters of the tractor's propulsion systems]. *Sel'skokhozyaystvennyye Mashiny i Tekhnologii*, (4), 24–30.

Laboratory of the Integrated Ecological and Geographical Research of the Arctic. (n.d.). http://www.ael-msu.org/

Ladányi, L., Ralphs, T. K., & Trotter, L. E. Jr. (2001). Branch, cut and Price: Sequential and Parallel. In N. Naddef & M. Jüenger (Eds.), *Computational Combinatorial Optimization* (Vol. 223). Berlin: Springer. doi:10.1007/3-540-45586-8_6

Land, A. H., & Doig, A. G. (1960). An Automatic method for solving discrete programming problems. *Econometrica, 28*(3), 497–520. doi:10.2307/1910129

Larochelle, H., Bengio, Y., Louradour, J., & Lamblin, P. (2009). Exploring Strategies for Training Deep Neural Networks. *Journal of Machine Learning Research, 10*, 1–40. Retrieved from https://dl.acm.org/citation.cfm?id=1577069.1577070

Lee, H.-H. (2016). Effective power dispatch capability decision method for a wind-battery hybrid power system. *IET Generation, Transmission & Distribution, 10*(3), 661-668. Retrieved from https://digital-library.theiet.org/content/journals/10.1049/iet-gtd.2015.0078

Lee, T., Johnson, J., & Cheng, S. (2016). *An Interactive Machine Learning Framework*. arXiv preprint arXiv:1610.05463

Lee, K. Y., Sode-Yome, A., & Park, J. H. (1998). Adaptive Hopfield neural networks for economic load dispatch. *IEEE Transactions on Power Systems, 13*(2), 519–526. doi:10.1109/59.667377

Leicht, A. S., Hirning, D. A., & Allen, G. D. (2003). Heart rate variability and endogenous sex hormones during the menstrual cycle in young women. *Experimental Physiology, 88*(3), 441–446. doi:10.1113/eph8802535 PMID:12719769

Lemmens, M. (2013). *Indoor Positioning*. https://www.gim-international.com/content/article/indoor-positioning-2

Lenz, R. (2016). Generalized Pareto Distributions—Application to Autofocus in Automated Microscopy. *IEEE Journal of Selected Topics in Signal Processing, 10*(1), 92–98. doi:10.1109/JSTSP.2015.2482949

Leskovec, J., Krause, A., Guestrin, C., Faloutsos, C., VanBriesen, J., & Glance, N. (2007, August). Cost-effective outbreak detection in networks. In *Proceedings of the 13th ACM SIGKDD international conference on Knowledge discovery and data mining* (pp. 420-429). ACM.

Li, B. (2012). How feasible is the use of magnetic field alone for indoor positioning? In *International Conference on Indoor Positioning and Indoor Navigation (IPIN)*. IEEE. 10.1109/IPIN.2012.6418880

Li, B., Gallagher, T., Rizos, C., & Dempster, A. G. (2013). Using geomagnetic field for indoor positioning. *Journal of Applied Geodesy, 7*(4), 299–308. doi:10.1515/jag-2013-0016

Liben-Nowell, D., & Kleinberg, J. (2007). The link-prediction problem for social networks. *Journal of the American Society for Information Science and Technology, 58*(7), 1019–1031. doi:10.1002/asi.20591

Lin, C. E., & Viviani, G. L. (1984). Hierarchical economic dispatch for piecewise quadratic cost functions. *IEEE Transactions on Power Apparatus and Systems, PAS-103*(6), 1170–1175. doi:10.1109/TPAS.1984.318445

Lin, J., & Wang, Z. J. (2019). Multi-area economic dispatch using an improved stochastic fractal search algorithm. *Energy, 166*, 47–58. doi:10.1016/j.energy.2018.10.065

Li, S., & Zhang, X. (2011). The study of PAFSSB on RO pre-treatment in pulp and paper wastewater. *Procedia Environmental Sciences, 8*(November), 4–10. doi:10.1016/j.proenv.2011.10.003

Liu, H., Mi, X., & Li, Y. (2018). Smart deep learning based wind speed prediction model using wavelet packet decomposition, convolutional neural network and convolutional long short term memory network. *Energy Conversion and Management, 166*, 120–131. doi:10.1016/j.enconman.2018.04.021

Liu, L., Tang, J., Han, J., Jiang, M., & Yang, S. (2010, October). Mining topic-level influence in heterogeneous networks. In *Proceedings of the 19th ACM international conference on Information and knowledge management* (pp. 199-208). ACM.

Li, W. X., Li, Z. M., Ni, Y. X., & Wu, F. F. (2003). Available transfer capability calculation with post-contingency generation rescheduling/load curtailment. In *2003 Sixth International Conference on Advances in Power System Control, Operation and Management ASDCOM 2003 (Conf. Publ. No. 497)* (Vol. 2, pp. 563–568). 10.1049/cp:20030648

Li, Y., Shi, H., Han, F., Duan, Z., & Liu, H. (2019). Smart wind speed forecasting approach using various boosting algorithms, big multi-step forecasting strategy. *Renewable Energy*, *135*, 540–553. doi:10.1016/j.renene.2018.12.035

Li, Y., Wu, H., & Liu, H. (2018). Multi-step wind speed forecasting using EWT decomposition, LSTM principal computing, RELM subordinate computing and IEWT reconstruction. *Energy Conversion and Management*, *167*, 203–219. doi:10.1016/j.enconman.2018.04.082

Lohtander, M., Ahonen, N., Lanz, M., Ratava, J., & Kaakkunen, J. (2018). Micro Manufacturing Unit and the Corresponding 3D-Model for the Digital Twin. *Procedia Manufacturing*, *25*, 55–61. doi:10.1016/j.promfg.2018.06.057

Lu, X. (2019). Time-periodic model of wind speed and its application in risk evaluation of wind-power-integrated power systems. *IET Generation, Transmission & Distribution, 13*(1), 46-54. Retrieved from https://digital-library.theiet.org/content/journals/10.1049/iet-gtd.2018.5619

Lu, L., Yang, H., & Burnett, J. (2002). Investigation on wind power potential on Hong Kong islandsan analysis of wind power and wind turbine characteristics. *Renewable Energy*, *27*(1), 1–12. doi:10.1016/S0960-1481(01)00164-1

Luna, R. R., Trejo, P. M., Vargas, D., & Os-Moreno, G. (2012). Op-timal sizing of renewable hybrids energy sys-tems: A review of methodologies. *Solar Energy*, *88*(4), 1077–1088. doi:10.1016/j.solener.2011.10.016

Lynch, C., O'Mahony, M. J., & Scully, T. (2014). Simplified Method to Derive the Kalman Filter Covariance Matrices to Predict Wind Speeds from a NWP Model. *Energy Procedia, 62*, 676–685.

Lysen, H. (1983). Introduction to wind energy. Consultancy Services. *Wind Energy (Chichester, England)*, 82–91.

Madhu, N. M., Mohan, V., Singh, J. G., & Suresh, R. (2018). Risk Adjusted Co-optimization of ATC in High-Low Voltage Interconnected Power Systems. In *2018 IEEE International Conference on Power Electronics, Drives and Energy Systems (PEDES)* (pp. 1–6). 10.1109/PEDES.2018.8707502

Mahabuba, A., & Khan, M. A. (2009). Small signal stability enhancement of a multi-machine power system using robust and adaptive fuzzy neural network basd power system stabilizer. *International Transactions on Electrical Energy Systems*, *19*(7), 978–01.

Maheta, H. H., & Dabhi, V. K. (2014, February). An improved SPEA2 Multi objective algorithm with non dominated elitism and Generational Crossover. In *Issues and Challenges in Intelligent Computing Techniques (ICICT), 2014 International Conference on* (pp. 75-82). IEEE. 10.1109/ICICICT.2014.6781256

Manish, D., Upadhyay, N., & Silakari, S. (2015). An exhaustive survey on nature An exhaustive survey on nature inspired optimization algorithms. *International Journal of Software Engineering and Its Applications*, *9*(4), 91–104.

Manoharan, P. S., Kannan, P. S., Baskar, S., & Iruthayarajan, M. W. (2008). Penalty parameter-less constraint handling scheme based evolutionary algorithm solutions to economic dispatch. *IET Generation, Transmission & Distribution*, *2*(4), 478–490. doi:10.1049/iet-gtd:20070423

Mante, O. D. (2018). *Sub-Saharan Africa Is Lighting Up: Uneven Progress on Electrification*. Academic Press.

Marjoram, R., St-Clair, K., McMahon, B., Goelz, A., & Achen, A. (2008). Integrated MR cabin suspension for reduction of human exposure to whole-body vibration in heavy-duty vehicles. Bremen, Germany: Messebremen-hvghanseatisch-everanstaltungs-gmbh. WOS:000270938400110

Marmolejo, J. A., & Rodriguez, R. (2015). Fat tail model for simulating test systems in multiperiod unit commitment. *Mathematical Problems in Engineering*, *2015*, 2015. doi:10.1155/2015/738215

Marmolejo, J. A., Velasco, J., & Selley, H. J. (2017). An adaptive random search for short term generation scheduling with network constraints. *PLoS One*, *12*(2), e0172459. doi:10.1371/journal.pone.0172459 PMID:28234954

Marugán, A. P., Márquez, F. P. G., Perez, J. M. P., & Ruiz-Hernández, D. (2018). A survey of artificial neural network in wind energy systems. *Applied Energy*, *228*, 1822–1836. doi:10.1016/j.apenergy.2018.07.084

Meng, A., Li, J., & Yin, H. (2016). An efficient crisscross optimization solution to large-scale non-convex economic load dispatch with multiple fuel types and valve-point effects. *Energy*, *113*, 1147–1161. doi:10.1016/j.energy.2016.07.138

Minsky, M. (1961). Steps toward artificial intelligence. *Proceedings of the IRE*, *49*(1), 8–30. doi:10.1109/JRPROC.1961.287775

Mirjalili, S. (2015). The ant lion optimizer. *Advances in Engineering Software*, *83*, 80–98. doi:10.1016/j.advengsoft.2015.01.010

Mironyuk, S. S., Smirnov, A., Sokolov, A. V., & Proshkin, Y. (2019). Optimization of Spectral Composition and Energy Economy Effectiveness of Phyto-Irradiators With Use of Digital Technologies. In Handbook of Research on Energy-Saving Technologies for Environmentally-Friendly Agricultural Development (pp. 191–212). doi:10.4018/978-1-5225-9420-8.ch008

Mitchell, J. E. (2002). Cutting plane algorithms for combinatorial optimization problems. In Handbook of Applied Optimization. Oxford University Press.

Mitchell, J. E. (2001). Branch and cut algorithms for integer programming. In C. A. Floudas & P. M. Pardalos (Eds.), *Encyclopedia of Optimization*. Kluwer Academic Publisher. doi:10.1007/0-306-48332-7_215

Modiri-Delshad, M., Kaboli, S. H. A., Taslimi-Renani, E., & Rahim, N. A. (2016). Backtracking search algorithm for solving economic dispatch problems with valve-point effects and multiple fuel options. *Energy*, *116*, 637–649. doi:10.1016/j.energy.2016.09.140

Mohamed, M. A., Eltamaly, A. M., & Alolah, A. I. (2016). PSO-based smart grid application for sizing and optimization of hybrid renewable energy systems. *PLoS One*, *11*(8), e0159702. doi:10.1371/journal.pone.0159702 PMID:27513000

Mohan, V., Singh, J. G., & Ongsakul, W. (2015). An efficient two stage stochastic optimal energy and reserve management in a microgrid. *Applied Energy*, *160*, 28–38. doi:10.1016/j.apenergy.2015.09.039

Mokhtar, N. M., Lau, W. J., Ismail, A. F., & Veerasamy, D. (2015). Membrane distillation technology for treatment of wastewater from rubber industry in Malaysia. *Procedia CIRP*, *26*, 792–796. doi:10.1016/j.procir.2014.07.161

Moncayo–Martínez, L. A., & Mastrocinque, E. (2016). A multi-objective intelligent water drop algorithm to minimise cost of goods sold and time to market in logistics networks. *Expert Systems with Applications*, *64*, 455–466. doi:10.1016/j.eswa.2016.08.003

Montello, D. (2015). *Spatial Cognition. International Encyclopedia of the Social & Behavioral Sciences*. doi:10.1016/B978-0-08-097086-8.72057-5

Moore, J. M. (1968). An n job, one machine sequencing algorithm for minimizing the number of late jobs. *Management Science*, *15*(1), 102–109. doi:10.1287/mnsc.15.1.102

Moriasi, D. N., Arnold, J. G., Van Liew, M. W., Bingner, R. L., Harmel, R. D., & Veith, T. L. (2007). Model Evaluation Guidelines for Systematic Quantification of Accuracy in Watershed Simulations. *Transactions of the ASABE*, *50*(3), 885–900. doi:10.13031/2013.23153

Motsatsi, L. (2015). The Development of a Post-Occupancy Evaluation tool for Primary Schools: Learner Comfort Assessment Tool (LCAT). *Proceedings of the Smart And Sustainable Built Environments Conference*.

MoWIE. (2019). *National electrification program 2.0; Intgrated planning for universal access*. Academic Press.

Možina, M., Demšar, J., Kattan, M., & Zupan, B. (2004, September). Nomograms for visualization of naive Bayesian classifier. In *European Conference on Principles of Data Mining and Knowledge Discovery* (pp. 337-348). Springer.

Munapo, E. (2016). Solving the Binary Linear Programming Model in Polynomial Time. *American Journal of Operations Research*, *6*(01), 1–7. doi:10.4236/ajor.2016.61001

Munapo, E. (2020). Network Reconstruction – A New Approach to the Traveling Salesman Problem and Complexity. In P. Vasant, I. Zelinka, & G. W. Weber (Eds.), *Intelligent Computing and Optimization. ICO 2019. Advances in Intelligent Systems and Computing, 1072*. Cham: Springer. doi:10.1007/978-3-030-33585-4_26

Nadef, D. (2002). Polyhedral theory and branch and cut algorithms for the symmetric TSP. In The traveling salesman problem and its variations. Kluwer.

Nafeh, A. E. S. A. (2011). Optimal economical sizing of a PV-wind hybrid energy system using genetic algorithm. *International Journal of Green Energy*, *8*(1), 25–43. doi:10.1080/15435075.2010.529407

Nakajima, J., Kunihama, T., & Omori, Y. (2017). Bayesian modeling of dynamic extreme values: Extension of generalized extreme value distributions with latent stochastic processes. *Journal of Applied Statistics*, *44*(7), 1248–1268. doi:10.1080/02664763.2016.1201796

Naresh, G., Raju, M. R., & Narasimham, S. V. L. (2016). Coordinated deign of power system stabilizers and TCSC employing improved harmony search algorithm. *Swarm and Evolutionary Computation*, *27*, 169–179. doi:10.1016/j.swevo.2015.11.003

Nasiruddin, I., Bhatti, T. S., & Hakimuddin, N. (2015). Automatic generation control in an interconnected power system incorporating diverse source power plants using bacteria foraging optimization technique. *Electric Power Components and Systems*, *43*(2), 189–199. doi:10.1080/15325008.2014.975871

Nematzadeh, H., Enayatifar, R., Motameni, H., Guimarães, F. G., & Coelho, V. N. (2018). Medical image encryption using a hybrid model of modified genetic algorithm and coupled map lattices. *Optics and Lasers in Engineering*, *110*, 24–32. doi:10.1016/j.optlaseng.2018.05.009

NERC. (1996, June). Available Transfer Capability Definitions and Determination. *North American Electric Realiability Council*, 1–42.

Neufeld, Z., Vicsek, M., & Vicsek, T. (1996). Complex spatiotemporal patterns in two lattice models with instability. *Physica A*, *233*(3-4), 754–766. doi:10.1016/S0378-4371(96)00188-4

Neufeld, Z., & Vicsek, T. (1995). Spatiotemporal chaos in a coupled map lattice with unstable couplings. *Journal of Physics. A, Mathematical and General*, *28*(18), 5257–5266. doi:10.1088/0305-4470/28/18/015

Niknam, T., Mojarrad, H. D., & Meymand, H. Z. (2011). Non-smooth economic dispatch computation by fuzzy and self adaptive particle swarm optimization. *Applied Soft Computing*, *11*(2), 2805–2817. doi:10.1016/j.asoc.2010.11.010

Niknam, T., Mojarrad, H. D., Meymand, H. Z., & Firouzi, B. B. (2011). A new honey bee mating optimization algorithm for non-smooth economic dispatch. *Energy*, *36*(2), 896–908. doi:10.1016/j.energy.2010.12.021

Nimal Madhu, M., Sasidharan, N., & Singh, J. G. (2016). A droop control incorporated dc equivalent power flow method for distribution and low voltage systems. *Electric Power Systems Research*, *134*, 56–65. doi:10.1016/j.epsr.2015.12.033

Ning, B., & Bloomfield, P. (2017). *Bayesian inference for generalized extreme value distribution with Gaussian copula dependence*. arXiv preprint arXiv:1703.00968

Nireekshana, T., Kesava Rao, G., & Sivanaga Raju, S. (2016). Available transfer capability enhancement with FACTS using Cat Swarm Optimization. *Ain Shams Engineering Journal*, *7*(1), 159–167. doi:10.1016/j.asej.2015.11.011

Noman, N., & Iba, H. (2008). Differential evolution for economic load dispatch problems. *Electric Power Systems Research*, *78*(8), 1322–1331. doi:10.1016/j.epsr.2007.11.007

Nourbakhsh, H., Emam-djomeh, Z., Omid, M., Mirsaeedghazi, H., & Moini, S. (2014). Prediction of red plum juice permeate flux during membrane processing with ANN optimized using RSM. *Computers and Electronics in Agriculture*, *102*, 1–9. doi:10.1016/j.compag.2013.12.017

Orjubin, G. (2007). Maximum field inside a reverberation chamber modeled by the generalized extreme value distribution. *IEEE Transactions on Electromagnetic Compatibility*, *49*(1), 104–113. doi:10.1109/TEMC.2006.888172

Ozkan, M. B., Küçük, D., Buhan, S., Demirci, T., & Karagoz, P. (2020). Large-Scale Renewable Energy Monitoring and Forecast Based on Intelligent Data Analysis. In *Handbook of Research on Smart Computing for Renewable Energy and Agro-Engineering* (pp. 53–77). Hershey, PA: IGI Global. doi:10.4018/978-1-7998-1216-6.ch003

Padberg, M., & Rinaldi, G. (1991). A branch and cut algorithm for the resolution of large-scale symmetric traveling salesman problems. *SIAM Review*, *33*(1), 60–100. doi:10.1137/1033004

Padiyar, K. R. (1996). *Power system dynamics stability and control*. Bangalore: Interline Publishing Private Ltd.

Panchenko, V. A. (2018). Solar Roof Panels for Electric and Thermal Generation. Applied Solar Energy, 54(5), 350-353. DOI: 10.3103/S0003701X18050146

Panchenko, V., & Kharchenko, V. (2019). Development and Research of PVT Modules in Computer-Aided Design and Finite Element Analysis Systems. Handbook of Research on Energy-Saving Technologies for Environmentally-Friendly Agricultural Development, 314-342. Doi:10.4018/978-1-5225-9420-8.ch013

Panchenko, V., Chirskiy, S., & Kharchenko, V. (2019). Application of the Software System of Finite Element Analysis for the Simulation and Design Optimization of Solar Photovoltaic Thermal Modules. Handbook of Research on Smart Computing for Renewable Energy and Agro-Engineering, 106-131. DOI: 10.4018/978-1-7998-1216-6.ch005

Panchenko, V., Chirskiy, S., & Kharchenko, V. V. (2019). Application of the Software System of Finite Element Analysis for the Simulation and Design Optimization of Solar Photovoltaic Thermal Modules. In Handbook of Research on Smart Computing for Renewable Energy and Agro-Engineering (pp. 106–131). doi:10.4018/978-1-7998-1216-6.ch005

Panchenko, V.A., Strebkov, D.S. & Persits, I.S. (2015). Razrabotka solnechnyh modulej s uvelichennym srokom nominal'noj raboty [Development of solar modules with extended nominal life]. *Nanostrukturirovannye materialy i preobrazovatel'nye ustrojstva dlya solnechnoj energetiki*, 91-94.

Panchenko, V.A., Strebkov, D.S., Polyakov, V.I. & Arbuzov, Yu.D. (2015). Vysokovol'tnye solnechnye moduli s naprya-zheniem 1000 V [High-voltage solar modules with a voltage of 1000 V]. *Al'ternativnaya energetika i ekologiya, 19*(183), 76-81.

Panchenko, V., Kharchenko, V., & Vasant, P. (2019). Modeling of Solar Photovoltaic Thermal Modules. *Intelligent Computing & Optimization. Advances in Intelligent Systems and Computing, 866*, 108–116. doi:10.1007/978-3-030-00979-3_11

Panda, S., Yegireddy, N. K., & Mohapatra, S. K. (2013). Hybrid BFOA-PSO approach for coordinated design of PSS and SSSC-based controller considering time delays. *International Journal of Electrical Power & Energy Systems, 49*, 221–233. doi:10.1016/j.ijepes.2013.01.006

Pandey, S. N., Pandey, N. K., Tapaswi, S., & Srivastava, L. (2010). Neural Network-Based Approach for ATC Estimation Using Distributed Computing. *IEEE Transactions on Power Systems, 25*(3), 1291–1300. doi:10.1109/TPWRS.2010.2042978

Panigrahi, B. K., Yadav, S. R., Agrawal, S., & Tiwari, M. K. (2007). A clonal algorithm to solve economic load dispatch. *Electric Power Systems Research, 77*(10), 1381–1389. doi:10.1016/j.epsr.2006.10.007

Papadimitriou, A., Symeonidis, P., & Manolopoulos, Y. (2012). Fast and accurate link prediction in social networking systems. *Journal of Systems and Software, 85*(9), 2119–2132. doi:10.1016/j.jss.2012.04.019

Papadimitriou, C. H. (1977). The Euclidean traveling salesman problem is NP-complete. *Theoretical Computer Science, 4*(3), 237–244. doi:10.1016/0304-3975(77)90012-3

Park, J. B., Lee, K. S., Shin, J. R., & Lee, K. Y. (2005). A particle swarm optimization for economic dispatch with non-smooth cost functions. *IEEE Transactions on Power Systems, 20*(1), 34–42. doi:10.1109/TPWRS.2004.831275

Park, J. H., Kim, Y. S., Eom, I. K., & Lee, K. Y. (1993). Economic load dispatch for piecewise quadratic cost function using Hopfield neural network. *IEEE Transactions on Power Systems, 8*(3), 1030–1038. doi:10.1109/59.260897

Parouha, R. P., & Das, K. N. (2016). A novel hybrid optimizer for solving Economic Load Dispatch problem. *International Journal of Electrical Power & Energy Systems, 78*, 108–126. doi:10.1016/j.ijepes.2015.11.058

Parretta, A., Bombace, M., Graditi, G., & Schioppo, R. (2005). Optical degradation of long-term, field-aged c-Si Photovoltaic modules. *Solar Energy Materials and Solar Cells, 86*(3), 349–364. doi:10.1016/j.solmat.2004.08.006

Parthenios, P., Mania, K., Oikonomou, A., Mallouchou, F., Ragiam, L., Patsavos, N., ... Dimitriou, M. (2014). Using WEBGl to design an interactive 3D Platform for the monuments of Crete. *Proceedings of the 1st CAA GR Conference.*

Paul, S., & Roy, P. K. (2015). Optimal design of single machine power system stabilizer using chemical reaction optimization technique. *International Journal of Energy Optimization and Engineering, 4*(2), 51–69. doi:10.4018/IJEOE.2015040104

Peng, C. K., Havlin, S., Stanley, H. E., & Goldberger, A. L. (1995). Quantification of scaling exponents and crossover phenomena in nonstationary heartbeat time series. *Chaos (Woodbury, N.Y.), 5*(1), 82–87. doi:10.1063/1.166141 PMID:11538314

Peres, W., Junior, I. C. S., & Filho, J. A. P. (2017). Gradient based hybrid metaheuristics for robust tuning of power system stabilizers. *International Journal of Energy Optimization and Engineering, 95*, 47–72.

Piepenrock, C. (2014). Article. *Human Factors, 56*(5), 942–951. doi:10.1177/0018720813515509 PMID:25141597

Pincus, S. M. (1991). Approximate entropy as a measure of system complexity. *Proceedings of the National Academy of Sciences of the United States of America, 88*(6), 2297–2301. doi:10.1073/pnas.88.6.2297 PMID:11607165

Pinedo, M. (2002). *Scheduling: Theory, Algorithms, and Systems.* Prentice Hall.

Piryonesi, S. M., & Tavakolan, M. (2017). A mathematical programming model for solving cost-safety optimization (CSO) problems in the maintenance of structures. *KSCE Journal of Civil Engineering*, *21*(6), 2226–2234. doi:10.100712205-017-0531-z

Plucar, J., Grunt, O., & Zelinka, I. (2017, July). Case study: Optimizing fault model input parameters using bio-inspired algorithms. In AIP Conference Proceedings (Vol. 1863, No. 1, p. 070027). AIP Publishing.

Pobedin, A. V., Dolotov, A. A., & Shekhovtsov, V. V. (2016). *Decrease of the vibration load level on the tractor operator working place by means of using of vibrations dynamic dampers in the cabin suspension.* Paper presented at 2nd International Conference on Industrial Engineering (ICIE), Chelyabinsk, Russia. 10.1016/j.proeng.2016.07.136

Polprasert, J., Ongsakul, W., & Dieu, V. N. (2013). A new improved particle swarm optimization for solving nonconvex economic dispatch problems. *International Journal of Energy Optimization and Engineering*, 2(1), 60–77. doi:10.4018/ijeoe.2013010105

Pomeranz, B., Macaulay, R. J., Caudill, M. A., Kutz, I., Adam, D., Gordon, D., & … . (1985). Assessment of autonomic function in humans by heart rate spectral analysis. *The American Journal of Physiology*, *248*(1 Pt 2), H151–H153. PMID:3970172

Poulek, V., Strebkov, D. S., Persic, I. S., & Libra, M. (2012). Towards 50 years lifetime of PV panels laminated with silicone gel technology. *Solar Energy*, *86*(10), 3103–3108. doi:10.1016/j.solener.2012.07.013

Pradhan, M., Roy, P. K., & Pal, T. (2016). Grey wolf optimization applied to economic load dispatch problems. *International Journal of Electrical Power & Energy Systems*, *83*, 325–334. doi:10.1016/j.ijepes.2016.04.034

Prathiba, R., Moses, B. B., Devaraj, D., & Karuppasamypandiyan, M. (2015). Multi-output On-Line ATC Estimation in Deregulated Power System Using ANN BT - Advances in Intelligent Informatics. Cham: Springer International Publishing.

Price, K., Storn, R. M., & Lampinen, J. A. (2006). *Differential evolution: a practical approach to global optimization.* Springer Science & Business Media.

Qi, Q., Tao, F., Zuo, Y., & Zhao, D. (2018). Digital twin service towards smart manufacturing. *Procedia CIRP*, *72*, 237–242. doi:10.1016/j.procir.2018.03.103

Qu, Z., Mao, W., Zhang, K., Zhang, W., & Li, Z. (2019). Multi-step wind speed forecasting based on a hybrid decomposition technique and an improved back-propagation neural network. *Renewable Energy*, *133*, 919–929. doi:10.1016/j.renene.2018.10.043

Radhakrishnan, S., & Vijayarajan, V. (2019). Optimized Deep Learning System for Crop Health Classification Strategically Using Spatial and Temporal Data. In Deep Learning Techniques and Optimization Strategies in Big Data Analytics (pp. 233–250). doi:10.4018/978-1-7998-1192-3.ch014

Rahaman, M. S. A., & Vasant, P. (2020). Artificial Intelligence Approach for Predicting TOC From Well Logs in Shale Reservoirs: A Review. In Deep Learning Techniques and Optimization Strategies in Big Data Analytics (pp. 46-77). IGI Global. doi:10.4018/978-1-7998-1192-3.ch004

Rahim, K. N. K. A., Elamvazuthi, I., Vasant, P., & Ganesan, T. (2020). Robotic Assistive System: Development of a Model based on Artificial Intelligent Technique. In Robotic Systems: Concepts, Methodologies, Tools, and Applications (pp. 1688-1720). IGI Global.

Ramoji, S. K., Rath, B. B., & Kumar, D. V. (2014). Optimization of hybrid PV/wind energy system using genetic algorithm (GA). *Journal of Engineering Research and Applications*, *4*, 29–37.

Rashedi, E., Nezamabadi-Pour, H., & Saryazdi, S. (2009). A gravitational search algorithm. *Information Sciences*, *179*(13), 2232–2248. doi:10.1016/j.ins.2009.03.004

Ray, P. P. (2016). A survey of IoT cloud platforms. *Future Computing and Informatics Journal*, *1*(1-2), 35–46. doi:10.1016/j.fcij.2017.02.001

Reda, S. M. (2007). Stability and photodegradation of phthalocyanines and hematoporphyrin doped PMMA as solar concentrators. *Solar Energy*, *81*(6), 755–760. doi:10.1016/j.solener.2006.10.004

Richardson, M., & Domingos, P. (2002, July). Mining knowledge-sharing sites for viral marketing. In *Proceedings of the eighth ACM SIGKDD international conference on Knowledge discovery and data mining* (pp. 61-70). 10.1145/775047.775057

Richter, H. (2008). Coupled map lattices as spatio-temporal fitness functions: Landscape measures and evolutionary optimization. *Physica D. Nonlinear Phenomena*, *237*(2), 167–186. doi:10.1016/j.physd.2007.08.016

Rodrigues, A. B., & Da Silva, M. G. (2007). Probabilistic Assessment of Available Transfer Capability Based on Monte Carlo Method With Sequential Simulation. *IEEE Transactions on Power Systems*, *22*(1), 484–492. doi:10.1109/TPWRS.2006.887958

Roy, S., Bhattacharjee, K., & Bhattacharya, A. (2017). A modern approach to solve of economic load dispatch using group leader optimization technique. *International Journal of Energy Optimization and Engineering*, *6*(1), 66–85. doi:10.4018/IJEOE.2017010104

Sabri, N. M., Puteh, M., & Mahmood, M. R. (2013). A review of gravitational search algorithm. *Int. J. Advance. Soft Comput.*, *5*(3), 1–39.

Sacha, D., Kraus, M., Keim, D. A., & Chen, M. (2018). Vis4ml: An ontology for visual analytics assisted machine learning. *IEEE Transactions on Visualization and Computer Graphics*, *25*(1), 385–395. doi:10.1109/TVCG.2018.2864838 PMID:30130221

Sadeghi, J., Sadeghi, S., & Niaki, S. T. A. (2014). A hybrid vendor managed inventory and redundancy allocation optimization problem in supply chain management: An NSGA-II with tuned parameters. *Computers & Operations Research*, *41*, 53–64. doi:10.1016/j.cor.2013.07.024

Sainath, T. N., Mohamed, A., Kingsbury, B., & Ramabhadran, B. (2013). Deep convolutional neural networks for LVCSR. In *2013 IEEE International Conference on Acoustics, Speech and Signal Processing* (pp. 8614–8618). 10.1109/ICASSP.2013.6639347

Saito, K., Nakano, R., & Kimura, M. (2008, September). Prediction of information diffusion probabilities for independent cascade model. In *International conference on knowledge-based and intelligent information and engineering systems* (pp. 67-75). Springer. 10.1007/978-3-540-85567-5_9

Salkin, H. M. (1991). A note on Gomory fractional cuts. *Operations Research*, *19*(6), 1538–1541. doi:10.1287/opre.19.6.1538

Salman, A. G., Heryadi, Y., Abdurahman, E., & Suparta, W. (2018). Single Layer & Multi-layer Long Short-Term Memory (LSTM) Model with Intermediate Variables for Weather Forecasting. *Procedia Computer Science*, *135*, 89–98. doi:10.1016/j.procs.2018.08.153

Saltelli, A. (2002). Sensitivity analysis for importance assessment. *Risk Analysis*, *22*(3), 579–590. doi:10.1111/0272-4332.00040 PMID:12088235

Salvelsbergh, M. W. P. (1997). A branch and price algorithm to solve the generalized assignment problem. *Operations Research*, *45*, 381–841.

Sambariya, D. K., & Prasad, R. (2014). Robust tuning of power system stabilizer for small signal stability enhancement using metaheuristic bat algorithm. *International Journal of Electrical Power & Energy Systems*, *61*, 229–238. doi:10.1016/j.ijepes.2014.03.050

Sanders, R. (1987). The Pareto principle: Its use and abuse. *Journal of Services Marketing*, *1*(2), 37–40. doi:10.1108/eb024706

Sandnes, B., & Rekstad, J. (2002). A photovoltaic/thermal (PV/T) collector with a polymer absorber plate: Experimental study and analytic model. *Solar Energy*, *72*(1), 63–73. doi:10.1016/S0038-092X(01)00091-3

Sato, N., & Miyake, S. (2004). Cardiovascular reactivity to mental stress: Relationship with menstrual cycle and gender. *Journal of Physiological Anthropology and Applied Human Science*, *23*(6), 215–223. doi:10.2114/jpa.23.215 PMID:15599065

Sato, N., Miyake, S., Akatsu, J., & Kumashiro, M. (1995). Power spectral analysis of heart rate variability in healthy young women during the normal menstrual cycle. *Psychosomatic Medicine*, *57*(4), 331–335. doi:10.1097/00006842-199507000-00004 PMID:7480562

Savelsbergh, M. W. P. (1994). Preprocessing and Probing Techniques for Mixed Integer Programming Problems. *ORSA Journal on Computing*, *6*(4), 445–454. doi:10.1287/ijoc.6.4.445

Schafer, A., & Vagedes, J. (2013). How accurate is pulse rate variability as an estimate of heart rate variability? A review on studies comparing photoplethysmographic technology with an electrocardiogram. *International Journal of Cardiology*, *166*(1), 15–29. doi:10.1016/j.ijcard.2012.03.119 PMID:22809539

Schleich, B., Anwer, N., Mathieu, L., & Wartzack, S. (2017). Shaping the digital twin for design and production engineering. *CIRP Annals*, *66*(1), 141–144. doi:10.1016/j.cirp.2017.04.040

Schmidt, G., Malik, M., Barthel, P., Schneider, R., Ulm, K., Rolnitzky, L., ... Schomig, A. (1999). Heart-rate turbulence after ventricular premature beats as a predictor of mortality after acute myocardial infarction. *Lancet*, *353*(9162), 1390–1396. doi:10.1016/S0140-6736(98)08428-1 PMID:10227219

Secui, D. C. (2015). The chaotic global best artificial bee colony algorithm for the multi-area economic/emission dispatch. *Energy*, *93*, 2518–2545. doi:10.1016/j.energy.2015.10.012

Selvakumar, A. I., & Thanushkodi, K. (2007). A new particle swarm optimization solution to nonconvex economic dispatch problems. *IEEE Transactions on Power Systems*, *22*(1), 42–51. doi:10.1109/TPWRS.2006.889132

Senion. (2016). *Indoor Positioning System*. https://senion.com/indoor-positioning-system/

Senkevich, S. E., Sergeev, N. V., Vasilev, E. K., Godzhaev, Z. A., & Babayev, V. (2019). Use of an Elastic-Damping Mechanism in the Tractor Transmission of a Small Class of Traction (14 kN): Theoretical and Experimental Substantiation (Chapter 6). Handbook of Advanced Agro-Engineering Technologies for Rural Business Development. – Hershey, Pennsylvania (USA): IGI Global – P. 149-179. Doi:10.4018/978-1-5225-7573-3.ch006

Senkevich, S. E., Lavrukhin, P. V., Senkevich, A. A., Ivanov, P. A., & Sergeev, N. V. (2020). *Improvement of Traction and Coupling Properties of the Small Class Tractor for Grain Crop Sowing by Means of the Hydropneumatic Damping Device. Handbook of Research on Energy-Saving Technologies for Environmentally-Friendly Agricultural Development* (pp. 1–27). Hershey, Pennsylvania, USA: IGI Global; doi:10.4018/978-1-5225-9420-8.ch001

Senkevich, S., Duriagina, V., Kravchenko, V., Gamolina, I., & Pavkin, D. (2020b). Improvement of the Numerical Simulation of the Machine-Tractor Unit Functioning with an Elastic-Damping Mechanism in the Tractor Transmission of a Small Class of Traction (14 kN). In P. Vasant, I. Zelinka, & G. W. Weber (Eds.), *Intelligent Computing and Optimization. ICO 2019. Advances in Intelligent Systems and Computing* (Vol. 1072, pp. 204–213). Cham: Springer; doi:10.1007/978-3-030-33585-4_20

Senkevich, S., Kravchenko, V., Duriagina, V., Senkevich, A., & Vasilev, E. (2019). Optimization of the Parameters of the Elastic Damping Mechanism in Class 1,4 Tractor Transmission for Work in the Main Agricultural Operations. In P. Vasant, I. Zelinka, & G. W. Weber (Eds.), *Intelligent Computing & Optimization. ICO 2018. Advances in Intelligent Systems and Computing* (Vol. 866). Cham: Springer; doi:10.1007/978-3-030-00979-3_17

Senkevich, S., Kravchenko, V., Lavrukhin, P., Ivanov, P., & Senkevich, A. (2020a). *Theoretical Study of the Effect of an Elastic-Damping Mechanism in the Tractor Transmission on a Machine-Tractor Unit Performance While Sowing. Handbook of Research on Smart Computing for Renewable Energy and Agro-Engineering* (pp. 423–463). Hershey, Pennsylvania, USA: IGI Global; doi:10.4018/978-1-7998-1216-6.ch017

Serebrenny, V., Lapin, D., & Mokaeva, A. (2019, July). The Concept of Flexible Manufacturing System for a Newly Forming Robotic Enterprises. In *Lecture Notes in Engineering and Computer Science: Proceedings of The World Congress on Engineering 2019* (pp. 267-271). Springer.

Shaw, B., Banerjee, A., Ghosal, S. P., & Mukherjee, V. (2011). Comparative seeker and bio-inspired fuzzy logic controllers for power system stabilizers. *Elect. Power and Energy Syst*, *33*(10), 1728–1738. doi:10.1016/j.ijepes.2011.08.015

Shayeghi, H., Shayanfar, H. A., Jalilzadeh, S., & Safari, A. (2010). Multi-machine power system stabilizers design using chaotic optimization algorithm. *Energy Conversion and Management*, *51*(7), 1572–1580. doi:10.1016/j.enconman.2010.02.015

Sheng, W., Liu, K. Y., Liu, Y., Meng, X., & Song, X. (2013). A new DG multiobjective optimization method based on an improved evolutionary algorithm. *Journal of Applied Mathematics*, *2013*, 2013. doi:10.1155/2013/643791

Shilaih, M., Goodale, B. M., Falco, L., Kubler, F., De Clerck, V., & Leeners, B. (2018). Modern fertility awareness methods: Wrist wearables capture the changes in temperature associated with the menstrual cycle. *Bioscience Reports*, *38*(6), BSR20171279. doi:10.1042/BSR20171279 PMID:29175999

Shoults, R. R., & Mead, M. M. (1984). Optimal estimation of piece-wise linear incremental cost curves for EDC. *IEEE Transactions on Power Apparatus and Systems*, *PAS-103*(6), 1432–1438. doi:10.1109/TPAS.1984.318480

Shukla, D., Lakshmi, E. S., & Singh, S. P. (2017). Estimation of ATC using PS-NR. In *2017 6th International Conference on Computer Applications In Electrical Engineering-Recent Advances (CERA)* (pp. 111–116). 10.1109/CERA.2017.8343311

Shukla, A., & Singh, S. N. (2017). Pseudo-inspired CBA for ED of units with valve-point loading effects and multi-fuel options. *IET Generation, Transmission & Distribution*, *11*(4), 1039–1045. doi:10.1049/iet-gtd.2016.1360

Shukur, O. B., & Lee, M. H. (2015). Daily wind speed forecasting through hybrid KF-ANN model based on ARIMA. *Renewable Energy*, *76*, 637–647. doi:10.1016/j.renene.2014.11.084

Sidney, J. B. (1973) An Extension of Moore's Due Date Algorithm. *Lecture Notes in Economics and Mathematical Systems*, *86*, 393–398.

Sierla, S., Kyrki, V., Aarnio, P., & Vyatkin, V. (2018). Automatic assembly planning based on digital product descriptions. *Computers in Industry*, *97*, 34–46. doi:10.1016/j.compind.2018.01.013

Silayev, A. A. (1972). *Spektral'naya teoriya podressorivaniya transportnykh mashin* [Spectral theory of the suspension of transport vehicles]. Moscow, Russia: Mashgiz.

Singh, M., & Dhillon, J. S. (2016). Multiobjective thermal power dispatch using opposition-based greedy heuristic search. *Electrical Power and Energy Systems*, *82*, 339–353. doi:10.1016/j.ijepes.2016.03.016

Singh, N. J., Dhillon, J. S., & Kothari, D. P. (2016). Synergic predator-prey optimization for economic thermal power dispatch problem. *Applied Soft Computing*, *43*, 298–311. doi:10.1016/j.asoc.2016.02.042

Singh, N. J., Dhillon, J. S., & Kothari, D. P. (2017). Surrogate worth trade-off method for multi-objective thermal power load dispatch. *Energy*, *138*, 1112–1123. doi:10.1016/j.energy.2017.07.063

Singh, N. J., Dhillon, J. S., & Kothari, D. P. (2018). Multiobjective thermal power load dispatch using adaptive predator–prey optimization. *Applied Soft Computing*, *66*, 370–383. doi:10.1016/j.asoc.2018.02.006

Singh, R. P., Mukherjee, V., & Ghoshal, S. P. (2015). Optimal reactive power dispatch by particle swarm optimization with an aging leader and challengers. *Applied Soft Computing*, *29*, 298–309. doi:10.1016/j.asoc.2015.01.006

Singh, S., & Kaushik, S. C. (2016). Op-timal sizing of grid integrated hybrid PV-biomass energy system using arti cial bee colony algorithm. *IET Renewable Power Generation*, *10*(5), 642–650. doi:10.1049/iet-rpg.2015.0298

Sinitsyn, S., Panchenko, V., Kharchenko, V., Kovalev, A., & Vasant, P. (2019, October). The Concept of Information Modeling in Interactive Intelligence Systems. In *International Conference on Intelligent Computing & Optimization* (pp. 249-259). Springer.

Sinitsyn, S., Panchenko, V., Kharchenko, V., & Vasant, P. (2020). Optimization of Parquetting of the Concentrator of Photovoltaic Thermal Module. *Intelligent Computing & Optimization. Advances in Intelligent Systems and Computing*, *1072*, 160–169. doi:10.1007/978-3-030-33585-4_16

Söderberg, R., Wärmefjord, K., Carlson, J. S., & Lindkvist, L. (2017). Toward a Digital Twin for real-time geometry assurance in individualized production. *CIRP Annals*, *66*(1), 137–140. doi:10.1016/j.cirp.2017.04.038

Soleimani, R., Shoushtari, N. A., Mirza, B., & Salahi, A. (2013). Experimental investigation, modeling and optimization of membrane separation using artificial neural network and multi-objective optimization using genetic algorithm. *Chemical Engineering Research & Design*, *91*(5), 883–903. doi:10.1016/j.cherd.2012.08.004

Sparkias, H., & Gibbon, A. (1993). *Lecture Notes on Parallel Computations*. Cambridge University Press.

Stark, R., Kind, S., & Neumeyer, S. (2017). Innovations in digital modeling for next generation manufacturing system design. *CIRP Annals*, *66*(1), 169–172. doi:10.1016/j.cirp.2017.04.045

Strebkov, D. S., Kirsanov, A. I., & Panchenko, V. A. (2017). Solnechnye krovel'nye paneli dlya programmy "Odin million solnechnyh krysh v Rossii" [Solar roofing panels for the program "One Million Solar Roofs in Russia"]. *Upravlenie innovacionnym razvitiem arkticheskoj zony Rossijskoj Federacii. Sbornik izbrannyh trudov po materialam Vserossijskoj nauchno-prakticheskoj konferencii s mezhdunarodnym uchastiem*, 393-397.

Strebkov, D.S., Persits, I.S. & Panchenko, V.A. (2014). Solnechnye moduli s uvelichennym srokom sluzhby [Long life solar modules]. *Innovacii v sel'skom hozyajstve*, *3*(8), 154-158.

Strebkov, D., Panchenko, V., Irodionov, A., & Kirsanov, A. (2015). The development of roof solar panels. *Research in Agricultural Electric Engineering*, *3*(4), 123–127.

Sukhorukov, A. V. (2003). Upravleniye dempfiruyushchimi elementami v sisteme podressorivaniya bystrokhodnoy gusenichnoy mashiny (Diss. ... kand. tekhn. nauk. Spetsial'nost' 05.05.03 - Avtomobili i traktory). [*Control of damping elements in the sprung weigt of a high-speed tracked vehicle* (Diss. ... candidate of technical sciences. Specialty 05.05.03 - Automobiles and tractors)].

Sun, G., & Zhang, A. (2013). A hybrid genetic algorithm and gravitational search algorithm for image segmentation using multilevel thresholding. *Iberian Conference on Pattern Recognition and Image Analysis*, 707-714. 10.1007/978-3-642-38628-2_84

Sun, N., Zhou, J., Liu, G., & He, Z. (2019). A hybrid wind speed forecasting model based on a decomposition method and an improved regularized extreme learning machine. *Energy Procedia*, *158*, 217–222. doi:10.1016/j.egypro.2019.01.079

Tada, Y., Yoshizaki, T., Tomata, Y., Yokoyama, Y., Sunami, A., Hida, A., & Kawano, Y. (2017). The Impact of Menstrual Cycle Phases on Cardiac Autonomic Nervous System Activity: An Observational Study Considering Lifestyle (Diet, Physical Activity, and Sleep) among Female College Students. *Journal of Nutritional Science and Vitaminology*, *63*(4), 249–255. doi:10.3177/jnsv.63.249 PMID:28978872

Taha, H.A. (2011). *TORA Software*. Windows Version 1.00.

Taha, H.A. (2015). *TORA Software*. Windows Version 4.

Taha, H. A. (2017). *Operations Research: An Introduction* (10th ed.). Pearson Educators.

Talaq, J. (2012). Optimal power system stabilizers for multi machine systems. *International Journal of Electrical Power & Energy Systems*, *43*(1), 793–803. doi:10.1016/j.ijepes.2012.06.030

Tao, F., Cheng, J., Qi, Q., Zhang, M., Zhang, H., & Sui, F. (2018). Digital twin-driven product design, manufacturing and service with big data. *International Journal of Advanced Manufacturing Technology*, *94*(9-12), 3563–3576. doi:10.100700170-017-0233-1

Tao, F., Sui, F., Liu, A., Qi, Q., Zhang, M., Song, B., ... Nee, A. Y. C. (2019). Digital twin-driven product design framework. *International Journal of Production Research*, *57*(12), 3935–3953. doi:10.1080/00207543.2018.1443229

Tesfaye, Y. (2014). *Application of Micro-Hydro Pv/Battery off-Grid Hybrid Energy System for Ethiopian Rural Area* (Masters Thesis). Addis Ababa University.

Teso, S., & Kersting, K. (2019). *Explanatory Interactive Machine Learning*. http://www. aies-conference. com/accepted-papers/

Thang, N. T. (2013). Economic emission load dispatch with multiple fuel options using Hopfiled Lagrange Network. *International Journal of Advanced Science and Technology*, *57*, 9–24.

Thomas, J. J., Karagoz, P., Ahamed, B. B., & Vasant, P. (2020). *Deep Learning Techniques and Optimization Strategies in Big Data Analytics*. Hershey, PA: IGI Global. doi:10.4018/978-1-7998-1192-3

Thomas, J., & Cook, K. A. (2005). *Illuminating the path: The R&D agenda for visual analytics national visualization and analytics center*. National Visualization and Analytics Center-US Department of Homeland Security.

Three-dimensional modeling system. (n.d.). https://www.kompas.ru/

Tizhoosh, H. (2005). Opposition based learning: A new scheme for machine intelligence. *Proc. of Int. Conf. on Computational Intelligence for Modeling Control and Automation*, 695-701. 10.1109/CIMCA.2005.1631345

Tong, C. (2014). Gravitational search algorithm based on simulated annealing. *Journal of Convergence Information Technology*, *9*(2), 231.

Tönissen, D. D., Van den Akker, J. M., & Hoogeveen, J. A. (2017). Column generation strategies and decomposition approaches for the two-stage stochastic multiple knapsack problem. *Computers & Operations Research, 83*, 125–139. doi:10.1016/j.cor.2017.02.009

Tsiptsis, I. N., Liimatainen, L., Kotnik, T., & Niiranen, J. (2019). Structural optimization employing isogeometric tools in Particle Swarm Optimizer. *Journal of Building Engineering, 24*, 100761. doi:10.1016/j.jobe.2019.100761

Ugalde-Loo, C. E., Acha, E., & Liceaga-Castro, E. (2010). Fundamental analysis of the electromechanical oscillation damping control loop of the static VAR compensator using individual channel analysis and design. *Power Delivery on IEEE Transaction, 25*(4), 3053–3069. doi:10.1109/TPWRD.2010.2046341

Ummels, B. C., Gibescu, M., Pelgrum, E., Kling, W. L., & Brand, A. J. (2007). Impacts of Wind Power on Thermal Generation Unit Commitment and Dispatch. *IEEE Transactions on Energy Conversion, 22*(1), 44–51. doi:10.1109/TEC.2006.889616

Unni, A. C., Ongsakul, W., & Madhu, M. N. (2019). Fuzzy-based novel risk and reward definition applied for optimal generation-mix estimation. *Renewable Energy*.

UNSD & WHO. (2019). Tracking SDG7: The energy progress report 2019. Authors.

Vaithilingam, C., & Kumudini Devi, R. P. (2013). Available transfer capability estimation using Support Vector Machine. *International Journal of Electrical Power & Energy Systems, 47*, 387–393. doi:10.1016/j.ijepes.2012.10.054

van den Akker, J. M., Hoogeveen, H., & Stoef, J. J. (2018). Combining two-stage stochastic programming and recoverable robustness to minimize the number of late jobs in the case of uncertain processing times. *Journal of Scheduling, 21*(6), 607–617. doi:10.100710951-018-0559-z

Varanasi, J., & Tripathi, M. M. (2016). Artificial Neural Network based wind speed & power forecasting in US wind energy farms. In *2016 IEEE 1st International Conference on Power Electronics, Intelligent Control and Energy Systems (ICPEICES)* (pp. 1–6). 10.1109/ICPEICES.2016.7853622

Vasant, K., & Watada, J. (2017). Metaheuristic techniques in Enhancing the Efficiency and Performance of Thermo-Electric cooling devices. *Energies*.

Vasant, P., Zelinka, & Weber, G-W. (2018). Intelligent Computing & Optimization. In *International Conference on Intelligent Computing & Optimization*. Springer.

Vasant, P., Zelinka, I., & Weber, G. W. (2018). Intelligent Computing & Optimization. In *Conference proceedings ICO* (p. 804). Academic Press.

Vasant, P., Zelinka, I., & Weber, G. W. (2019). Intelligent Computing and Optimization. In *Proceedings of the 2nd International Conference on Intelligent Computing and Optimization, (ICO) 2019*. Springer International Publishing. 10.1007/978-3-030-00979-3

Vasant, P., Kose, U., & Watada, J. (1703). Metaheuristic (2017). Techniques in Enhancing the Efficiency and Performance of Thermo-Electric Cooling Devices. *Energies*, 10.

Vasant, P., Litvinchev, I., & Marmolejo-Saucedo, J. A. (Eds.). (2019). *Innovative computing trends and applications*. Springer International Publishing. doi:10.1007/978-3-030-03898-4

Vasant, P., Marmolejo, J. A., Litvinchev, I., & Aguilar, R. R. (2019). Nature-inspired meta-heuristics approaches for charging plug-in hybrid electric vehicle. *Wireless Networks*. doi:10.100711276-019-01993-w

Vaskov, A. G., Tyagunov, M. G., Shestopalova, T. A., Deryugina, G. V., & Ishchenko, I. (2018). Structure and parameter optimization of renewable- Based hybrid power complexes. In Handbook of Research on Renewable Energy and Electric Resources for Sustainable Rural Development (pp. 352–382). doi:10.4018/978-1-5225-3867-7.ch015

Vega-Barbas, M., & Pau, I. (2017). Interaction Patterns for Smart Spaces: A Confident Interaction Design Solution for Pervasive Sensitive IoT Services. *IEEE Access: Special Section on Human-Centered Smart Systems and Technologies.*

Venhovens, P. J. T. (1994). The Development and Implementation of Adaptive Semi-Active Suspension Control*. *Vehicle System Dynamics*, *23*(1), 211–235. doi:10.1080/00423119408969057

Vo, D. N., & Ongsakul, W. (2012). Economic dispatch with multiple fuel types by enhanced augmented Lagrange Hopfield network. *Applied Energy*, *91*(1), 281–289. doi:10.1016/j.apenergy.2011.09.025

Walshaw, D. (2013). Generalized extreme value distribution. Encyclopedia of Environmetrics.

Wang, J. S., & Song, J. D. (2017). A Hybrid Algorithm Based on Gravitational Search and Particle Swarm Optimization Algorithm to Solve Function Optimization Problems. *Engineering Letters, 25*(1).

Wang, L., & Li, J. (2016). Estimation of extreme wind speed in SCS and NWP by a non-stationary model. *Theoretical and Applied Mechanics Letters, 6*(3), 131–138.

Wang, J., Ye, L., Gao, R. X., Li, C., & Zhang, L. (2019). Digital Twin for rotating machinery fault diagnosis in smart manufacturing. *International Journal of Production Research*, *57*(12), 3920–3934. doi:10.1080/00207543.2018.1552032

Wang, L., & Singh, C. (2009). Multicriteria design of hybrid power generation systems based on a modi ed particle swarm optimization algorithm. *IEEE Transactions on Energy Conversion*, *24*(1), 163–172. doi:10.1109/TEC.2008.2005280

Wei, J., Li, G., & Zhou, M. (2015). Numerical bifurcation and its application in computation of available transfer capability. *Applied Mathematics and Computation*, *252*, 568–574. doi:10.1016/j.amc.2014.12.003

Weiser, M., & Brown, J. (1995). *Designing Calm Technology*. https://people.csail.mit.edu/rudolph/Teaching/weiser.pdf

Weldemariam, L. E. (2010). *Genset-solar-wind hybrid power system of off-grid power station for rural applications*. Academic Press.

Williams, R. A. (1994). Electronically controlled automotive suspensions. *Computing & Control Engineering Journal*, *5*(3), 143–148. doi:10.1049/cce:19940310

Winston, W. L. (2004). *Operations Research: Applications and Algorithms* (4th ed.). Thomson Brooks/Cole.

Wohlgemuth, J. H. (2003). Long Term Photovoltaic Module Reliability. *NCPV and Solar Program Review Meeting*, NREL/CD-520-33586.

Wolsey, L. A. (1980). Heuristics analysis, linear programming and branch and bound. *Mathematical Programming Study*, *13*, 121–134. doi:10.1007/BFb0120913

Wu, B., & Malik, O. (2006). Multivariable adaptive control of synchronous machine. *IEEE Power India Conference.*

Wu, Y.-K. (2007). A novel algorithm for ATC calculations and applications in deregulated electricity markets. *International Journal of Electrical Power & Energy Systems*, *29*(10), 810–821. doi:10.1016/j.ijepes.2007.06.014

Xingyuan, W., Le, F., Shibing, W., Zhang, C., & Yingqian, Z. (2018). Spatiotemporal chaos in coupled logistic map lattice with dynamic coupling coefficient and its application in image encryption. *IEEE Access: Practical Innovations, Open Solutions*, *6*, 39705–39724. doi:10.1109/ACCESS.2018.2855726

Yadav, M. R., Singh, K. G., & Chaturvedi, A. (2017). Short-term wind speed forecasting of knock airport based on ANN algorithms. In *2017 International Conference on Information, Communication, Instrumentation and Control (ICICIC)* (pp. 1–8). 10.1109/ICOMICON.2017.8279089

Yamamoto, Y., Hughson, R. L., & Peterson, J. C. (1991). Autonomic control of heart rate during exercise studied by heart rate variability spectral analysis. *Journal of Applied Physiology, 71*(3), 1136–1142. doi:10.1152/jappl.1991.71.3.1136 PMID:1757310

Yamashita, R., Nishio, M., Do, R. K. G., & Togashi, K. (2018). Convolutional neural networks: An overview and application in radiology. *Insights Into Imaging, 9*(4), 611–629. doi:10.100713244-018-0639-9 PMID:29934920

Yangg, H., Zhou, W., & Lou, C. (2008). Optimal siz-ing method for stand-alone hybrid solar{wind system with LPSP technology by using genetic algorithm. *Solar Energy, 82*(4), 354–367. doi:10.1016/j.solener.2007.08.005

Yeragani, V. K., Sobolewski, E., Jampala, V. C., Kay, J., Yeragani, S., & Igel, G. (1998). Fractal dimension and approximate entropy of heart period and heart rate: Awake versus sleep differences and methodological issues. *Clinical Science (London, England), 95*(3), 295–301. doi:10.1042/cs0950295 PMID:9730848

Yildirir, A., Kabakci, G., Akgul, E., Tokgozoglu, L., & Oto, A. (2002). Effects of menstrual cycle on cardiac autonomic innervation as assessed by heart rate variability. *Annals of Noninvasive Electrocardiology, 7*(1), 60–63. doi:10.1111/j.1542-474X.2001.tb00140.x PMID:11844293

Yoshida, Y., Furukawa, Y., Ogasawara, H., Yuda, E., Hayano, J., & Group, A. R. (2016). Longer lying position causes lower LF/HF of heart rate variability during ambulatory monitoring. *Proceedings of the 2016 IEEE 5th Global Conference on Consumer Electronics (GCCE).*

Yoshida, Y., Yuda, E., & Hayano, J. (2019). Machine-learning estimation of body posture and physical activity by wearable acceleration and heartbeat sensors. *Signal and Image Processing: an International Journal, 10*(3), 01–09. doi:10.5121ipij.2019.10301

Young, T., Palta, M., Dempsey, J., Skatrud, J., Weber, S., & Badr, S. (1993). The occurrence of sleep-disordered breathing among middle-aged adults. *The New England Journal of Medicine, 328*(17), 1230–1235. doi:10.1056/NEJM199304293281704 PMID:8464434

Young, T., Peppard, P. E., & Gottlieb, D. J. (2002). Epidemiology of obstructive sleep apnea: A population health perspective. *American Journal of Respiratory and Critical Care Medicine, 165*(9), 1217–1239. doi:10.1164/rccm.2109080 PMID:11991871

Yuan, G. X., Ho, C. H., & Lin, C. J. (2012). Recent advances of large-scale linear classification. *Proceedings of the IEEE, 100*(9), 2584–2603. doi:10.1109/JPROC.2012.2188013

Yuda, E., & Hayano, J. (2018). *Menstrual Cycles of Autonomic Functions and Physical Activities.* Paper presented at the 2018 9th International Conference on Awareness Science and Technology (iCAST), Fukuoka, Japan. 10.1109/ICAwST.2018.8517191

Yuda, E., Furukawa, Y., Yoshida, Y., & Hayano, J. (2018). Characteristics of basal heart rate during daily life: Influences of age, gender, and seasons. *Artificial Intelligence in Medicine, 18.*

Yuda, E., Yoshida, Y., & Hayano, J. (2018). Impacts of sleeping time during the day on the timing and level of basal heart rate: analysis of ALLSTAR big data. *Wireless Networks,* 1-5.

Yuda, E., & Hayano, J. (2020). *Changes in Heart Rate Dynamics with Menstrual Cycles* (Vol. 1072). Cham: Springer. doi:10.1007/978-3-030-33585-4_14

Yu, F., & Crolla, D. A. (1998). State Observer Design for an Adaptive Vehicle Suspension. *Vehicle System Dynamics*, *30*(6), 457–471. doi:10.1080/00423119808969461

Yuvaraj, D., Sivaram, M., Mohamed Uvaze Ahamed, A., & Nageswari, S. (2019). Some Investigation on DDOS Attack Models in Mobile Networks. *International Journal of Interactive Mobile Technologies*, *13*(10), 71–88. doi:10.3991/ijim.v13i10.11304

Zelinka, I., Tomaszek, L., Vasant, P., & Duy, V. H. (2018). A novel approach on evolutionary dynamics analysis – A progress report. *Journal of Computational Science*, *25*, 437-445.

Zhang, L., Liu, S., & Yu, C. (2014). Chaotic behaviour of nonlinear coupled reaction–diffusion system in four-dimensional space. *Pramana*, *82*(6), 995–1009. doi:10.100712043-014-0753-2

Zhang, Y. Q., He, Y., & Wang, X. Y. (2018). Spatiotemporal chaos in mixed linear–nonlinear two-dimensional coupled logistic map lattice. *Physica A*, *490*, 148–160. doi:10.1016/j.physa.2017.07.019

Zhang, Y. Q., & Wang, X. Y. (2014). Spatiotemporal chaos in mixed linear–nonlinear coupled logistic map lattice. *Physica A*, *402*, 104–118. doi:10.1016/j.physa.2014.01.051

Zhang, Y., Chen, G. P., Malik, O. P., & Hope, G. S. (1993). An artificial neural network based adaptive power system stabilizer. *IEEE Transactions on Energy Conversion*, *8*(1), 71–77. doi:10.1109/60.207408

Zhan, J., Wu, Q. H., Guo, C., & Zhou, X. (2015). Economic dispatch with non-smooth objectives—Part II: Dimensional steepest decline method. *IEEE Transactions on Power Systems*, *30*(2), 722–733. doi:10.1109/TPWRS.2014.2330618

Zhao, Y., Li, P., Li, R., & Li, X. (2019). Direct filtration for the treatment of the coagulated domestic sewage using flat-sheet ceramic membranes. *Chemosphere*, *223*, 383–390. doi:10.1016/j.chemosphere.2019.02.055 PMID:30784745

Zheng, E. L., Fan, Y. D., Zhu, R., Zhu, Y., & Xian, J. Y. (2016). Prediction of the vibration characteristics for wheeled tractor with suspended driver seat including air spring and MR damper. *Journal of Mechanical Science and Technology*, *30*(9), 4143–4156. doi:10.100712206-016-0826-x

Zheng, P., Lin, T. J., Chen, C. H., & Xu, X. (2018). A systematic design approach for service innovation of smart product-service systems. *Journal of Cleaner Production*, *201*, 657–667. doi:10.1016/j.jclepro.2018.08.101

Zhou, Y., Zhao, H., Bai, H., Zhang, L., & Tang, H. (2012). Papermaking Effluent Treatment: A New Cellulose Nanocrystalline/Polysulfone Composite Membrane. *Procedia Environmental Sciences*, *16*, 145–151. doi:10.1016/j.proenv.2012.10.020

Zong, W. G. (2012). Size optimiza-tion for a hybrid photovoltaic wind energy system. *Electrical Power and Energy Systems*, *42*(1), 448–451. doi:10.1016/j.ijepes.2012.04.051

Zwillich, C., Devlin, T., White, D., Douglas, N., Weil, J., & Martin, R. (1982). Bradycardia during sleep apnea. Characteristics and mechanism. *The Journal of Clinical Investigation*, *69*(6), 1286–1292. doi:10.1172/JCI110568 PMID:7085875

About the Contributors

Pandian Vasant is a senior lecturer at Department of Fundamental and Applied Sciences, Faculty of Science and Information Technology, University of Technology Petronas in Malaysia. He holds PhD (UNEM, Costa Rica) in Computational Intelligence, MSc (UMS, Malaysia, Engineering Mathematics) and BSc (2nd Class Upper-Hons in Mathematics, UM, Malaysia) in Mathematics. He has co-authored research papers and articles in national journals, international journals, conference proceedings, conference paper presentation, and special issues lead guest editor, lead guest editor for book chapters, edited books, and keynote lecture. (257 publications indexed in Web of Science, H Index 23). In the year 2009, Dr. Pandian Vasant was awarded top reviewer for the journal Applied Soft Computing (Elsevier), awarded outstanding reviewer in the year 2015 for ASOC (Elsevier) journal and Top reviewer for Sentinels of Science: Computer Science (Oct. 2015 - Sept. 2016). He has 26 years of working experience at the various universities from 1989-2017. Currently he is Editor-in-Chief of IJEOE and Member of American Mathematical Society.

Igor Litvinchev is a Head of Department at Computing Center Russian Academy of Sciences (CC RAS), Moscow and Professor at Nuevo Leon State University (UANL), Postgraduate Program in Systems Engineering,Monterrey, Mexico. Received his M.Sc. (1979) degree from Moscow Institute of Physics and Technology (Fizteh), Moscow, Russia, and Ph.D. (1984) and Dr.Sci. (1995) degrees in systems modeling and optimization from CC RAS. He has held positions at universities in Brazil, Mexico, and Norway, as well as positions at various universities and research centers in Russia. His research focuses on large-scale system modeling, optimization, and control. Dr. Litvinchev is a member of Russian Academy of Natural Sciences and Mexican Academy of Sciences.

* * *

Tarun Kanti Bandyopadhyay received his PhD, MTech, BTech in Chemical Engineering in 2011, 1997 and 1994 respectively. He also received BSc Hons. in Chemistry from Calcutta University in 1991. He joined NIT Agartala in 2010 as an Assistant Professor in Department of Chemical Engineering. He holds many administrative posts including HOD in the Department of Bio Engineering. He has guided one doctoral student, five students are pursuing PhD under his guidance, 14 Master's thesis, and 20 BTech projects. He has authored one book, three book chapters, and 30 international journal papers. He has reviewed two doctoral thesis and an NPTEL site. His research area includes CFD, fluid dynamics, membrane integrated separation and filtration process, bio fuel, enzyme extraction, etc. He has 22 years of academic, research and industrial experience.

Anirban Banik is pursuing his PhD in Civil Engineering Department from National Institute of Technology Agartala. He received his BE in Civil Engineering from G.H. Raisoni College of Engineering, Nagpur in 2014 and MTech in Civil Engineering from National Institute of Technology Agartala in 2016. His research interest are soft computing, Optimisation, membrane based separation process.

Olexander Barmak is Professor of the Department of Computer Science and Information Technology of Khmelnitskyi National University.

B. Bazeer Ahamed received a Bachelor of Technology in Vel Tech Engineering College, Affiliated to Anna University, Chennai, India and Master of Computer Science Engineering in Anna University of Technology, Tiruchirapalli, India. Ph.D From Sathyabama Institute of Technology and Science, Chennai India. He has published more than 20 peer reviewed international journals and participated in several high profile conferences. At present he is working as Lecturer in the Department of Information Technology, Al Musanna College of Technology, Sultanate of Oman. Prof. Bazeer research is mainly focused on Data Mining &Information retrieval; additionally his research includes Networks, Data bases, Big Data. He is a Member of IEEE, ISTE, IAENG, and CSTA. He chaired the several sessions at National and International Conference.

Sushant Kumar Biswal received his PhD from IIT Kanpur in 2012. He joined National Institute of Technology Agartala 2012 as an Assistant Professor in Civil Engineering Department. He is supervising PhD thesis of three students from 2016. He has six journal papers, eight international conference and five national conference papers. His research interest includes fluvial hydraulics, turbulence, and sediment transport. He has reviewed three international journal papers. He has 16 years of academic, research and industrial experience.

Balamurugan Easwaran having 20 years of enriching experience in the field of Teaching, Research and Project Coordination in the field of Computer Science & Applications. At present working as Director for ICT Services and faculty in Department of Mathematics and Computer Science, University of Africa, Toru – Orua, Nigeria . Prior to this worked as a Dean (Academics & Evaluation) & Assoc. Professor (IT & IS) at BlueCrest University College, Accra, Ghana. As an educationalist carrier is Associated with Gobi Arts & Science College, Bannari Amman Institute of Technology, Sathyamangalam, India as a Lecturer, Assistant Professor, Associate Professor in Computer Science & Applications. Proficient in preparing proposals for funding agencies, Affiliation process, Framing of Curriculum, Regulations and syllabus for Undergraduate and Postgraduate degree programmes like B.Sc. (CS / IT / CT), M.C.A., M.Sc. (CS/SE) Degree autonomous programs. Headed various positions as Professor In charges for numerous Societies, Associations, Coordinator of Cell's Includes SDC, IIPC etc., Skilled in arranging & coordinating Short term trainings, Consultancy Services, National & International level Conference, Workshops and Events related to academic's. Member of various technical societies includes IEEE, Ghana Chapter, Indian Society for Technical Education (ISTE), Computer Society of India (CSI), India, International Association of Engineers (IAENG), Canada, Science and Engineering Institute (SCIEI), Hong Kong, International Association of Computer Science and Information Technology (IACSIT), Singapore, Indian Academicians and Researchers Association (IARA), India. The areas of research includes Image Processing, Document Image analysis, Digital Forensics, Cloud Computing, GOOGLE Apps Development, Image In-painting, Information Security and Data Mining. Research publications includes numerous International, National level journals & conferences also Reviewer of various international conferences and journals. Delivered various special lectures in conferences & workshops also resource person for training Senior Secondary School Teachers in Ghana in collaboration with Ghana Education Service, Government of Ghana on Cloud Computing using Google APPS.

Olusegun Ewemooje is a lecturer at FUTA in Nigeria.

Diriba Kajela Geleta is a PhD candidate at Department of Mathematics, Punjabi University, Patiala, Punjab State, India.

Vasiliki (Valina) Geropanta is an Assistant Professor of Urban Planning with ICT at the Technical University of Crete, Faculty of Architecture. She holds a PhD in Smart City from the Università degli studi di Roma, Sapienza, Italy, an Ma in Housing & Urbanism from Architectural Association, School of Architecture, London, UK, and a Master's Degree of Architect-Engineer from the University of Patras. Her research interests are on innovation ecosystems, the smart - intelligent cities (digital and virtual innovation environments, smart clusters και hyperconnected - intelligent districts, living labs, smart environments and smart buildings), user-experience data, culutral heritage and climate change. Through these fields she stands for the persevation of cultural heritage and works to establish a new architectural educational agenda for the 4th industrial revolution. She has participated in European research projects in capacity building in the field of Smart Tourism and Smart City, while she has been awarded grants and funding for research and lecturing in Cambridge, India, Korea, Peru, Qatar, Italy and Lebanon. She has published extensively on urban and architectural development with ICT, smart cities, smart tourism and the use of ICT into international monuments, museums, housing, and contemporary education at international journals.

Zakhid Godzhaev is deputy director in FSAC VIM, Moscow, Russian Federation. In 1979 he received PhD (Candidate of Engineering Sciences) in Moscow. After that, in 1995 he became a doctor of Engeniiring Sciences. Finally, in 2007 he reached a status of "Full Professor". Scientific interests: Creation of mobile agricultural machines and tractors. Also, professor at Bauman Moscow State Technical University and The Educational Scientific-and-technical Center "MAMI". Godzhaev Z. has more than 150 scientific works. Corresponding Member, Russian Academy of Sciences.

Junichiro Hayano graduated Nagoya City University Medical School, Nagoya, Japan and received an M.D. degree in 1980. From 1981 to 1983, he received residency training in psychosomatic medicine at Kyushu University School of Medicine, Fukuoka, Japan. He obtained a Ph.D. degree (Dr. of Medical Science) in 1988 from Nagoya City University Graduate School of Medical Sciences. From 1990 to 1991, he was a visiting associate at the Behavioral Medicine Research Center, Duke University Medical Center, Durham, NC, USA. In 1984, he got a faculty position at Nagoya City University Medical School and has been a Professor of Medicine at Nagoya City University Graduate School of Medical Sciences since 2003. His current interests are applications of bio-signal processing to health sciences.

Anna Karagianni is a PhD researcher focusing on the spatial transformation of smart environments through the use of IoT systems and on the user-centered design of smart buildings. Since 2014, she has been teaching Architectural Design, Digital Media in Architectural Design and Advanced Digital Media at the School of Architecture of the Technical University of Crete. Her entrepreneurial project EcoMotivate, an innovative application aiming to reduce the energy footprint and increase eco-awareness at work, was selected among other 400 proposals at the e-nnovation International University Competition in 2015. In 2011, she co-founded ANT Design & Construction, a company undertaking medium-scale planning and construction projects in Greece. Prior to this, she worked as a designer in International Architectural Offices, such as OAB - Carlos Ferrater and Associates in Barcelona and Buerger - Katsota architects in Athens, where she participated in multiple award-winning projects, such as the Frontaura Winery in Zamora – ES. She received her Diploma in Architecture from the School of Architecture, University of Patras and a Master's Degree (MSAAD) from Columbia University. Since 2015, she serves as the General Secretary of the Columbia Alumni Association of Greece, an active Association with more than 500 members.

Viktor Kuzmin, 2009-2015 studied at the Volgograd State Technical University, the city of Volgograd. 2015-2018 he studied at the graduate school of the Federal Scientific Agroengineering Center VIM. Currently a researcher in the mobile energy department. Research interests of the dynamics of machines. 2019 defended his thesis and received the degree of candidate of technical sciences.

Tanawat Laopaiboon completed his Masters from Asian Institute of Technology, Thailand in 2017. He is now working with the Metropolitan Water Authority of Thailand.

Nimal Madhu M. received his Doctoral degree from Asian Institute of Technology (AIT), Thailand, in 2016 and Masters from Indian Institute of Technology Bombay (IITB), India, in 2012. He is currently a Post-Doctoral Research Associate at AIT. His research interests include Power System Optimization, Power System Restructuring and Deregulation, Smart Grid, Microgrids, and AI and Deep Learning Applications in Power Systems.

Mrinmoy Majumder received his PhD from Jadavpur University in 2010. Received his ME and BE from Jadavpur University and Biju Patnaik University of Technology in 2006 and 2003 respectively. He joined National Institute of Technology Agartala in 2010 as an Assistant Professor in the School of Hydro-Informatics (now under Civil Engineering). He is supervising the PhD thesis of eight students from 2012 and two students from 2011. He has authored 10 international monographs, 01 book, 26 international journals, 01 national paper, three international conference papers and four national conference papers. His research interest includes hydro-informatics engineering, nature inspired optimisation algorithms, decision science, water based renewable energy resources, etc.

Mukhdeep Singh Manshahia is working as Assistant Professor in the Department of Mathematics, Punjabi University Patiala. He has published more than 30 research paper in journals of national and international repute and edited a book. He has presented more than 20 papers in national and international seminars/ conferences. He has also participated in many international workshops on wireless communication and sensor networks. His areas of interest are Modelling and optimization of RES, Internet of Things (IoT), Wireless Sensor Networks, Pervasive computing, Artificial intelligence and ICT.

Lais-Ioanna Margiori is an undergraduate university student at TUC in the department of architecture. Currently in the 3rd year (2019). From Heraklion, Crete, Greece.

Sofia Mavroudi holds a M.Eng in Architecture from the Aristotle University of Thessaloniki (A.U.TH.). She currently attends the post-graduate Master Program in "Tourism, Culture and the Mediterranean: Contemporary Design Approaches" at the Technical University of Crete (TUC), aiming to enrich her studies in the field of innovative digital technologies and the use of ICT and its impact on cities. She is a member of the Technical Chamber of Greece and speaks fluent English with a good knowledge of German and French.

Izzet Melikov, 1989-1994 student of the Agroengineering University Zernograd city Rostov region. 1994-1997 postgraduate-course student of the Department of Tractors and Cars. 2001 defended thesis for the degree of candidate of technical sciences. 2003 awarded the title of associate Professor of the Department of Tractors and Cars. From 1997-2004, associate professor of the Department of Tractors and Cars of the Agroengineering University Zernograd city Rostov region . From 2004 to the present, Associate Professor of the Department of Technical Operation of Cars of the Dagestan State Agrarian University named after M.M. Dzhambulatova.

Vivek Mohan is with Dept. of Electrical and Electronics Engineering at National Institute of Technology Tiruchirappalli, India. He has his B.Tech degree in Electrical and Electronics Engineering from Amrita Vishwa Vidyapeetham, M.Tech degree in Power systems from Govt. Engineering College Thrissur and Doctor of Engineering (D.Eng) degree in Electric power system management from Asian Institute of Technology Thailand. He was a post-doctoral fellow in Dept. of electrical engineering at The Hong Kong Polytechnic University (PolyU HK) during 2017-18. His areas of research include risk management, deregulation, reliability and stability of power systems.

Jayakumar N., Ph.D., has published more than 10 Scopus research articles. His research interests include the applications of soft computing techniques to power system problems.

Weerakorn Ongsakul has completed his Master's and Doctoral studies from Texas A&M University. Currently, he is a Professor at the Department of Energy, Environment and Climate Change, Asian Institute of Technology, Thailand.

Balachandar P. has published 5 research articles in reputed international journals. His research interests include Economic dispatch and soft computing techniques.

Vladimir Panchenko, from 2003 to 2009 studied at "Bauman Moscow State Technical University" at the faculty "Power Engineering " with qualification of engineer. From 2009 to 2012 studied at the graduate school of the ''All-Russian Scientific Research Institute of Agriculture Electrification'', specializing in "Renewable Energy Installations". Passed an internship program on the International Energy Program in the Solar Energy Center (SEC), the Certificate under the Indian Technical and Economic Cooperation (ITEC), the Program of the Ministry of External Affairs (MEA), the Government of India - India, Delhi, November-December 2011. Participated in the International Conference-Exhibition Intersolar Europe 2012, Germany, Munich, June 11 - 15, 2012 as part of the delegation of the Ministry of Energy of the Russian Federation. Ph.D. thesis "Development and research of solar photovoltaic thermal module with a paraboloid-type concentrator" was protected on July 2, 2013. From 2013 to 2017 head of the laboratory ''Solar photovoltaic modules with an experimental - technological site''. The Council of

Young Scientists of the VIESH was elected the Chairman of the Council of Young Scientists and was appointed a member of the Academic Council of the VIESH in 2014 - 2017. In 2014 was appointed as a contact person and a member of the International Solar Energy Society of the Russian section of the community. Since September 2015 associate professor at the ''Russian University of Transport''. Since 2017, a senior researcher at the ''Solar Energy Laboratory'' of "Federal Scientific Agroengineering Center VIM". Passed the advanced training in 2017 in ''Russian University of Transport'' the program "Using e-learning and distance educational technologies in the implementation of educational programs on distance learning". In 2016 awarded the grant ''Young lecturer of Russian University of Transport''. In 2018 awarded the Scholarship of the President of the Russian Federation for 2018 - 2020 for young scientists and graduate students engaged in advanced research and development in priority areas of the modernization of the Russian economy. Since 2019 member of the Council of Young Scientists of the Russian University of Transport MIIT. Has 3 medals of exhibitions, 3 diplomas of the winner of competitions, participated in more than 90 conferences and exhibitions of various levels, has more than 170 publications, including patents, monographs and educational publications. Area of interest includes three-dimensional modeling, use of solar energy and electrical transport.

Panagiotis (Panos) Parthenios is an Associate Professor at the School of Architecture, Technical University of Crete, where he teaches Architectural Design with Digital Media since 2009. He received a PhD in Digital Media for Architects from Harvard GSD in 2005, and has conducted research at the Harvard Center for Design Informatics. He serves as the Director of the Digital Media Lab at the School of Architecture since 2010. Since 2016 he serves as the Director of the Post Graduate Studies Program, where he teaches Information Technologies and Cultural Heritage and Advanced Digital Media. His research interests focus on smart and sustainable buildings, interactive media, music and architecture, advanced 3D visualizations for architecture and cultural heritage. A 2000 Fulbrighter, he has been awarded the 2001 Harvard Digital Media Prize and the 2002 National Internet and Digital Media Prize in Greece. He has been selected among Europe's 40 most important emerging young architects for 2009, receiving the "Europe 40 Under 40" Award from the European Centre for Architecture Art Design and Urban Studies. He is the co-founder of Parthenios architects+associates, an Athens based architectural firm, which focuses on design and construction-supervision of a wide range of projects, including hotels, hospitals, health centers, private residences, office buildings, museums, theaters, housing complexes and industrial and retail spaces in Greece and abroad. Their work has been published on several magazines and books, participated in exhibitions and distinguished on international architectural competitions.

Sourav Paul was born in 1987 at Dhanbad, Jharkhand, India. He completed his B-Tech in Electrical Engineering from Dr. B. C. Roy Engineering College, Durgapur, West Bengal, India in 2005; M-Tech in Power Systems from Dr. B. C. Roy Engineering College, Durgapur, West Bengal, India in 2012. Presently he is working as Assistant Professor of Electrical Engineering at Dr. B. C. Roy Engineering College, Durgapur, West Bengal, India. His field of interest includes Power System Stabilizer, Automatic Generation Control and Soft computing Techniques.

Provas Kumar Roy was born in Mejia, West Bengal, India in 1973. He obtained PhD degree in Electrical Engineering from National Institute of Technology Durgapur in 2011. He received his Master degree in Electrical Machine in 2001 from Jadavpur University. He finished his Engineering studies in Electrical Engineering from Regional Engineering College (Presently Known as National Institute of

Technology) Durgapur. Presently, he is working as a Professor in Electrical Engineering Department at Kalyani Government Engineering College, West Bengal, India. He was the recipient of the Outstanding Reviewer Award for International Journal of Electrical Power and Energy Systems (Elsevier) in 2018, Engineering Application of Artificial Intelligence (Elsevier) in 2017, Renewable Energy Focus (Elsevier) in 2018, Ain Shams Engineering Journal (Elsevier) in 2017. He has published more than 150 research papers in National/International Journals and conference. He has more than 60 Journals published in reputed SCI and Scopus indexed Journals. Moreover, he has more than 10 book chapters and one book of international standard. Six research scholars have obtained/submitted their Ph.D. degree under his guidance and 8 students are perusing their Ph.D. under his guidance. His research interest includes economic load dispatch, optimal power flow, FACTS, automatic generation control, radial distribution network, power system stabilizer, image processing, machine learning, evolutionary computing techniques, etc.

Subramanian S., Ph.D., is working as Professor and Head in the Department of Electrical Engineering, Annamalai University, Chidambaram, Tamilnadu, India. His contributions in the research include 50 SCI articles and 22 Ph.Ds. awarded under his guidance. He is a senior member in IEEE (SM 90389250) and also a Fellow of Institution of Engineers (India) (F1130982).

R. Thamarai Selvi did her Master of Computer Science in Bishop Heber College, Bharathidasan University, Trichirappalli, Tamilnadu, India in the year 1992. She has finished her M.hil in Computer science followed by her Ph. D in 2018. She has also qualified SET and NET(UGC) in the year 2012. She is currently working as an Associate Professor and Head in the Department of Computer Applications, at Bishop Heber College, Tiruchirapalli. She holds professional membership of quite a few advisaries such as National Advisary Member of techresearchnet.com. She has published more than 20 research papers in reputed journals. Her main research work focuses on Data Mining, Information Retrieval, Big Data Analytics, Blockchain and Compiler Design. She has earned over 27 years of teaching experience.

Sergey Senkevich, 1997-2002 student of the Agroengineering University. 2002-2005 postgraduate-course student of the Department of Tractors and Cars. 2006 defended thesis for the degree of candidate of technical sciences. 2009 awarded the title of associate Professor of the Department of Tractors and Cars. 2005-2013 associate Professor of the Department of Tractors and Cars. 2013-2019 senior researcher, Laboratory of Mobile Energy Facilities. From 2019 to the present, head of the laboratory for the automated drive of agricultural machinery.

Jai Govind Singh has obtained his doctoral, master and bachelor degrees in Electrical Engineering from IIT Kanpur (2008), IIT Roorkee (2003) and MNNIT Allahabad (2001), respectively. Currently, he is working as an Associate Professor in the Department of Energy, Environment and Climate Change at SERD, Asian Institute of Technology Bangkok. He has worked as a postdoctoral research associate at KTH, Stockholm (2008/2009), followed by the UQ, Brisbane (2009), as a Postdoctoral Research Fellow. His research and teaching interest include power system dynamics, operation, and control; Smart Grid and Microgrid; Renewable energy integration; deregulation, FACTS controllers, power distribution system planning, electric and hybrid vehicles.

Ganesan Sivarajan (Senior Member IEEE) is presently working as Assistant Professor in the Department of Electrical and Electronics Engineering, Government College of Engineering, Salem, Tamil

Nadu, India. He has 16 years / 12 years of teaching / research experience. He has authored 12 SCI research articles. He is an active member in various professional bodies like the Institution of Engineer (India) and Indian Society for Technical Education.

Periakaruppan Sudhakaran was born in 1979, Trichy, Tamilnadu, India. He received his B.E. degree specialized in Computer Science and Engineering from Bharathidasan University, Tiruchirappalli, M.E. degree specialized in the field of Computer Science and Engineering from Anna University, Chennai, Tamilnadu, India, M.B.A degree specialized in the field of Information Technology from Sikkim Manipal University, Karnataka, India. Then he received his Doctoral - Ph.D. in Computer Science Engineering in the year 2016 from Anna University, Chennai, India. He has published over more than 10 Research papers in different International Journals and Conferences, more in the area of opinion mining and data mining, Information Retrieval, Web mining.

Emi Yuda was born in Tokyo, Japan. She studied informatics at M. V. Lomonosov Moscow State University until 2003 and then received M.S. degree from Tsukuba University, Japan. She received Ph.D. from Nihon University in 2019. From 2013 to 2014 she was a research assistant at Santa Monica College in California, USA. From 2015 to 2019, she was a NEDO project researcher in Nagoya City University Graduate School of Medical Sciences. Since 2019, she has been an assistant professor at Tohoku University Graduate School of Engineering. Her current research is Medical Informatics, Data Science and Bio-signal Processing. She has many achievements in the field of Informatics and Big Data.

Index

T

U

V

W

www.ingramcontent.com/pod-product-compliance
Lightning Source LLC
LaVergne TN
LVHW062113300926
849161LV00010B/2449